Proceedings

of the 5th International Yellow River Forum on Ensuring Water Right of the River's Demand and Healthy River Basin Maintenance

Volume Ⅱ

Yellow River Conservancy Press

图书在版编目(CIP)数据

第五届黄河国际论坛论文集/尚宏琦,骆向新主编. —郑州:黄河水利出版社,2015.9
ISBN 978-7-5509-0399-9

Ⅰ. ①第… Ⅱ. ①尚… ②骆… Ⅲ. ①黄河-河道整治-国际学术会议-文集 Ⅳ. ①TV882.1-53

中国版本图书馆 CIP 数据核字(2012)第 314288 号

出 版 社:黄河水利出版社
地址:河南省郑州市顺河路黄委会综合楼 14 层 邮政编码:450003
发行单位:黄河水利出版社
发行部电话:0371-66026940、66020550、66028024、66022620(传真)
E-mail:hhslcbs@126.com
承印单位:河南省瑞光印务股份有限公司
开本:787 mm×1 092 mm 1/16
印张:149.75
印数:1—1 000
版次:2015 年 9 月第 1 版 印次:2015 年 9 月第 1 次印刷

定价(全五册):960.00 元(US$155.00)

Under the Auspices of

Ministry of Water Resources, People's Republic of China

Sponsored & Hosted by

Yellow River Conservancy Commission(YRCC), Ministry of Water Resources, P. R. China

China Yellow River Foundation(CYRF)

Editing Committee of Proceedings of the 5th International Yellow River Forum on Ensuring Water Right of the River's Demand and Healthy River Basin Maintenance

Welcome

(preface)

The 5th International Yellow River Forum (IYRF) is sponsored by Yellow River Conservancy Commission (YRCC) and China Yellow River Foundation (CYRF). On behalf of the Organizing Committee of the conference, I warmly welcome you from over the world to Zhengzhou to attend the 5th IYRF. I sincerely appreciate the valuable contributions of all the delegates.

As an international academic conference, IYRF aims to set up a platform of wide exchange and cooperation for global experts, scholars, managers and stakeholders in water and related fields. Since the initiation in 2003, IYRF has been hosted for four times successfully, which shows new concepts and achievements of the Yellow River management and water management in China, demonstrates the new scientific results in nowadays world water and related fields, and promotes water knowledge sharing and cooperation in the world.

The central theme of the 5th IYRF is "Ensuring water right of the river's demand and healthy river basin maintenance". The Organizing Committee of the 5th IYRF has received near one thousand paper abstracts. Reviewed by the Technical Committee, part of the abstracts are finally collected into the Technical Paper Abstracts of the 5th IYRF.

An ambience of collaboration, respect, and innovation will once again define the forum environment, as experts researchers, representatives from national and local governments, international organizations, universities, research institutions and civil communities gather to discuss, express and listen to the opportunities, challenges and solutions to ensure the sustainable water resources management.

We appreciate the generous supports from the co - sponsors, including domestic and abroad governments and organizations. We also would like to thank the members of the Organizing Committee and the Technical Committee for their great supports and the hard work of the secretariat, as well as all the experts and authors for their outstanding contributions to the 5th IYRF.

Finally, I would like to present my best wishes to the success of the 5th IYRF, and hope every participant to have a good memory about the forum!

Chen Xiaojiang
Chairman of the Organizing Committee, IYRF
Commissioner of YRCC, MWR, China
Zhengzhou, September 2012

Contents

D. Adaptation and Measures of Water Resources Management in River Basins Due to Global Climate Change

E. Ecological Protection and Sustainable Water Utilization in River Basins

D. Adaptation and Measures of Water Resources Management in River Basins Due to Global Climate Change

Hydrological Response to Adapt Climate Changes

Yang Hanxia

Hydrology Bureau of YRCC, Zhenzhou, 450004, China

Abstract: The Yellow River is a very difficulty river to harness because it is one of the most complex rivers in the world. Its unique determines its hydrological characteristics of heavier task, higher requirements and more service items than that of other rivers. The Yellow River has four flood seasons. There are peach flood in spring, major flood in summer, fall flood in autumn, and ice-jam flood in winter. So it is necessary to undertake the hydrological gauging and forecasting tasks throughout the whole year for flood controlling. The main features of the Yellow River are less water and more sediment, different water and sediment origination and unmatched water and sediment relationship. Accurate and timely hydrometry is required to actualize water and sediment regulation. Meanwhile, the Yellow River is the only river in China authorized to annually allocate available water and integrally regulate water resources of the whole river due to its sharp contradiction between supply and demand of water resources. Therefore, for the Yellow River, water resources monitoring and forecasting are especially important. Moreover, along with hydrological regime changes owing to climate change plus more and more human activities, the Yellow River hydrological work will confront much more new challenges in the future.

Key words: water resources regime changes, human activities, climate change, system

1 Changes of hydrological regime in the Yellow River

The latest scientific research shows that the average global surface temperature had increased 0.74° C in the last century (1906 ~ 2005), and is expected to raise 1.1 ~ 6.4 ℃ by the end of the 21st century.

The YRCC research indicates that the annual average temperature of the Yellow River Basin had risen 1.3 ℃ from 1961 to 2005, average 0.3 ℃ per ten-year. While the river's average annual precipitation showed a decreasing trend. It is 468 mm in 1961 ~ 1979, is 450 mm from 1961 to 2005. By the combined influence of climate change and human activities, the Yellow River water and sediment regime changed a lot. That is, the average annual runoff and sediment volume is significantly reduced. The former is 58×10^9 m^3 at Lijin across-section in 1919 ~ 1975 and reduced to 53.5×10^9 m^3 in 1956 ~ 2000. The latter is 1.6×10^9 in 1919 ~ 1958 and decreased to 1.2×10^9 in 1956 ~ 2000.

The average natural annual runoff of Huayuankou station is 58.7×10^9 m^3 in 1950 – 1970, 54.6×10^9 m^3 in 1970 ~ 1990 and 43.8×10^9 m^3 in 1990 ~ 2010, showed a decreasing trend.

The ice regime characteristics of the reach from Ningxia to Inner Mongolia in the Yellow River also changed a lot. The average winter air temperature is −7 ℃ in the 1950s and 1960s, up to −4.5 ℃ in 1991 ~ 2010, increased more than 2 ℃. Low flow at Toudaoguai cross-section lasted 20 ~ 30 days in 1987 ~ 1998, 30 ~ 60 d in 1999 ~ 2008. The cross-section flow capacity reduced and low flow lasted longer. While the channel detention capacity increased from 1.14×10^9 in 1970 ~ 2000 to 1.53×10^9 t in 2001 ~ 2010 in the river course from Shizuisan to Toudaoguai hydrological station. Over the past decade, affected by climate change, reservoir regulation, and water usage, bridge construction and river siltation, the river channel detention capacity significantly increased and ice regime became more and more complex. The ice disasters have occurred from time to time.

Available statistical data shows that the extreme weather events obviously increased in the Yellow River Basin that regional storm, extreme high temperature cropped up more and more suddenly, frequently and simultaneously. In July 2012, continuous local strong storm happened at Sanxi-Shaanxi area with rainfall intensity in many stations not seen for 30 years and rainfall volume in several stations setting a new record

Water resources development and utilization has a long history in the Yellow River. There are various water supply projects built, such as water storage projects, water diversion works and Water lifting engineering, and so on, which undertake heavy water supply tasks for agriculture, industry, and life, ecological usage etc inside and outside the river basin. Related researches figure out that, along with the economic and social development, the total annual water supply increased from 44.6×10^9 m^3 to 51.2×10^9 m^3 from 1980 to 2010, an increase of 6.6×10^8 m^3 (Among it, 6.3×10^9 m^3 happened inside the basin) in the Yellow River. In particular, the total water supply of the Yellow River in 2011 reached a new record of 53.5×10^9 m^3. In a word, total water resource continues to decrease while water demand continues to increase, leading to sharper and sharper contradiction between water supply and demand in the Yellow River.

2 Challenges brought about by changes of hydrological regime

2.1 Implementation of the most strict water resources management requires finer water resources gauging and forecasting

Affected by climate change, rainfall and runoff in the Yellow River showed a decreasing trend, while more and more water resources are needed by the watershed economic and social development, resulting in sharper and sharper contradiction between water supply and demand. In order to effectively manage water resources, the State Council in the 3rd file of 2012 clearly put forward the most strict water management system, made control indicators and phases targets of water resources development and utilization, water usage efficiency and "three red lines" of water function zone for limiting pollutant. To achieve the expected goal, comprehensive, timely and accurate hydrology and water resources gauging and forecasting are absolutely essential. Scientific allocation of water resources and effective control of the total water amount require exact runoff changes situation, which put a higher demand on the runoff forecasting accuracy. Likewise, the performance of annual water allocation and the implementation of real-time water resources scheduling require higher monitoring and gauging capacity, not only in gauging runoff process in real-time, but also in providing solid support for the implementation of various water regulation directives. In order to well implement the indicators of water function zone “three red lines” for limiting pollutant, there are unprecedented new requirements for water quality monitoring and gauging at the cross-section of important river courses, provincial boundaries, key water function zones.

2.2 Flood control and drought relief propose new challenges for hydrology work

With economic development, wealth accumulation, and raised economic and social sensitivity to natural disasters, flood and drought are declined by whether government or public. Influenced by climatic factors and human activities, basin flood reduced while regional flood and local storm flood occurred frequently in the Yellow River. Local storm flood of strong outburst, short lead time, and turbulent flow, high flood peak has brought great difficulties to the hydrological gauging and forecasting work, which is a serious threat to local safety. Since late July 2012, heavy rain storm process continuously occurred in the upper and middle reaches of the Yellow River, maximum precipitation observed at Shengjiawan station in Jiluhe river since there are data and No. 1, No. 2, No. 3, and No. 4 flood peak of the 2012 year formed with maximum discharge at some stations in 40 to 50 years.

Besides, in recent years the Yellow River constantly suffered droughts, such as moderate drought in the upper reaches, severe drought in the middle and lower reaches in 2008 ~ 2009. The drought issue has become more and more highlight. More new requirements have been put forward to meet the need of the flood control and drought relief tasks due to continuously aggravating floods and droughts, not only on the timely and accurate hydrological gauging and forecasting work, but also on monitoring and gauging of groundwater, soil and so on.

2.3 Water and sediment regulation system put forward higher demands for hydrological monitoring and forecasting.

Less water, more sediment and uncoordinated water and sediment relationship, are the main reasons why the Yellow River is difficulty to harness. The key is to shape the coordinated water and sediment relationship which not only needs accurate and timely flood and runoff forecasting, but also needs real-time sediment gauging and forecasting. Currently, the means of sediment gauging lag behind and the gauging frequency is too low and sediment particle size distribution data can not be obtained in time. Sediment forecasting technique has been researched at main stations in the middle and lower reaches of the Yellow River. Affected by the sediment gauging and many other factors, its forecasting accuracy and timeliness can not meet the demand for water and sediment regulation scheduling. Hydrological gauging and forecasting of the Yellow River is facing new challenges.

2.4 Ice flood control work urgently needs hydrological gauging and forecasting in the Yellow River

Last decade, due to the impact of climate change, river siltation, human activities and other factors, the ice regime characteristics of the reaches from Ningxia to Inner Mongolia in the Yellow River significantly change. The ice run date and ice cover time postponed; ice freeze-up and break-up are unstable; low flow process lasts for a long time at Toudaoguai cross-section in ice flood season; the channel detention capacity significantly increased; water level remained high in the critical period of ice freeze-up and break-up; Ice regime is more complex; ice disasters have occurred from time to time; all mentioned above make ice flood control work face a huge challenge and highly require ice gauging and forecasting. Precisely calculating river course storage increment and real-time forecasting ice break-up process in the reach from Ningxia to Inner Mongolia are required to meet the demand of the ice flood control and water regulation in the Yellow River. Now, there are only following forecasting items: ice run date, ice freeze-up and break-up time, maximum water level and largest discharge in the period of ice break-up in current ice forecasting projects. Ice freezing and melting law should be exploited to establish ice forecasting mathematical models with physical mechanism based on increasing frequency of ice observations and river cross-section measurements, to provide solid support for the Yellow River ice flood control scheduling.

2.5 Comprehensive harness of water and soil conservation urgently needs hydrological support.

With the implementation of the new Water and Soil Law, on the basis of existing hydrological data, the strictest system of supervision and monitoring of soil and water conservation needs supplementing station network in the area of plenty and coarse sediment as well as enriching gauging content and conducting dynamic monitoring. At the same time, the research on soil erosion discipline, on evaluating and forecasting of sediment reduction effects as well as water and soil conservation are also in dire need of the support of the hydrological data.

3 Hydrological response measures in the Yellow River

Impacted by climate change and human activities, the hydrological regime of the Yellow River is changing. It needs an in-depth exploration and research on its inherent laws to cognitive the relationship between each other. Ceaselessly changing hydrological regime, however, has been generated a direct impact on the basin' s economic and social development, river ecology and water resources regulation and so on. In order to adapt to such changes, to better serve the development and management in the Yellow River, in the aspect of automatic monitoring, online sediment monitoring and gauging, and sediment forecasting, tributary runoff forecasting, the Loess Plateau Water and Sediment dynamic gauging, effective work have been done and made some progress. Nevertheless, to meet the deeper needs of rational and scientific water resources scheduling and water and sediment regulation system construction, more efforts and exploration in more aspects should be made.

3.1 To supplement and improve the hydrological monitoring network

On the basis of the current station network, comprehensively considering the needs of flood control and drought relief, of the allocation and management and of water resources, of water emergency response, of water and sediment regulation practice, station network in headwaters regions, in ice flood prevention reaches from Ningxia to Inner Mongolia, and in the middle reaches of rainstorm flood-prone areas, in the Loess Plateau of plenty and coarse sediment area, in watershed or provinces (autonomous regions) boundary, large-scale water control projects region et al. should be mainly enriched and improved.

3.2 To vigorously carry out hydrological scientific and technologic innovation

Online sediment gauging technology, automated discharge measurement technology under movable bed conditions and suitable hydrological monitoring equipments in the alpine region are still the difficult technological problems of the hydrological development. It is essential to further expand the applications of new equipments such as microwave discharge meter and ADCP and so on and strive to achieve breakthroughs in key technologies of the online sediment monitoring, sediment forecasting and ice forecasting. Recently, substantial progress is expected in sediment gauging equipment through using the strong field pole method and isotopic dating method.

3.3 To focus on hydrological basic research

Strengthen the analysis and research on basic situation, basic data and the basic law of the Yellow River hydrology. Deeply study the situation and the variation law of water and sediment; strive to explore the river ice freezing and melting law; carefully analyze the causes of ecological changes in the headwaters region; find out the driving force factors and human control effect about sharp decrease of runoff in the middle reaches of the Yellow River. Actively carry out the topographic mapping work of river courses, reservoirs, lakes and coastal area and other basic hydrological work to safeguard the quality of hydrological gauging and forecasting with reliable basic data and basic results.

3.4 To accelerate the construction of rainstorm flood warning and forecasting system in the He-San region (from Hekou Town to Sanmenxia Reservoir)

Late July of 2012, floods happened one after another, more hydrological gauging and forecasting problems exposed. There is an urgent need to establish rainstorm flood warning and forecasting system in the He-San region through enriching and improving the hydrological station network, strengthening the real-time sediment monitoring and gauging means at the major stations in the sediment source area, and optimizing hydrological data collection, transmission and application programs, strengthening research on flow generation and accumulation law, quantitative precipitation forecasting coupled with flood forecasting measures, and developing flood forecasting model and so on.

3.5 To expand runoff forecasting work

In non-flood season, the long-term runoff forecasting work mainly includes the total runoff volume and ten-day and monthly runoff forecasting at control stations in the main headwaters area of the Yellow River with statistical correlation and time-series analysis as main forecasting methods. To meet the requirements of the water regulation, medium-and long-term runoff forecasting of the Yellow River Basin, it is need to expand the runoff forecasting from non-flood season to flood season, from headwaters area to important tributaries and from single-model to multi-model ensemble forecasting technology.

3.6 To comprehensively improve monitoring and surveying capabilities

We should actively put into application of advanced equipment such as automatic discharge gauging system, microwave discharge measurement instrument, and the radar gun, ADCP etc, expand the application of measuring means at representative point and line in medium and small rivers, And optimize measuring mode to improve the efficiency of measurement; We should improve applicable technical performance of vibrating sediment measuring instrument, OPUS, and QICPIC to improve the ability of sediment measurement; We should strengthen the construction of low-water measuring equipment to further improve the ability of low-water monitoring and measurement.

The Yellow River belongs to China, also belongs to the world. We are willing to work with domestic and foreign counterparts to deeply discuss the new issues and common challenges raised from climate change in the Yellow River Hydrology. We hope the co-operation on the online sediment monitoring, sediment forecasting and flood forecasting and other aspects, and will exert unremitting efforts to safeguard the healthy life of the Yellow River.

Water and Climate Change "If the Greenhouse – effect Gases are Responsible for Climate Change, Fresh Water is the First Victim"! Integrated and Sound Water Resources Management at the Level of River Basins is Obviously Essential Worldwide!

Jean – François DONZIER

Permanent Technical Secretary of the International Network of Basin Organizations (INBO), General Manager of the International Office for Water, International Office for Water, Paris France

Abstract: Global warming now seems to be unavoidable so adaptation of water management to climate change is urgently needed worldwide. Therefore quick action is required, for reducing costs and damage with regard to floods and droughts in particular.

Integrated and sound water resources management at the level of river basins is obviously essential, based on integrated information systems, on management plans or master plans and on the development of Programs of Measures.

Cooperation between riparian Countries should be strengthened for good management of transboundary rivers, lakes and aquifers with the participation of stakeholders and the civil society.

Key words: climate change, water resources management, river basin, cooperation, stakeholders' participation, basin management plan, non – regret measures, adaptation, flooding, drought, natural phenomenons, water saving, warning systems

1 Introduction

Floods, shortages, pollution, wastage, destruction of ecosystems: the seriousness of the situation in many countries requires that comprehensive, integrated and consistent management of water resources is implemented to preserve the future and the human heritage.

Water has no national or administrative boundary. It is thus necessary to take into account the specific situation of the 276 rivers or lakes and several hundreds of aquifers over the world, whose resources are shared by at least two riparian countries or sometimes much more. Their joint management is thus strategic and a priority.

2 Adaptation of water management to climate change is urgently needed worldwide

Global warming now seems to be unavoidable. One of the first consequences will be a change in the hydrological cycles. Should ambitious measures be globally taken by all the countries to appreciably reduce their emissions of greenhouse – effect gases, the effect on climate would only be perceptible at best at the end of the century.

Changes in precipitations and hydrological cycles have already started and will undoubtedly be appreciable by 2040 or 2050: in less than a generation! It is necessary to react quickly, before it is not too late and it is clear that the control of gas discharges alone is insufficient to change this evolution within these deadlines.

During the past forty years, the number and intensity of floods and droughts have already increased, sometimes in a spectacular way. The melting of glaciers in particular, has an effect on water supply, especially in periods of low water level, and on the increase of flood hazards.

It is thus essential to adapt to the consequences of climate change, and, as regards basin organizations, water resources management policies in particular, by taking into account the new elements of climate change. It is especially necessary to quickly evaluate the hydrological consequences of this change, according to various scenarios.

Freshwater resources will be directly affected in the coming years, with for consequences, in particular and according to the regions:

(1) Changes in the intensity and frequency of floods and droughts, increase of extreme hydrological and hydrogeological phenomena with the risk of human losses, destructions and catastrophic economic damage;

(2) Modification of the flows of rivers coming from mountains, because of the melting of glaciers and reduction of the snow cover, and the corresponding impacts on the associated aquifers in particular;

(3) increased erosion caused by the modification of plant and soil cover;

(4) Higher plant evapotranspiration leading to changes in ground and surface waters, decrease even in agricultural production, regarding irrigation in particular;

(5) Changes in the flows to the river mouths, as well as salt water intrusion inland and in coastal aquifers, because of the increase of sea and ocean level and the reduction in the piezometric levels during the dry season;

(6) *Changes in the relations between surface flows and the groundwater of subjacent aquifers (supply of aquifers or flows of groundwater into the river)*;

(7) *A strong regional impact on the demand and production of energy, hydropower in particular* …

(8) *Changed and unforeseeable conditions of water supply to large wetlands of local and international interest*;

(9) *A risk of increased prevalence of water - related diseases* …

Indeed, these effects will cumulate with the significant pressures linked to demographic growth, urbanization and development. The social, economic and ecological consequences are likely to be very significant. It is thus essential to work now to adapt water resources management policies.

"If the greenhouse - effect gases are responsible for climate change, fresh water is the first victim"!

It is thus essential to adapt water resources management policies and mechanisms to face climate change. We must thus learn how to anticipate the damage and take the necessary measures to prevent or, at the very least, to minimize their negative effects, in short to adapt us!

3 Quick action will allow reducing costs and damage

The International Network of Basin Organizations (INBO) is worried about the "no - action cost"!

Indispensable adaptation policies and mechanisms should be developed in each river basin, by urgently developing adapted research and knowledge acquisition programs to measure the hydrological and hydrogeological consequences of these changes according to various scenarios, and by drafting ambitious basin management plans or master plans and the programs of measures needed.

It should well be admitted that the urgency of launching programs for adaptation to climate change, in which water management is a central element, the core, has not yet reached the political world and has not been systematically introduced, as evidence, into the plans of most countries nor in the projects supported by international organizations nor in the Conference of Parties in Copenhagen in December 2009.

The implementation of regularly updated planning processes is well adapted to the uncertainty that remains on the forms that the phenomenon will take in each basin. This means that it is important to regularly update the management plans to face, in a pragmatic way, the various situations encountered and their evolution in the coming years, while relying on observation and increasingly fine projections of the climate effects.

The basin management and planning process is the best suited mechanism by which the demands on available water resources could be adjusted in the long term, in order to avoid persistent shortage and to give a clear response to the need for also managing the increasing flood hazards in most areas of the world.

In a first phase at least, it is possible that the effect of climate change is interfered with or especially amplified by the continuation of human activities which have the highest impact on water resources, such as the increase in water abstractions for irrigation, the building of infrastructures which modify hydrology, the creation of obstacles to flows, the destruction of wetlands and the increase in pollution of any origin, etc.

The development of hydropower in addition to other forms of renewable energy will allow contributing effectively to the adaptation to climate change, while improving the living conditions of the poorest populations.

3.1 With regard to floods

It is, first, necessary to make the "upstream – downstream" common cause a main item of consistent management on the scale of basins and sub – basins.

Protection against floods must pass through a coordinated approach, combining:

(1) The protection of people and properties.

(2) The protection of groundwater intakes which can ensure supply water during floods.

(3) The reduction of vulnerabilities.

(4) The restoration of the open flows of rivers.

(5) The conservation and the re – building of the natural flood storage areas.

(6) The forecasting of events.

(7) The identification of zones at risk.

(8) The identification of protected emergency water intakes.

(9) The publication of "atlases" of easily flooded zones, including the flooded areas caused by the rising of aquifers.

● the control of urbanization.

● warning and education.

In the transboundary basins in particular, cooperation between riparian States, for jointly looking for coordinated solutions and for sharing information and responsibilities, should be promoted.

3.2 With regard to droughts

Situations of water shortages, too often ignored, are a growing problem in an increasing number of areas and are likely to worsen in the future.

The availability of fresh water – in sufficient quantity and quality – may become, in a generation from now, one of the main limiting factors of the economic and social development in many countries.

Climate change will worsen the structural problems which already lead to water scarcity in many areas: on this subject it is useful to make a distinction between drought and scarcity, the latter being initially related to a permanent and structural imbalance between available resources and abstractions.

The prevention of recurring droughts can, no more, be done on a case – by – case basis, but must be planned in the long term, by solving the structural problems which occur, in order to prevent, in the best possible way, their effects and to avoid the total degradation of water resources.

It is essential to intensify efforts for better managing water demand and thus reducing the pressures on the resources especially in period of drought, by reducing, in particular, abstractions for irrigation which is the reason for the most significant abstractions in many areas.

Very often, the rarefaction of the resource will require looking for water saving, by, first of all, managing the demand but also by mobilizing non – conventional water and by reusing water, by systematically fostering ecologically sound solutions, socially acceptable and economically reasonable.

Mobilizing new resources and creating reserves should be planned after rationalizing water demands and only when it will be ecologically acceptable and economically reasonable. The economic incidence of water reuse on its efficient use should not be forgotten, in period of drought in particular.

Building new dams will not be enough without the implementation of water saving and recycling programs: the solutions will pass by proactive water management together with constant incentive measures for more rational uses facilitated by innovation and new technologies.

Plans for the Management of Water Scarcity should prioritize drinking water supply, making sure that water is equitably and soundly shared between the various uses, ensuring a better optimization of water and avoiding wastages.

They must ensure a better optimization of existing water resources, before planning the launching of projects for the mobilization of new resources.

Water saving, leak detection, recycling, the reuse of treated water, groundwater recharge, the desalination of sea water, research on low – consumption uses, must become priorities.

4 A new approach to water uses in agriculture should be looked for

In a context of increased pressure on water resources and lands, the importance of the agricultural component should be stressed, as continuing the "business as usual" scenario would be irresponsible.

Feeding the world population today and in the future (9×10^9 inhabitants foreseen in 2075) implies using, in all the countries, an agriculture which is less water – consuming and less sensitive to climate hazards: to a very large extent that will require effective irrigation.

The farmers will be among the first victims of the fluctuations of water supply due to the variations of the climate.

Accompaniment to the changes in agricultural practices towards water saving should be planned with good dissemination of innovation in general, thanks to education, training, research and development.

Financial incentive mechanisms should be gradually developed for respecting water resources allocation and water saving (pricing, quotas, subsidies) and for facilitating access to credits for the modernization of plot irrigation.

The reduction of non – point pollution, as regards the use of fertilizers and pesticides, is also a prerequisite to maintain or recover good water status.

The issues of agricultural water should thus be clearly included into the Integrated Water Resources Management (IWRM) approaches in transboundary and national river basins.

5 Integrated and sound water resources management at the level of river basins is obviously essential worldwide

Adaptation to the effects of climate change is initially a problem of better water management.

Admitting this obvious fact and implementing the necessary reforms are real development opportunities.

The basins of rivers, lakes and aquifers are the relevant natural geographical territories in which to organize this integrated and sound management.

Indeed, river basins are the natural territories in which water runs on the soil or in the sub soil, whatever are the national or administrative boundaries or limits crossed.

Significant progress has already been made since the 1990s:

River basin management experienced a quick development in many countries, which made it the basis of their national legislation on water or experimented it in national or transboundary pilot basins.

It is now widely recognized that water resources management should be organized:

(1) On the scale of local, national or transboundary basins of rivers, lakes and aquifers, as well of their related coastal waters;

(2) With the participation in decision – making of the concerned Governmental Administrations and local Authorities, the representatives of different categories of users and associations for environmental protection or of public interest. Indeed, this concerted participation will ensure the social and economic acceptability of decisions taking into account the real needs, the provisions to be ac-

ted upon and the contribution capabilities of the stakeholders in social and economic life;

(3) Based on integrated information systems, allowing knowledge on resources and their uses, polluting pressures, ecosystems and their functioning, the follow - up of their evolutions and risk assessment. These information systems will have to be used as an objective basis for dialogue, negotiation, decision - making and evaluation of undertaken actions, as well as coordination of financing from the various donors;

(4) Based on management plans or master plans that define the medium and long - term objectives to be achieved;

(5) Through the development of Programs of Measures and successive multiyear priority investments;

(6) With the mobilization of specific financial resources, based on the polluter - pays principle and "user - pays" systems; by looking for geographical and inter - sectoral equalizations to gather the necessary amounts;

Legal and institutional frameworks should allow the application of these six principles in each country and at the regional level.

6 Cooperation between riparian Countries should be strengthened for good management of transboundary rivers, lakes and aquifers in particular

For several centuries, many agreements have certainly been signed between riparian countries to ensure freedom of navigation or the sharing of flows and, since the end of the 19th century, for the building of hydropower dams.

It is now essential that cooperation agreements, conventions or treaties on pollution control, environmental protection or the prevention of floods and integrated management of these shared basins be initiated or signed between the riparian countries of these transboundary resources to achieve indispensable common cause at the basin level and develop a common vision of the future.

Although the United Nations Convention of 21 May 1997, on the uses other than navigation on the international rivers, did not yet come into effect, its principles are now more and more recognized as a basis for relations among the riparian States concerned.

Resolution A/RES/63/124, adopted in December 2008 by the General Assembly of the United Nations, offers to the States the framework for joint management their transboundary aquifers.

In addition, the European Water Framework Directive of 2000 (WFD) lays down an objective of good status in the national or international river basin districts of the 27 current Member States and the Countries applying for accession to the European Union.

Agreements on transboundary aquifer management should also be developed, taking into account their fragility and the time needed for the restoration of degraded situations.

Improving knowledge of water resources, aquatic environments and of their uses is essential to allow decision - making.

It is recommended to promote the creation of information systems on water resources and their uses in each basin.

Systems for warning against floods, droughts and pollution should be improved, developed and coordinated for better facing the natural disasters caused by water and for protecting human lives and properties.

It is necessary to promote the emergence in this field of means and competences for specific engineering and to support any work aiming at defining common standards and nomenclatures for data administration in order to allow exchanges, comparisons and syntheses of information between partners at all the relevant levels of observation.

If climate change can no more be doubted, significant uncertainties remain regarding its local impact and the best way of facing it in each situation. It is clear that acting now and very quickly is needed but also that it is necessary to reinforce research on climate in each large basin or areas.

In a practical way, it is essential to test the sensitivity of each basin and the relevance of the management plans using various projection assumptions provided by climate models, in order to establish as finely as possible the combinations of measures to be taken with best cost - effectiveness, especially in the case of transboundary basins which require increased coordination and exchanges between riparian countries.

7 The participation of stakeholders and the civil society should be organized for a real mobilization of partners

INBO recommends that this participation be organized in Basin Committees or Councils.

These Basin Committees should be involved in the decision – making related to water policy in the basin, with procedures that clearly define their role. In particular, they should be associated to the definition of long – term objectives, to the preparation of Management Plans or master plans, to the selection of development and equipment priorities and to the implementation of Programs of Measures and multiyear priority investment programs, as well as to the setting of financing principles and to the calculation of water taxes that concern them.

If the various partners are involved at the earliest possible time, the more will be chance of good acceptance of all the measures which will have to be taken and of a definition of a real intersectoral adaptation strategy. Each sector must be well informed on the possible effects of climate change on its activity.

It is necessary to establish intersectoral links to foster exchanges of information and experience and coordination of actions in each basin.

Moreover, it is necessary that users, professionals or not, such as farmers, tree growers, fishermen, environmental associations, producers of hydropower, managers of navigation… adopt administrative, corporative or associative structures in basins and sub – basins.

Finally, significant means should be devoted to public awareness and participation, women and young people in particular, and to the training of their representatives regarding decision – making.

The transfers of research outcomes to water managers and decision makers, regarding socioeconomics and prospective analysis in particular, should allow improving and providing the basis of these decision – makings.

The investments necessary for the sustainable management, conservation and control of water resources and ecosystems and for the exploitation and maintenance of public utilities and the renewal of installations require huge financial resources.

This adaptation will also require additional financial resources that will undoubtedly have to be found by adopting new mechanisms such as basin taxes, insurance systems or market instruments.

It is necessary to set up everywhere complementary financing systems that are based on the users' participation and common cause.

It is thus necessary to consider specific and additional financial resources by combining national or local administrative taxes, the pricing of community services, the creation of geographic and inter – sectoral equalization mechanisms and taxes specific to objectives retained through dialogue.

These arrangements should be an incentive to limiting wastage and to removing pollution by changing the users' behavior.

8 Conclusions

Integrated and sound water resources management is more than ever a priority when this scarce resource is already a limiting factor for sustainable development in many countries of the world.

Mobilization is essential for humanity to win the water battle and prepare the future. Organizing this management on a basin scale is an effective solution which deserves to be developed, fostered and supported.

INBO intends to actively contribute to the efforts for adapting to the effects of climate change:

(1) By supporting programs for identification of the threats, allowing anticipation, thanks, for example, to the development of integrated information systems,

(2) By allowing the populations to be better warned and informed on the evolutions and behaviors that are likely to overcome the difficulties,

(3) By protecting natural spaces in basins and adapting infrastructures within the framework of

basin management plans,

(4) By supporting the development of better coordinated agricultural and forestry policies regarding deforestation control, irrigation and water storage in particular.

It is especially necessary:

(5) To improve the collection of information allowing a modeling of the phenomena and the development of scenarios leading to an identification of the most vulnerable black spots, to giving priority to the actions to be carried out and to a suitable answer;

(6) To reinforce the water management institutions to guarantee a long – term and rational meeting of the needs of the populations, industry, hydropower power, agriculture and fresh water fish farming, tourism and of the ecosystems.

Investing in water management is profitable! This produces immediate advantages but also creates a social, economic and environmental strength in the long term.

INBO member organizations have experience and expertise which they intend to pool at the disposal of all the countries and institutions which would like to follow them in an effective basin management approach.

Response of Runoff to Climate Change in the Weihe River Basin

Zuo Depeng and ***Xu Zongxue***

Key Laboratory of Water and Sediment Sciences, Ministry of Education,
College of Water Sciences, Beijing Normal University, Beijing, 100875, China

Abstract: The Weihe River Basin (WRB) is the major source of water supply for the Guanzhong Plain, and the economic hub of Western China. Many researchers found that the runoff had decreased significantly in the WRB since 1990s. Therefore, studies on impact of climate change on runoff in the WRB have important significance in utilizing water resources in an efficient and sustainable way and maintaining good health of the river ecosystem in Western China. In this study, Soil and Water Assessment Tool (SWAT) was selected to set up a hydrological model in the WRB, calibrated and validated with Sequential Uncertainty Fitting program (SUFI-2) based on river discharge. Sensitivity and uncertainty analysis were also performed to improve the model performance. Results indicate that the model performance criteria were quite satisfactory for this case. Then future daily precipitation, maximum and minimum air temperature data series at each station, generated by the Statistical Downscaling Method (SDSM), were inputted to drive the SWAT model to analyse the spatiotemporal characteristics of runoff during the future periods (2046 ~ 2065 and 2081 ~ 2100) under three climate scenarios including CSIRO, INM and MRI. Two emission scenarios (SRES A2 and SRES B1) were also included. The results show that average values of mean annual runoff in the periods of 2046 ~ 2065 and 2081 ~ 2100 were 80.4×10^8 m^3 and 104.3×10^8 m^3, which were greater than runoff in the base period by 12.4% and 45 %, respectively. In both of future periods, low flows would be much lower, while high flows tend to be much higher in the period of 2046 ~ 2065 than that in the base period. In other words, there would be more extreme events (droughts and floods) in the future. For the spatial distribution of runoff over the WRB, it showed consistency for runoff changes under most combined scenarios, with runoff decreasing at some areas of the upstream and the upstream of Beiluo River, while increasing at mid-lower stream of the WRB.

Key words: climate change, hydrological response, SWAT, GCM, Weihe River

1 Introduction

Many researchers have found that pervasive human interferences, such as construction of soil and water conservation projects, reservoirs and irrigation infrastructures etc., have contributed a lot to the reduction of water resources in the Yellow River Basin (Xu et al., 2002; Ringler et al., 2010). However, climate change, including reduction of precipitation and increase of temperature, have also played an important role on runoff changes in the second largest river in China (Wang et al., 2003; Liu et al., 2011). Hydrological cycle, as the key linkage of the atmosphere and biosphere, is inevitably influenced by the natural forces resulted by the climate changes (Vörösmarty et al., 2000; Xu et al., 2005a). Regarding the water resources management at present and in the future, investigations on hydrological response to climate change are becoming more and more important (Christensen et al., 2004; Xu et al., 2005b).

The Weihe River is the largest tributary of the Yellow River. It is the major source of water supply for the Guanzhong Plain, the economic hub of Western China. The WRB (Weihe River Basin) plays an important role for the management and development of water resources in the Yellow River Basin, and has a great strategic significance in the regional economic development and the development of Western China. Therefore, studies on climate change in the WRB have important significance in utilizing water resources in an efficient and sustainable way and maintaining good health of the river ecosystem in Western China (Zuo et al., 2012).

At present, assessment of hydrological response to climate change is basically following a paradigm that "climate scenario generation - hydrological simulation- hydrological response assessment". Coupling GCMs and hydrological models is considered as the most promising way to assess impact of climate change on water resources, because not only the GCMs can provide more credible information on the future climate, but also the hydrological modeling can obtain more knowledge on local hydrological processes and changes (Bronstert et al., 2002). Although distributed hydrological models with finer resolution are inherently incompatible with GCMs on spatial and temporal scales, statistical downscaling bridges two different scales by establishing empirical (statistical) relationships between large-scale features simulated reliably by GCMs and regional or local climate variables (Xu et al., 2009a).

In this study, coupling GCMs and hydrological model approach is adopted to evaluate hydrological response to future climate change scenarios. A unidirectional linkage system, i. e. GCMs outputs - statistical downscaling method - distributed hydrological model (SWAT) coupling system, is developed in the WRB. Firstly, scenarios of climate change in the future are generated based on GCMs output using statistical downscaling method. Secondly, the distributed hydrological model SWAT is calibrated and validated to simulate catchments discharge in the baseline. Finally, the climate scenarios are used to drive the SWAT model and possible changes of flow regime are projected in the future periods. The results can be used as a reference for water resources management for local water authorities.

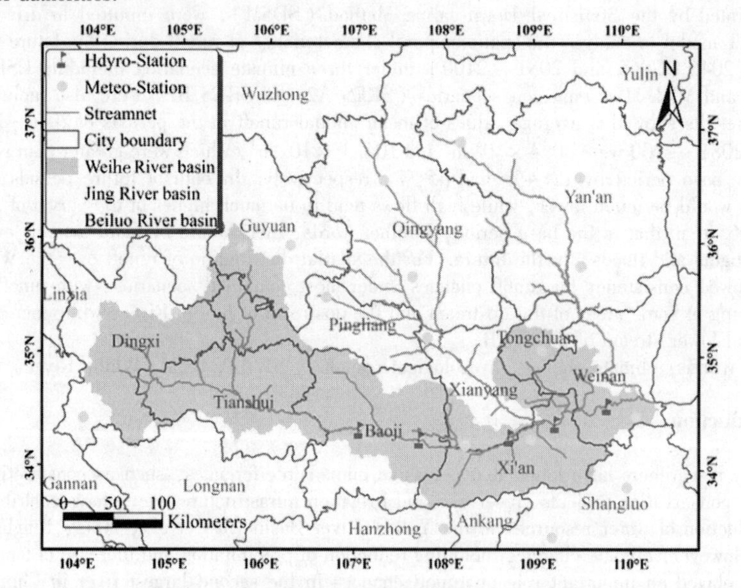

Fig. 1 Sketch map of the Weihe River Basin

2 Study Area Description

The Weihe River (Fig. 1) originates from the Niaoshu Mountain, goes through the provinces of Gansu, Ningxia and Shaanxi, and runs into the Yellow River at Tongguan. The basin is located between 104°00′ E ~ 110°20′ E and 33°50′ N ~ 37°18′ N with a length of 818 km and a drainage area of 1.35×10^5 km^2. Northern part of the basin is the Loess Plateau and southern part is the Qinling Mountains, western part is geographically high while eastern part is low. The Weihe River has many tributaries. The Jing River, the largest tributary of the Weihe River, goes across 455.1 km and covers an area of approximately 45,400 km^2 accounting for 33.7% of the total area of WRB;

the Beiluo River, the second largest tributary of the Weihe River, has a length of 680 km with a catchment area of 26,900 km^2 occupying 20% of the whole WRB.

The WRB is located in a semi-arid area with the temperate continental monsoon climate. The climate is cold, dry, and rainless in winter controlled by the Mongolia high pressure system, while hot and rainy in summer affected by the West Pacific subtropical high pressure system. The precipitation, temperature, evaporation, and runoff vary greatly at inter-annual and intra-annual timescales. The mean air temperature is 7.8 ~ 13.5 ℃ (decrease from the main channel toward the north and south tributaries), mean annual precipitation is 400 ~ 800 mm (decrease from south to north), potential evapotranspiration is 800 ~ 1000 mm (decrease from east to west), mean runoff is 195 m^3/s, and runoff coefficient varies within 0.1 ~ 0.2 (He et al., 2009).

3 Hydrological simulation

3.1 The hydrological model SWAT

The physically-based distributed hydrological model, SWAT (ver. 2009) developed within ArcGIS 9.3, is adopted in this study. SWAT is a continuous hydrological model that was developed by the US Department of Agriculture, Agricultural Research Service (USDA ARS) in the early 1990s to assess the impact of land management practices on water, sediment and agricultural chemical yields in large complex watersheds with varying soils, land uses and management conditions over long periods of time. The main components of SWAT include hydrology, climate, nutrient cycling, soil temperature, sediment movement, crop growth, agricultural management and pesticide dynamics (Neitsch et al., 2005). It has been used extensively and worldwide (Arnold and Allen, 1996; Moon et al., 2004; Xu et al., 2009b).

Spatial parameterization for the SWAT model is performed by dividing the watershed into sub-basins on the basis of topography, which is further subdivided into a series of Hydrologic Response Units (HRU) based on unique soil, land use, and slope characteristics. The responses of each HRU in terms of water and nutrient transformations and losses are determined individually, aggregated at the sub-basin scale and routed to the associated reach and catchment outlets through the channel network. SWAT represents the local water balance through four storage volumes: snow, soil profile, shallow aquifer, and deep aquifer. The soil water balance equation is the basis of hydrological modeling. The simulated processes include surface runoff, infiltration, evaporation, plant water uptake, lateral flow, and percolation to shallow and deep aquifers. Surface runoff is estimated by a modified Soil Conservation Service (SCS) curve number equation using daily precipitation data based on soil hydrologic group, land use/cover characteristics and antecedent soil moisture. Further technical details on the model are given by Neitsch et al. (2005).

3.2 Calibration and uncertainty analysis procedure SUFI-2

The SUFI-2 procedure (Abbaspour, 2007) was used for parameter optimization. In this procedure, all sources of uncertainties such as driving variables, conceptual model, parameters, and measured data are depicted onto the ranges of parameters, which are calibrated to bracket most of the measured data in the 95% prediction uncertainty. Overall uncertainty in the output is quantified by the 95% prediction uncertainty (95PPU) calculated at the 2.5% (L95PPU) and 97.5% (U95PPU) levels of the cumulative distribution of an output variable obtained through Latin hypercube sampling. The SUFI-2 starts with large but physically meaningful parameter ranges that bracket most of the measured data within the 95PPU, and then decreases the parameter uncertainties iteratively. After each iteration, new and narrower parameter uncertainties are calculated where the more sensitive parameters find a larger uncertainty reduction than the less sensitive parameters. Two indices are used to quantify the goodness of calibration/uncertainty performance: *P-factor*, which is the percentage of measured data bracketed by the 95PPU band, and *R-factor*, which is the average width of the band divided by the standard deviation of the corresponding measured variable.

Theoretically, the value of *P-factor* ranges between 0 and 100% , while that of R-factor ranges between 0 and infinity. A *P-factor* of 1 and R-factor of 0 is a simulation that exactly corresponds to measured data. As a larger *P-factor* can be found at the expense of a larger R-factor, a trade-off between the two indices must be sought (Abbaspour, 2007).

3.3 The hydrological simulation in the Weihe River Basin

Input data needed in the SWAT model include Digital Elevation Model (DEM), land use map, soil map, digital stream network, meteorological and hydrological observed data, etc. In this study, DEM (1:250,000), land-use map (1:250,000) and soil map (1:1,000,000) were obtained from the Data Center for Resources and Environmental Sciences, Chinese Academy of Sciences (RESDC). Daily meteorological data series (precipitation, maximum and minimum air temperature, relative humidity, wind speed, etc.) were obtained from China Meteorological Data Sharing Services System Network for 21 stations from 1961 to 2008. Daily discharge data series were obtained from the Yellow River Conservancy Commission (YRCC) for 5 gauging stations (Linjiacun, Weijiabao, Xianyang, Lintong, and Huaxian) during the same period of 1961 ~ 2008, which were used to calibrate and validate the SWAT model.

On the basis of DEM and digital stream network, a minimum drainage area of 8,000 hm^2 was chosen to discretize the WRB into 106 sub-basins. The geomorphology, stream parameterization, and overlay of soil and land use were automatically done within the SWAT model. These 106 sub-basins were further subdivided into 565 HRUs based on unique combinations of soil and land use.

Sensitivity analysis, calibration, validation and uncertainty analysis were performed for the hydrological model using river discharges. The period for calibration was 1990 ~ 2008, considering the first 3 years as warm-up period. Validation period was 1961 ~ 1989, the first 3 years were also excluded from the analysis in order to mitigate the unknown initial conditions. As hydrology in SWAT involves a large number of parameters, a sensitivity analysis was performed to identify the key parameters across different regions. For the sensitivity analysis, 28 parameters integrally related to stream flow were initially selected. They are referred to as "global" parameters in this study. Sensitivity analysis results showed that most of the 28 global parameters related to hydrology were sensitive to river discharge. As expected, parameters such as CN2 (SCS runoff curve number), ALPHA_BF (Baseflow alpha factor) were the most sensitive parameters.

The Nash and Sutcliffe (1970) coefficient of efficiency (E_{NS}), the coefficient of determination (R^2), and a modified version of the efficiency criterion Φ (Krause et al., 2005) were used to measure the model performance. The closer the values of E_{NS}, R^2, and Φ are to 1, the more successful the model calibration/validation.

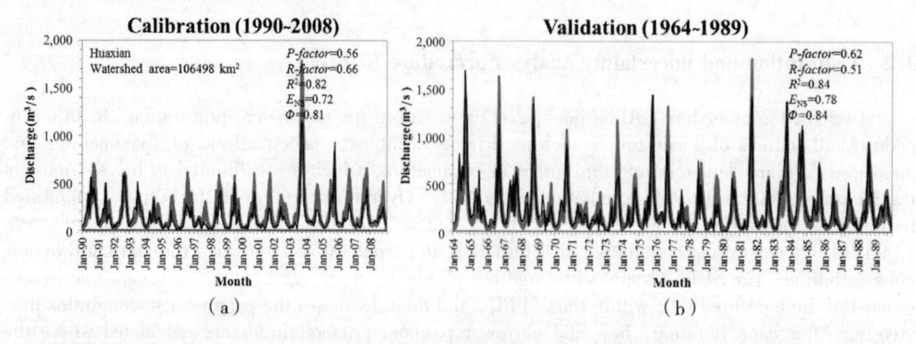

Fig. 2 Comparison between the observed and simulated prediction uncertainty band monthly hydrographs at Huaxian stations for both calibration and validation period

Fig. 2 presents the calibration and validation results for Huaxian station. *R*-factor for both calibration and validation at five stations (Linjiacun, Weijiabao, Xianyang, Lintong, and Huaxian) is quite small. However, the P-factor for the calibration at Linjiacun and Weijiabao is small, which indicates that the actual uncertainty is larger than expected. In calibration (validation), Φ and ENS values at most of the stations were higher than 0.6 (0.7), and R^2 at all stations was higher than 0.7 (0.8). Generally speaking, the model performance criteria were quite satisfactory for this case. Some stations with relative poor simulation were Linjiacun and Weijiabao, located in the upstream region. Based on the information from local experts and previous researches in the study area (Shangguan et al., 2002; Wang et al., 2007), possible reasons for large uncertainties in some stations include the construction or operation of reservoirs and insufficient accounting of agricultural water uses.

4 Future runoff under climate change scenarios

Daily precipitation, maximum and minimum air temperature data series at all the metrological stations, which were generated by the statistical downscaling method, were inputted into the calibrated SWAT model to simulate hydrological processes under climate change scenarios in the WRB. The daily meteorological data series are with two periods, 2046 to 2065 and 2081 to 2100. There were three GCMs selected for statistical downscaling, which are CSIRO: MK30, INM: CM30, and CGCM2.3.2, respectively. It will be described as CSIRO, INM and MRI in the following. It also includes two emission scenarios (SRES A2 and SRES B1). Runoff changes will be calculated based on the stream flow at Huaxian station.

4.1 Runoff driven by climate change scenarios

Mean values of the annual runoff in the periods of 2046 ~ 2065 and 2081 ~ 2100 were shown in the Tab. 1. It was 72×10^8 m^3 for mean annual runoff in the base period. In the period of 2046 ~ 2065, the highest mean annual runoff scenario would be 97.6×10^8 m^3, which was under the climate change scenario downscaled from CSIRO-B1; while the lowest mean annual runoff would be 68 $\times10^8$ m^3, which was driven by MRI-B1. In the period of 2081 ~ 2100, the highest mean annual runoff scenario would be 145.6×10^8 m^3, which was under the climate change scenario downscaled from INM-B1; while the lowest mean annual runoff would be 68.4×10^8 m^3, which was driven by CSIRO-B1. Therefore, it was interesting that the highest and lowest runoff scenarios would be under B1 scenarios in both periods of 2046 ~ 2065 and 2081 ~ 2100. Overall, it showed increasing trends for runoff of the WRB in the future, with the magnitude of these trends greater in 2081 ~ 2100 than 2046 ~ 2065. Because average values of mean annual runoff in the periods of 2046 ~ 2065 and 2081 ~ 2100 were 80.4×10^8 and 104.3×10^8 m^3, which were greater than runoff in the base period by 12.4% and 45%, respectively.

Tab. 1 also described the 10th, 50th, and 90th percentiles (described as R_10, R_50, and R_90, respectively) of the annual runoff data series in the periods of 2046 ~ 2065 and 2081 ~ 2100. The 10 th and 90 th percentiles of annual runoff indicated the dry and wet years, respectively. The normal runoff could be described as mean or the 50 th percentile values of annual runoff. The driest R_10 was under CSIRO-B1, with the value 37.6×10^8 m^3; while the highest R_10 would be 100×10^8 m^3, which was under the climate change scenario downscaled from INM-B1. It was also true that the highest and lowest *R*10 would be under the combined scenarios of INM-B1 and CSIRO-B1 in the periods of 2046 ~ 2065 and 2081 ~ 2100 respectively, which was consistent with mean annual runoff mentioned above. That was also consistent with mean annual runoff. In generally, average R10 under all of scenarios was 52.5×10^8 m^3 and 64.7×10^8 m^3 in the periods of 2046 ~ 2065 and 2081 ~ 2100, respectively.

It could be seen from Tab. 1 that the most wet R_90 was under CSRIO-B1 in the period of 2046 ~ 2065, with the value 154×10^8 m^3; while the lowest R_90 would be 77.6×10^8 m^3, which was under the climate change scenario downscaled from INM-B1 in the period of 2046 ~ 2065. In generally, average R_90 under all of scenarios was 108.3 $\times 10^8$ and 138.3 $\times 10^8$ m^3 in the periods of 2046 ~ 2065 and 2081 ~ 2100, respectively.

Tab. 1 Quantile and mean values for annual runoff in the periods of 2046 ~ 2065 and 2081 ~ 2100

(Unit: 10^8 m^3/a)

Periods	Quantile	A2			B1		
		CSIRO	INM	MRI	CSIRO	INM	MRI
2046 - 2065	R_10	413.3	539.1	678.3	344.4	376.5	723.9
	R_50	479.2	607.3	762.2	447.3	440.1	798.3
	Mean	500.2	620.4	764.6	470.5	485.6	798.1
	R_90	591.6	703.4	806.1	558.2	604.3	870.6
2081 - 2100	R_10	517.0	528.2	721.4	472.3	489.4	705.7
	R_50	553.1	562.8	743.8	534.2	567.7	774.3
	Mean	584.3	592.9	748.5	543.3	603.9	775.1
	R_90	679.3	688.8	776.2	582.3	733.0	861.7

Besides mean and percentiles values for annual runoff data series, the percentiles daily streamflow were also analyzed in this study. The 10th and 90th percentiles of the daily streamflow data series (described as $Q10$ and $Q90$) in the base and future periods (2046 ~ 2065 and 2081 ~ 2100) were shown in the Tab. 2. In the period of 2046 ~ 2065, it showed decreasing trends for $Q10$ under all of climate change scenarios except for INM-A2, and exhibited increasing trends for $Q90$ under all of climate change scenarios. That indicated low flows would be much lower, while high flows tend to be much higher in the period of 2046 ~ 2065 than that in the base period. In other words, there would be more extreme events (droughts and floods) in the future. The lowest $Q10$ in this period would be lower by 27.9% than that in the base period with the value 11.5 m^3/s, which was driven by MRI-B1. The greatest $Q90$ in this period would be increased by 82% with the value 837.7 m^3/s, which was under the climate change scenario downscaled from CSIRO-B1.

In the period of 2081 ~ 2100, it showed increasing trends for $Q90$ under all of climate change scenarios, and exhibited no obvious trends for $Q10$ under different climate change scenarios. It was partially consistent with that in the period of 2046 ~ 2065, which indicated there would also be more extreme flood events in the period of 2081 ~ 2100. The lowest $Q10$ in this period would decrease by 38% with the value 9.9 m^3/s, which was under the climate change scenario downscaled from the CSIRO-B1. The greatest Q90 in this period would be increased by about one time with the value 964.9 m3/s, which was under the climate change scenario downscaled from MRI-B1.

4.2 Spatial distribution of future scenario runoff

Spatial distribution of mean annual runoff under climate change scenarios downscaled from CSIRO was shown in Fig. 3. Overall, the spatial distribution of mean annual runoff driven by climate change scenarios downscaled from CSIRO is consistent with that in the base period. It tended to be the highest runoff in the some areas of upstream and midstream regions of the basin, while showed the most poorly runoff in the most areas of upstream region and Beiluo River Basin. Runoff under B1 scenario was smaller than that under A2 scenario in both of periods, because area with runoff more than 50 mm under B1 scenario was smaller than that under A2 scenario. The differences of runoff between the periods of 2046 ~ 2065 and 2081 ~ 2100 were not as obviously as that between two emission scenarios. There was little difference for the highest and lowest runoff between two periods.

Tab. 2 Quantile values for daily streamflow in the periods of 2046 ~ 2065 and 2081 ~ 2100

			Base period	A2			B1		
				CSIRO	INM	MRI	CSIRO	INM	MRI
2046 ~ 2065	*Q*10	Flow (m^3/s)	390	377.2	335.3	298.1	319.5	251.2	293.2
		Changes (%)		-3.3	-14.0	-23.6	-18.1	-35.6	-24.8
	Q90	Flow (m^3/s)	3,070	3,166.3	3,841.6	4,020.8	2,976.3	3,159.9	4,093.4
		Changes (%)		3.1	25.1	31.0	-3.1	2.9	33.3
2081 - 2100	Q10	Flow (m^3/s)	390	395.1	377.3	256.7	431.1	296.0	316.4
		Changes (%)		1.3	-3.2	-34.2	10.5	-24.1	-18.9
	Q90	Flow (m^3/s)	3,070	3,520.1	3,602.6	3,870.9	3,261.4	3,880.8	4,131.3
		Changes (%)		14.7	17.3	26.1	6.2	26.4	34.6

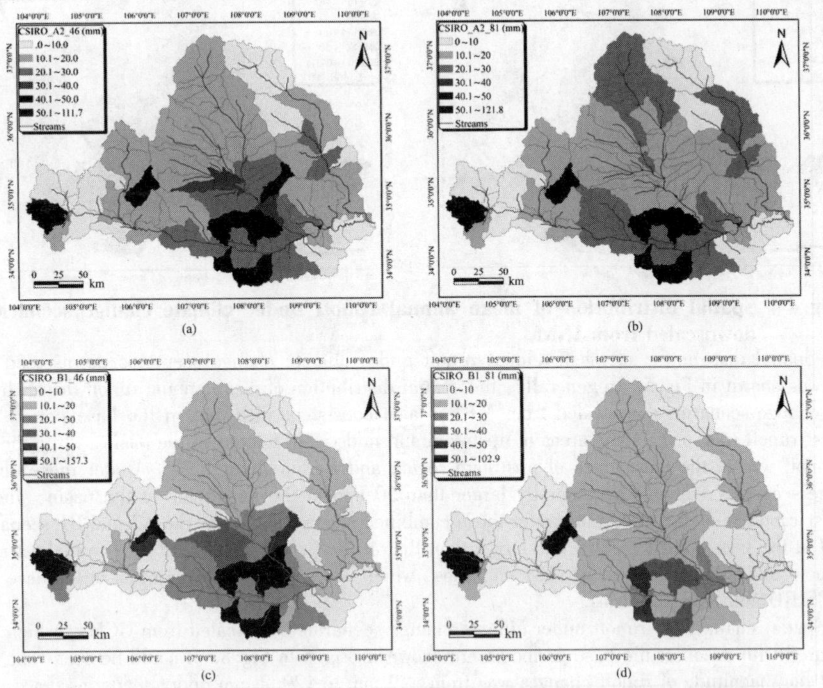

Fig. 3 Spatial distribution of mean annual runoff under climate change scenarios downscaled from CSIRO

Fig. 4 described spatial distribution of mean annual runoff under climate change scenarios downscaled from INM. In generally, the spatial distribution of mean annual runoff driven by climate change scenarios downscaled from INM is also consistent with that in the base period. The highest runoff was in the some areas of upstream and midstream regions of the basin, while the lowest runoff was at the most areas of upstream region and Beiluo River Basin. Under these climate change scenarios, runoff was generally larger than 40 mm at the midstream of the basin, and the highest runoff was about 200 mm under these combined scenarios. As for the different climate change scenarios, it showed consistency for spatial distribution among combined scenarios of A2 in 2046 ~ 2065 and

2081 ~ 2100, B1 in 2046 ~ 2065, while runoff under B1 scenario in 2081 ~ 2100 was obviously greater than other scenarios, and the highest value in the midstream was more than 200 mm.

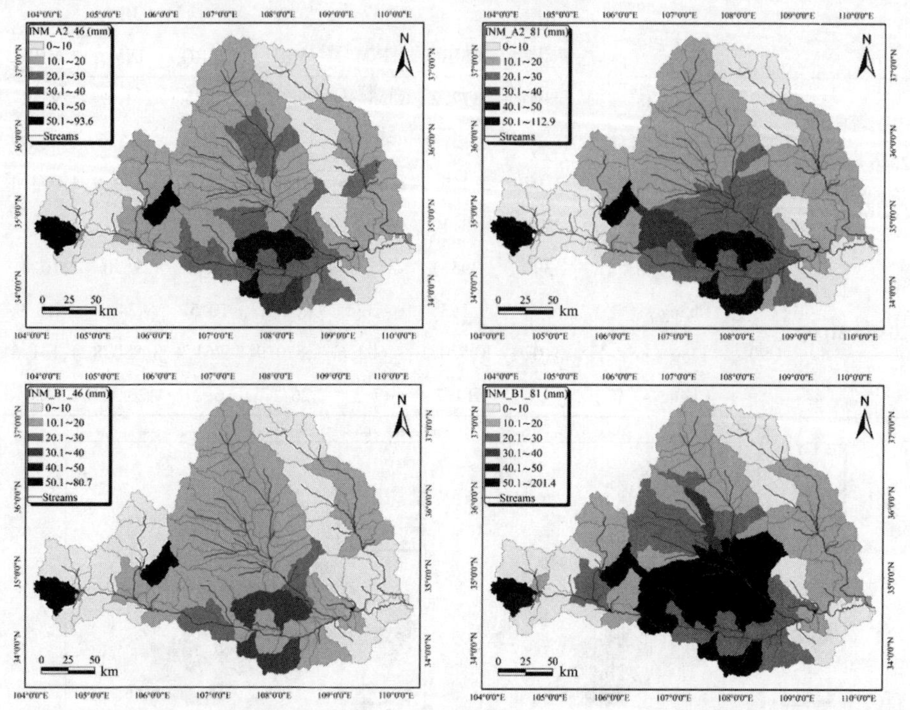

Fig. 4 Spatial distribution of mean annual runoff under climate change scenarios downscaled from INM

Spatial distribution of mean annual runoff under climate change scenarios downscaled from MRI was shown in Fig. 5. In generally, the spatial distribution of mean annual runoff driven by climate change scenarios downscaled from MRI is also consistent with that in the base period. The highest runoff was in the some areas of upstream and midstream regions of the basin, while the lowest runoff was at the most areas of upstream region and Beiluo River Basin. Under these climate change scenarios, runoff was generally larger than 50 mm at the midstream of the basin, and the highest runoff was about 200 mm under these combined scenarios. Under the A2 and B1 scenarios, runoff at the period 2081 ~ 2100 was larger than that at the period 2046 ~ 2065. The highest runoff was more than 220 mm driven by MRI scenarios, which were much larger than that at the base period, CSIRO and INM scenarios.

Spatial changes for runoff under climate change scenarios downscaled from GCMs, which were compared with runoff at the base period, were shown in Fig. 6 to Fig. 8. It could be seen from Fig. 6 that the magnitude of runoff changes was from −9 mm to 129.2 mm under scenarios downscaled from CSIRO, and the change of runoff in the most region showed positive trend. In generally, it showed consistency for runoff changes under different combined scenarios, with runoff decreased at the headwater catchment of WRB and upstream of the Beiluo River basin, while increased at the middle and downstream catchment of WRB. The magnitude of decreasing trends was less than 10 mm, while obviously increasing trends at the downstream region of the catchment, especially under B1 scenario with the largest magnitude of these trends about 100 mm. Overall, it showed decreasing trends at the headwater catchment and increasing trends at the midstream catchment for runoff under CSIRO climate change scenarios, with the magnitude of increasing runoff obviously greater under B1 scenario than that under A2 scenario.

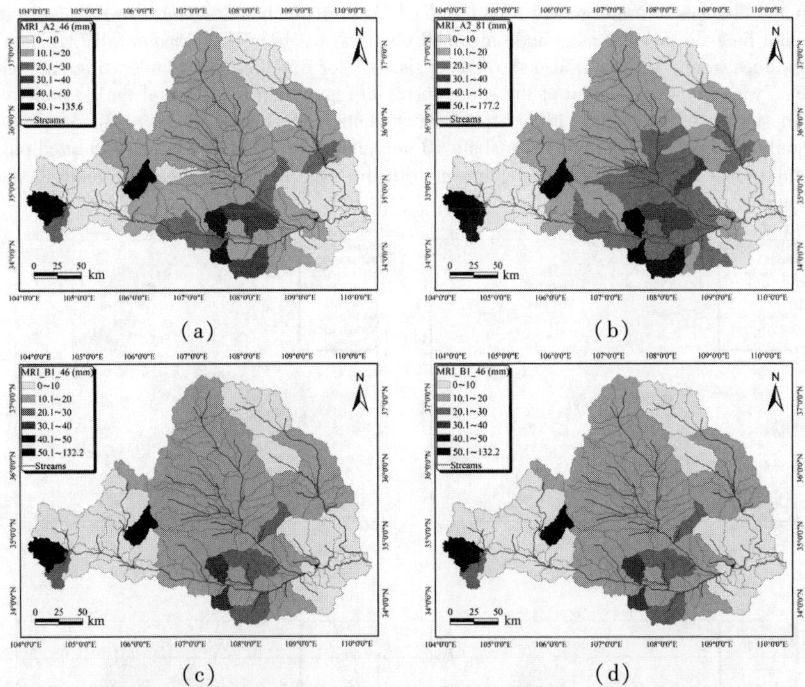

Fig. 5 Spatial distribution of mean annual runoff under climate change scenarios downscaled from MRI

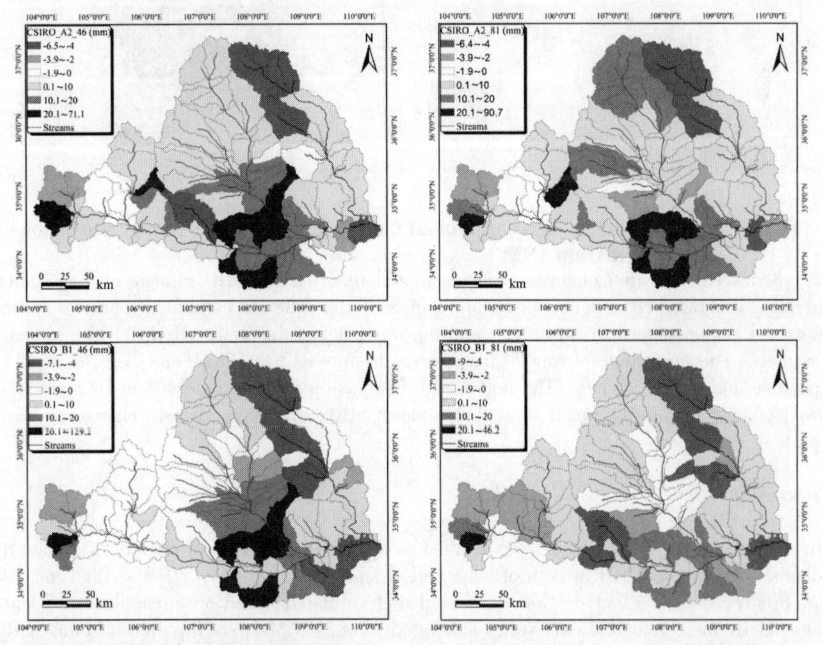

Fig. 6 Spatial changes of mean annual runoff under climate change scenarios downscaled from CSIRO

Spatial changes of mean annual runoff under INM climate change scenarios were shown in Fig. 7. It could be seen that the magnitude of runoff changes was from −28.5 mm to 173.4 mm, with most of values were positive. It also showed consistency for runoff changes under different combined scenarios, with runoff decreased at the some upstream region and upstream of the Beiluo River basin, while increased at the mid-lower stream catchments. The region where runoff decreased with the magnitude of decreasing trends was about 10 mm and up to about 30 mm. It exhibited obviously increasing trends at the downstream catchment with the magnitude of increasing trends larger than 20 mm and up to about 173.4 mm.

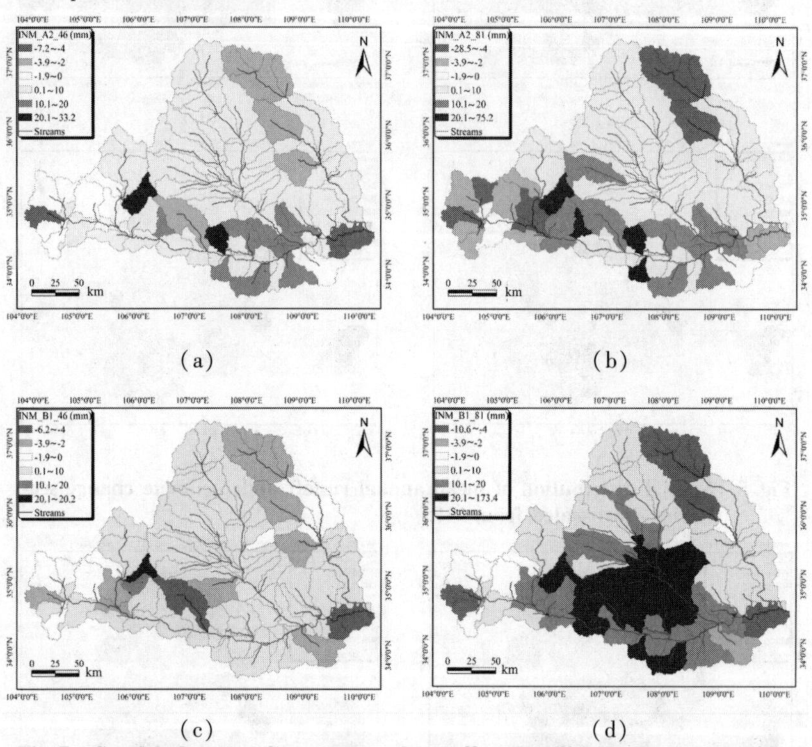

Fig. 7 Spatial changes of mean annual runoff under climate change scenarios downscaled from INM

Fig. 8 described spatial changes of mean annual runoff under MRI climate change scenarios. In generally, it showed consistency for runoff changes under different combined scenarios, with runoff decreased at the some upstream area, and the upstream of the Beiluo River, while increased at other regions. The magnitude of runoff changes was from −8.2 to 198.1 mm, with most of values were positive and up to 190 mm. The region with decreasing runoff was less than 10 mm. Overall, it showed greater increasing runoff than that under CSIRO and INM climate change scenarios at most parts of the basin.

5 Conclusions

In this study, the SWAT model was applied as the detail hydrological model to simulate hydrological processes in the WRB in both of base period and future periods (2046 ~ 2065 and 2081 ~ 2100). Firstly, the SWAT model was calibrated and validated based on streamflow of 5 stations. Secondly, daily meteorological data series generated from 3 GCMs were inputted into the calibrated SWAT model to simulate hydrological processes under these climate change scenarios. Finally, future runoff was analyzed based percentiles values and spatial distributions. More specifically, the following conclusions were made:

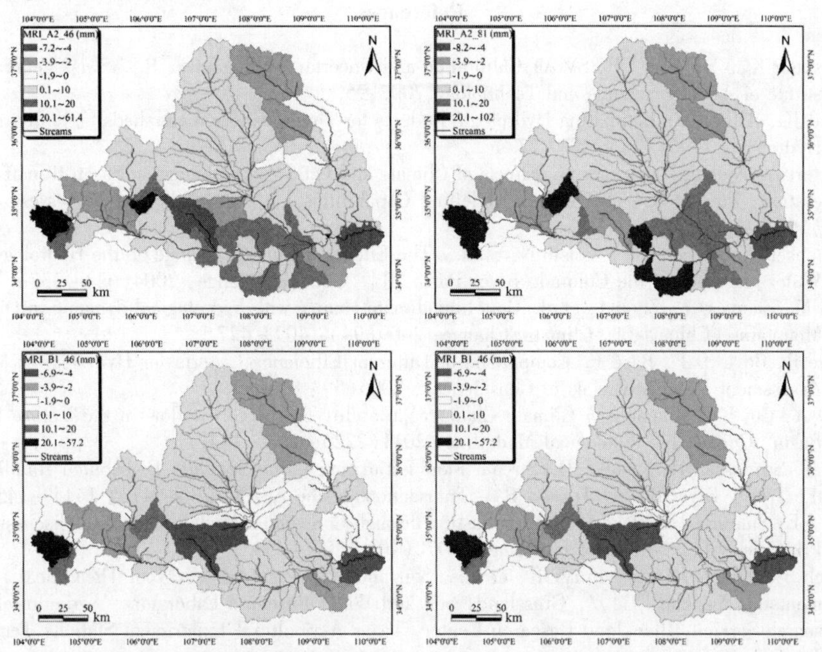

Fig. 8 Spatial changes of mean annual runoff under climate change scenarios downscaled from MRI

The calibrated SWAT model was satisfied for application. The model could capture both temporal and spatial distribution of runoff of the basin, with water balance and hydrograph shape were well simulated.

Average values of mean annual runoff in the periods of 2046 ~ 2065 and 2081 ~ 2100 were 80.4×10^8 m^3 and 104.3×10^8 m^3, which were greater than runoff in the base period by 12.4% and 45 %, respectively. In a word, it showed increasing trends for runoff of the WRB in the future, with the magnitude of these trends greater in 2081 ~ 2100 than 2046 ~ 2065.

In both of future periods, it generally showed decreasing trends for $Q10$ under most of climate change scenarios, while exhibited increasing trends for $Q90$ under most of climate change scenarios. That indicated low flows would be much lower, while high flows tend to be much higher in the period of 2046 ~ 2065 than that in the base period. In other words, there would be more extreme events (droughts and floods) in the future.

For the spatial distribution of runoff over the WRB, it showed consistency for runoff changes under most combined scenarios, with runoff decreased at some areas of the upstream regions and the upstream of Beiluo River basin, while increased at middle and downstream of the basin. Although there was no significant difference for the magnitude of decreasing trends under different combined scenarios with values about −10 mm, the highest increasing runoff exhibited at the midstream region of basin with value about 100 mm.

Acknowledgements

This study is jointly supported by EU-China River Basin Management Programme, Yellow River Climate Change Scenario Development-Phase 2, the Sino-Swiss Science and Technology Cooperation Project (2009DFA22980), the Ministry of Science and Technology, P. R. China, and the Major Special S&T Project on Water Pollution Control and Management (2009ZX07012 - 002 - 003), the Ministry of Environmental Projection, P. R. China.

References

Abbaspour KC. SWAT-CUP, SWAT Calibration and Uncertainty Programs [R]. Swiss Federal Institute of Aquatic Science and Technology,2007:95.

Arnold JG, Allen PM. Estimating Hydrologic Budgets for Three Illinois Watersheds [J]. Journal of Hydrology, 1996(176): 57 – 77.

Bronstert A, Niehoff D, Burger G. Effects of Climate and Land – use Change on Storm Runoff Generation: Present Knowledge and Modelling Capabilities [J]. Hydrological Processes, 2002 (16): 509 – 529.

Christensen NS, Wood AW, Voisin N, et al.. The Effects of Climate Change on the Hydrology and Water resources of the Colorado River Basin [J]. Climatic Change, 2004(62): 337 – 363.

He H M, Zhang Q F, Zhou J, et al. Coupling climate change with hydrological dynamic in Qinling Mountains, China [J]. Climatic Change, 2009(94): 409 – 427.

Krause P, Boyle D P, Bäse F. Comparison of Different Efficiency Criteria for Hydrological Model Assessment [J]. Advances in Geosciences, 2005(5): 89 – 97.

Liu Q A, Cui B S. Impacts of Climate Change/Variability on the Streamflow in the Yellow River Basin, China [J]. Ecological Modelling, 2011(222): 268 – 274.

Moon J, Srinivasan R, Jacobs JH. Stream Flow Estimation Using Spatially Distributed Rainfall in the Trinity River Basin, Texas [R]. Transactions of the ASAE, 2004(47): 1445 – 1451.

Nash J E, Sutcliffe J V. River Flow Forecasting through Conceptual Models Part I – A discussion of Principles [J]. Journal of Hydrology, 1970(10): 282 – 290.

Neitsch S L, Arnold J G, Kiniry JR, et al.. Soil and Water Assessment Tool Theoretical Documentation Version [C]//. Grassland, Soil and Water Research Laboratory, Agricultural Research Service, Blackland Research Center, Texas Agricultural Experiment Station: Temple, Texas.

Ringler C, Cai X M, Wang J X, et al.. Yellow River Basin: Living with Scarcity [J]. Water International, 2010(35): 681 – 701.

Shangguan Z P, Shao M, Horton R, et al.. A Model for Regional Optimal Allocation of Irrigation Water Resources under Deficit Irrigation and its applications [J]. Agricultural Water Management, 2002(52): 139 – 154.

Vörösmarty C J, Green P, Salisbury J, et al.. Global Water Resources: Vulnerability from Climate Change and Population Growth [J]. Science, 2000(289): 284.

Wang G Q, Li H B. Causes of Run – off and Sediment Variation in Weihe River Basin [J]. Journal of Experimental Botany, 2003(54): 55 – 55.

Wang Z Y, Wu B S, Wang G Q. Fluvial Processes and Morphological Response in the Yellow and Weihe Rivers to Closure and Operation of Sanmenxia Dam [J]. Geomorphology, 2007(91): 65 – 79.

Xu C Y, Widén E, Halldin S. Modelling Hydrological Consequences of Climate Change—progress and Challenges [J]. Advances in Atmospheric Sciences, 2005(22): 789 – 797.

Xu Y, Gao X J, Shen Y, et al.. A Daily Temperature Dataset over China and its Application in Validating a RCM Simulation [J]. Advances in Atmospheric Sciences, 26: 763 – 772.

Xu Z X, Takeuchi K, Ishidaira H, et al.. An Overview of Water Resources in the Yellow River Basin [J]. Water International, 2005(30): 225 – 238.

Xu Z X, Takeuchi K, Ishidaira H, et al.. Sustainability Analysis for Yellow River Water Resources using the System Dynamics Approach [J]. Water Resources Management, 2002 (16): 239 – 261.

Xu Z X, Zhao F F, Li J Y. Response of Streamflow to Climate Change in the Headwater Catchment of the Yellow River Basin [J]. Quaternary International, 2009(208): 62 – 75.

Zuo D P, Xu Z X, Yang H, et al.. Spatiotemporal Variations and Abrupt Changes of Potential Evapotranspiration and its Sensitivity to Key Meteorological Variables in the Weihe River Basin, China [J]. Hydrological Processes, 2012(8): 1149 – 1160.

Relation Analysis of Cold Wave Weather and Ice Regime of the Yellow River Basin

Wang Chunqing[1,2,3], *Arthur E. Mynett*[2,3], *Zhang Yong*[1] and *Zhang Fangzhu*[1]

1. Bureau of Hydrology, Yellow River Conservancy Commission, Zhengzhou, 450004, China
2. UNESCO-IHE, Institute for Water Education, 2601DA, Delft, the Netherlands
3. Delft University of Technology, Faculty of CiTG, 2600GA, Delft, the Netherlands

Abstract: Using the winter monthly air temperature data (November of this year to March of next year) from 1951 to 2008, statistically analyze the cold wave weather process of the last 50 years of the Yellow River, and described the temporal and spatial distribution, variation characteristics, intensity and influence scope of the cold wave. Analyze the cold wave weather with the relation of the ice regime and also its influence on the Yellow River, which lays the foundation to improve the ice regime forecast standard.

Key words: cold wave, ice regime, Ning-Meng reach, the lower Yellow River

In high latitude area which low solar height, short sunshine, low solar radiation in winter, the surface of continent cooling down quickly. Large amount cold air mass accumulated at the high latitude area. Guided by the northwest air stream behind the upper east Asian trough, cold air mass often flow from the source area to low latitude warm and humid area, which cause the strong wind cooling weather, with accompany occurring the snow and rainfall, freezing fog and frozen rain in front of the cold front, which is cold air activity. If the cooling intensity reaches certain standard, then the cold air activity can be called cold wave. The cold wave can lead to river frozen, freeze-up length increase. So for the river system, the cold wave weather in winter is one the important disastrous weather.

There are three reaches occur the ice flood in the Yellow River basin (see Fig. 1), one is the Ning-Meng reach (including Ningxia Hui Nationality Autonomous Region reach and Inner Mongolia Autonomous Region reach) of the upper Yellow River, the second is Hequ reach (from Tianqiao reservoir to Longkou) of the middle Yellow River, the last is the lower Yellow River. The Ning-Meng reach and the lower Yellow River are most severe reaches with the high frequency occurrence of the ice flood disasters for the Yellow River, so carry on the analysis and research on these two reaches.

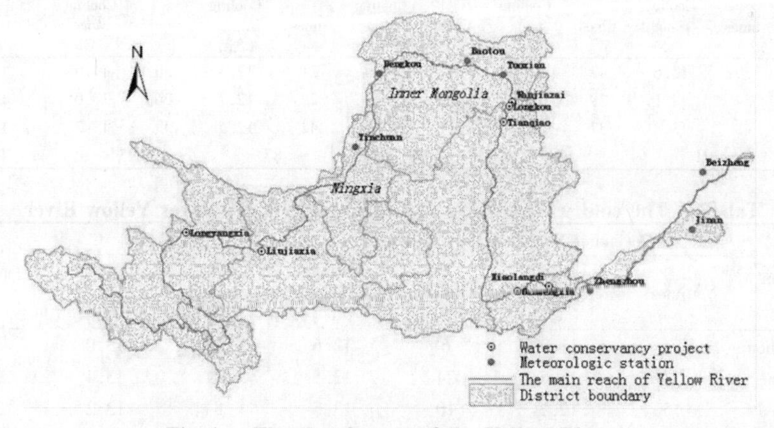

Fig. 1 The sketch map of the Yellow River

1 Standard of the cold wave classification

According to the requirement of the ice regime and weather and climatic character, and based on the China national cold air grades standard, take the weather process of air temperature decrease larger than 8 ℃ in 24 h and larger than 10 ℃ in 48 h and total process larger than 12 ℃ with minimum air temperature less than 4 ℃ as ordinary cold wave process; the weather process of air temperature decrease larger than 12 ℃ in 24 h with minimum air temperature less than 4 ℃ as moderate intensity cold wave process; the weather process of air temperature decrease larger than 16 ℃ in 24 h with minimum air temperature less than 0 ℃ as strong intensity cold wave process. During the statistic of the cold wave process, if occur the adjacent months process, then take the more days month as the cold wave process occur.

2 Climatic character of the cold wave

2.1 Spatial and temporal distribution

2.1.1 Occurrence frequency

According to the cold wave standard mentioned above, carry on the statistical analysis of daily average air temperature of Dengkou, Baotou and Tuoxian for the Ning-Meng reach from November of this year to March of next year and the data series is from 1954 to 2008 and of Zhengzhou, Jinan, Beizhen for the lower Yellow River from December of this year to February of next year and the data series is from 1951 to 2008 respectively. The statistical results are shown as Tab. 1 and Tab. 2. From the occurrence frequency of the cold wave, in recent 50 years, the Ning-Meng reach occur 2 to 4 cold wave weather processes in every year and the lower Yellow River occur 0.5 to 2 cold wave weather processes. The reason of the occurrence difference for the two reaches is the Ning-Meng reach located at the north part and every year the times of cold wave occur next to northeast of China, middle and north part of Inner Mongolia Autonomous Region and Xinjiang Uyghur Autonomous Region, and belongs to frequent occurrence region of cold wave, the lower Yellow River is located at the mid-latitude and the cold wave occurrence frequency is lower than that of the Ning-Meng reach.

Tab. 1 The cold wave process statistical table of the Ning-Meng reach

Station	November		December		January		February		March		Total
	times	Cooling range (℃)	times	Cooling range (℃)	times	Cooling range (℃)	times	Cooling range (℃)	times	Cooling range (℃)	
Dengkou	39	13.6	42	12.8	25	12.4	21	12.6	30	11.9	157
Baotuo	34	14.1	29	13.6	24	13.0	22	12.7	26	13.0	135
Tuoxian	48	13.8	37	14.3	29	12.9	42	12.5	39	12.2	195
Total	121		108		78		85		95		487

Tab. 2 The cold wave process statistical table of the lower Yellow River

Station	December		January		February		Total
	times	Cooling range (℃)	times	Cooling range (℃)	times	Cooling range (℃)	
Zhengzhou	14	11.7	6	12.6	13	13.0	33
Jinan	34	13.5	24	12.5	38	13.4	96
Beizhen	15	12.7	10	13.5	8	13.2	33
Total	63		40		59		162

Besides the above difference, the occurrence frequency of the cold wave is different in the same reach. For the Ning-Meng reach, the east is larger than the west of the cold wave occurrence frequency, and for the lower Yellow River, the Jinan station is located at the north part of the Jinan city before 1999, and after that moved to the Gui mountain which is located at the south part of Jinan, due to the influence of the hilly land, the cold wave occurrence frequency is evident larger than that of the other two stations, so has the character of hilly land is larger than the plain area. According to the cold wave occurrences, the geographical distribution of the two reach can be divided into the following:

High frequent area: Tuoxian of the Ning-Meng reach and Jinan of the lower Yellow River which occur the cold wave frequent most in winter. Tuoxian has totally occurred 195 cold wave s since 1951, and annual mean is about 4 times and one more time larger than that of Dengkou and Baotou. Jinan has totally occurred 96 cold waves since 1954, and annual mean is about 2 and nearly 3 times than that of Zhengzhou and Beizhen.

Low frequent area: Mainly as Zhengzhou and Beizhen, the annual average cold wave occurs less than one time.

2.1.2 Spatial and temporal variation

From Tab. 1 and Tab. 2, there are two frequency maximum values which occurred in November and December of the cold wave. And this takes 48.4% of the total cold wave times, March takes the second place. So November and December is the cold wave most active period of the Yellow River, and January and February is the cold wave relatively quiet period, which is because at that time the air temperature is very low and ordinary it is difficult to cool down 8 ℃, so the number of cold wave is relative less than the other months.

Through the statistics of cold wave decadal occurrence times in winter, the cold wave occur more prevail in 1960's and 1970's, and occur less prevail since 1990. So in the recent 50 years, the cold wave activities more frequent in 1960's and 1970's, in 1950's takes the second the place, and since 1980 it appears the tendency of decrease.

The cold wave decadal variation of the upper and lower stream appears different characteristics. The variation tendency of the three stations for the Ning-Meng reach is almost same, in recent 50 years, the cold wave occurred most frequent in 1971 ~ 1972 and 1965 ~ 1966, which total occurred 19 times cold waves both. Next is in 1970 ~ 1971 and 1966 ~ 1967, and occurred 18 and 17 times respectively. The minimum is in 1983 ~ 1984, which no cold wave process occurred. So the decadal variation of the Ning-Meng reach is great.

The variation tendency of the three stations for the lower Yellow River is incompletely same. The Zhengzhou and Jinan is almost same, but the cold wave annual variation of Jinan is larger than the other two stations, the cold wave annual variation of Beizhen is comparable stable which means difference of the maximum year and minimum year is only one time. The cold wave occurred most frequent is 1978 ~ 1979 which is 10 times, and no cold wave process occurred year is in 1984 ~ 1985 and 1990 ~ 1991.

2.2 Intensity and scope

2.2.1 Intensity

Tab. 3 shows the 24 h maximum air temperature cooling value of the cold wave in recent 50 years. From Tab. 3, there is total 3 moderate intensity cold wave occurred in Jinan, and mainly in February and December. The number of the moderate intensity cold wave and above of the Ning-Meng reach is larger than that of the lower Yellow River, and Tuoxian occurred 11 times, and Dengkuo 5, Baotou is the last and only 1, the occurrence time mainly January, and then March. The reason is the Ning-Meng reach is located at the high latitude and the cold air mass route is from north to south, and the intensity of the cold air mass weaker as moves to south, so the cold air mass intensity of the Ning-Meng reach is stronger than that of the lower Yellow River.

From Tab. 3, the strong intensity cold wave occurred less, besides Dengkuo occurred 3 times, the other stations not occur, and only takes 2% of the total cold wave times. The 24 hours maximum cooling range is 18.4℃, comparing with the that of other place of China, the cooling range is only less than that of the northeast of China and middle and north part of Inner Mongolia Autonomous Region.

Tab. 3 24 h maximum air temperature cooling value of the cold wave for the Yellow River

Dengkou		Baotou		Tuoxian		Jinan	
Date (Y. M. D)	Range (℃)	Date (Y. M. D)	Range (℃)	Date (Y. M. D)	Range (℃)	Date (Y. M. D)	Range (℃)
				1968. 11. 6 ~ 9	12. 6		
1954. 12. 30 ~ 1955. 1. 2	17. 4					1963. 12. 22 ~ 25	12. 7
1955. 1. 3 ~ 6	18. 4			1956. 1. 5 ~ 6	13. 2		
1971. 1. 18 ~ 22	12. 5			1962. 1. 8 ~ 9	13. 9		
				1971. 1. 19 ~ 21	14. 7		
				1972. 1. 23 ~ 27	13. 1		
				1980. 1. 25 ~ 29	12. 0		
						1979. 2. 20 ~ 22	12. 9
						2001. 2. 22 ~ 23	13. 5
1978. 3. 8 ~ 9	12. 6	2005. 3. 9 ~ 11	14. 4	1956. 3. 5 ~ 6	13. 5		
2005. 3. 9 ~ 11	16. 1			1960. 3. 10 ~ 12	12. 9		
				1978. 3. 8 ~ 10	12. 2		
				1987. 3. 18 ~ 20	12. 0		
				2005. 3. 9 ~ 11	12. 8		

2. 2. 2 Scope

Due to the difference of cold wave route and its geological conditions, the influence scope of the cold wave exist significance difference, sometime can cover the whole basin, sometime only influence the partial area. Tab. 4 shows the monthly cold wave times take the percentage of the total. In recent 50 years, there are 12 time all 3 stations of the lower Yellow River reach the cold wave standard, and takes 36. 4% of the total, the cold wave occur at the same time of Zhengzhou and Jinan are 28 times and takes 84. 8% of the total, and Beizhen and Jinan are 26 times and takes 78. 8% of the total, Zhengzhou and Beizhen are 11 times and takes only 33. 3% of the total. While for the Ning-Meng reach, all 3 stations reach the cold wave standard as 26 times in November, 16 in December, 11 in January, 9 in February and 16 in March. In November the number of all 3 stations reaches the cold wave standard is the highest, which means in this month the influence area of the cold wave is also large, in March is next to that. The cold wave scope covers 2 stations and above takes 74. 1%, and in January and February the influence scope is minimum, which takes 56. 4% and 72. 6% respectively. And both the Ning-Meng reach and the lower Yellow River occur the cold wave at the same time is only 5 times.

Tab. 4 The monthly cold wave times take the percentage of the total

Month	Ning-Meng reach (%)	Lower Yellow River (Y. M. D)	Both two reaches (Y. M. D)
11	67. 2		
12	48. 9	1963. 12. 22 ~ 25	1963. 12. 22 ~ 25
		1965. 12. 13 ~ 16	1985. 12. 3 ~ 8
		1975. 12. 3 ~ 12	
		1982. 12. 3 ~ 6	
		1985. 12. 3 ~ 8	
1	42. 3	1958. 1. 12 ~ 15	1958. 1. 12 ~ 15
		1979. 1. 25 ~ 31	
		1980. 1. 27 ~ 30	
2	32. 1	1966. 2. 19 ~ 22	1966. 2. 19 ~ 22
		1969. 2. 11 ~ 16	1996. 2. 13 ~ 18
		1979. 2. 19 ~ 24	
		1993. 2. 5 ~ 7	
		1996. 2. 13 ~ 18	
3	51. 6		

Through the above analysis, the cold wave occurrence of the Yellow River influenced by latitude, cold wave route and topography, so the spatial and temporal distribute uneven. The Ning-Meng reach which located at the high latitude area, the frequency of the cold wave occurrence is larger than that of the low Yellow River, and also intensity and scope.

3 Cold wave and ice regime

3.1 Relation with the river freeze-up and breakup time

3.1.1 Influence on freeze-up time

Firstly statistically analyze 15 d before the beginning of the freeze-up whether the cold wave process occurred or not, and then classify the cold wave influence the Yellow River within 5 d as direct influence and without 5 d as indirect influence. There are 31 times which take 57.41% cold wave occurred before river freeze-up since 1954, and direct influence are 23 times which takes 42.59% for the Ning-Meng reach. And for the lower Yellow River, from 1954 to 2008 there are 8 years no river freeze-up, besides this 8 years, the cold wave result in the river freeze-up directly 22 years which takes 44.9% , the direct and indirect influence the river freeze-up takes 51% and 25 times. Therefore, this proves the important of cold wave to the river freeze-up.

In order to analyze the relation of sold wave and ice regime, using the characteristic value of cold wave occurrence time, duration, intensity, minimum air temperature and influence area to calculate the correlation with the freeze-up time. For the freeze-up time representation, let the December as the base, if the freeze-up time earlier than December as negative, otherwise as positive, for example, December 5th take as 5 and November 28th as -2. And for the breakup time, the method is same. The results show in the Ning-Meng reach, the correlation of the duration of the cold wave and freeze-up time is best, the correlation coefficient $R = -0.326,4$ ($\alpha_{0.05} = 0.28$), and next is the cold wave intensity, $R = -0.253,1$ ($\alpha_{0.10} = 0.24$). Due to the less cold wave processes of Zhengzhou and Beizhen for the lower Yellow River, can not make the linear correlation analysis, and the relation of the cold wave characteristic value of Jinan in December with the freeze-up time do not reach the statistical reliability, so the relation of cold wave and freeze-up is not simple linear relation. For the relation of cold wave process duration and freeze-up time, Dengkuo is best and $R = 0.301,2$, Tuoxian $R = 0.275,8$ and Baotou $R = 0.251,6$.

3.1.2 Influence on breakup time

Use the same method mentioned above, and the results show the cold wave occurrence frequency before the 15 days of the river breakup time is 33.3% and 16 times, and 7 years which occurs cold wave during the breakup period for the Ning-Meng reach. For the lower Yellow river, there are 10 times and take 20.4% before the 15 d of the river breakup time and only 3 years occurs cold wave. Through the contrast analysis of the air temperature data, found the cold wave occurred before the 15 d has the postpone influence on the river breakup time. The cold wave occurred before the river breakup time usually the result of the air temperature increasing, although this kind of cold wave can reach the standard, but the minimum air temperature is not very low and after the cold air mass pass, the air temperature rising sharply.

Another situation is some of the years which the cold wave break before the river breakup within 1 d to 10 d, this kind of river breakup mainly caused by the air temperature rising sharply before the cold wave break. With the Ning-Meng reach as example, the Tab. 5 shows the characteristics value of the cold wave break before the river breakup, besides 1994 ~ 1995 the mean air temperature below 2.0 ℃ of 10 d before the river breakup, the other years are all above 2.0 ℃. The river breakup mainly caused by the air temperature rising sharply before the cold wave break.

Tab. 5 The characteristics value of the cold wave break before the river breakup for the Ning-Meng reach

Year	Pre-10 d Mean air temperature (℃)	Pre −3 d air temp. temperature (℃)	Pre −2 d air temp. temperature (℃)	Pre −1 d air temp. temperature (℃)	Daily air temp. temperature (℃)	Max. (Min) air temp. temperature (℃)	Duration (D)	Breakup time (M. D)
1957 ~ 1958	4.9	5.4	5.4	8.2	11.5	11.5 (−2.1)	23 −25	3.23
1965 ~ 1966	2.8	2.7	−5.6	−0.7	2.6	7.0 (−1.9)	21 −22	3.20
1968 ~ 1969	3.2	0.8	6.9	11	12.7	12.5 (0.5)	25 −27	3.25
1970 ~ 1971	3.8	7.2	3.9	2	8.7	8.7 (−1.3)	30 −31	3.30
1973 ~ 1974	2.9	−0.8	0.3	4.4	7.5	9.2 (−1.8)	29 −31	3.28
1980 ~ 1981	5.4	6.5	6.5	5	8.7	6.9 (−0.7)	23 −25	3.21
1981 ~ 1982	2.4	0.2	0	2.3	6.2	9.4 (1.3)	22 −23	3.20
1994 ~ 1995	1.2	0.1	4	5	1	5.0 (−8.9)	14 −16	3.15
1997 ~ 1998	3.5	4.7	3	2.4	3.7	7.3 (−8.8)	17 −19	3.12
2003 ~ 2004	4.7	1.4	3.3	2.2	8.3	10.1 (−4.2)	15 −17	3.14

3.2 Relation with freeze-up duration

Through the correlation analysis of cold wave process times, duration, intensity, cooling range, minimum air temperature and influence area with the freeze-up duration of the Ning-Meng reach, and results show the cold wave process times has close relationship with freeze-up duration, and which is positive correlation. The correlation coefficient is 0.317,3, 0.201,7 and 0.284,9 for Dengkou, Baotuo and Tuoxian respectively, and Dengkou and Tuoxian reach the reliability ($\alpha_{0.05}$ = 0.28), and the cold wave times of Dengkou has close relationship with the freeze-up duration.

The freeze-up duration of the lower Yellow River has close relationship with minimum air temperature in December of the cold wave, R = −0.400,9 ($\alpha_{0.01}$ = 0.34), the cooling range also has close relationship with that of R = 0.259,5 ($\alpha_{0.05}$ = 0.26).

3.3 Relation with freeze-up length

Through the correlation calculation of cold wave process times, duration, intensity, cooling range minimum air temperature and influence area with the freeze-up length, to get the correlation coefficient of cold wave characteristics and freeze-up length which pass the statistical test for the lower Yellow River as show in Tab. 6. Besides the correlation of cold wave influence area and freeze-up length not reach the statistical reliability, other factors all have the correlation with the freeze-up length, and the cold wave minimum air temperature has close relationship. So in winter when the more cold wave processes, large cooling range, long duration and low minimum air temperature, the freeze-up length is longer than that of the normal year for the lower Yellow River. The same result can be obtained for the Ning-Meng reach.

Tab. 6 The correlation coefficient of cold wave characteristics value and freeze-up length which pass the statistical test for the lower Yellow River

Month	Process times	Cooling range(℃)	Duration	Minimum air temperature (℃)
12			0.287,5	−0.349,8
1				−0.267,9
2				−0.311,8
12 ~ 2	0.344,8	0.333,2	0.273,7	−0.443,7

4 Conclusions

The spatial and temporal distribution of the cold wave for the Yellow River is uneven, and the cold wave occurrence frequency of the Ning-Meng reach is higher than that of the lower Yellow River, and also for cold wave intensity and influence area.

The ordinary cold wave weather process is most of the Yellow River, the strong intensity cold wave seldom occur, and besides Dengkou occurred 3 times, the other stations not occurred. Since 1980, the number of cold wave activities has the tendency of decrease, the winter cold wave occurrences tend to quiet.

Cold wave has the positive relation with freeze-up time, freeze-up duration and freeze-up length, and reach the requirement of the correlation test ($r = 0.05$). Almost 50% of the freeze-up of the Yellow River caused by cold wave weather, more cold wave processes and longer the freeze-up length.

References

Flood Control office of Yellow River Conservancy Commission, Hydraulic Engineering Department of Tsinghua University. The Ice Flood of the Lower Yellow River [M]. Beijing: China Science Press, 1979.

Wang Wencai. Ice Regime Analysis of the Lower Yellow River in the 1980's [J]. Journal of Yellow River, 1992 (5).

Xi Xiufeng ,Zheng Shifang. Contrast Analysis of the Air Temperature Data on Present and Old Station of Jinan City [J]. Journal of Shandong Meteorology, 2003, 23(4).

Peng Meixiang, Wang Chunqing, Wen Liye et al.. Ice Flood Causing Analysis and Forecasting Research of the Yellow River [M]. Beijing: China Meteorological Press, 2007.

Estimation of Vulnerable Areas Around Rivers of the Republic of Armenia Caused by Global Climate Change

V. H. Tokmajyan, B. P. Mnatsakanyan, V. S. Sargsyan

Yerevan State University of Architecture and Construction, 105/1a Teryan Street, Yerevan, 0009, Republic of Armenia

Abstract: Analysis of observations data recorded by weather stations shows that snowfall and river flow during the last decades has decreasing tendency. Taking into account the observation base period snow precipitation values their expected changes have been predicted for 2030, 2070, and 2100. The obtained results enable modeling and mapping of vulnerable areas. In general, the most vulnerable areas located on and over 1800m above sea-level where rivers originating receive a significant portion of their flow.

Key words: vulnerable area, low-water, snow cover, base period of time, forecasting

Over the last 100 years climate observations have shown the Earth's climate global change. The territory of Armenia is not an exception and certain deviations of such elements of the whole weather picture, particularly, the average air temperature and atmospheric precipitations are observed compared with the accepted base period of time between 1961 and 1990. Studies showed that especially the above said elements fluctuations cause essential change of the country water resources, particularly river flow.

To estimate vulnerability of our water resources caused by the climate change we have studied the degree of both the annual and seasonal distribution and low – water flow. Toward this end the flow was considered for both until 2008 and long – term predictions after 2030, 2070, and 2100, for which values of rivers' flow changes obtained by applying IPCC scenarios.

Hydrological regime of Armenia's rivers until 2008 depending on weather conditions – intensive increase of evaporation and decrease of precipitations – has been subjected to essential changes. Recently some rivers in warm seasons began dry out at their estuaries of which one the main causes, except climate, is increased water intake to meet grown water requirements.

It should be noted that the role of melted snow water is prominent in formation of rivers flow in Armenia. The main source feeding most of rivers is melting snow which ranges from 30% to 60% of annual flow at the average. It is clear that global climate change affects on formation of snow cover and the latter in its turn affects both the amount of rivers flow and the hydrological regime.

To estimate the snow cover in Armenia and, naturally, water resource contained in snow, taking into account incomplete information available on actual snow cover in high – level regions, we have made use of both the data provided by high – level weather stations' winter records (from December to February) on snowfall and snow measuring routes.

Analysis of observations data recorded by weather stations' during 70 ~ 80 years shows that snowfall between 1961 and 2008, except Lake Sevan coastal area, Dzoraget and Maghri rivers water catchment reservoirs, recorded precipitation has decreasing tendency which is reflected in Tab. 1.

Having analysed fluctuations of snowfall during observations period until 2008 we can arrive to a conclusion that the most decrease tendency is expected in areas located above altitude of 1,700 m which is from 20 mm to 70 mm (with the exception of Lake Sevan reservoir).

Taking into account the observation base period snow precipitation values their expected changes have been predicted for 2030, 2070, and 2100, findings are as follows: snowfall in Armenia will decrease by 7% in 2030, 20% ~30% in 2070, and 30% ~40% in 2100.

The obtained results enable modeling and mapping vulnerable areas, and bringing them together vulnerability degrees have been defined. Fig. 1 illustrates weak, moderate, average, and strong vulnerable areas in terms of water resources until 2008 for different parts of the country.

Tab. 1 Weather stations records on snow precipitation values in mm and predictions in %

	Observation period	Station altitude (m)	Change in observation period(mm)	Change in one observation year(mm)	Base period (1961 ~ 1990) (mm)	2030 (%)	2070 (%)	2100 (%)
Amasia	1961 ~ 2008	1,876	-76	-1.58	132	-11	-21	-30
Giumri	1961 ~ 2008	1,556	-16	-0.33	108	-11	-21	-30
Aparan	1961 ~ 2008	1,881	-44	-0.92	213	-11	-21	-30
Hrazdan	1961 ~ 2008	1,765	-52	-1.08	231	-11	-21	-30
Fantan	1961 ~ 2008	1,798	-20	-0.42	205	-11	-21	-30
Gavar	1924 ~ 2008	1,961	6	0.07	63	1	5	9
Masrik	1935 ~ 2008	1,940	7	0.09	66	1	5	10
Shorzha	1936 ~ 2008	1,914	5	0.07	51	1	5	9
Martuni	1926 ~ 2008	1,945	6	0.07	134	1	3	7
Garni	1935 ~ 1988	1,422	-18	-0.33	109	-11	-21	-30
Jermuk	1961 ~ 2008	2,066	-50	-1.04	292	-11	-22	-30
Nameless mountain pass	1961 ~ 2008	…	-60	-1.25	210	-11	-22	-30
Sisian	1961 ~ 2008	1,580	-22	-0.46	93	-15	-29	-40
Goris	1935 ~ 2008	1,398	-23	-0.31	110	-15	-29	-40
Vorotan mountain pass	1961 ~ 2008	…	-20	-0.42	195	-15	-29	-40
Meghri	1935 ~ 2008	627	3	0.04	15	1	3	6

The most vulnerable are mountainous areas located on and over 1,800 m above sea - level where rivers originating receive a significant portion of their flow from snowmelt. Such areas are in the Akhuryan, Kasakh, Hrazdan, Azat, and Arpa rivers' basins, where compared with the observation base period a decrease is expected in 2080 from 10 mm to 30 mm, in 2070 from 20 mm to 60 mm, and in 2100 from 30 mm to 80 mm.

As for the northeastern and southern parts of Armenia snowfall increase is expected in 2030 from 0.2 mm to 1.5 mm, in 2070 from 0.5 mm to 3.5 mm, and in 2100 from 0.9 mm to 7.0 mm.

Summarizing the investigation it can be stated that in Armenia especially vulnerable to climate change are areas located over 1,800 m above sea - level and where the main sources feeding summer stream flows are formed.

The map clearly shows that currently the river flow is also strongly vulnerable in Ararat valley. Namely here are raised the best part of cultivated plants and it has become imperative to take pressing measures to provide further development this branch of agriculture. To this end it is suggested to regulate rivers water which is envisaged by RA law On Water National Program and install modern irrigation technologies shifting from widely used in Armenia traditional furrow irrigation to drip irrigation.

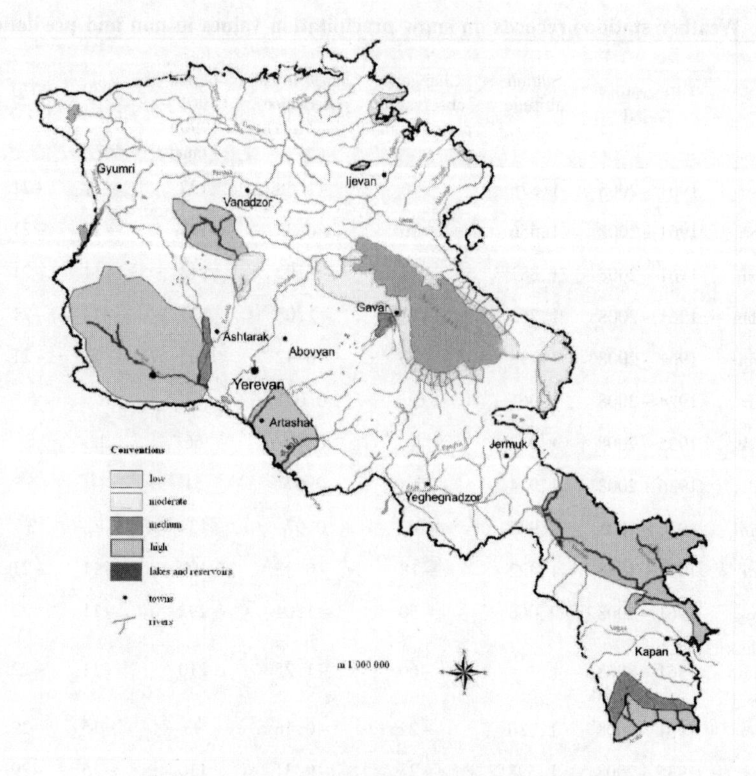

Fig. 1　RA river flow vulnerability until 2008

References

First National Report of the Republic of Armenia according to UNO Climate Change Framework Convention. Brief summary. Yerevan, 1999:372.

Armenia. Problems of climate change. Transactions 2nd issue. Yerevan, 2003: 362.

Second National Report on Climate Change according to UNO Climate Change Framework Convention. Yerevan, 2010:112.

Characteristics of Climate Change over the Yellow River Basin

Liu Jifeng[1], *Zhang Haifeng*[1], *Zhang Huaxing*[2], *Zhao Su*[1] and *Li Mingzhe*[1]

1. Hydrology Bureau of the Yellow River Conservancy Commission, Zhengzhou, 450004, China
2. Yellow River Conservancy Commission, Zhengzhou, 450003, China

Abstract: The Yellow River Basin (YRB for short) is in the arid or semi - arid regions of China, which water resources system is very sensitive to climate change. Temperature and precipitation in YRB have been experiencing obviously changing in recent decades. Since the middle of 1980s, temperature has been eventually rising, especially in winter, and the north basin is the remarkable warming zone. In 1990s, precipitation in YRB was decreasing apparently, while slightly increasing after entering the 21st century. Climate warming and precipitation decreasing are the important reasons of runoff sharp reduction over YRB. On the bases of climate change model, the temperature would apparently increase and precipitation would have a slightly increment in 2050, which aggravates the contradiction between the demand and supply of water resources in YRB. So, adaptive countermeasures should be taken to reduce the negative effects from climate change.

Key words: climate change, the Yellow River Basin, water resources

1 Introduction

According to the 4th report of IPCC, the global climate has been experiencing remarkable change which main characteristic is warming. The average linear warming rate in recent 50 years is about 0.13 ℃/10 a, doubling than the recent 100 years. After entering the 21st century, the climate warming caused by greenhouse gases and climate natural variation is becoming more and more obvious. Eleven of the last twelve years (1995 ~ 2006) are ranked among the twelve warmest years in the instrumental record of global surface temperature since 1850. China is the remarkable region of global warming by the speed 0.22 ℃/10 a, especially in arid area and Tibet Plateau, though the change trend of precipitation in China is not so obviously.

YRB locates in the middle north China, influenced by the continental monsoon climate with less precipitation that mostly occurs from June to September. The Yellow River is the biggest sediment laden river in the world. Annual average runoff and sediment load in the Yellow River is about 53.5×10^9 m^3 and 1.25×10^9 t, respectively. Lanzhou hydrology station in the upper watershed is the main water - yield area with about 62% total runoff of YRB; and the zone between Hekouzhen and Longmen station is the main sediment - yield area with 54% total sediment. The downstream reach is the suspended river because of serious sediment deposition. So, "Less water with excessive sediment, different original source of sediment and water, unbalanced sediment discharge relationship" are the remarkable features of the Yellow River water resources.

These features imply that the water resources system of YRB is sensitive and fragile to climate change. Since 1980s, with the development of social - economic along the river, contradictions between water supply and demand are becoming more and more serious, Water resources management is becoming the principal issue of basin managements. In this situation, people pay more attention on researches of climate change effects on water resources over YRB.

Based on the last observation data, climate change characteristics in recent 50 years are analyzed to support river basin management and master plan of YRB.

2 Data and methods

Temperature and precipitation data come from the National Climate Center of China Meteorology Bureau. 70 stations with data years of 1961 ~ 2005 were selected from 86 meteorology stations in YRB by considering the data series integrity. Fig. 1 shows the stations distribution. Single missing data in some stations were interpolated by proximal point interpolation method.

Series on basin level: area weighted Thiessen – Plogon method was used to obtain the time series at basin level and each sub – basin level. Climate abrupt change points were detected by Mann – kendal method. Linear trend of time series were filtered by conic fitting, then change period of stationary series can be obtained by wavelet transform.

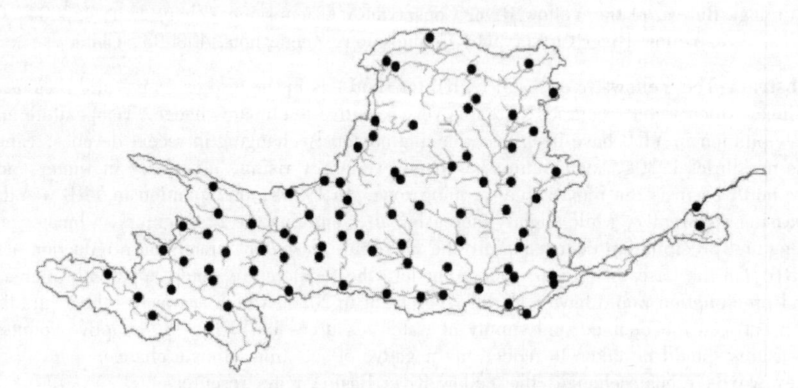

Fig. 1 Meteorology station distribution in the YRB

3 Basic characteristics of climate

By the statistics, the average annual temperature of YRB is about 6.95 ℃ (in Tab. 1), increasing from northwest to southeast with amplitude in the range of minus 8 ~ 14 ℃. Temperature in YRB has obvious seasonal variation with lowest monthly temperature minus 8 ℃ in January and highest 20 ℃ in July. The lowest annual temperature is 5.88 ℃ in 1967 and highest 8.36 ℃ in 1998. The average annual precipitation is about 446 mm, with quite uneven spatial distribution and decreasing from southeast to northwest in general. Influenced by continental monsoon climate, the annual precipitation is concentrated, and precipitation during May to October accounts for 87%. The highest amount of annual precipitation is 628 mm in 1964 and lowest 332 mm in 1965.

Tab. 1 Statistics of temperature and precipitation in YRB

(Units: Temperature, ℃ Precipitation, mm)

	Jan.	Feb.	Mar.	Apr.	May	Jun.	Jul.	Aug.	Sep.	Oct.	Nov.	Dec.	Year
Temperature	-7.91	-4.41	1.75	8.62	14.08	18.04	19.96	18.64	13.66	7.43	-0.17	-6.24	6.95
Precipitation	3.3	5.1	12.8	25.1	43.5	58.7	96.5	91.3	65.6	31.8	9.6	2.8	446

4 Trend of climate change

4.1 Trend of temperature

4.1.1 General trend of temperature

Temperature in YRB has increasing trend by 0.307 ℃/10 a in general. Since the middle of 1980s, temperature has been obviously increasing by 0.6 ℃/10 a. Both YRB and the north hemisphere have the same increasing tendency with the speed 0.307 ℃/10 a and 0.241 ℃/10 a respectively. (see Fig. 2, Fig. 3) Comparing with the global warming, climate warming in YRB is more violent with larger amplitude.

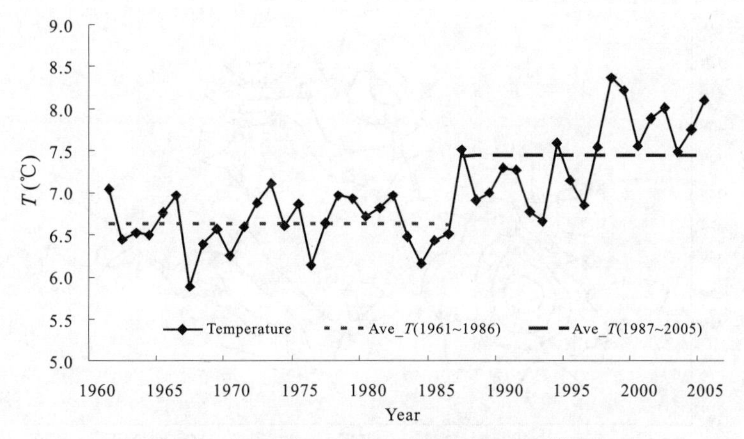

Fig. 2 Annual average temperature in YRB

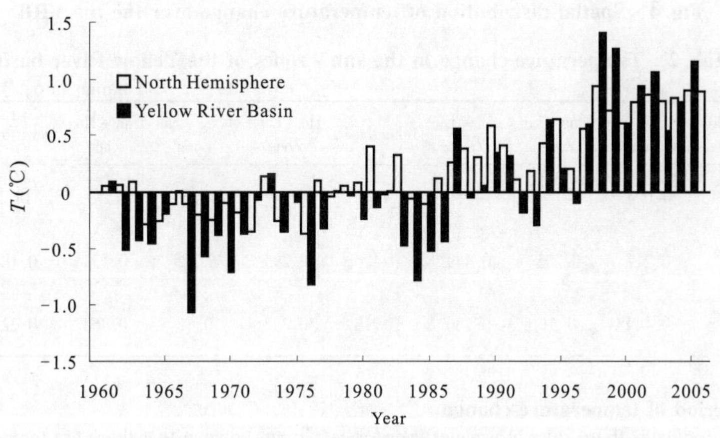

Fig. 3 Annual temperature in YRB and the north hemisphere
(heavy line is 10 – year moving average)

4.1.2 Spatial feature of temperature change

Most area of YRB shows the warming trend with the feature of "high warming speed in north and low in south" (Fig. 4). Ningxia and Inner – Mongolia provinces in YRB have obvious warming trend, and Linhe station is the temperature increasing center with warming rate 0.754 ℃/10 a; Qinghai province and west of Gansu province in YRB show less amplitude of climate warming, and Henan station of Qinghai province is the low value center with the warming rate minus 0.411 ℃/10 a. Temperature increasing in downstream is not obviously. There are more remarkable temperature increasing in winter season, especially in Ningxia and Inner – mongolia provinces in YRB, and the warming rate in Linhe station reached to 0.979 ℃/10 a. As the key reach of ice regime, the persistent winter warming has important impact on the ice disaster of YRB.

Tab. 2 shows the temperature change in each water yield zone of YRB. Firstly, all zones show warming trends. Among, interior drainage area and Lan – Tuo zone are the apparent warming regions with 0.486 ℃/10 a and 0.445 ℃/10 a; San – Hua zone shows small increasing speed of 0.132 ℃/10 a. Secondly, climate warming has been being extraordinarily fast since 1980s in all zones. Among, amplitudes in both Lan – Tuo and He – Long zone are above 0.7 ℃/10 a.

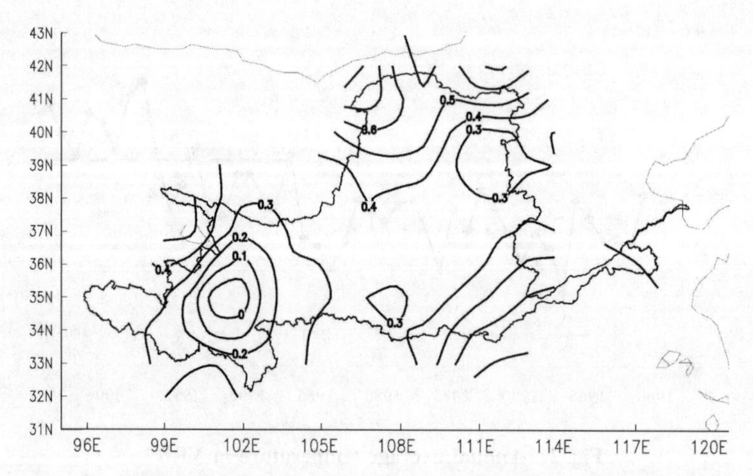

Fig. 4 Spatial distribution of temperature change over the the YRB

Tab. 2 Temperature change in the sub – zones of the Yellow River basin

(Units:T, ℃ Change rate, ℃/10 a)

Water – yield Zones	Up Longyangxia	Long – Lin Zone	Lan – Tuo Zone	Inlandbasin	He – Long Zone	Long – San Zone	San – Hua Zone	Downstream	YRB
Temperature	−0.4	4.4	7.5	7.4	8.4	4.4	12.3	14.1	6.95
Change rate since 1961	0.217	0.205	0.449	0.486	0.295	0.205	0.132	0.185	0.307
Change rate after 1980	0.449	0.51	0.736	0.718	0.732	0.51	0.464	0.225	0.6

4.1.3 Period of temperature change

Fig. 5 indicates the results of wavelet transforms. It can be seen that there are temperature periods of 10 ~ 13 years in YRB, espccially since the middle of 1980s, 12 – year is the main period. Climate warming period of YRB is in accordance with the north hemisphere, which implies that the global warming is the background of temperature increasing of YRB and YRB warming is also the component of global warming.

4.2 Features of precipitation

4.2.1 Trend of precipitation

Precipitation in YRB has been experiencing decreasing with average speed of 11.7 mm/10 a in recent 50 years, with obvious decrease since 1990s and slight increase after entering into 21st century (Fig. 6). Precipitation in 2003 reached to 551 mm, 24% more than the normal.

4.2.2 Spatial distribution of precipitation

Fig. 7 shows the distribution of trend coefficients of precipitation in YRB. Except the north of Ningxia – Inner Mongolia reach, precipitation in most area of YRB shows the decreasing trend. Precipitation decreased obviously in the sources region and He – Long zone (including Jinghe River, Weihe River and Luohe River). Precipitation decreases in these regions have important impact on runoff yield, and zero runoff sometimes appeared in some tributaries even in flood season.

Since 1980s, there were more obvious decreases of precipitation in YRB (Fig. 3). Among,

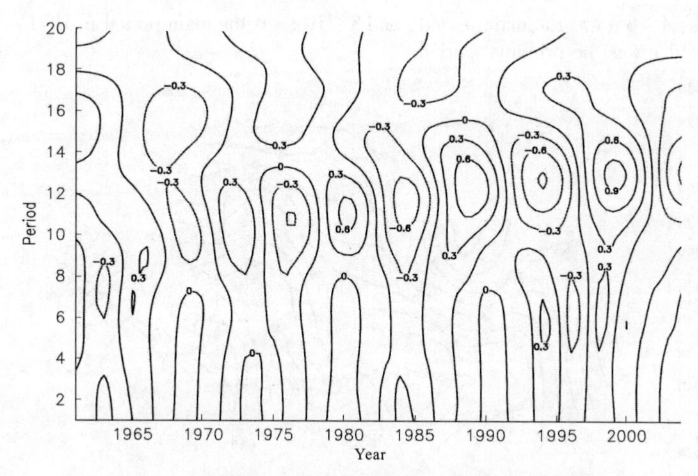

Fig. 5 Wavelet analysis of temperature in YRB

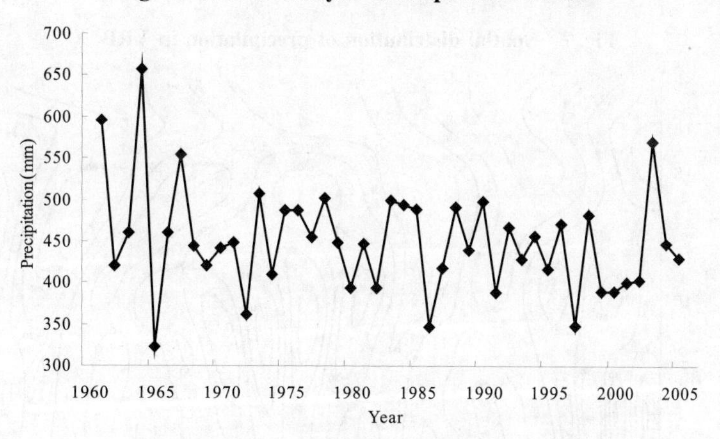

Fig. 6 Annual average precipitation of YRB

the source region decreasing speed was 32 mm/10 a, other important water yield zones, such as Long – Lan, He – Long and Long – San, increased by the speed above 10 mm/10 a. The decreases of precipitation in YRB, especially in the source region, were the main reason of runoff reduction (See Tab. 3).

Tab. 3

Zone	Up Longyangxia	Long – Lan	Lan – Tuo	He – Long	Long – San	San – Hua	Interior drainage area
Precipitation	490	446	274	447	446	640	283
Change rate	–6.7	–7.0	–2.4	–19.5	–7.0	1.6.3	–2.4
Rate after 1981	–32.2	–10.0	10.9	–13.6	–10.0	–8.1	12.1

4.2.3 Features of precipitation period

Fig. 8 shows the wavelet transform of precipitation in YRB. Influenced by sunspots, ocean and atmosphere interaction, the change periods of precipitation are more complicated than temperature.

Before 1970s, 4 ~ 6 a was the main period, and 8 ~ 10a was the main period in 1971 ~ 1990. Since 1990s, 12 ~ 14 a was the principal period.

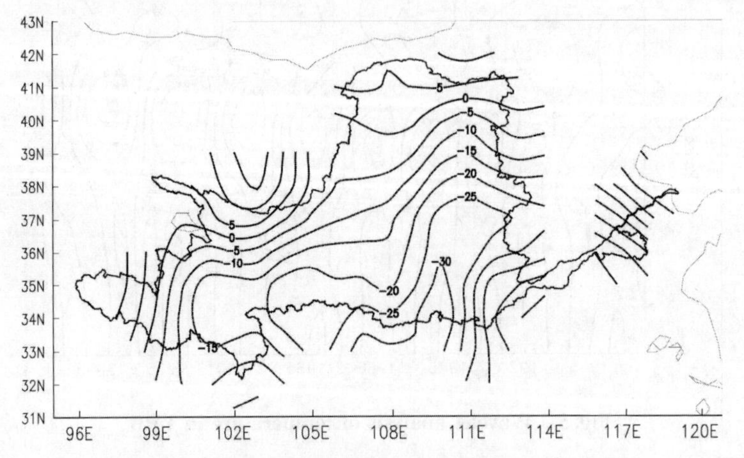

Fig. 7 Spatial distribution of precipitation in YRB

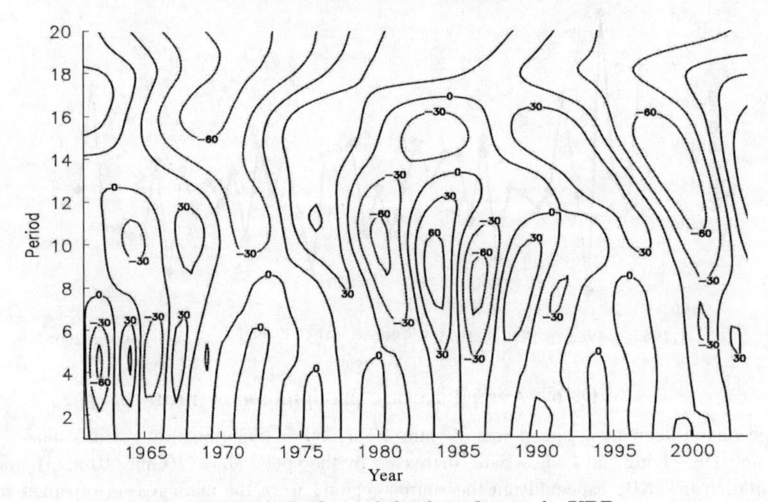

Fig. 8 Period of precipitation change in YRB

5 Climate change scenarios

There were many climate change evolution models used in scenarios researches of YRB. Results from seven climate models indicated that temperature in YRB would increase about 3 ℃ and precipitation only 5% . The statistics downscaling results under A2/B2 scenarios showed the precipitation increment was about minus 81.9 ~ 60.4 mm, with the rate minus 18.1% ~ 13.4%. The future temperature and precipitation under A2/B2 scenarios in YRB were obtained with dynamic downscaling method by using general circulation model (HadCM3) to drive region climate model (PRECIS) of Hadley Center of UK. The results indicated that temperature in 2050 would increase apparently, and compared with 1961 ~ 2005, the average temperature would increase 1.6 ℃ (A2) and 1.4 ℃ (B2), respectively. Precipitation in 2050 would increase slightly by 3.5 % (A2) and 0.4 % (B2), compared with 1961 ~ 2005.

The forth report of IPCC pointed out that the general circulation models were provide as a kind of very reliable tools in the predictions of future climate change, and the reliability of temperature was better than precipitation. Synthesizing these research results, it can be believed that the temperature would increase apparently and precipitation increase slightly in 2050, which can led to the decreases of water resources in YRB with the sharp increase of evaporation caused by temperature. Certainly, effected by the uncertain of future greenhouse gases emission, development level of climate model, precision of observed data, there was large uncertainty. Despite all this, departments of water resources management should establish some adaptive countermeasures to climate change to relive the disadvantage effects, and keep the river healthy life and achieve the accord development of natural environment, economic and society in river basin.

6 Discussions and conclusions

(1) YRB is the remarkable region of global warming with the apparently temperature increasing and precipitation decreasing in recent 45 years. Since 1980s, temperature in YRB has been experiencing most obvious increase. There is the most evidence reduction of precipitation in 1990s and slight increase after 21st century. Climate change in YRB is the response to global warming and the important component element of global change.

(2) There are regional differences of climate change in interior of YRB. The north basin (around Ningxia and Inner – Mongolia provinces) is the most apparent warming area and Qinghai province and the west Gansu province are the least warming regions. There is consistent climate warming trend in different seasons, and the most warming season is winter, then spring, autumn and summer.

(3) There is obvious 11 ~ 13 a periods of temperature changes, which is in accordance with the north hemisphere. Periods of precipitation are complicated and 8 ~ 12 a are the main periods.

(4) Based on the climate model predictions, the future temperature would rise apparently and precipitation increase slightly. The results imply that water resources in YRB would reduce because of the evaporation sharp decrease caused by climate warming. Though there is large uncertainty of future climate change, none can afford to neglect the impacts of climate on water resources in YRB, and some countermeasures should be taken to relive the disadvantage of climate change of YRB.

References

IPCC. Climate Change 2007: The Physical Science Basis. In: Solomon S, Qin D, Manning M, et al. Contribution of Working Group I to the Fourth Assessment Report of the Intergovernmental Panel on Climate Change [M]. Cambridge and New York: Cambridge University Press, 2007: 117 – 118.

Qing Dahe, Ding Yihui, Su Jilan, et al. Climate and environment changes in China (Vol. 1): Evolution and prediction of climate and environment [M]. Beijing: China Science Press, 2007.

Zhang Xueqin, Sun Yang, Mao Weifeng, et al. China Drought Area of Air Temperature Change on Regional Response to Global Warming [J]. Arid Zone Research, 2010, 27(4): 592 – 599.

Xu Xiaoling, Yan Junping, Liang Xunfeng. Sanjiang Source Region Changes in Runoff and the Effect of Human Activities [J]. Arid Zone Research, 2009, 26(1): 88 – 94.

Li Guoying. The Mmajor Problems and its Countermeasures in the Yellow River [J]. China Water Resources, 2002(1): 21 – 23.

Yao Chensheng, Ding Yuguo. Climatic statistics [M]. Beijing: China Meteorological Press, 1990.

Wei Fengying. Modern Climatic Statistical Diagnosis and Prediction Technology [M]. Beijing: China Meteorological Press, 1999.

Lin Zhenshan, Deng Ziwang. Climatic Diagnosis Technology Based on Wavelet [M]. Beijing:

China Meteorological Press, 1999.

Zhu Likai, Meng Jijun. Temporal and Spatial Variation of Precipitation in Centre Inner - Mongolia Province in Recent 40 Years[J]. Arid Zone Research, 2010, 27(4):536 - 545.

Shao Xiaomei, Xu Yueqing, Yan Changrong. Wavelet analysis of rainfall variation in the Yellow River Basin[J]. Acta Scientiarum Naturalium Universitatis Penkinensis, 2006, 42(4): 503 - 509.

Xu Ying, Ding Yihui, Zhao Zongci. Scenarios Analysis on Temperature and Precipitation under Human Activities over the Yellow River Basin[J]. Advances in Water Science, 2003, 14 (supplement): 34 - 40.

Liu Lvliu, Liu Zhaofei, Xu Zongxue. Trends of Climate Change for the Upper - Middle Reaches of the Yellow River Basin in the 21st century[J]. Advances in Climate Change Research, 2008, 4(3): 167 - 172.

Characterizing Uncertainty of Hydro – climatic Variables for Water Management in the Yellow River Basin: Learning from the Past

Congli Dong, ***Gerrit Schoups*** and ***Nick van de Giesen***

Section of Water Resources Management, Faculty of Civil Engineer and Geosciences, Delft University of Technology, Delft, the Netherlands

Abstract: Unpredictable variability such as caused by climate change, and imperfect knowledge of hydrological processes introduce uncertainty in water resources planning and management. As a result, analysing and studying the variability and uncertainty of hydro – climatic variables help to better understand the variations of these processes and provide information to water planning and management. The paper focuses on characterizing the variability and uncertainty of climatic and hydrological variables such as temperature, precipitation, and runoff in the Yellow River Basin. Historical information and data is used to analyse the trend of these variables, and the variability and uncertainty of these hydro – climatic processes. Two statistical approaches, including the bootstrap method and the maximum likelihood (MLE) method, are used to estimate uncertainties given historical data. We apply the analysis to both seasonal and annual data, and to various sub – regions of the Yellow River Basin.

Key words: uncertainty, bootstrap method, maximum likelihood method, Yellow River Basin, water management

1 Introduction

Climate variability directly influences the availability of water resources (Yang, Li et al., 2004). Global warming related to climate change has been widely discussed over the past two decades. The impacts of climate change on water resources have been studied using the projection of the General Circulation Models (GCMs); however, climate variations are not possible to be assessed precisely by these models due to unpredictable variability and limited understanding of the natural climate system. Moreover, when considering how the water system responses to the climate system, uncertainties are introduced, for example, from the imperfect knowledge of hydro – climate interactive system and hydrological processes. As a result, analysing and studying the variability and uncertainty of hydro – climatic variables can help to better understand the variations of these processes and provide information to water resources management.

Variability represents variation or heterogeneity in a well characterized population; and uncertainty arises due to incomplete or imperfect knowledge about poorly – characterized phenomena or models (Frey and Burmaster 1999). Climate change and hydrological process are subject to both variability and uncertainty. Variability refers to the unpredictable natural behaviours, for example, the natural variation of climate system and water system. Uncertainty rises when lack of knowledge or understanding regarding the true climate and hydrological process, or unreliable available data of climatic and hydrological variables.

Historical information and data is used to analyse the trend of these variables, more importantly, the variability and uncertainty of these hydro – climatic processes. Two statistical approaches, including the bootstrap method and the maximum likelihood (MLE) method, are used to estimate uncertainties given historical data. In this paper, the hydro – climatic variability and uncertainty is analysed annually and sub – regionally in the Yellow River Basin, China.

2 Study area

The Yellow River is the second longest river in China. It flows around 5,500 km in north China, originating from the Tibetan plateau, going through the northern semiarid region, the loess plateau, the eastern plain, and finally discharging into the Bohai Sea. (see Fig. 1). It accumulates about 573,000 km^2 drainage basin, where has 1.00×10^8 people and 12.00×10^8 hm^2 farmland, and half of the farmland is irrigated by Yellow River. In the Yellow River Basin, annual evaporation is much higher than the annual precipitation, with 200 ~ 700 mm/a precipitation while 850 ~ 1,600 mm/r evaporation. With the increasing pressure from the population growth, economic development and climate change, water shortage has been a major issue of water management in the Yellow River Basin.

The whole basin is divided into three reaches, in which the upper reach is from the source to Toudaoguai gauging station, the middle reach is until Huayankou station, and the downstream is below the Huayuankou station (Fig. 1). The lower reach of the Yellow river is suspended, as a result, the natural discharge of the Huayuankou gauge station can be taken as the natural runoff for the whole Yellow River Basin. Besides, the area from the river source to Lanzhou gauging station in the upper stream contributes 56% of the total runoff for the whole basin. Therefore, this paper focus on analysing variability and uncertainty of the two main hydro – climatic variables, precipitation and river discharge, in the two regions: river source – Lanzhou, and river source to Huayuankou station. For the purpose, 48 years of precipitation and river discharges data have been collected.

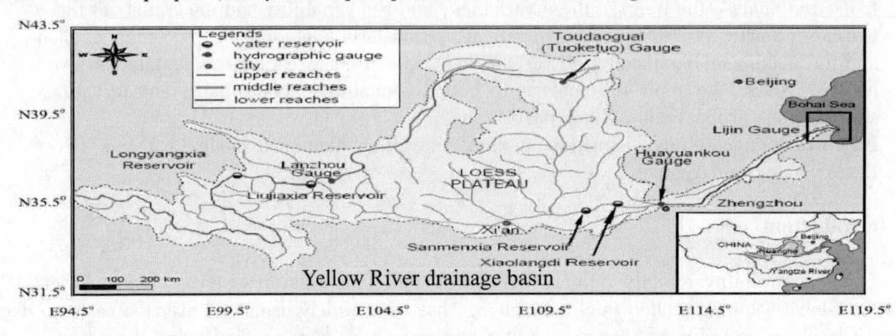

Fig. 1 The Yellow River and Yellow River drainage basin

3 Methods

There has been a trend to the use of probabilistic methods to quantify uncertainty and variability. A two – dimensional framework was proposed to represent variability and uncertainty using probabilistic methods (Frey and Burmaster 1999). The steps of the framework is described as below:

(1) Cumulative density function (CDF) of a parametric probability distribution is used to best fit the data to represent variability of climate and hydrological variables. Parametric CDF is used instead of empirical CDF, because parametric probability distributions extrapolate the unobserved tails of the unknown population, while empirical distributions are likely to limit the range of the observed data.

(2) Parameters of the probability distribution are estimated using methods such as maximum likelihood (MLE) method or Method of Matching Moments (MOMM).

(3) Confidence intervals of the distributions are constructed to represent uncertainty of these variables. Bootstrap simulation can be used to quantify uncertainty in the fitted distribution under the assumption that the estimated probability distribution is the best estimate of the true but unknown population distribution.

3.1 Maximum likelihood estimation

In this study, method of maximum likelihood estimation (MLE) is used to estimate the parameters of the probability distributions. Maximum likelihood method was developed by R. A. Fisher (Fisher, 1925), and is one of the most widely used methods of statistical estimation (Le Cam, 1990). Generally speaking, the method of maximum likelihood estimation picks up values of pa’ ters in given statistical models, which can maximize the probability of the observed data.

Suppose a data set has independent samples $\{x_1, x_2, \cdots, x_n\}$ which are from a probability density function (PDF) $\{f(X/\theta), \theta\epsilon\Theta\}$, $\theta$is the true but unknown value of the PDF. The task is to find the estimator $\hat{\theta}$ which would be close to the true value θ_0 as much as possible. To use MLE method, the joint probability density function for all samples is generated first:

$$f(x_1, x_2, \cdots, x_n/\theta) = f(x_1/\theta)f(x_2/\theta)\cdots f(x_n/\theta)$$

The likelihood function for the independent samples is:

$$L(\theta/x_1, x_2, \cdots, x_n) = f(\theta/x_1)f(\theta/x_2)\cdots f(\theta/x_n) = \Pi_{i=1}^{n} f(\theta/x_i)$$

In practice, it is often convenient to use the logarithm of the likelihood function, called the log - likelihood:

$$\ln L(\theta/x_1, x_2, \cdots, x_n) = \sum_{i=1}^{n} \ln f(\theta/x_i)$$

The principle of the method is to find the estimator $\hat{\theta}$ by maximize the likelihood (log likelihood) function $L(\theta/x)$:

$$\hat{\theta}_{mls} = \mathrm{argmax} L(\theta/x_1, x_2, \cdots, x_n) \qquad \theta\epsilon\Theta$$

3.2 Bootstrap simulation

Bootstrap simulation was introduced by Efron in 1979 with the purpose of deriving robust estimates of standard errors and confidence intervals of statistics such as mean, median, percentile, etc. (Elfron, 1979). It is a statistical method for estimating the sample distribution of a statistic by sampling with replacement of the original sample set from an empirical distribution, and each random replacement is called a bootstrap sample. The approach is referred to as resampling. Suppose the original data set is

$$X = \{x_1, x_2, \cdots, x_n\}$$

All data in the data set have the same probability to be resampled into a bootstrap sample. A bootstrap sample of size n from this original data set is denoted by:

$$X^* = \{x_1^*, x_2^*, \cdots, x_n^*\}$$

Form each bootstrap sample, a bootstrap replication of a statistic can be calculated :

$$\hat{\theta}^* = s(X^*)$$

where, $s(X^*)$ represents a statistical estimator applied to a bootstrap replication of the original data set. To estimate the uncertainty of the statistic, bootstrap samples can be replicated N times.

$$\hat{\theta}_i^* = s(X_i^*), \text{where } i = 1, 2, \cdots, N$$

The number of the bootstrap replications depend on the information desired. Efron and Tibshirini suggest that $N = 200$ or less in order to calculate the standard error of a statics (Efron and Tibshirani, 1993). Frey and Burmaster resampled more than 1,000 times to estimate confidence intervals (Frey and Burmaster, 1999).

4 Results

Two parametric probability distribution, lognormal distribution and gamma distribution types, are chosen to fit the average annual precipitation and discharge data. Normal distribution is not considered due to its implausible predictions of negative values. Fig. 2 shows the comparison between empirical cdf of the data and the best - fit parametric cdf whose parameters were obtained using MLE approach. It suggests that both type fit the data well. In this paper, lognormal distribution is chosen to fit the data.

Bootstrap simulation is used to resample the data and generate the confidence interval. In this research, 2,000 samples of the data set were made. For each bootstrap sample, the statistics , named bootstrap replications of the statistics, are calculated using MLE approach. Each bootstrap replications of these statistic can be used to generate a sampling distribution describing the variability of the data set. 2,000 replications of the statistic simulate 2,000 possible distributions. Due to

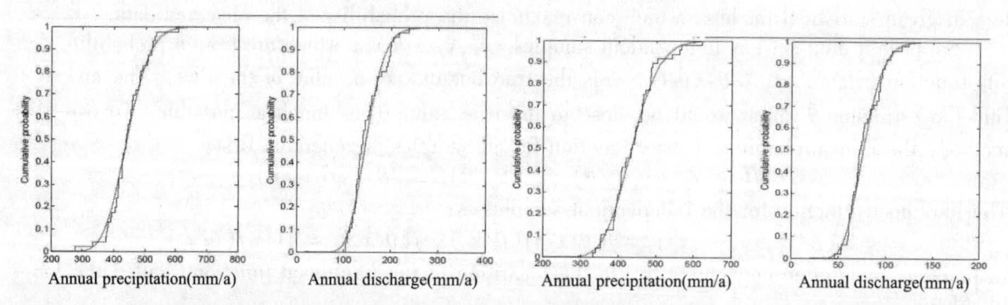

Fig. 2 Empirical cdf (blue stairs) , lognormal (red line) fit and gamma(dark line) fit of the annual precipitation and discharge data above lanzhou(left) and Huayuankou (right)

the error of the bootstrap sampling associated with a finite sample size, the uncertainty in the selected distribution can be quantified in order to characterize both the variability and uncertainty of the variables of interest in a two – dimensional framework. Fig. 3 shows the two – dimensional simulation of variability and uncertainty of annual precipitation and discharge above Lanzhou and Huayuankou station.

Two – dimensional analysis of variability and uncertainty can be used to produce point estimation, which is useful to decision makers(Frey and Burmaster, 1999). On one hand, it is possible to know the confidence interval regarding the probability when ' data values less than or equal to a given number. For example, the annual precipitation less than 400 mm/a is between 12% to 38% within a 95% confidence interval above Lanzhou station; while between 18% to 40% above Huayuankou station.

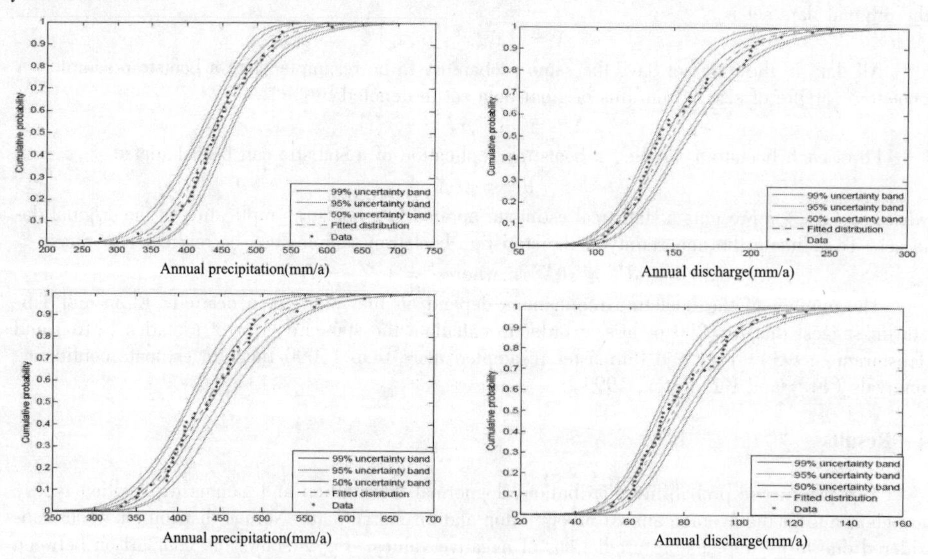

Fig. 3 Two – dimensional simulation of variability and uncertainty of annual precipitation and discharge above Lanzhou (top) and Huayuankou (bottom) stations

On the other hand, it is also possible to know the uncertainty regarding the true value of the random variable at a given percentile of the population of interest. Tab. 1 shows the 95% confidence interval for the 5th ,95th and 99th percentile of variability of annual precipitation and discharge above Lanzhou and Huayuankou gauging station. For example, in the 95th percentile of the variability, the range of annual discharge is between 139 mm/a and 157 mm/a in 95% confidence interval above Lanzhou , while between 72 mm/a and 81 mm/a above Huayuankou. Compared with the same percentile, the annual precipitation range in 95% confidence interval has less difference than the annual discharge between the two areas.

Tab. 1 95% confidence interval of 5%, 95% and 99% percentile of precipitation and discharge data above Lanzhou and Huayuankou stations

Statistic	Annual precipitation (mm/a)		Annual discharge (mm/a)	
	Lanzhou	Huayuankou	Lanzhou	Huayuankou
5th percentile	(323,382)	(332,367)	(99,113)	(50,59)
95th percentile	(423,457)	(415,449)	(139,157)	(72,81)
99th percentile	(513,582)	(508,577)	(187,227)	(98,118)

5 Conclusions

This paper demonstrated how to characterize both the variability reflected by the probability of the population of interest, and the uncertainty represented by confidence interval using a two – dimensional probabilistic framework. MLE approach was used to estimate parameters of the best – fit probability distribution to reflect the variability of the variables and the bootstrap simulation was used to produce the uncertainty band based on the sampled distributions.

The probabilistic method was applied to analyse the variability and uncertainty of hydro – climatic variables in the areas above Lanzhou and Huayuankou gauging stations of the Yellow River Basin. Lognormal distribution was found to be a good fit to the datasets. The paper also analysed the information learned from the two – dimensional simulation of variability and uncertainty of the annual precipitation and discharge based on historical data, which can become reference for the decision makers and water managers to plan and manage water issues at present and in the future.

References

Efron B, R. J. Tibshirani, Eds. An Introduction to the Bootstrap. CRC Monographs on Statistics & Applied Probability[R]. New York, Chapman & Hall, 1993.

Elfron B. Bootstrap Methods: Another Look at the Jackknife[J]. The Annals of Statistics, 1979,7(1): 1 – 26.

Fisher R. A. Theory of Statistical Estimation. Proc. Cambridge Phil. Soc., 1925(22): 700 – 725.

Frey H. C., Burmaster D. E. Methods for Characterizing Variability and Uncertainty: Comparison of Bootstrap Simulation and Likelihood – Based Approaches[J]. Risk Analysis, 1999,19(1): 109 – 130.

Le Cam L. Maximum likelihood – an Introduction[J]. ISI Review, 1990,58(2): 153 – 171.

Yang D. W., Ni. C, et al. Analysis of Water Resources Variability in the Yellow River of China during the Last Half Century Using Historical Data[J]. Water Resources Research, 2004,40(6).

Analysis of Runoff Variation and Causes in the Yellow River Source Region

Yang Libin[1,2], *Wang Yu*[1], *Yang Guoxian*[1] and *Cui Changyong*[1]

1. Yellow River Engineering Consulting Co., Ltd., Zhengzhou, 450003, China
2. Xi'an University of Technology, Xi'an, 710048, China

Abstract: The Yellow River source region is located in the hinterland of Qinghai-Tibet Plateau, it is the import source area of the Yellow River because of plenty of water systems. The runoff of the source region occupies 38% of the total source of the whole basin, it plays a decisive role in the security of water supply and environment. In recent years, the runoff decreased obviously in the source region of the Yellow River, the value of runoff was 21.6×10^9 m^3 in 1960s, 17.2×10 m^3 in the last 10 years. On the basis of the research on the runoff and its temporal and spatial distribution and change characteristics, the change and the cause of the runoff was analyzed from temperature, precipitation, vegetation condition and other natural factors and human activities impact, so as to provide the necessary technology in support of water conservation and ecological protection in the source region of the Yellow River. The results showed that: the runoff variation was mainly influenced by the precipitation. At the same time, temperature rise, permafrost layer thinning and even disappeared, making soil water seep to the deep soil, also led to the runoff decrease. Vegetation degradation and desertification, rainfall infiltration increase, and soil evaporation ability improvement, also led to runoff decrease. Lakes, marshes and other wetlands atrophy, increasing the early loss of precipitation in value, made the runoff function decline too.

Key words: the source region of the Yellow River, runoff, variation, cause analysis

1 Characteristics of the Yellow River source region

1.1 Natural geography

The Yellow River source region is located in the northeast part of Qinghai-Tibet Plateau, between latitude 32°09′ ~ 36°33′, longitude 95°53′ ~ 103°25′, area of 131,000 km^2, occupied 16.4% of the Yellow River drainage area. There lies the Xiqing mountain, Animaqing mountain and Bayankela mountain from northeast to southwest in turn in the Yellow River source region, constituted the basic terrain skeleton, average height of 4,079 m, terrain elevation change is large, the highest peak is Maqinggangre of Animaqing mountain, the elevation is 6,282 m, and the lowest valley is 2,700 m. Influenced by the plateau monsoon and the geographical environment, the climate characteristics of the Yellow River source region are as follows: it is cold and long period of low temperature in the winter; it is cool in summer and large temperature difference between day and night; low humidity, dry and wet trenchant; much gale hail frost thunderstorms and snow disaster weather; low pressure, and low oxygen content.

1.2 Precipitation

The average annual rainfall is 487 mm in the Yellow River source region, and it is higher in the northwest than that in the southeast, the average annual rainfall of the Maduo station in the source area is only 314mm, but it is high up to 762.4 mm of the southeastern station named Hongyuan. The precipitation has obviously seasonal, mainly in May to October (occupied 85% of the whole year). The rain area is very large, the period is long, and the intensity is low. The rain area sometimes covers the entire region above Longyangxia, the rain process can be continued for more than 30 d; and the daily precipitation generally is less than 50 mm, the small intensity continuous rain is very common.

1.3 River system

The river system is developed in the source region of the Yellow River, there are 56 distributaries directly flowing into the main river, there are 24 distributaries whose watershed area are more than 1,000 km^2, the drainage area of the Duoqu River, the Requ River, the Baihe River, the Heihe River, the Zequ River, the Qiemuqu River, the Bagou River, the Qushian River, the Daheba River, and the Mangla River is more than 3,000 km^2, and there are a lot of snow, glacier, lakes, wetlands in the source region of the Yellow River.

1.4 Runoff

The average annual runoff of Tangnaihai station is 20×10^9 m^3 according to the serial from 1956 to 2009, it accounts for about 38% of the total amount of runoff of the Yellow River, and it is known as the "Water Tower" of the Yellow River Basin. The runoff mainly concentrates in June to October, accounting for more than 70% of the annual runoff, the maximum runoff is 3.1 times the minimum ones. The runoff of the source region of the Yellow River is mainly from the reaches between Jimai and Jungong, accounting for 66.6% of the runoff above Tangnaihai, with only 43.8% of the area. The area above the Huangheyan station occupies over 17.2% of the total area above Tangnaihai, but the runoff only occupies 3.6%.

1.5 Frozen soil

According to the frozen soil remote sensing investigation and study, the high plains, Animaqing mountain, Bayankela mountain and plateau region in the source region of the Yellow River are patchy permafrost zone, the southeast and northeast areas are almost the seasonal frozen soil, except individual mountains have sporadic island permafrost. The elevation of the lower bound of the permafrost is about 3,840 ~ 4,000 m in Xinghai, Zeku, and Tongde counties, and the permafrost depth can be 20 ~ 49 m. According to the survey of Wang Shaoling (1991), the changes of the permafrost lower bound was closely related to the elevation and latitude, it was 3,850 ~ 4,000 m in Ela mountain, 4,000 ~ 4,050 m in Animaqign mountain, 4,150 ~ 4,200 m from Dawu to Gander and Bayankela mountain, and 4,150 ~ 4,200 m in Jiuzhi county.

2 Runoff and its change in the source region of the Yellow River

2.1 Spatial variation of the runoff

Affected by the precipitation and natural conditions, the runoff changed significantly in the source region in the Yellow River.

2.1.1 The area above Huangheyan station

The precipitation was scarce, the average annual precipitation was 200 ~ 300 mm. the areas of lakes and marshes are very large, the function of water regulation and control is very strong, the evaporation is very high, and runoff condition was poor, the annual mean runoff depth is 32.8 mm, and amount of the runoff is 0.7×10^9 m^3, accounting for 3.4% of the total runoff above the Tangnaihai station, accounting for 17.2% of the areas, it was the weak area of the runoff in the source region of the Yellow River.

2.1.2 The area between Huangheyan and Jimai

The precipitation increased from 313.4 mm of Maduo station to 543.4 mm of Jimai station. The annual mean runoff depth was 132.3 mm, and the amount of the runoff was 3.2×10^9 m^3, accounting for 15.9% of the total runoff above Tangnaihai station, and accounting for 19.7% of the areas, the runoff was 4 times more than that in the areas above Huangheyan station.

2.1.3 The area between Jimai and Maqu

The precipitation increased from 543.4 mm of Jimai station to 753.1 mm of Hongyuan station. The annual mean runoff depth was 253.5 mm, and the amount of the runoff was 10.4×10^9 m^3, accounting for 52% of the runoff above Tangnaihai station, accounting for 33.6% of the areas, it was the most water production area of the source region of the Yellow River.

2.1.4 The area between Maqu and Jungong

The runoff decreased gradually from south to north, the annual mean runoff depth was 225.8 mm, the amount of the runoff was 2.8×10^9 m^3, accounting for 14% of the runoff above Tangnaihai station, accounting for 10.1% of the areas, it was the better region of water production.

2.1.5 The area between Jungong and Tangnaihai

The precipitation gradually reduced from north to south, the annual rainfall is only 252.1 mm in Tangnaihai station. The annual mean runoff depth is only 124.4 mm, and the amount of the runoff was 2.8×10^9 m^3, accounting for 14% of the runoff above Tangnaihai station, accounting for 19.3% of the areas.

From the view of spatial distribution, the area between Jimai and Jungong was the main source of runoff in the source region of the Yellow River, the annual runoff was 13.2×10^9 m^3, accounting for 66% of the runoff of Tangnaihai station, which was due to the abundant rainfall and good conditions.

2.2 Temporal variation of the runoff

2.2.1 The inter-annual variation

The annual runoff was small changed largely between different years in the area above Huangheyan station, the inter-annual coefficient of variability was high up to 0.96, the maximum and minimum flow ratio was high up to 126. From Jimai towards the downstream, with the catchment area increased, the annual runoff increased, the inter-annual coefficient of variability reduced gradually, reached 0.26 in Maqu station, which was similar with Jungong station and Tangnaihai station. From the runoff process of each hydrological station, the runoff temporal variation of the stations below Jimai was basically consistent and had good synchronization.

According to the analysis of the annual runoff difference curves, the runoff had periodic change between wet years and dry years in the Yellow River source regions. The dry periods were 1956 ~ 1960, 1969 ~ 1974, 1994 ~ 2004, and the wet periods were 1961 ~ 1968, 1972 ~ 1984, and the normal periods were 1985 ~ 1993. The average annual runoff of the three dry periods were respectively 16.2×10^9 m^3, 17.5×10^9 m^3 and 15.9×10^9 m^3, about 12.4% ~ 20.6% less than the mean annual runoff, especially, in 2002, the runoff was 10.6×10^9 m^3, only 53% of the average annual runoff, which was the driest in recent fifty years. The average annual runoff of the two wet periods was respectively 23.1×10^9 m^3 and 24.7×10^9 m^3, 15.4% ~ 23.3% more than the mean annual runoff; the wettest year was 1989, which annual runoff was 32.8×10^9 m^3, occurred in 1989. The hydrological characteristics of the main stations in the Yellow River source region showed as Tab. 1.

2.2.2 Variation within the year

In the Yellow River source region, the runoff distribution in a year was similar with the precipitation, which had obviously seasonal change, the runoff was mainly coming from the precipitation in the flood season. The runoff month distribution proportion of the main hydrologic stations was substantially similar except Huangheyan station. The amount of the runoff from June ~ October accounted for 71% ~ 73% of the annual runoff, and the proportion of each month was greater than 10%; the runoff from June ~ October runoff of Huangheyan station accounted for 65.5% of the annual runoff, the maximum monthly runoff accounted for 14.8% of the annual runoff, the minimum monthly runoff accounted for 4%, the difference was only 10.8%, which was minimal among the hydrologic stations.

Tab. 1 The hydrological characteristics of the main stations in the Yellow River source region

Station name	Huangheyan	Jimai	Mentang	Maqu	Jungong	Tangnaihai
Catchment area (km^2)	20,930	45,019	59,655	86,048	98,414	121,972
Mean annual runoff($\times 10^9$ m^3)	0.7	3.9	7.3	14.3	17.1	20.0
Average annual runoff depth(mm)	32.8	86.04	122.0	165.9	173.4	163.9
Maximum annual runoff($\times 10^9$ m^3)	2.5	8.3	14.9	22.3	28.2	32.8
Minimum annual runoff($\times 10^9$ m^3)	0.02	2.0	4.1	7.2	8.1	10.6
Ratio of maximum and minimum runoff	126	4.2	3.6	3.1	3.5	3.1
Inter-annual coefficient of variation (*Cv*)	0.96	0.38	0.35	0.26	0.27	0.27

2.2.3 The runoff variation of Tangnaihai hydrologic station

The average annual runoff of Tangnaihai station was 20.0×10^9 m^3, the average annual runoff was 21.6×10^9 m^3 in 1960s, 20.4×10^9 m^3 in 1970 s, 24.1×10^9 m^3 in 1980 s, 17.6×10^9 m^3 in 1990s, and 17.2×10^9 m^3 from 2000 to 2009, it showed that the runoff decreased from 1990 s. the runoff variation of Tangnaihai hydrological station from 1956 to 2010 shown in Fig. 1.

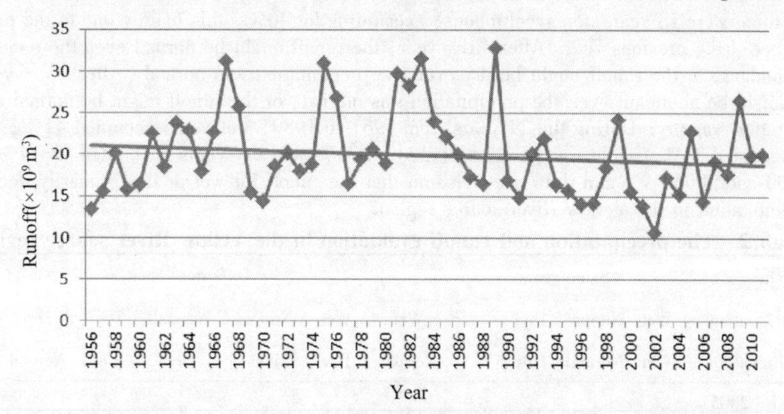

Fig. 1 The runoff variation of Tangnaihai hydrological station from 1956 to 2010

3 Analysis of the factors affecting runoff

3.1 Precipitation effect on Runoff

According to the measured data of Jimai, Zeku, Henan, Maqin, and Tangnaihai rainfall station, calculated the arearl mean rainfall by Tyson polygon method. The average annual precipitation was 487 mm, the maximum was 613 mm and the minimum was 405 mm in the Yellow River source region. The average precipitation was 490 mm in 1960 s, 482 mm in 1970 s, 507 mm in 1980 s, 478 mm in 1990 s, and 480 mm from 2000 to 2009. It was dry since 1990 in the Yellow River source region. The precipitation variation of the Yellow River source region from 1956 to 2009 was shown as Fig. 2.

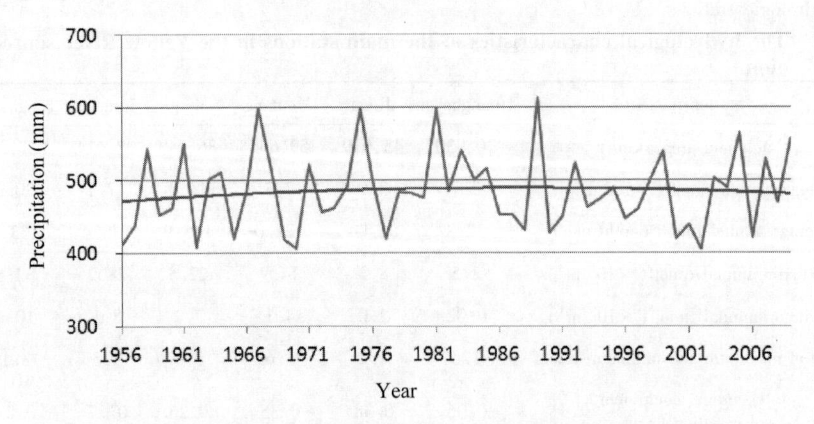

Fig. 2 The precipitation variation of the Yellow River source region from 1956 to 2009

The analysis of the change of relationship between the runoff and the precipitation showed strong synchronization. The rate defined as follow, $P < 37.5\%$ was wet year, $37.5\% < P < 62.5\%$ was normal year, $P > 62.5\%$ was dry year, the precipitation and runoff evaluation in the Yellow River source region was shown as Tab. 2. From the table, we could see that there were 27 years that the runoff and precipitation were synchronous during 1961 and 2005, accounting for 60%, there were 18 years non synchronous, accounting for 40%, this mainly due to the precipitation effect of the previous year. After a dry year, the runoff might be normal even the precipitation was abundant, or the runoff might be dry even the precipitation was normal. After a wet year, the runoff might be abundant even the precipitation was normal, or the runoff might be normal even the precipitation was dry. During the 29 years from 1961 to 1989, wet years occupied 41%, and dry years occupied 34%; while wet years accounted for 31% and dry years accounted for 44% during the 1990 and 2005. We can draw a conclusion that the runoff for wet or dry primarily decided by the precipitation in the Yellow River source region.

Tab. 2 The precipitation and runoff evaluation in the Yellow River source region

Item	Same years				Different years						
Precipitation	Wet	Normal	Dry	Total	Wet	Wet	Normal	Normal	Dry	Dry	Total
Runoff	Wet	Normal	Dry		Normal	Dry	Wet	Dry	Wet	Normal	
1961 ~ 2005 (45 years)	13	3	11	27	3	1	5	3	1	5	18
1961 ~ 1989 (29 years)	11	3	4	18	1	0	4	0	1	5	11
1990 ~ 2005 (16 years)	2	0	7	9	2	1	1	3	0	0	7

3.2 Temperature effect on Runoff

From the observation data of Maduo station, the temperature was low from 1960s to 1970s, increased from 1980s, and increased more obviously after 1998 in the Yellow River source region. The average temperature was −3.7 ℃. It was −4.1 ℃ in 1960 s, −4.3 ℃ in 1970 s, −3.9 ℃ in 1980 s, −3.5 ℃ in 1990 s, and −2.6 ℃ from 2000 to 2007. The annual mean temperature changes of Maduo station from 1956 to 2007 was shown as Fig. 3.

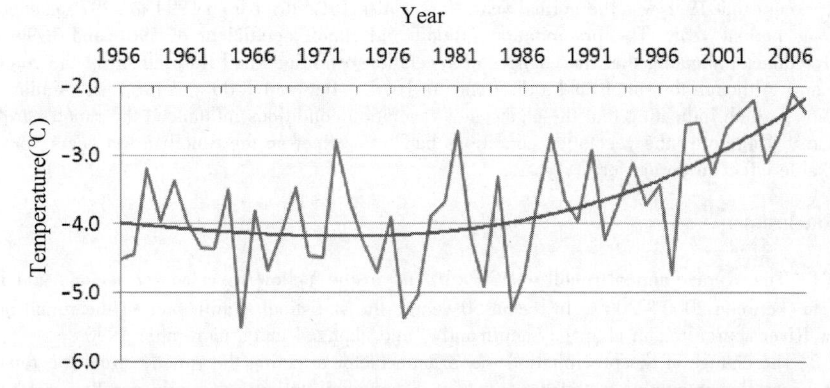

Fig. 3 The annual mean temperature changes of Maduo station from 1956 to 2007

The ground temperature observation data of Qinghai-Tibet Plateau showed that, the temperature continued warming since the 1970 s, especially since 1980 s, which affected the ground temperature above 40 m, the shallow ground temperature above 20 m increased the most obviously. The ground temperature data of recent 15 ~ 20 years showed that, the annual ground temperature increased 0.3 ~ 0.5 ℃ in the plateau of the seasonal frozen soil area, melting rivers and island permafrost zone with less ice (water) , 0.1 ~ 0.3 ℃ in the large continuous permafrost zone.

The study on the change of the frozen soil depth in Jimai and Maqu station by the Cold and arid regions Environmental and Engineering Research Institute, Chinese Academy of Sciences showed that, the maximum frozen soil depth was decreased continuously since 1980s, and changed the most obviously from January to March, the frozen soil depth decreased about 11 cm during every 10 years, and the upper position in permafrost descended greatly in Maqu station, descended 6.7 cm every 10 years. As a result of maximum frozen soil depth becomes shallow and permafrost upper position downward, the permafrost layer to thin, even many years frozen earth disappear, seasonal frozen soil layer thickens, permafrost degradation is a regional phenomenon. As the permafrost layer position downward, the frozen soil layer level down to the deep soil, caused by soil water leakage, resulting in surface runoff weakening. There were more than 90% area of permafrost in the Yellow River source region, the permafrost degradation had a certain influence on the runoff.

3.3 Vegetation condition effect on runoff

Remote sensing interpretation results showed that high coverage grassland was 46,832 km^2 in 1989 and 46,066 km^2 in 2005, decreased 766 km^2; the desert soil was 12,123 km^2 in 1989 and 12,346 km^2 in 2005, increased 223 km^2.

If the vegetation condition was poor, the vegetation degradation and desertification appeared, the capacity of rainfall infiltration and soil evaporation increased, but the capacity of water conservation reduced; rivers, lakes, marshes and other wetlands reduced too, which increased the initial loss of the precipitation, decrease the function of runoff. Conversely, if the vegetation condition was good, the vegetation, small lakes, marsh wetland filling, the capacity of water conservation was enhanced, the runoff would be relatively stable.

The degree of the effect on runoff by the vegetation condition could be defined through different periods of precipitation and runoff in comparative analysis. Selected 1961 and 1962 as good vegetation condition typical years in 1960s, 1999 and 2000 as the poor vegetation condition typical years in 1990s, which had similar precipitation and antecedent precipitation, so that to analyze comparatively. The 1961 and 1999 were the similar wet years with similar precipitation and antecedent precipitation, and 1962 and 2000 were similar dry years with similar precipitation and antecedent precipitation. The precipitation was 18% less than the average annual precipitation from 1956 to

1960, except that 1958 was the normal year. It was also 18% drier from 1994 to 1997, except that 1998 was normal year. The precipitation, runoff and runoff coefficient of 1961 and 1999 were roughly equal, indicating that the changes of vegetation conditions had little effect on the runoff in wet years. Although the runoff had a difference of 16.4% between 2000 and 1962 with similar precipitation, which indicating that the changes of vegetation conditions influenced the runoff hardly in dry years. Therefore, the vegetation conditions had little effect on the runoff in wet years, but had remarkable effect in normal or dry years.

4 Conclusions

(1) The average annual runoff was 20×10^9 m^3 in the Yellow River source region, and 17.2×10^9 m^3 between 2000 ~ 2009. In recent 50 years, the statistical significance of the runoff of the Yellow River source region changed significantly, and changed much more after 1990s.

(2) The change of the precipitation was a main factor affecting the runoff. From the relationship between the precipitation and the runoff of Tangnaihai station, we could see that wet or dry runoff was mainly determined by the precipitation, the amount of the runoff not only depended on the precipitation in the same year, also influenced by the precipitation of the year before.

(3) The vegetation condition had a significant impact on the runoff in the dry year. The temperature rising, resulted that permafrost depth increases, and the surface soil drying, vegetation degradation, land desertification, evapotranspiration capacity enhancement, infiltration increase, and runoff decrease at last. The rainfall decreased 1.8% in 1990s than the annual average value, but the runoff decreased 12.0%. The comparative analysis of the typical year in different periods showed that the change of vegetation condition had little influence on the runoff in wet years, but had remarkable influence in normal or dry years.

(4) The reason of the reduction of the runoff in the Yellow River source region in 1990s was the evapotranspiration capacity increase, infiltration increase and the rainfall reduction.

References

Yellow River Conservancy Commission. The integrated planning of water resources of the Yellow River Basin [R]. Zhengzhou: Yellow River Conservancy Commission, 2009.

Li Rong, et al.. The influence on the runoff by the climate change in the reaches between Sanmenxia and Huayuankou in the Yellow River Basin [J]. Yellow River, 2007(10).

Song Yudong, et al.. Research on water resources and environment in Talimu River in China[M]. Xinjiang: Xinjiang People's Publishing House, 2000.

Li Yanbin, et al.. Wavelet analysis of the runoff variation of Yellow River [J]. Water Resources and Hydropower Engineering, 2012, 43(1).

Causes and Countermeasures of Ice Flood Disasters in the Yellow River Xiaobeiganliu Reaches

Zheng Shixun and ***Guo Guili***

Yellow River Shanxi Bureau, YRCC, Yuncheng, 044000, China

Abstract: Since 1996, large or small ice flood disasters have been occurring frequently in the Yellow River Xiaobeiganliu reaches, causing serious economic losses. Through an analysis of the characteristics of ice flood disasters occurring on the left bank Shanxi side of Xiaobeiganliu since 1996, it was found that temperature changes in the river channel, the effect of water control projects on the upper reaches, worsening of the channel pattern of Xiaobeiganliu and imperfection of engineering and non - engineering measures for ice flood control are the main causes of ice flood disasters. Ice flood control of Xiaobeiganliu should be carried out in the following aspects: strengthen engineering harnessing in Xiaobeiganliu and accelerate non - engineering measures construction to gradually improve the ice flood control and disaster reduction system; clear away barriers in the river to effectively reduce obstacles to ice discharge; control the discharge of the Wanjiazhai Reservoir in a scientific way and carry out research on the effect of operation of upper and lower reservoirs on the ice run in Xiaobeiganliu to give full play to the role of reservoir regulation in ice flood control; pay close attention to temperature change in the channel and plan in advance ice flood control countermeasures; and improve reservoir system construction in the Baiganliu reaches in accordance with the planning of YRCC to effectively improve the river regime of Xiaobeiganliu, intercept ice from the upper reaches to relieve the ice flood control pressure on Xiaobeiganliu.

Key words: ice flood disaster; Xiaobeiganliu; cause; countermeasure

1 Overview

1.1 Reach profile

Xiaobeiganliu, namely the reaches of the middle Yellow River between Yumenkou and Tongguan, as the boundary river between Shanxi and Shaanxi Provinces, is 132.5 km long and is located at 34°35′ ~ 35°49′N and 110°15′ ~ 110°38′E. The channel of the reaches is 3 ~ 18 km wide, and is a typical accumulation wandering channel. On the left bank are 5 counties (cities) Hejin, Wanrong, Linyi, Yongji and Ruicheng under Yuncheng City, Shanxi Province, and on the right bank are the 4 counties (cities) Hancheng, Heyang, Dali and Tongguan under Weinan City, Shaanxi Province. On the left bank Shanxi side of Xiaobeiganliu there is 42.29 $\times 10^4$ mu of floodplain, where there are 8 large - scale electric pumping stations, and within the floodplain are distributed 34 enterprises and institutions and historical sites such as the Guanque Tower, the Iron Bull of the Tang Dynasty and so on. On the right bank Shaanxi side there is 62.49 $\times 10^4$ mu of floodplain. In the reaches, construction of flood control works began in 1968. Up to now, a total of 41 flood control works have been built on both banks, with length of 160.32 km, of which 27 works are on the left bank Shanxi side, whose length is 91.65 km.

1.2 Overview of ice run and disasters

The Xiaobeiganliu reaches of the Yellow River belong to semi - arid temperate continental monsoon climate, and are controlled by Mongolian high pressure in winter, where the climate is dry and cold, and there is little rainfall. Usually the reaches only had floating ice but did not freeze, and in special years parts of the reaches might be frozen up. In the Xiaobeiganliu reaches, ice usually appeared in late November or early December, and disappeared in mid - February of the following year, and the main ice run situations were shore ice and floating ice.

In the Yellow River Xiaobeiganliu reaches there were no records of freezing before the Song and Yuan Dynasties. In the 392 years between Wanli 48th year of the Ming Dynasty (1620 AD) and 2011, there were 32 years when freezing – up occurred in Xiaobeiganliu, and among the freezing – up years there were 8 years when disasters occurred on the left bank Shanxi side. The earliest disaster took place in the first month of Guangxu 20th year of the Qing Dynasty (1894 AD), when ice damming occurred in Xiaoshizui area of Henin, the dammed water flooded Cangtouzhen Port where 120 families resided and turned thousands of mu of wheat field into an ice field. After founding of P. R. China in 1949, from 1949 to 2011, it is recorded that 7 years have seen ice flood disasters. In December 1965, spreading of dammed water occurred in Hejin reaches, inundating 6,658 mu of wheat field. From January to February 1996, a 60 – year rare ice flood disaster occurred in Xiaobeiganliu, which hit a population of 100 thousand in 61 villages under 8 townships (towns) of Yongji and Ruichang counties on Shanxi side along the river, inundated 80 $\times 10^4$ mu of farmland, damaged wells, roads, electric appliances and flood control projects in the floodplain, causing a direct economic loss of 0.469,9 $\times 10^8$ yuan. In February 2000, an once in 100 – years extraordinary ice flood disaster occurred in Qingjianwan and Xiaoshizui reaches of Hejin. The 10 – km river reaches from Yumenkou to Daguotou were all covered with huge ice blocks, and within the 40 – km^2 channel about 60 $\times 10^8$ m^3 of ice was piled up, nearly ten thousand mu of floodplain behind the dam was inundated, and more than 60 water source wells of Shanxi Aluminum Manufacturer and local wells and electric network and equipment were besieged, with economic loss of 30 million yuan. In January 2004, ice overbank and project danger occurred in Hanjiazhuang reaches of Yongji, over 64 $\times 10^4$ mu of floodplain land was inundated. In February 2008, Qingjianwan bend – regulating downward – extending project of Hejin was in danger, causing an economic loss of 0.42 $\times 10^6$ yuan. In January of the same year, Hanjiazhuang Project of Yongji was in danger, losing 0.27 $\times 10^6$ yuan of investment. In the end of December 2009, overbank occurred due to ice jam in Qucun reaches of Linyi, flooding 0.39 $\times 10^4$ mu of floodplain land. In January 2010, ice jam occurred in Hejin reaches, causing overtopping at Qingjianwan bend – regulating downward – extending project and danger at the Fenhekou Project, and the rescue cost 0.56 $\times 10^6$ yuan (see Tab. 1).

Tab. 1 Statistics of ice flood disasters on the left bank of the Xiaobeiganliu reaches of the Yellow River

Time		Place	Ice Run Situation	Loss
In history	First month of Guangxu 20th year (1894 AD)	Xiaoshizui area of Hejin	Ice damming occurred in Xiaoshizui area	Cangtouzhen Port where 120 families resided was completely flooded and thousands of mu of wheat field was turned into an ice field.
	Dec. 19 ~ 20, 1965	Hejin reaches	Water spreading occurred due to ice damming in Hejin reaches, a large quantity of ice and water flowed to the east bank, and, by Lianbo Village, rushed directly to the south – east of old town of Hejin	Ice damming occurred at Fenhekou, villages of Lianbotan, yongan and Taiyang of Hejin were stricken by the disaster, and 6,658 mu of wheat field was inundated
In recent years	Jan. 18 ~ Feb. 17, 1996 (extraordinary ice flood disaster)	Ruicheng and Yongji reaches	Affected by freezing at the end of the Sanmenxia Reservoir area, the river froze at Tongguan. Affected by ice damming at Tongguan, the Xiaobeiganliu channel could not discharge smoothly, thus freezing gradually extended upstream to the reaches west of Yongji Town. The frozen river was as long as 27 km and the ice cover was over 0.6 m thick	8.0 $\times 10^4$ mu of farmland was inundated, 61 villages and a population of over 100 thousand under 8 townships (towns) of Ruicheng and Yongji counties were stricken, with economic loss of 45.23 million yuan. The backwards reconstructed works at the west of Yongji and Fenghuangzui works of Ruicheng lost 1.76 $\times 10^4$ yuan

Continued Tab. 1

Time		Place	Ice Run Situation	Loss
	Feb. 7 ~ 9, 2000 (extraordinary ice flood disaster)	Hejin reaches	In early February, the discharge of Wanjiazhai Reservoir on the middle Yellow River increased, along with it a great quantity of floating ice rushed downstream, and then ice jam occurred at Daguotou of Hejin. 0.6×10^8 m^3 of ice piled up at Qingjianwan and Xiaoshizui	Burst occurred at the location 5 + 400 of the Qingjianwan Project with length of 62 m, and more than 1.0×10^4 mu of farmland was flooded. The flood besieged Shanxi Aluminum Manufacturer, over 60 water source wells and local wells, more than 1,000 electric poles, causing an economic loss of up to 0.3×10^8 yuan.
	Jan. 28 ~ 29, 2004	Yongji reaches	The main channel of the reach between the west of Yongji and Hanjiazhuang froze, the dammed water overflowed through by bypass channels, water entered Hanjiazhuang floodplain, the bypass channel at the end of Hanjiazhuang project was jammed up by ice, and the bypass channel water flowed backwards and spread along the back of the project	The back slope of Hanjiazhuang project collapsed, Hanjiazhuang project and the river engineering section were besieged, and over 64×10^4 mu of floodplain farmland was inundated
In recent years	Jan. 21 ~ 24, 2008	Yongji reaches	In January, the floating ice in the reach between Shundi and Hanjiazhuang of Yongji became denser and the water stage was dammed higher. Affected by the bend of the main stream of the Yellow River, the No. 18 ~ 20 dams of Hanjiazhuang project encountered danger of root stone sinking in succession.	Hanjiazhuang project was in danger for 160 m, and lost 1,114 m^3 of stone, with investment loss of $0.271,8\times10^6$ yuan
	Feb. 5, 2008	Hejin reaches	Parts of Hejin reaches froze, the water did not discharge smoothly; with increase of freeze area, the water stage in front of the project dam kept rising, the flood overflowed through the first back – silting opening of Qingjianwan bend – regulating emergency flood control project, and then discharged along the back side of the project	The back – silting opening was destroyed for 80 m, and 120 – m earth base on the back side of the main work was destroyed, with total investment loss of 0.42×10^6 yuan
	Dec. 31, 2009	Linyi reaches	Affected by the bend of the main stream and ice jam, ice overbank occurred at Qucun reaches of Linyi	3,900 mu of floodplain land was inundated
	Jan. 14 ~ 18, 2010	Hejin reaches	Affected by lasting low temperature and the discharge of ice from the upper reaches, the main channel from Yumenkou to Fenhekou partly froze, a great amount of ice jammed and accumulated, the main stream did not discharge smoothly and the water stage was raised high rapidly, causing project danger	Qingjianwan bend – regulating and downward – extending project and Fenhekou project together lost 4,200 m3 of earth and 750 m3 of stone. The Fenhekou project had 190 m of water penetration and 170 m of crest cracks, and the breach in the right embankment of the warping test project was about 30 m long. The rescue cost 0.56×10^6 yuan.

2 Analysis of the characteristics of ice flood disasters

Most ice flood disasters in Xiaobeiganliu by Shanxi side occurred in January to February, and the ice flood occurring frequency has been increasing since 1996: in the 376 years before 1996 only two years saw ice flood disasters, while in the 16 years since 1996 there have been 6 years when ice flood disasters occurred. The reaches subject to disasters also have changed. The two disasters before 1996 both occurred in Hejin reaches where the Yellow River comes out of Longmen, while after 1996, Hejin, Yongji and Linyi reaches encountered ice flood disasters one after another. It can be seen from the ice flood disasters in the past years that the ice flood disaster tends to occur abruptly and is difficult to predict. Once the channel is jammed by ice, the water stage will rise rapidly within a few hours, causing large area of floodplain to be inundated, or even causing projects to be overflowed or dams to be destroyed, especially, once ice dams form, the dammed water will extend a long distance, it is difficult to predict where danger may occur, and the projects in danger are difficult to guard.

Geographically, the Xiaobeiganliu of the Yellow River is a reach of the middle Yellow River at the lowest latitude, where the temperature is relatively high, so it should not have been the channel with the most serious ice flood disasters. Yet, with special channel characteristics, plus the flow direction of it together with Dabeiganliu parallel to the longitude line while crossing the latitude line, the 857.5 – km channel crosses nearly 6 degrees latitude, the huge ice blocks from the high – cold areas in the north of Dabeiganliu are transferred to the Xiaobeiganliu channel, which may bring about serious ice flood disasters in the Xiaobeiganliu channel. Therefore, it is necessary to analyze the causes of the ice flood disasters in Xiaobeiganliu and put forward corresponding defensive countermeasures.

3 Causes of ice flood disasters

Based on analysis of the ice flood disasters occurred since 1996 on the left bank at Shanxi side of Xiaobeiganliu, it is considered that the formation of ice flood disasters was mainly affected by four factors: the first is by temperature factor, the second by the effect of water control projects, the third by the effect of river regime, and the fourth by imperfection of the ice flood control and disaster reduction system.

3.1 Temperature factor

Channel ice is the outcome of low temperature, and temperature change is an important factor of ice flood disasters.

3.1.1 Steady drop in temperature causes river freezing, the incoming water from the upper reaches is dammed high and finds other flow routes, thus forming ice flood disasters

When the daily lowest temperature is lower than –5 ℃ (and daily average temperature is lower than 0 ℃), shore ice begins to appear, and when the shore ice increases to some extent, river freezing forms. For example, during the ice flood disaster in 1996, the 3 d average temperature dropped abruptly about 11 ℃ in Ruicheng, with the daily lowest temperature below –10 ℃; the 3 d average temperature dropped 10 ℃ suddenly, and the daily lowest temperature was –7.8 ℃ in Yongji. The frozen channel between Ruicheng and Yongji was up to 27 km long. During the 2007 ~ 2008 ice flood period, the temperature lower than 0 ℃ lasted 80 d, with the lowest temperature –10 ℃, the reach between Yumenkou and Fenhekou of Hejin partly froze for 42 d, and the Shundi reach of Yongji froze for 19 d, which were both longer than the previous years. From January to February 2008, affected by low temperature and river freezing, ice flood disasters occurred in Hanjiazhuang reach of Yongji and Qingjianwan reach of Hejin in succession. In January 2010, in Hejin reach the temperature kept being lower than 0 ℃, with the lowest temperature –8 ℃. Affected by lasting low temperature, the main channel of the Yumenkou – Fehekou reach partly froze, the main channel could not discharge smoothly and the water stage was dammed high, causing project danger.

3.1.2 Great change and abrupt rise and drop in temperature are also likely to cause disasters

For example, in February 4 ~ 7, 2000, the temperature of Hejin rose, with highest temperature 7 ℃; on February 8, the temperature dropped abruptly, the lowest temperature was as low as −7.8 ℃, causing the melt ice in the reaches upstream of Yumenkou to discharge in great quantity. After coming out of Yumenkou, the river suddenly widened and became shallower, a great quantity of ice blocks jammed in the wide and shallow channel, plus abrupt drop in temperature, the ice blocks froze to form an ice dam, raising water level and causing ice flood disasters. Thus it can be seen that the change in temperature is an important factor of causing ice flood disasters.

3.2 Effect of water control projects

On the upper reaches of Xiaobeiganliu there are Wanjiazhai, Longkou and Tianqiao Reservoirs, and on the end reaches there is the Sanmenxia Reservoir. As for the Wanjiazhai Water Control Project, which began to store water on October 1, 1998, under normal operation of original design, the water level should be 970 ~ 975 m in November, 975 ~ 977 m from December to March of the following year, and lower to 970 m during the thawing period. Yet, due to the requirement of ice flood control in the upper reaches, the Wanjiazhai Project had to discharge in great quantity in January and February, its flow rate increased and became unsteady, which was likely to cause ice flood disasters in Xiabeiganliu reaches. For instance, in early February, 2000, the daily discharge of the Wangjiazhai Reservoir changed greatly (from 0 to 1,000 m^3/s), the daily discharge was up to 650 m^3/s, and the discharge at Longmen station was 500 ~ 600 m^3/s. With the increase of the river discharge, the large amount of floating ice within the Shanxi – Shaanxi Gorge rushed downward, causing an extraordinary ice flood disaster in Hejin reaches. Having weaker regulation capacity, the Tianqiao Reservoir, which is located downstream of the Wanjiazhai and Longkou reservoirs, discharged the ice as soon as it comes, so the reservoir could not relieve the flood control pressure on Xiaobeiganliu.

3.3 Effect of river regime

River regime is another important factor of ice flood occurrence. The Xiaobeiganliu reaches are like a dumbbell in shape, and the river is 8.5 km wide on average. (see Tab. 2) Coming out of Yumenkou, the gorge stream less than 100 m wide suddenly widens to several km, and when it flows to Tongguan, the river narrows to 850 m, and this special river regime makes the Xiaobeiganliu reaches a natural flood and sand detention area. Before 1977, the Xiaobeiganliu reaches encountered "bottom tearing" scouring 7 times, the channel was narrow and deep, thus ice flood disasters were relatively rare. Up to now since 1977, no wide – range bottom tearing scouring has occurred, thus channel silting has become serious, the channel has been keeping shrinking and the difference in elevation between floodplain and channel has reduced, numerous sandbanks and trenches appear here and there, causing frequent occurrence of ice flood disasters. Due to worsening of the river regime, the huge ice blocks from the gorge channel upstream of Yumenkou could not but get stranded after going into the Xiaobeiganliu reaches, those ice blocks that manage to float would jam up in narrow reaches or at nodal points to form ice dams. For example, in 2000, the jammed ice blocks formed an ice dam at Daguotou of Hejin, causing an extraordinary ice flood disaster. Due to river silting, the flood discharge capacity of the channel has decreased sharply, and the protective capacity of the projects has become weaker and weaker, thus ice run are likely to cause water spreading and project overflowing. For instance, in 2008 and 2010, the danger of overflowing occurred at Qingjianwan bend – regulating project of Hejin.

In recent years, structures across the river on the Xiaobeiganliu reaches have increased year after year, at present, there are altogether 6 bridges and floating bridges that have been completed and are being built. These bridges are mostly built at narrow points of the river, and their piers, floating pontoons and the temporary construction roads have narrowed the flood discharge cross – section, which has further aggravated river regime worsening and increased the chance of ice flood disaster occurrence.

Tab. 2 Characteristics of basic river regime of Xiaobeiganliu reaches

Reaches	Length (km)	River width (km)			Average slope (‰)
		Widest	Narrowest	Average	
Yumenkou—Miaoqian	42.5	13.0	3.5	6.6	0.57
Miaoqian—Jiamakou	30.3	6.6	3.5	4.73	0.47
Jiamakou—Tonghuan	60.0	18.8	3.0	11.59	0.31
Yumenkou—Tongguan	132.5	18.8	3.0	8.87	0.41

3.4 Imperfection of ice flood control and disaster reduction system

In ice flood control projects, firstly the existing projects on the left bank of Xiaobeiganliu were mostly built in the 1970s, with river bed rising due to silting, the protective standards have decreased, and the projects are subject to ice overtopping. For instance, during the ice flood disaster period in Hejin in 2000, overtopping occurred in succession at Xiaoshizui, Qingjianwan and Yumenkou projects. Secondly, in some reaches, the space between projects is too large so that the river regime can not be controlled effectively, which is liable to cause overbank. For example, during the ice flood period in 1996, the dammed water flowed into Puzhou floodplain through the project space west of the town, causing a large area of overbank; in January 2004, since the main channel froze, the dammed water overflowed through bypass channels, overbank occurred in Hanjiazhuang reaches of Yongji, inundating 64 thousand mu of floodplain land; in 2009, in Qucun reaches of Linyi, affected by the bend of the main stream and ice damming, 3.9×10^4 mu of floodplain land was submerged.

As for non – engineering measures, firstly the ice monitoring means in the Xiaobeiganliu reaches are backward, basically at the state of manual monitoring; secondly there are no specialized ice flood control rescue teams, and mass protection teams are difficult to set up, thus optimum opportunities for rescue would be missed; thirdly there are no special facilities to break ice and dredge the river, once ice damming occurs, it is difficult to deal with; and fourthly, construction of information technology has lagged behind.

4 Ice flood control countermeasures

4.1 Advance construction of ice flood control engineering and non – engineering measures

Strengthen engineering harnessing in the Xiaobeiganliu reaches, actively lay up construction projects and try to get funds for construction, speed up heightening and reinforcement of old projects, advance construction of new projects and gradually eliminate spaces between projects so as to improve the ability of flood control and ice flood control. The "Feasibility Report on Recent Harnessing of the Yumenkou—Tongguan Reaches of the Yellow River" has planned seven newly – built or continued projects with length of 10.55 km, which are to be constructed in the Twelfth Five – Year Plan period. The fulfillment of the report on recent harnessing will further improve the flood control and ice flood control ability of the Xiaobeiganliu reaches. As for non – engineering measures, it is planned to gradually establish and perfect emergency plans for ice flood control, organize specialized ice flood control rescue teams, strengthen training of ice flood control teams to ensure that the teams are well - trained and skilled; strengthen ice run monitoring, allocate facilities special for ice flood control, advance construction of information technology and perfect forecast and monitoring system for the Yellow River ice flood control so as to improve the level of dealing with emergencies caused by ice run in an all - round way.

4.2 Remove barriers in the channel for ice flood control

The woods, production dykes and floating bridges are significant factors that help form ice jam and cause ice flood disasters. To remove barriers in the channel effectively may ensure ice flood control safety. We must draw up plans for barrier removal before floating ice discharge, and completely remove barriers that hinder ice discharge in the channel to ensure smooth ice discharge. Strengthen inspect and supervision of river - related projects that are being built, and in case of special situations, take powerful measures to eliminate obstacles that hinder ice discharge so as to prevent jammed ice from forming ice dams and causing ice flood disasters. Remove floating bridges before floating ice discharge to ensure smooth ice discharge of the channel and to prevent jammed ice from forming dams to cause dangerous situations.

4.3 Give full play to the role of reservoir operation in ice flood control

Collect ice run date of the Dabeiganliu reaches and establish an ice run database. On the basis of the ice run of the upper reaches and incoming water, control the discharge of the Wanjiazhai Reservoir in a scientific way to minimize ice flood disasters in the Xiaobeiganliu reaches. Carry out research on the effect of the reservors on the upper and lower reaches on ice run in Xiaobeiganliu, that is, the relation between the operation of the Sanmenxia Reservoir and ice run in Xiaobeiganliu, and the effect of joint operation of the Wanjiazhai, Longkou and Tianqiao reservoirs on ice flood control of Xiaobeiganliu, and put forward reservoir operation mode favorable for ice flood control of Xiaobeiganliu and suggestions on improvement.

4.4 Pay close attention to temperature change and plan countermeasures for ice flood control in advance

Temperature change is an important factor that causes ice run. During ice run period, flood control and ice flood control offices at various levels should earnestly implement responsibility system of ice flood control, strengthen watch for ice flood control, pay close attention to the change in channel temperature, based on the local weather forecast, keep abreast of the weather characteristics in advance, carry out dynamic monitoring of river temperature and ice run situation. In case of abrupt drop or unsteady change in temperature, hold consultations for ice flood control, analyze and predict the development trend of the ice run situation and plan protective measures in the light of the actual situation to ensure ice flood control safety.

4.5 Perfect reservoir system construction

At present, there are three water control projects on the Beiganliu reaches of the Yellow River: the Wanjiazhai Reservoir is 619 km from Longmen Hydrological Station, the Longkou Hydropower Station is 25.6 km downstream of Wanjiazhai, and the Tianqiao Water Control Project is 70 km downstream of Longkou. The nearest Tianqian project is over 500 km from Longmen. As a key water control projects on the Dabeiganliu reaches, the Guxian Water Control Project has been listed in the Twelfth Five - Year Plan. The Guxian Water Control Project is to be located on the Qikou - Yumenkou reaches of the middle Yellow River's Beiganliu, about 10.1 km upstream of the Hukou Waterfall. After completion, the Guxian project can reduce silting in the Xiaobeiganliu reaches, improve the river regime, and at the same time intercept ice from the upper reaches, carry out operation for ice flood control. The construction of the Guxian Reservoir will further perfect the regulation system of ice flood control reservoirs with major reservoirs taking the leading position, improve the ability to control ice flood and greatly relieve the ice flood control pressure of the Xiaobeiganliu reaches.

5 Conclusions

Based on analysis of the characteristics of ice flood disasters on the left bank of Xiaobeiganliu, the authors believe that the ice flood disasters in the Xiaobeiganliu reaches were mainly affected by four factors: temperature, operation of water control projects, river regime and imperfection of the ice flood control and disaster reduction system. We suggest that, through construction of engineering and non - engineering measures for ice flood control, we further perfect ice flood control and disaster reduction system to lay a solid foundation for ice flood control; that, by clearing away barriers in the river, we effectively reduce barriers that hinder ice discharge in the river to ensure ice flood control safety; that we pay attention to change in river temperature, hold consultations for ice flood control earlier and plan in advance countermeasures for ice flood control; that, by perfecting construction of reservoir system, we give full play to the role of water operation in ice flood control, relieve ice flood control pressure, reduce silting and improve the river regime of Xiaobeiganliu so as to minimize ice flood disasters.

References

Yuan Baoping. Annals of Xiaobeiganliu of the Yellow River in Shanxi Province[M]. Zhengzhou: Yellow River Conser vancy Press, 2002.

Ke Sujuan, Wang Min, Rao Suqiu et al.. Research on the Yellow River Ice[M]. Zhengzhou: Yellow River Conser vancy Press, 2002.

Yellow River Conservancy Commission. Implementation Program for Construction of Comprehensive Ice Flood Control Capacity of the Yellow River[R]. 2008.

Yellow River Engineering Consulting Co., Ltd.. Feasibility Report on Recent Harnessing Project of Yumenkou - Tongguan Reaches of the Yellow River[R]. 2011.

Zhai Jiarui. How to Optimize Reservoir Operation in Ice Flood Control[J]. Yellow River, 2005.

PMP/PMF Estimation over Consideration of Global Warming

Wang Guoan[1], ***Li Chaoqun***[1] and ***Wang Chunqing***[2]

1. Yellow River Engineering Consulting Co., Ltd., Zhengzhou, 450003, China
2. Hydrology Bureau of Yellow River Conservancy Commission, Zhengzhou, 450004, China

Abstract: Under the precondition of agreement to (1) global warming, (2) climate warming of the design region, (3) the climate warming of the design region during the flood periods and (4) the surface dew point (Td) increases during the efficiency storm occurs in flood periods, methods to obtain PMP/PMF over consideration of global warming were exploited in this paper. The 4 hypothesizes mentioned above should be considered together, otherwise any methods will be meaningless. The main framework of calculating PMP/PMF is as High - Efficiency storm→Moisture maximization→Transposition→Enveloping →PMP→Runoff generation and confluence calculation→PMF. The major influences of climate warming on PMP/PMF estimating lie in Moisture maximization and Runoff generation and confluence calculation respectively. Considering the design attributes and prediction attributes of PMP/PMF, methods to obtain PMP/PMFs should be recapitulative. The method to estimate PMP/PMFs of design watersheds over consideration of global warming was presented, only considering the influence of the air temperature increased on the moisture maximization ratio. In details, the PMP moisture maximization ratio was calculated using the similar method to obtain the precipitable water from surface dew point. However, considering that the dew point of high efficiency storm will increase due to climate warming, and the relationship of the dew point and the precipitable water lies in that while the dew point increases by 1℃, the precipitable water increases by about 10%, and accordingly the PMP increases by about 10%. Subsequently, the PMF increases by about 10%, ignoring the influence of climate warming on the Runoff generation and confluence for the project safety. Results can be obtained using the same method under other conditions.

Key words: global warming, extreme hydro - meteorological events, storm transposition, probable maximum precipitation (PMP), probable maximum flood (PMF)

1 Introduction

In recent 20 and 30 years, due to the influence of human activities, the amount of greenhouse air is continuing to increase, the global climatic change and extreme hydro - meteorological event frequently occurs, and heavy casualties and serious economic losses in many countries are caused. So impel the world large quantities experts and scholars to carry on the research and study on that.

Nowadays, the hydro - meteorological engineers of water conservancy planning and designing departments mainly focus on a problem that what is the influence of global warming on PMP/PMF. It is very complicated, and this paper attempts from the viewpoint of the project to propose a simple method to approximately solve this problem. In order to understand this influence, it is firstly necessary to take seriously research and study, and then for the following 4 questions to give positive answer. Otherwise, if one answer of the following questions is negative, which unnecessary to considering the influence of climate change on the PMP/PMF. The 4 questions are as following:

The first one is the global warming is true or not? The second is the climate is warming or not in the design region including the design watershed and its surrounding meteorologically homogeneous regions? The third is during flood periods of the design region the climate is warming or not? This because the flood control requirements of the projects are mainly during flood periods. The fourth one is under the high - efficiency storms which occur during flood periods, the air temperature or dew point is increasing or not? This because the air temperature or dew point is increasing which mean has the influence on PMP/PMF.

The pupose of this paper is, based on taking the positive answers of 4 questions mentioned above, to explorer the PMP/PMF estimation method for discussion.

2 Concepts of PMP/PMF

2.1 Definition of PMP/PMF

Generally, PMP/PMF is the approximately physical upper limited precipitation/flood for the design of a given project in the target watershed under current meteorological conditions.

2.2 The purpose of using PMP/PMF

Traditionally, PMP/PMF is considered as the design storm and flood of the highest flood control standard in the world, which is used as the design for the reservoir projects and nuclear power plants as the risks of disastrous consequences such as loss of lives, huge social and economic loss and seriously ecological and environmental damage.

2.3 The basic attributes of PMP/PMF

According to the above mentioned of section 2.1 and 2.2, the basic attributes of PMP/PMF as the following:

(1) Upper limit attribute: The PMP/PMF is the approximately physical upper limited precipitation for the design of a given project in the target watershed, and has the attribute of maximization. If encounter such kind of storm flood which can guarantee the safety operation of the project built according to its design during its useful life period.

(2) Design attribute: The PMP/PMF is obtained according to the storm and flood characteristics of the design watershed and combined with the specific requirement of the project and using statistical and generalized to design the possible maximum precipitation and flood by hydro - meteorological engineers. In other words, the PMP/PMF is coming from the reality and higher than reality.

(3) Prediction attribute: PMP/PMF is same as the all design storm and flood, which all are results of is super long (several years and hundred years) storm and flood prediction, which belongs to prediction attribute, and never know its specific occurrence date and totally different with the real storm and flood forecast results of the reality.

3 Estimation of PMP/PMF

3.1 Basic idea

Due to the PMP/PMF has design attribute and prediction attribute, its estimation methods have the principle of broad rather than detailed, which is focus on the main factors which affects on the magnitude of storm and flood and neglect the other secondary factors, and at the end the estimation results of PMP/PMF (mainly flood peak and flood volume) which is right qualitatively and basic right quantitatively.

3.2 The basic structure of estimation method

Up to now, the structure of the main method to estimate the PMP/PMF in the world is as following:

High - efficiency storm→Moisture maximization→Transposition→Enveloping→PMP→Runoff generation and confluence calculation→PMF.

Here, high? efficiency storm, simply speaking, is the maximum storm of a certain duration and in a certain area with the assumption that its precipitation efficiency has reached its maximum in the design region which has long period observed storm data.

According to the above structure, the influence of the climate change, for PMP mainly indicated as moisture maximization, for PMF from PMP, totally indicated as runoff generation and confluence calculation.

3.3 PMP estimation

How to solve the moisture maximization is the new problem caused by climate warming. We propose to use precipitable water maximization method for solving this problem.

3.3.1 The concept of precipitable water

For PMP estimation, precipitable water is the indicator to represent the water vapor amount in the air mass; its definition is the corresponding water depth which condensed at the bottom of air column from the water vapor, not including the liquid water and solid water in the vertical air column.

The specific value of precipitable water can be calculated from the aerological observation data. However, due to the less distribution of the areological observation stations, and short observation period, and much less the useful radar observation data during the rainstorm and no this kind data in historical rainstorm, the surface dew point generally used to derive the precipitable water. The surface dew point data observed simply, more observation stations and long observation period.

Using the surface dew point to derive precipitable water is based on the assumption of from the surface to upper the whole air column saturated, and air temperature (dew point) varies with the height of pseudo – wet adiabatic lapse rate, which is assuming the whole layer air is saturated and mixed very well. Under this assumption, the precipitable water of the air column for a given location from surface to each height is the uniform function of the surface dew point for that given location. Therefore it is very directly according to the surface dew point to derive the precipitable water which can be queried from the specific table from the reference.

Some researches both at home and abroad indicate that the rainstorm under the condition of long duration and large range, the above basic assumption basically coincide with the reality.

3.3.2 Moisture maximization ratio calculation

Without consideration of the climate change, the moisture maximization ratio is calculated using the following formula:

$$K = \frac{W_m}{W}$$

where, K is moisture maximization ratio; W is precipitable water of high – efficiency storm (mm); W_m is possible maximum precipitable water in design condition (mm).

Obviously, over consideration of climate warming, the moisture maximization ratio should be calculated using the above formula, just by replacing W_m in the formula as the possible dew point after climate warming.

According to our research experiences of several years, in middle and low latitude areas, the surface dew point increase by 1℃, correspondingly the precipitable water will increase by about 10%. To be on the safe side, recently we specially made the comparison of calculation. The detail results can be seen Tab. 1 to Tab. 3.

From Table 1 to Tab. 3, whatever for the sea surface level (SSL, 1,000 hPa) and other different surface level (such as 600 m, 1,000 m) all have the following principles:

While the surface dew point (T_d) increase by 1℃, the precipitable water (W) increase by about 10%;

While the surface dew point increase by 2 ℃, the precipitable water increase by about 20%;

While the surface dew point increase by 3 ℃, the precipitable water increase by about 30%.

Tab. 1 The design precipitable water ratio between possible maximum dew point (T_{dm}) and storm representative dew point (T_d) in middle and low latitude regions

Dew point T_d (T_{dm}) (℃)		20	21	22	23	24	25	26	27	28	29	30
Precipitable water W (mm)		52	57	62	68	74	81	88	96	105	114	123
Maximization ratio $K=\frac{W_m}{W}$	T_d (℃) 22			1.00	1.10	1.19	1.31					
	23				1.00	1.09	1.19	1.29				
	24					1.00	1.10	1.19	1.30			
	25						1.00	1.09	1.19	1.30		
	26							1.00	1.09	1.19	1.30	
	27								1.00	1.09	1.19	1.28
	28									1.00	1.09	1.17

Note: The precipitable water is the theory value from sea surface level (1,000 hPa) to tropopause (200 hPa).

Tab. 2 The design precipitable water ratio between possible maximum dew point (T_{dm}) and storm representative dew point (T_d) in middle and low latitude regions

Dew point T_d (T_{dm}) (℃)		20	21	22	23	24	25	26	27	28	29	30
Precipitable water W (mm)	①Sea surface level	52	57	62	68	74	81	88	96	105	114	123
	②SSL to 600 m	10	10	11	11	12	13	14	15	15	16	17
	③ΔW = ① - ②	42	47	51	57	62	68	74	81	90	98	106
Maximization ratio $K=\frac{W_m}{W}$	T_d (℃) 22			1.00	1.12	1.22	1.33					
	23				1.00	1.09	1.19	1.30				
	24					1.00	1.10	1.19	1.31			
	25						1.00	1.09	1.19	1.32		
	26							1.00	1.09	1.22	1.32	
	27								1.00	1.11	1.21	1.31
	28									1.00	1.09	1.18

Note: The precipitable water W ③ is the theory value from 600 m to tropopause (200 hPa).

Tab. 3 The design precipitable water ratio between possible maximum dew point (T_{dm}) and storm representative dew point (T_d) in middle and low latitude regions

Dew point T_d (T_{dm}) (℃)		20	21	22	23	24	25	26	27	28	29	30
Precipitable water W (mm)	①Sea surface level	52	57	62	68	74	81	88	96	105	114	123
	②SSL to 1,000m	15	16	17	18	20	21	22	23	25	26	28
	③ΔW = ① - ②	37	41	45	50	54	60	66	73	80	88	95
Maximization ratio $K=\frac{W_s}{W}$	T_d (℃) 22			1.00	1.11	1.20	1.33					
	23				1.00	1.08	1.20	1.32				
	24					1.00	1.11	1.22	1.35			
	25						1.00	1.10	1.22	1.33		
	26							1.00	1.11	1.21	1.33	
	27								1.00	1.10	1.21	1.30
	28									1.00	1.10	1.19

Note: The precipitable water W③ is the theory value from 1000m to tropopause (200 hPa).

3.3.3 Estimation of PMP over consideration of climate warming

According to the above mentioned variation principle of T_d and W, the influence of climate warming on PMP can be easily solved.

If we through research can determine the designed reservoir project within its useful life period, which the future air temperature of occurring high – efficiency storm in flood period increase by 2 ℃ compare with the current air temperature, to enlarge 20% of the present PMP of the project as the PMP after climate warming, the other conditions are the same.

3.4 Estimation of PMF

The influence of climate warming on PMF, from PMP to PMF is the influence of runoff generation and confluence factors.

The factors influence on runoff generation and confluence are topography, physiognomy, geology, soil, vegetation, basin shape, drainage density, channel morphology and human activities influence on the watershed underlying surface conditions etc. However, for a specific watershed, the climate change only affects the vegetation, and for the other factors affects less. Although the vegetation variation is gradually varied, needs several years to change significantly. Also, the exact date when the PMF occurs is not clear and maybe it is a short time after the project is built. So according to the principle of broad rather than detailed and take account of the project safety, there is no need to consider the influence of vegetation variation on PMF. Therefore, the method using the current moisture maximization ratio (K) to maximum the present PMF to derive the PMF considering climate warming is recommended.

4 Conclusions

PMP/PMF estimation method was proposed in this paper based on the agreement to ① global warming, (2) climate warming of the design region, ③ the climate warming of the design region during the flood periods and ④ the surface dew point (T_d) increases during the efficiency storm occurs in flood periods. If the 4 factors mentioned above are possible close to the reality for the design conditions, the proposed PMP/PMF method is the simplest and easiest way, though this method is coarser, the derived results can reach the applicable accuracy.

Acknowledgments

The authors are very grateful to Prof. Wang Jiaqi, Prof. Ding Jing, Prof. Fu An and Prof. Ma Xiufeng, whose suggestions helped to clarify and improve the paper.

References

WMO – No. 168, Guide to Hydrological Practices (Sixth edition), Volume II Management of Water Resources and Application of Hydrological Practices [S]. Geneva, WMO, 2009: 11.5 – 24 – 11.5 – 27.

WMO – No. 1045, Manual on Estimation of Probable Maximum Precipitation (PMP) [S]. Geneva, WMO, 2009: 1 5, 211 – 217.

Wang Guoan, Calculation Principle and Method of PMP/PMF [M]. Beijing: Chinese Water Conservancy and Hydropower Press; Zhengzhou: Yellow River Conservancy Press, 1999: 46 – 50.

Responses to Climate Change of Hydrology and Water Resources in the Yellow River Source Region

Jin Junliang[1,2], ***Zhang Jianyun***[1,2], ***Wang Guoqing***[1,2], ***Gu Ying***[1], ***Liu Cuishan***[1,2] and ***He Ruimin***[1,2]

1. The State Key Laboratory of Hydrology – Water Resources and Hydraulic Engineering, Nanjing Hydraulic Research Institute, Nanjing, 210029, China
2. Research Center for Climate Change, MWR, Nanjing, 210029, China

Abstract: Variation of Precipitation, Temperature and runoff in Yellow River Source Region was analyzed with Mann – Kendall and Spearman method over past 60 years. Based on the ten climate scenarios, responses of hydrological process to climate change were simulated using the Variable Infiltration Capacity model. Results indicated that recorded Temperature in the Yellow River Source Region presented significant increasing trend during over past 60 years. The daily minimum temperature presents higher increasing trend than the daily maximum temperature. Yearly gross precipitation presents minor increasing, and the year runoff present minor decreasing. The runoff and soil moisture will likely undergo decreasing trend and the heterogeneity on Runoff Process will be increase in the future. Much more drought may threaten the social development in this region in the future.

Key words: Yellow River source region, climate change, climate scenario, hydrological responses, drought

Global warming is a major atmospheric issue that has been damaging our precious environment for many years. Water cycle is accelerated and changed by global warming. Water holding capacity of atmosphere is enhanced because of the global warming. Precipitation, evaporation, runoff and soil moisture will be changed under the global change environments. In recent years the issue of global warming has become an increasingly popular matter among society. Both the causes and effects of global warming have been the topics of heated debate not only among scientists, but among politicians, businesses, and general members of society. The World Meteorological Organization (WMO), United Nations Educational, Scientific and Cultural Organization (UNESCO) and International Association of Hydrological Sciences (IAHS) prompted a series of science projects, such as World Climate Research Programme (WRCP), International Geosphere – Biosphere Program (IGBP) and International Hydrological Programme (IHP) in order to study the impact of climate change to water resources. Many Chinese scientists have been doing the research of the impact of climate change to water resources. Such as, Wang Guoqing (Wang Guoqing, Li Mi, et al., 2012) gives the impact of climate change to Fujiang river basin and hydrological sensitivity. The accuracy of 22 Precipitation and temperature of GCMs were evaluated by Ju Qin (Ju Qin, Hao Zhenchun et al., 2011). Furthermore, lots of hydrological scientists (Andreadis and Lettenmaier, 2006; Hao Zhenchun Hao, Wang Jiahu, et al., 2006; Zhang Jianyun and Wang Guoqing, 2007) have been doing relevant study of impact of climate change to water resources.

Forcing the meteorological data into hydrological model and analyze the results of model is the main method to assess the impact of climate change to district water resources. The research was always focusing on runoff and frequency but short in soil moisture. Nowadays, soil moisture is a most important physical element in hydrological cycle. Soil moisture can change the albedo and the latent heat flux to influence the weather system. The research shows that the 65% precipitation in global come from the evapotranspiration of land surface, the proportion of inland may much bigger. So the soil moisture was the second hydrological cycle element that is only less important than sea temptation in climate change research, and we should give much more concentrations in the study of climate change.

The Yellow River is always called the mother river of China. The source region of Yellow River is the most important water sources conservation area. It also is the major part of the water tower of China. The dataset shows that (Niu Yuguo and Zhang Xuecheng, 2005) the runoff of river basin from the source region accounts for 40% of the total Yellow River Basin. So the hydrological cycle change of source region will directly affect the water supply in the downstream of the Yellow River Basin. This paper analyzed the trend and characteristics of hydro-metro data over the past 60 years. The latest GCMs scenarios were used to forcing the hydrological model in order to predict the hydrological process in the further.

1 Dataset and methodology

1.1 Study area

The Yellow River source region was the district which the hill catchment areas of the Tangnaihai cross section. It is located in the northeast of the Tibet plateau. The watershed of the Yellow River source region is about 122,000 km^2. The length of the main river is over 300 km, and the average slope is about 1.2‰. High mountains, canyon and basins, lakes and swamp and complicated topography exist in this region. West high and east low terrain was in the Yellow River source region. As the long statistic data showing, the mean annual gross precipitation is 485.9 mm. Mean discharge runoff is 2.052×10^{10} m^3, and it accounts for the 38% of total natural runoff. The Yellow River source region is located the subfrigid zone of Tibetan plateau with high temperature in southeast and lower temperature in northwest. This region has the feathure of inland plateau weather characteristics.

The 21 meteorological stations and 23 hydrological stations in this region are shown in Fig. 1. The daily precipitation, daily maximum and minimum temperature and daily discharge were collected in this study. Almost all the stations have longterm observed data from 1950.

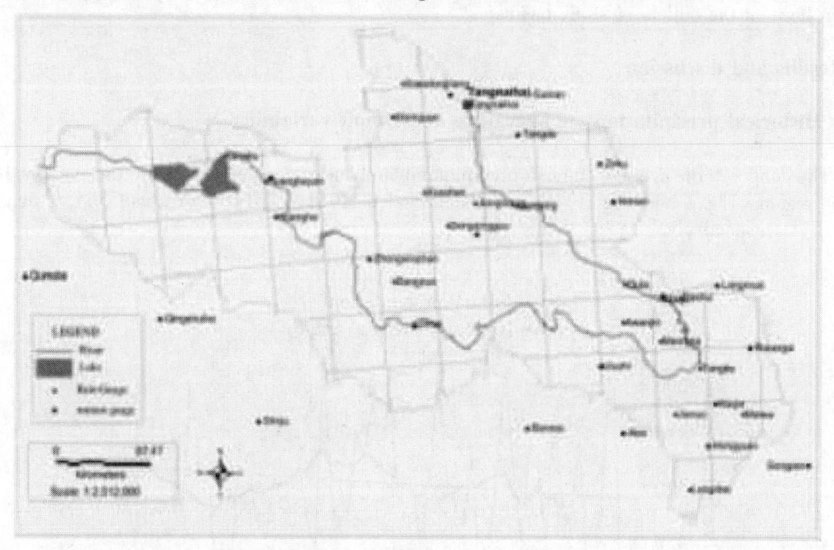

Fig. 1 Hydro-meteorological stations in the Yellow River source region

1.2 Description of the VIC model

In this study, Variable Infiltration Capacity (VIC) model (Liang et al., 1994, 1996) was used to evaluate the hydrological process change over the Yellow River source region. VIC is a macroscale hydrologic model that solves full water and energy balances, originally developed by Liang Xu at the University of Washington. And in its various forms it has been applied to many watersheds including the Columbia River, the Ohio River, the Arkansas-Red Rivers, and the Upper Mississippi Rivers, as well as being applied globally (Abdulla et al., 1996; Cherkauer and Lettenmaier, 1999; Hamlet and Lettenmaier, 1999; Liang and Xie, 2001). The model was designed both for inclusion in GCMs as a land atmosphere transfer scheme, and for use as a stand-alone macroscale hydrology model.

The VIC model divides a study catchment into grid cells, and then divides the soil column in each grid into three layers. The upper two layers, which are usually treated as one layer, are designed to represent the dynamic response of soil to rainfall events, while the lower soil layer is used to characterize seasonal soil moisture behavior. Three types of evaporation are considered: evaporation from wet canopy, evapotranspiration from dry canopy, and evaporation from bare soil. Stoma resistance is used to reflect the effects of radiation, soil moisture, vapor pressure deficiency, air temperature. etc., when calculating transpiration from the canopy.

The total runoff estimates to consist of surface flow and base flow. Surface flow, including infiltration excess flow and saturation excess flow, is generated in the two top layers only. In order to consider the heterogeneity of soil properties, the soil storage capacity distribution curve and infiltration capacity curve are employed. The double curves are individually described as a power function with a B exponent. Base flow occurs in the lowest layer only, and is described by the Arno method (Habets F, et al, 1999) using the one-dimensional Richards' equation to describe the vertical soil moisture movement.

The VIC model was established over the Yellow River source region at a resolution of 50 km. The model domain consists of 74 computational grid cells. A 50 km × 50 km river network based on 1 km DEM was developed over the entire Yellow river soure region for purposes of defining the model's river routing scheme using the method of Lohmann et al. (Lohmann et al., 1996; Lohmann D, 1998) which takes daily VIC surface and subsurface runoff as input to obtain model simulated stream flows at the outlets of study basins.

2 Results and discussion

2.1 Historical precipitation, temperature and runoff variability

The long – term average annual precipitation and mean temperature over the Yellow River source region (Fig. 2 and Fig. 3) during the period of 1951 ~ 2011 were about 505.9 mm and −1.96 ℃ respectively.

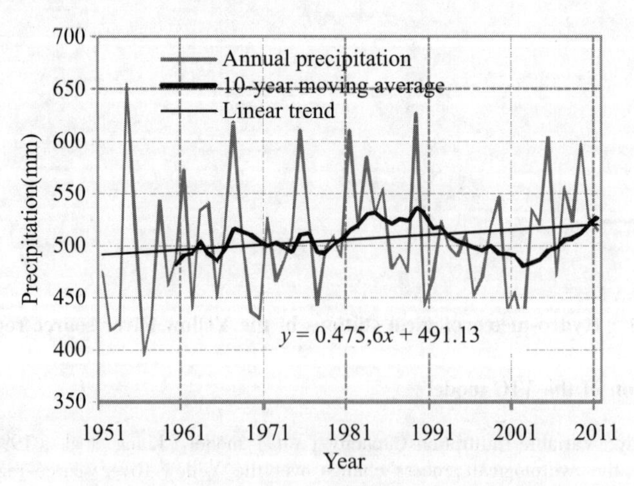

Fig. 2 Yearly gross precipitation in study area

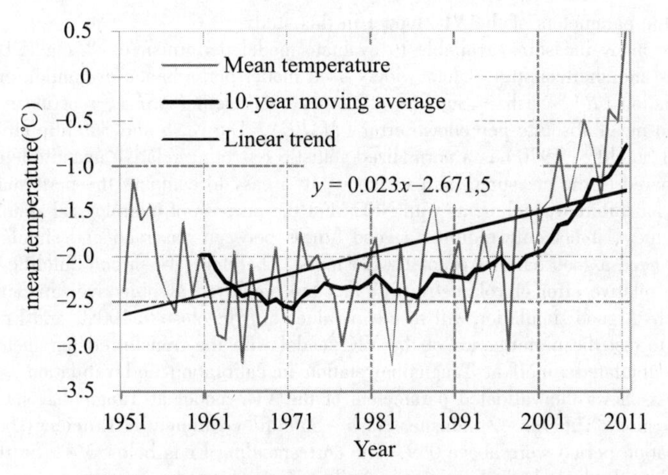

Fig. 3 Yearly average temperature in study area

Both the annual precipitation and mean temperature had increasing trends of +0.475 mm/a and +0.023 ℃/a respectively, Mann - Kendall test for the both series are 1.39 and 3.93, indicating the variation trends of the temperature series not significant increasing. The quickly temperature increasing period occurred after 1980.

Tab. 1 indicates that the results of meteorological data trend test over Yellow River source region. Annual precipitation, mean maximum temperature and mean minimum temperature had increasing trends over past sixty years. The annual runoff had decreasing trends. The daily mean minimum temperature had much obvious than daily mean maximum temperature. The annual precipitation and annual runoff had slightly increasing and decreasing trend respectively. The two all have not excess the 0.05 threshold. It indicates the trend is not obvious.

Tab. 1 Trends detection for hydro-meteorological data in study area

Hydro-meteorological data	slope (/10 a)	Spearman Statistics	Mann-Kendall Statistics	Trend	Significant Test
Annual precipitation	4.76 mm	-1.39	1.39	Increase	No significant
Daily maximum temperature	0.17 ℃	-2.65 *	2.54 *	Increase	Significant
Daily minimum temperature	0.29 ℃	-5.30 *	4.85 *	Increase	Significant
mean temperature	0.23 ℃	-4.18 *	3.93 *	Increase	Significant
Annual runoff	-6.85×10^{8} m^3	1.78	-1.71	decrease	No significant

Note: sign * represents the significant test bigger than 0.05 confidence levels.

2.2 Discharge simulation

The meteorological forcing dataset from year 1951 to 2011 was interpolated by the inverse distance to a power gridding method to 50 km × 50 km spatial resolution grids. The VIC model was driven to run by forcing to calculate through each grid. Dataset series of Tangnaihai station from 1961 ~ 1990 were divided into two periods: a calibration period from 1961 to 1980, and a validation period from 1981 to 1990. The start input forcing data of model is ahead of 1 year in order to preheating the model's initial parameters, such as soil moisture.

The rosenbrock optimization method (Zhang Jianyun, Xuan Yunqin, et al., 1999) was used to calibrated the parameters of the VIC model in this study.

There are many measures available to evaluate model performance. Wang (2007) analyzed the advantages and shortcomings of four widely used model performance evaluation criteria, coefficient of correlation (R), Nash - Sutcliffe efficiency coefficient (NSE), root mean squared error ($RMSE$), and mean absolute percentage error ($MAPE$). The Nash and Sutcliffe efficiency criterion (Nash and Sutcliffe, 1970) is a normalized statistic reflecting relative magnitude of the residual variance compared to the measured data variance. It's easy to compare the performance of hydrological model for different catchments with NSE. For the purpose of hydrological simulation and climate change study, it not only requires a good fitness between observed and simulated runoff series, but also need a good balance of total water mass. Therefore, Nash and Sutcliffe efficiency criterion and the relative error of volumetric fit (Er) are employed as objective functions to calibrate the VIC model. A good simulation will result in values of NSE close to 100% and Er close to 0%.

In order to test the performance of the VIC model with the transferred parameter values, the observed and simulated runoff at Tangnaihai station in calibration and validation is compared in Fig. 4. Tab. 3 shows the validated parameters of the VIC model at Tangnaihai station of Yellow River source region. The Tab. 2 indicates Nash – Sutcliffe coefficient of Efficiency both in calibration and validation period were above 0.9. The corresponding Er is below 3%. So the VIC model can well simulate the hydrological process at Yellow River source region.

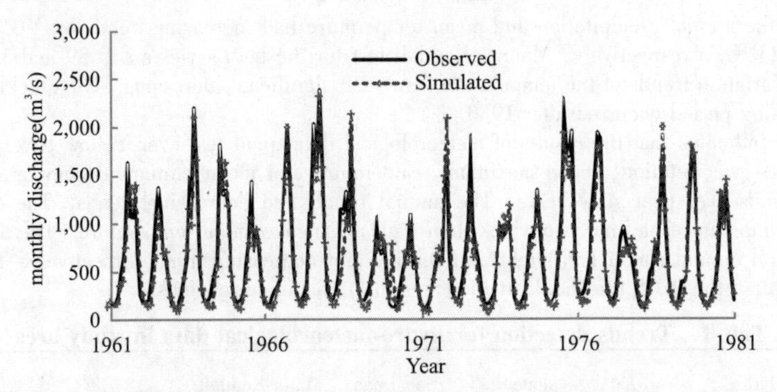

Fig. 4 The simulated and recorded monthly discharge 1961 ~ 1980 of Tangnaihai station in calibration

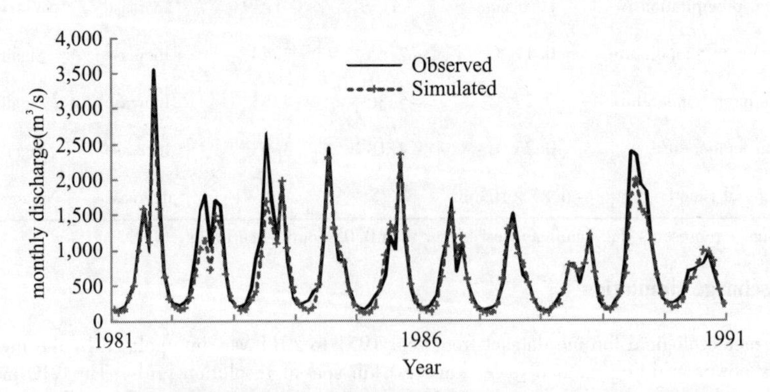

Fig. 5 The simulated and recorded monthly discharge 1981 ~ 1990 of Tangnaihai station in validation

Tab. 2 Validated parameters of the VIC model at Tangnaihai station of Yellow River source region

Parameters						Performance			
						Calibration (1961 ~ 1980)		Validation (1981 ~ 1990)	
B	*Ds*	*Dm*	*Ws*	*D2*	*D3*	*NSE*	*Er*(%)	*NSE*	*Er*(%)
0.38	0.98	30	0.41	0.60	0.59	0.91	-2.42	0.93	0.92

2.2.1 Climate change scenarios

General Circulation Model (GCM) is the most important tool to predict the climate in future. Also because the accuracy of the GCM is not higher, but at present it is the major means to predict the precipitation and temperature. The four latest General Circulation Models (GCMs) in current research were used in this study, and there are Commonwealth Scientific and Industrial Research Organization (CSIRO), National Center for Atmospheric Research (NCAR), Max - lanck - nstitut für Meteorologie (MPI) and Hadley center coupling with PRECIS (PRECIS). The four GCMs have the metrological data over the whole China under SRES (Special Report on Emission Scenarios) scenarios A2, B2, B1 and A1B in the 21st century. Tab. 3 shows the information of scenario data.

Tab. 3 The information of scenario data

GCM	Abbreviations	Emission scenarios	Spatial/temporal scale
Commonwealth Scientific and Industrial Research Organisation	CSIRO	A1B, A2, B1	2001 ~ 2100, daily, 0.5°
Max-Planck-Institut für Meteorologie	MPI	A1B	2001 ~ 2100, daily, 0.5°
National Center for Atmospheric Research	NCAR	A1B, A2, B1	2000 ~ 2099, daily, 0.5°
Hadley center coupling with PRECIS	PRECIS	A1B, A2, B2	2001 ~ 2100, daily, 50km

Metrological forcing data from Scenario data including daily precipitation, daily maximum temperature and daily minimum temperature from the period 2020 to 2050 are used to drive the VIC model. The model predicts the hydrological process in 2020s (year 2021 ~ 2030), 2030s (year 2031 ~ 2040) and 2040s (year 2041 ~ 2050) over the Yellow River source region.

2.2.2 Runoff

The change of projected averaged annual runoff over the Yellow River source region under the four emissions scenarios are shown in Tab. 4. Tab. 4 indicates the runoff over the Yellow River source region will decrease in the further thirty years. Runoff under A2 emissions scenarios in 2040s will reduce the most. The trend under B1, B2 and A1B emissions scenarios almost the same. The result shows that the runoff under these emissions scenarios will reduce in the further over the Yellow River source region. And it decreased severely with time prolonged.

Tab. 4 The change of runoff in each decade comparing with baseline (1961 ~ 1990)

Emission scenarios	Change of runoff compared with baseline (%)		
	2020s	2030s	2040s
A2	-8.40	-7.78	-11.78
B1	-0.63	-5.96	-8.63
B2	-2.11	1.91	-5.77
A1B	-3.39	-4.32	-5.61

Fig. 6 shows the change of runoff in each decade in the period of years from 2020 to 2040 comparing with baseline. Fig. 6 indicates that the runoff under different GCMs is not identical. For example, In 2020s, the runoff with NCAR under A1B emission scenario will increase 1.9% compared with baseline, but the results with other nine scenarios are opposite with it. The runoff under other nine scenarios in 2020s will decrease. The change is from −0.16% to −9.8% compared with the baseline. The average of ten scenarios is −4.21%. So it means the runoff will most likely decrease in the further. And the water resources in the Yellow River region will most likely decrease severely with time prolonged.

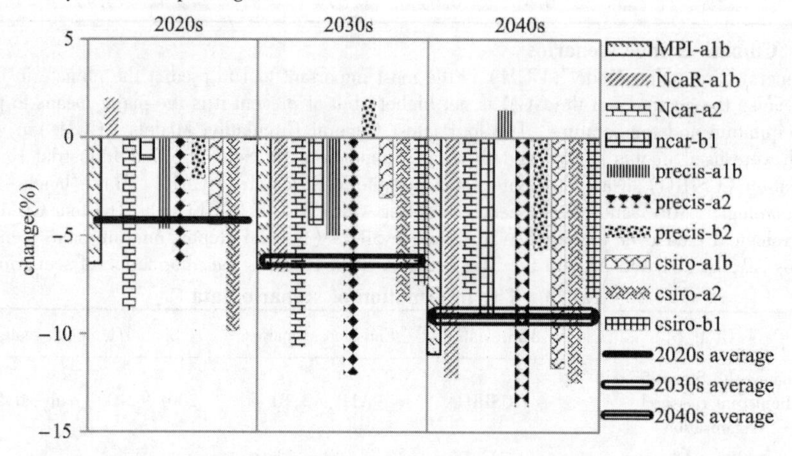

Fig. 6 The change of runoff in each decade comparing with baseline

2.2.3 Soil moisture

The change of soil moisture in each decade comparing with baseline is shown in the Fig. 7. Fig. 7 indicates that the soil moisture in the middle Yellow River source region will decrease, meanwhile, it will increase in the downstream region in 2020s. In the decades 2030s and 2040s, the region of soil moisture decrease will enlarge with the time prolonged. So it means that the drought will be much possible in the further. The climate change will enhance the situation over the Yellow River source region. The short of water resources may threaten the agricultures and animal husbandry over this region.

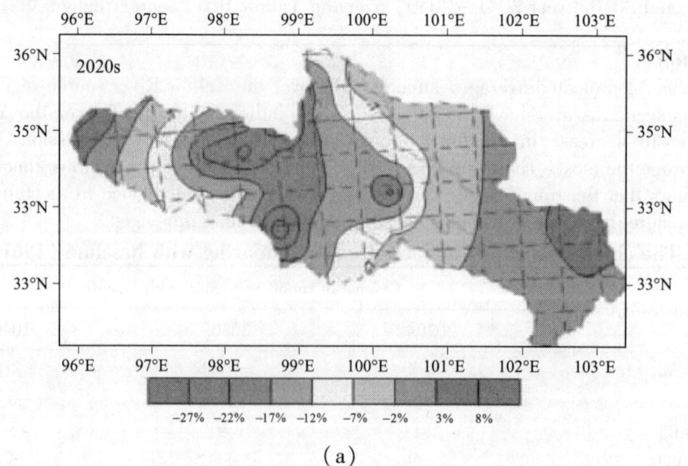

(a)

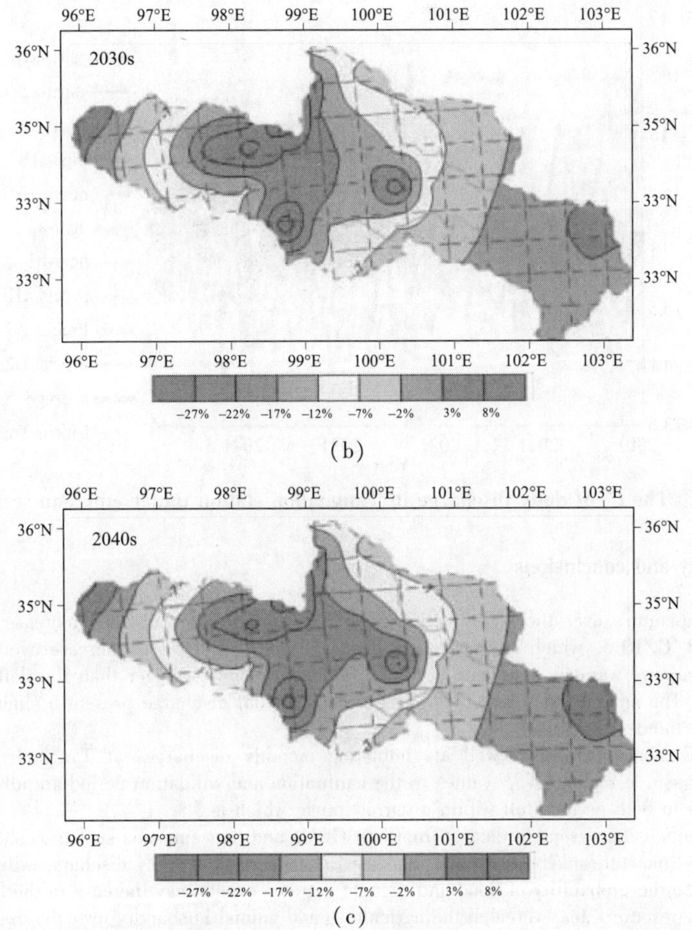

(b)

(c)

Fig. 7 The change of soil moisture in each decade comparing with baseline

2.2.4 Hydrologic frequency

The variation of runoff in year can be described by the coefficient of variation.

$$C_v = \sigma \sqrt{R} \tag{1}$$

$$\sigma = \sqrt{\frac{1}{365}\sum_{i=1}^{365}(R_i - \overline{R})^2} \tag{2}$$

$$\overline{R} = \frac{1}{365}\sum_{i=1}^{365} R_i \tag{3}$$

Where, R_i is the daily discharge; $\overline{R}$ is the average annual discharge.

Fig. 8 shows the coefficient of variation of Daily Discharge in Tangnaihai Station under emission scenario. We can know that from Fig. 8, the coefficient of variation increase with the time prolonged. It means the inhomogeneity of daily discharge will increase in the further. So, the probability of flood and drought extreme events may increase in the further.

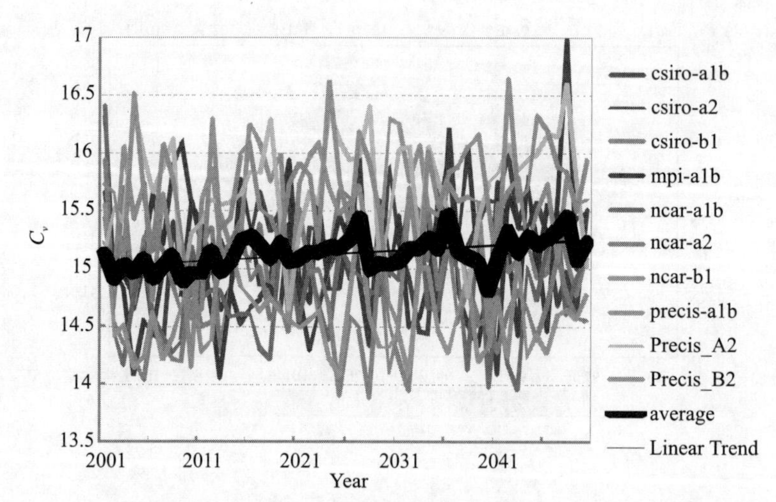

Fig. 8 The C_v of daily discharge in Tangnaihai station under emission scenarios

3 Summary and conclusions

The temperature over the Yellow River source region presents obvious increase trend. The slope is 0.23 ℃/10 a. which is bigger than the slope 0.13 ℃/10 a of increase over global land surface temperature warmed. The rate at which daily minimum is bigger than the daily maximum temperature. The annual gross precipitation and mean annual discharge present a slightly increase and decrease trend, respectively.

The VIC model performed well at simulating monthly discharges at Tangnaihai station in Yellow River source region. *NSE* values in the calibration and validation period are all above 0.9, the Er values in both periods fell within a narrow range which is 5%.

The results using ensemble dataset from four GCMs and four emission scenarios show that, the runoff and soil moisture maybe decrease, and the inhomogeneity of daily discharge will increase in the further. So the probability of flood and drought extreme events may increase in the further. The short of water resources may threaten the agriculture and animal husbandry over this region. Effective strategies for adaptation to climate change are essential for the sustainable devclopment of water resources in the Yellow River source region.

Acknowledgements

This study has been financially supported by the National Basic Research Program of China (grant no. 2010CB951103), the National Key Technology R&D Program (grant no. 2012BAC19B03), the International Science & Technology Cooperation Program of China (grant no. 2010DFA24330), the Non-profit Industry Program of the Ministry of Water Resource (grant no. 201001042), the Jiangsu Planned Projects for Postdoctoral Research Funds (grand no. 1101044C), the Fund on Basic Scientific Research Project of Nonprofit Central Research Institutions (grand no. Y11008, Y511009), the ACCC project funded by DFID, SDC and DECC. Thanks also to the anonymous reviewers and editors.

References

Wang Guoqing, Li Mi, Jin Junliang, et al. Variation Trend of Runoff in Fujiang River Catchment and Its Responses to Climate Change[J]. Journal of China Hydrology, 2012,32(1): 2-28.

Ju Qin, Hao Zhen - chun, Yu Zhong-bo, el al. Runoff Prediction in the Yangtze River Basin Based on IPCC AR4 Climate Change Scenarios[J]. Advances in Water Science, 2011,22(4): 462 -469.

Zhang Jian - Yun, Wang Guoqing. The Impacts of Climate Change to Water Resources[M]. Beijing: China Science Press,2007.

Hao Zhen-chun, Wang Jia - hu, Li Li, el al. Impact of Climate Change on Runoff in Source Region of Yellow River[J]. Journal of Glaciology and Geocryology, 2006,28(1): 1-7.

Abdulla F A, Lettenmaier D P, Wood E F, et al. Application of a Macroscale Hydrologic Model to Estimate the Water Balance of the Arkansas—Red River Basin[J]. J. Geophys. Res., 1996, 101, 7449 -7459.

Cherkauer K A, Lettenmaier D P. Hydrologic Effects of Frozen Soils in the Upper Mississippi River Basin[J]. J. Geophys. Res.,1999, 104, 19599 -19610.

Hamlet A F, Lettenmaier D P. Effects of Climate Change on Hydrology and Water Resources in the Columbia River Basin[J]. Am. Water Res.,1999, Assoc 35, 1597 -1623.

Andreadis K M, Lettenmaier D P. Trends in 20th Century Drought Over the Continental United States[J]. Geophysical research letters.,2006, 33(10): 1 -4.

Ma Zhuguo, Fu Congbin, Xie Li, el al. Some Problems in the Study on the Relationship Between Soil Moisture and Climatic Change[J], Advance In Earth Sciences,2001, 16(4): 563 -568.

Niu Yu-guo, Zhang Xue-cheng. Preliminary Analysis on Variations of Hydrologic and Water Resources Regime and Its Genesis of the Yellow River Source Region[J]. Yellow River, 2005, 27(3): 31 -36.

Habets F, et al. The ISBA Surface Scheme in a Macroscale Hydrological Model Applied to the Hapex - Mobilhy area—Part I: Model and Database[J]. Journal of Hydrology, 1999, 217, 75 -89.

Liang X, Lettenmaier D P, Wood E F. A Simple Hydrologically Based Model of Land Surface Water and Energy Fluxes for General Circulation Models[J]. Journal of Geophysical Research., 1994,99(7): 14415-14428.

Jin Junliang, Lu Guihua, Wu Zhiyong. The Applicability Study of VIC Model in Arid Region of Northwestern China[J]. Water Resources and Power., 2010,28(1): 12 -14.

Lu Guihua, Jin Junliang, Wu Zhiyong, et al. Obtaining Method of Vegetation Parameter for Hydrological Model and Its Application[J]. Hydro - Science and Engineering, 2009 (4): 1 -6.

Wu Zhiyong, Lu Guihua, Zhang Jianyun, et al. Simulation of Daily Soil Moisture Using VIC Model [J]. Scientia Geographica Sinica, 2007,27(3): 359 -364.

Zhang Jianyun, Xuan yunqin, Li jian. Hydrological Model Optimization Method and Its Applications[J]. Journal of China hydrology, 1999(sppl): 61 -66.

Nash J E, Sutcliffe J V. River Flow Forecasting Through Conceptual Models: Part1 - a Discussion of Principles[J]. Journal of Hydrology, 1970,10(3): 282 -290.

Wu Zhiyong, Lu G, Wen L, et al. Thirty - Five Year (1971 -2005) Simulation of Daily Soil Moisture Using the Variable Infiltration Capacity Model Over China[J]. Atmosphere-Ocean, 2007,45(1): 37 -45.

Sensitivity Study of Hydrological Process in Source Area of the Yellow River

Wang Jinhua[1], *Liu Jifeng*[2] and *Zhang Ronggang*[2]

1. Yellow River Institute of Hydraulic Research, Zhengzhou, 450003, China
2. Hydrology Bureau, Yellow River Conservancy Commission, Zhengzhou, 450004, China

Abstract: The sensitivity test has been done by improved and rated SWAT model in order to analyzing the affection of climate change and land use/land cover change to runoff. The results of sensitivity test show that, climate change is the dominant factor in affecting runoff change of the source regions of the Yellow River. The impact of climate change on runoff change accounts for 70%, and land use/land cover change is about 30%. 25 different combinations of water and heat have been designed according to the temperature and precipitation fluctuations of the source regions of the Yellow River in recent 40 years. All the combinations of water and heat was to be sued to explore the impact of climate factors change on runoff. The combinations of water and heat were designed by two principles, one is precipitation in five hypotheses (constant, add/subtract 10% and add/ subtract 20%) while temperatures in constant, the other is temperatures in five hypotheses (constant ascending/descending 1 ℃, ascending / descending 2 ℃) while precipitation in constant. The simulation results show that the impact of precipitation on runoff of the source regions of the Yellow River is more obvious than temperatures, that is to say precipitation is the main factor to runoff change. It is the worst things to the water resources of source area of the Yellow River if the temperature ascending 2 ℃ and precipitation decreased 20% in the future. The runoff from source area of the Yellow River will be reduced by 31.19% in that case.

Key words: runoff, land use/land cover change, climate change, sensitivity test, source area of the Yellow River

Source region of the Yellow River, above the Tang Naihai hydrometric station, is located in the northeastern part of the Tibetan Plateau, across Qinghai, Gansu, Sichuan provinces. Between 95°55′E and 102°50′E, 32°10′N and 36°05′N, the catchments area of the source region of the Yellow River is about 0.12×10^6 km^2, which is 15.35% accounting for the total area of the Yellow River basin. The ratio of runoff from source region of the Yellow River to the whole runoff is 38%. It is well known to us that the source region of the Yellow River is the tower of the Yellow River.

The occurrence probability of flood and drought will change with the global warming. The unique hydrology and water resources system of the source region of the Yellow River is very sensitive to climate change because of its special geography and fragile ecological. The results of sensitivity test showed that runoff from upper reaches of the Yellow River increased with precipitation, and decreased with temperature. That is to say, the runoff is far greater sensitivity to precipitation than to temperature . The evaporation will increase and runoff will reduce in the upper reaches of the Yellow River if the temperature continues to rise . The influences of human activities on runoff is more obvious than before with the rapid development of social economy. Wang Hao. etc have taken the research on the reasons of runoff change and take the Point of view that the climate and human activities are taking the same important role in runoff reduction. Liu Xiaoyan thought that the land use/land cover pay more affections on the relationship between precipitation and runoff. But it is difficult to clear the climate change and human activities, which make more influence on the underlying surface change.

The improved SWAT model has been sued to analyze the trend of runoff from source region of the Yellow River with the different climate scenarios and human activities. The spatial interpolation module and weather generator module have been improved. The simulation capability of improved model was more powerful in cold areas than the original model. Different land use and climate change scenario in the future had been designed in order to studying the sensitivity of hydrological process to human activities and climate change. The results may provide a scientific basis to ecological protection and reasonable utilization of water resources.

1 Simulation study in influencing factors of the hydrology change in source region of the Yellow River

The characteristic of runoff was analyzed according with the measured annual runoff data, from 1955 to 2006, Tang Naihai hydrological station. Average value of the measured runoff from Tang Naihai hydrological station is about 1.9890 × 10^{10} m^3, and variation coefficient (C_v) is about 0.27. The runoff's changing process has been showed in Fig. 1. It can be seen from the graph that the runoff from source area of the Yellow River decreased significantly since 1990s. That is lesser than coming from 1960 to 1980s. Compared with the runoff in 1960 to 1989, the reduction amplitude of measured runoff in 1990 to 1999 is about 11.1%, and the reduction amplitude of measured runoff in 2000 to 2006 is about 19.8%.

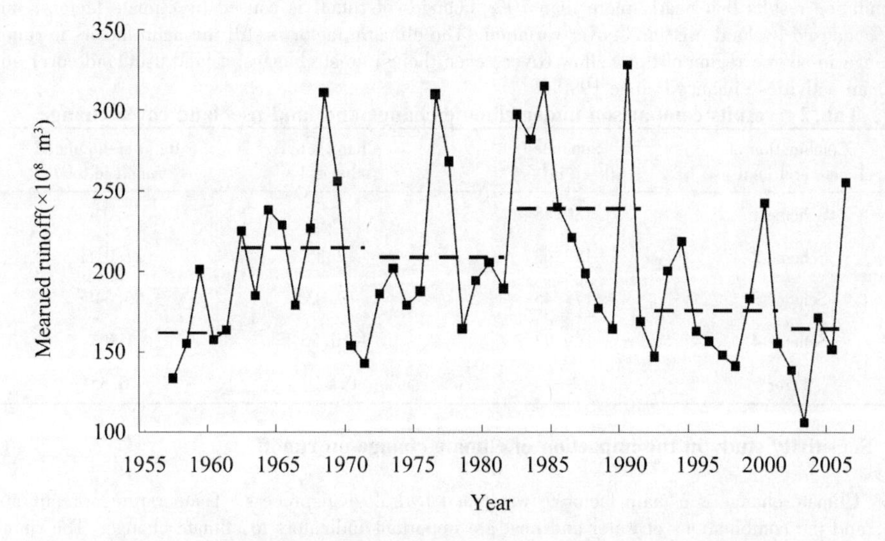

Fig. 1 Measured runoff changes process in Tangnaihai Station

The runoff from source regions of the yellow River is changed by many factors. Climate change and land use/land cover change are two important factors. Balance of water and heat is controlled by climate change, and hydrological cycle mode and process are changed by land use/land cover change. It is important to make clear respective the contribution of climate change and land use/land cover change to runoff variation. The main influential factor has been researched by sensitivity test basing different period of climate and land use type combinations.

1.1 Test schemes

Test schemes were designed to analyze the affection of climate change and land use/land cover change on runoff since 1990s. Land use/land cover data in 1995 and 2000 were employed. Two series data from 1991 to 1997 and 1998 to 2004 are used for climate change. The combinations of climate change and land use/land cover change used in this research were shown in tab. 1.

Tab. 1 Combinations of climate change and land use/land cover change

Scheme	Combinations
1	1991 ~ 1997s' climate and 1995s' land use/land cover
2	1991 ~ 1997s' climate and 2000s' land use/land cover
3	1998 ~ 2004s' climate and 1995s' land use/land cover
4	1998 ~ 2004s' climate and 2000s' land use/land cover

1.2 Main factors analysis of runoff variation

The simulation test was carried out by improved, rated SWAT model according four kinds of different combination schemes. The results (Tab. 2) show that flow was reduced approximately 10.55 m^3/s since 1990s. The results from scheme 2 show the affections caused by land use/land cover change compared with results from scheme 1. And the results from scheme 3 show the affections caused by climate change compared with results from scheme 1. It is not difficult to see from the simulation results that nearly more than 70% changes of runoff is caused by climate factors, and 30% caused by land use/land cover variation. The climatic factor is still the main factors to runoff change in source region of the Yellow River, even though great changes of land use/land cover and human activities enhanced since 1990s

Tab. 2 results comparison under climate change and land use/land cover change

Combination of climate and land use	Simulated flow (m^3/s)	Changes to scheme 1	Ratio of simulated runoff to total
Scheme 1	180.48	0	0
Scheme 2	177.63	-2.85	27.01%
Scheme 3	173.45	-7.03	66.64%
Scheme 4	169.93	-10.55	100%
Error	—	0.67	-6.35%

2 Sensitivity study in the impaction of climate change on runoff

Climate change is a main factor to watershed hydrological process. Temperature, precipitation, and the combinations of water and heat are important indicators to climate change. The runoff variation caused by climate change has been analyzed according the assumed combination of water and heat.

2.1 Designed climate scenario

It is the consensus of scientists that climate is warming in future, but the precipitation's change has obvious regional feature and uncertainty in future. The assumption of future temperature and precipitation changes in amplitude is respectively ΔT and ΔP. The variations of runoff have been simulated according with different combinations of water and heat by hydrologic simulation model in source region of the Yellow River. 25combinations of water and heat have been designed according to the characteristics of temperature in recent 40 years. Those combinations are as below:

$\Delta T = -2℃, \Delta P = -20\%$; $\Delta T = -2℃, \Delta P = -10\%$; $\Delta T = -2℃, \Delta P = 0\%$; $\Delta T = -2℃, \Delta P = 10\%$; $\Delta T = -2℃, \Delta P = 20\%$; $\Delta T = -1℃, \Delta P = -20\%$; $\Delta T = -1℃, \Delta P = -10\%$; $\Delta T = -1℃, \Delta P = 0\%$; $\Delta T = -1℃, \Delta P = 10\%$)。($\Delta T = -1℃, \Delta P = 20\%$; $\Delta T = 0℃$, $\Delta P = -20\%$; $\Delta T = 0℃, \Delta P = -10\%$; $\Delta T = 0℃, \Delta P = 10\%$; $\Delta T = 0℃, \Delta P = 20\%$; $\Delta T = 1℃$, $\Delta P = -20\%$; $\Delta T = 1℃, \Delta P = -10\%$; $\Delta T = 1℃, \Delta P = 0\%$; $\Delta T = 1℃, \Delta P = 10\%$; $\Delta T = 1℃$,

$\Delta P = 20\%$; $\Delta T = 2℃$, $\Delta P = -20\%$; $\Delta T = 2℃$, $\Delta P = -10\%$; $\Delta T = 2℃$, $\Delta P = 10\%$; $\Delta T = 2℃$, $\Delta P = 20\%$.

In this study, the sensitivity of runoff to precipitation and temperature were the major consideration. So the human activities, land utilization condition etc. were content.

2.2 Changes of runoff

The runoff was simulated by the improved SWAT model based on 2000's land use/land cover, precipitation from 1986 to 2004 and temperature from 1986 to 2004 data.

From the results of simulation test, temperature and precipitation are the main factors to runoff. The influence of rainfall on runoff is more obvious than that of temperature. Runoff increased as the temperature decreased, and also increased as precipitation increasing. The variation of runoff is not significant under the conditions of precipitation keeping constant and temperature increasing / decreasing. The amplitude variation is not more than 5%. But the amplitude variation is above 15% under the conditions of temperature keeping constant and precipitation increasing / decreasing 10%. What's more, the amplitude variation of runoff will exceed 30% under the conditions of temperature keeping constant and precipitation increasing / decreasing 20%. The most favorable cases are that the temperature will decrease 2 ℃ and precipitation will increase 20% , in this condition the runoff may be 35.42% increasing. On the contrary, if the temperature increase 2 ℃ and precipitation decrease 20% , the runoff will decrease 31.19%. (in Tab. 3)

Tab. 3 Ratio of runoff changes under different combinations of water and heat (%)

ΔT	ΔP				
	-20%	-10%	0	+10%	+20%
-2℃	-24.81	-11.97	2.7	15.47	35.42
-1℃	-25.47	-13.35	1.01	14.1	31.95
0℃	-26.53	-14.92	0	12.28	26.94
+1℃	-28.82	-16.25	-1.82	11.52	24.46
+2℃	-31.19	-17.87	-3.08	10.33	23.15

3 Conclusions

(1) Sensitivity test of runoff variation has been carried out by improved SWAT model. The results showed that climate change has more affection on runoff than land use/land cover change. Runoff changes of 70% caused by climate change, and 35% caused by land use/land cover change.

(2) Precipitation has great impact on the runoff change in source region of the Yellow River. The simulated tests have been carried out to analyze the response of runoff to climate change according 25 combinations of heat and water. The results showed that if the temperature increase 2 ℃ and precipitation decrease 20% , the runoff will decrease 31. 19%. It is serious for the whole Yellow River basin if the water resources from source region of the Yellow River decreased.

(3) Climate change and human activities have great impact on the hydrological process and water source in source region of the Yellow River. The impaction research of climate change and human activities on runoff is a very complex issue. It is difficult to exactly identify the affections of melting permafrost, seasonal permafrost, tourism resources development, and other human activities on runoff. Further researches should be carried out in those areaes

References

Qing Dahe. The Truth about Climate Change, Impaction and Countermeasure [A] //. Climate

Change and Ecological environment Symposium [C]. Beijing: China Meteorological Press, 2004.

Wang Jinhua, Kang Lingling, Yu Hui, et al.. Analysis in Affection of Olimate Change on Natural Runoff from Upper Reaches of the Yellow River [J]. Arid Land Geography, 2005(6): 288 – 291.

Gao Xuejie, Zhao Zongci. Numercal Simulation of Greenhouse Temperature to Climate Change in Northwest China by Regional Climate Model [J]. Permafrost and Glacier, 2003(2):165 – 169.

Wang Zhao, Chen Dehua. Analysis of the Factors Effecting the Runoff from Source Regions of the Yellow River [J]. Engineering Geology, 2001(3):35 – 37.

Liu Xiaoyan, Chang Xiaohui. Research on Changes of the Runoff from Source Regions of the Yellow River Change [J] . Yellow River, 2005(2):6 – 8, 14.

Research on Runoff Variation in Yellow River Headwater Area under Climate Change

Liu Meng, ***Cai Ming*** and ***Qiao Mingye***

Yellow River Engineering Consulting Co. , Ltd. , Zhengzhou, 45003, China

Abstract: Due to comprehensive impact from climate change and other factors, the water resource quantity of Yellow River valley in recent years has dropped sharply, and the event of cutout of Yellow River resource and downstream happened occasionally. The climate change causes change of a series of hydrographic circulation elements, like rainfall, evapotranspiration, river flow, subsurface flow and the like so as to further change the current status of the global hydrographic circulation to cause redistribution of water resource in space and time. Therefore, taking the Yellow River headwater area with fragile ecological environment as the research object and adopting wavelet analysis method to analyze temperature variation and runoff variation characteristics in the Yellow River headwater area may provide help for constantly improving the ecological environment of the Yellow River headwater area. The research has made main achievements as follows: ①taking analysis on the climate change characteristics of the Yellow River headwater area in recent 50 years, and result: the temperature of the Yellow River headwater area will increase constantly in the next few years; ②in accordance with annual average flow time series of Tangnaihe station (1959 ~ 2009), after adopting the wavelet analysis method to analyze the periodic variation of time series, finding that variation of two periods, 14 ~ 16 years and 9 ~ 12 years is the most distinguishable; and then the variation of the period of 3 ~ 6. 15 years is defined as the first period of runoff variation by calculating wavelet variance of runoff time series.

Key words: the Yellow River headwater area, climate change, runoff

The Yellow River headwater area indicates the area above Tangnaihe hydrometric station of main stream of Yellow River and is located in the northeast of Qinghai - Tibet Plateau between longitude 95°50′ ~ 103°30′ east and latitude 32°20′ ~ 36°10′north, with drainage area of 121,972.00 km^2. The drainage area above the Yellow River is 20,930.00 km, the length of the main stream is 270 km, lake, swamp and wet land in the area occupies 2,000 km^2, of which NgoringHu and Gyaring Lake occupy more than 1,000 km^2, they play significant role in adjustment and storage of runoff; the drainage area of the Yellow River along Jimai district occupies 24,089.00 km^2, and the length of the main stream is 325 km; the drainage area from Jimai district to Machu occupies 41,029.00 km^2, the length of the main stream is 585 km; the drainage area from Machu to Tangnaihe district occupies 35,924 km^2, and the length of the main stream is 373 km^2.

The drainage area in the headwater area occupies 15% of the drainage area of the Yellow River, the average natural runoff of many years takes up 38% of the natural runoff of the Yellow River, it serves as the main water production area and called "water tower of the Yellow River" iconically. However, as the global temperature increases gradually, the temperature of the Yellow River headwater area also increases gradually so as to produce a series of climate change and ecological and environment degradation problems, therefore, it is necessary to take research on runoff variation of the Yellow River headwater area for providing scientific basis for protecting water resource and ecological environment in headwater area.

1 Climate change characteristics

1.1 Global climate change trend

It is speculated in the fourth assessment report from IPCC that the global carbon dioxide concentration had increased to 379ppm in 2005 from 280ppm before Industrial Revolution due to human factor and driving effect of natural factor on climate change. It is proved by ice core research that the impact caused due to increment of air carbon dioxide concentration in 2005 was far more than the variation scope (180 ~ 300 ppm) caused due to natural factor in the past 650,000 years. The increment rate of carbon dioxide in the last 10 years is 1.9ppm/a, but the increment rate since continuous direct measurement record is 1.4 ppm/a. Since 1750, human activities have prompted climate warming with net effect of +1.6 W/m^2 (+0.6 ~ +2.4 W/m^2).

In accordance with recent direct observation of climate change, both average global atmospheric temperature and ocean temperature increase to cause ice and snow to melt and global sea level to rise in large range. From the land area and sea basin scale, large long-term climate change evidences have been observed, including Arctic Ocean temperature and ice change, and variation of rainfall, sea salinity, wind model and extreme climate in large range. The warming trend in the last 50 years is increasing by 0.13 ℃/10 a (0.10 ~ 0.16 ℃), which is almost twice of that in the last one hundred years. Comparing 2001 ~ 2005 with 1850 ~ 1899, the overall temperature has increased by 0.76 ℃ (0.57 ~ 0.95 ℃).

1.2 Analysis on temperature variation in headwater area

Since instruments have been provided for observation in the Yellow River headwater area, the general temperature change characteristics are as follows: the temperature was high in 1950s, reduced in 1960s, raised again in 1970s, entered high period in 1980s, continued rising in 1990s, and kept on rising in 2000s in comparison with 1990s. It can be seen by analyzing temperature data of ten stations, including Madoi, Darlag, Jigzhi, Tongde, Xinghai, Zekog, Maqen, Hongyuan, Zoige and Machu in the headwater area in last 50 years that the annual average temperature in the Yellow River headwater area was in rising trend with fluctuation, its climate inclination rate is 0.41 ℃/10 a (Fig. 1), which is obviously higher than average global temperature increment value 0.13 ℃/10 a and the average national temperature increment value 0.16 ℃/10 a. Since 1987, the temperature in the headwater area has been in rising trend, the average temperature in 1960s was -0.66 ℃, up to 0.18 ℃ in 1990s, which is 0.53 ℃ higher than that in 1960s. The temperature increment in recent 10 years since 2000 has been particularly distinguishable, and the temperature was 0.85 ℃ higher than that in 1990s.

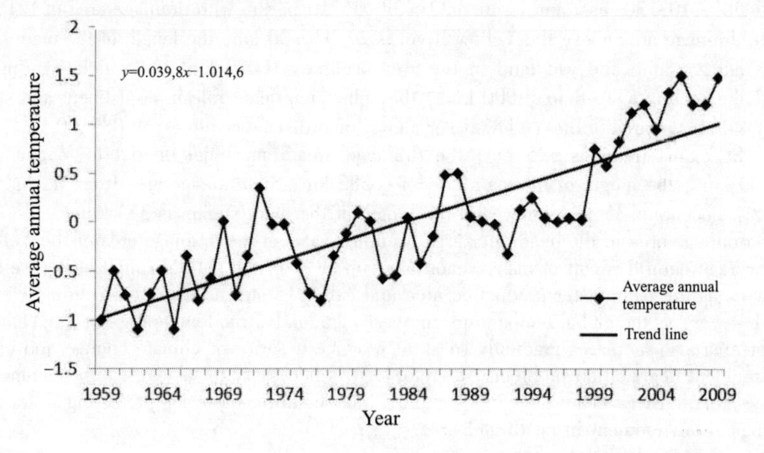

Fig. 1 Annual average temperature variation in the Yellow River headwater area

Our country is one of the climate change sensitive areas and fragile areas in the world, and the first to be affected by climate warming. Scientists have reached a consensus on the future global warming trend, however, they have different forecast results on the future global warming amplitude. The relatively rational forecast is that the global average temperature will increase by 1 ~ 2 ℃

in comparison with the current in 2050, and further increase by 1.5 ~ 3.5 ℃ in 2100. In addition, it is pointed out in IPCC report that the temperature increment in almost all lands may be greater than the global average level. Therefore, there are reasons to believe that the Yellow River headwater area will be one of the most obvious area of temperature increment in the world in next few years.

2 Analysis on runoff variation in headwater area

2.1 Annual runoff change

Although runoff in the Yellow River headwater area is seriously affected by atmospheric general circulation, its annual change is still stable in comparison with other large basin in our country. Real annual runoff data of station in the area along the Yellow River, and Guimet, Machu and Tangnaihe stations is calculated, the runoff variation coefficient of areas distributed along the Yellow River and in Guimet station is affected by adjustment and storage of the lake and relatively high, the value of Machu and Tangnaihe stations is below 0.24, which indicates that the annual runoff variation of the control station in the headwater area is relatively stable. See Tab. 1 for annual runoff value in each year. The average flow value from 1990 to 2003 in the Yellow River headwater area is 28.8% less than the average value of many years from 1959 to 1989; it can be seen from the annual average flow variation duration curve of Tangnaihe station (Fig. 2) that the average flow in the Yellow River headwater area in 40 years has passed through the variation trend from low to high and then to low, in addition, the average flow at the end of 1990s started to fall back gradually and has risen slightly since 2000s.

Tab. 1 Table of annual runoff C_v value in Yellow River headwater area

Station name	Area along Yellow River	Guimet	Machu	Tangnaihe
C_v	0.78	0.34	0.22	0.24

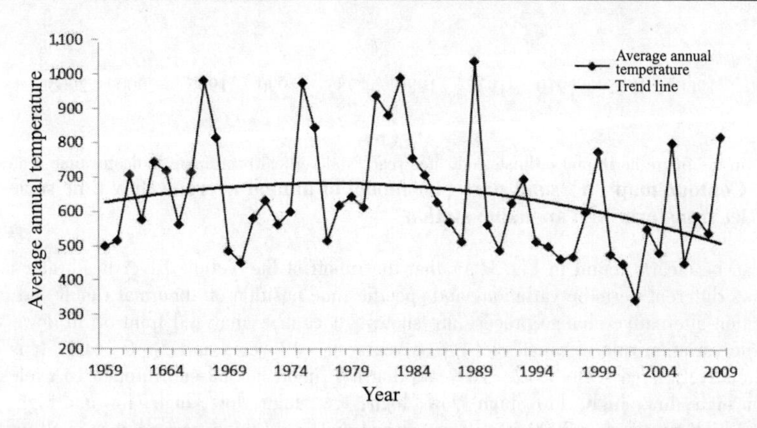

Fig. 2 Annual average flow variation duration curve of Tangnaihe station

2.2 Runoff variation periodic analysis

2.2.1 Wavelet transform of runoff time series

Complex Matlab wavelet is adopted for continuous wavelet transform in accordance with annual average flow time series of Tangnaihe station to obtain solid part and model of wavelet transform coefficient, wavelet transformer contour map should be drawn, respectively, see Fig. 3 (a) and

Fig. (b). When the real number of wavelet coefficient is positive, it indicates the runoff is high, it is drawn with solid line in the figure; and when the real number is negative, it indicates the runoff is low, it is drawn with dotted line in the figure; when the wavelet coefficient is zero, it corresponds to a catastrophe point.

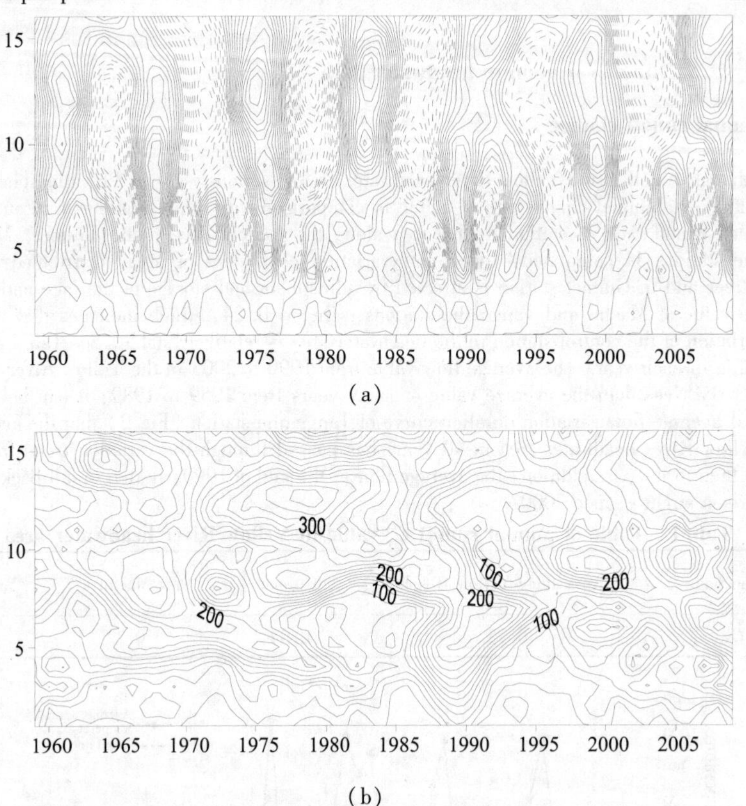

(Note: in the figure horizontal ordinate indicates year, and vertical coordinate indicates time scale/year)

Fig. 3 Contour map (a) solid part (b) model of annual average flow time series wavelet transform of Tanghaihe station

It can be clearly found in Fig. 3(a) that the runoff of the Yellow River headwater area in 50 years shows different periodic variations and specific time position of abnormal runoff and increment and reduction alternative change process are shown. It can be analyzed from up to down to obtain the periodic variation rule of runoff of 14 ~ 16 years, 9 ~ 12 years and 3 ~ 6 years. It is analyzed from the relatively large scale of 12 ~ 16 years that the runoff has passed through 13 cycles alternatively from high, low, high, low, high, low, high, low, high, low, high, low and high, the contour line of high runoff after 2009 is still not closed and at the high period; it is analyzed from the scale of 6 ~ 12 years that the rainfall precipitation also has passed through 13 cycles alternatively, the contour line of high runoff after 2009 is still not closed and at high period; the periodic variation of the two scales is stable during the whole analysis period and global. And it has passed through more cycles alternatively on the scale of 3 ~ 6 years.

Model of Matlab wavelet coefficient indicates energy density, and the model contour line represents the distribution condition of periodic variation on various time scales in time domain. If the wavelet coefficient model is larger, it indicates the corresponding time and scale periodicity are more obvious. It can be seen from Fig. 3(b) that the wavelet coefficient models on the scale of

14 ~ 16 years and 9 ~ 12 years are large, it indicates that the two periodic variations are most obviously; and then is the periodic variation on the scale of 3 ~ 6 years.

2.2.2 Wavelet variance of runoff time series

Main period of each series can be determined by calculating wavelet variance of runoff time series, see Fig. 4 for wavelet variance. There are six peak values in the figure, corresponding to the time scales of 3, 5, 7, 10, 12 and 15, respectively, the first peak value is on the scale of 15 years, it indicates that the periodic oscillation of about 15 years is the strongest, it is the first period of runoff variation, the second period is 12 years, and then 7, 5, 10 and 3 years in sequence.

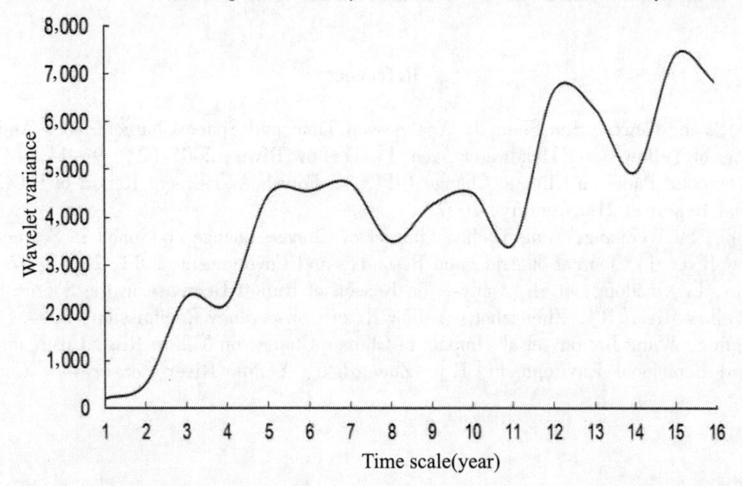

Fig. 4 Wavelet variance of runoff time series in the Yellow River headwater area

3 Impact of climate change on runoff

3.1 Calculation formula of natural runoff

Before analysis on impact of climate change on runoff, usually a natural runoff calculation formula should be created. Therefore, in accordance with the incoming runoff formation mechanism of Longyangxia, both the impact of precipitation in the same period and temperature on runoff formation and action of early precipitation on runoff in the current year are considered. And on the basis of calculating the relationship between the natural annual runoff of Tangnaihe station from 1956 to 1989 and precipitation and temperature of each station, related obvious representative station will be taken, and the precipitation in the same period, average temperature from May to October and precipitation from September to December in the last year are taken as correlation factors to finally obtain the calculation formula of natural annual runoff as follows:

$$W = -207.117 + 0.855,276x_1 - 19.007x_2 + 0.600,983x_3$$

In the formula: W indicates the natural annual runoff; x_1 indicates precipitation in the same period; x_2 indicates average temperature from May to October in the same year; and x_3 indicates precipitation from September to December in the last year.

With simulation analysis on annual flow from 1956 to 1989, the result indicates that the natural runoff calculation value perfectly fits the real value, variation trends of the two values are approximately consistent, and the main peak value and valley value are almost the same. After calculation, multiple correlation coefficient of the formula is up to 0.946, and average relative error is 9.1%, there are 28 years in which the relative error is less than 15%, occupying 84.8% of the total samples. It indicates that the impact of climate change on the natural annual runoff calculated by the formula is highly reliable.

3.2 Calculation analysis on impact quantity of climate change

Correlation factors of Tangnaihe station since 1990s should be substituted to calculate the natural runoff of each year since 1990s by using above formula. After calculation, the recent average annual runoff is 17.24×10^9 m^3, compared with the average natural annual runoff before 1990s (21.61×10^9 m^3), the runoff has been reduced by 4.37×10^9 m^3, it indicates that the reduction amplitude of the natural runoff caused by climate change is up to 20.2%.

References

Che Qian, Wang Genxu, Sun Shengli. Analysis on Time and Space Characteristics And Relevant Factors of Yellow River Headwater Area[J]. Yellow River, 2005 (2): 9 – 11.

Intergovernmental Panel on Climate Change (IPCC). Fourth Assessment Report of IPCC[R]. Financial Report of 21st Century, 2007.

Liu Caihong, Su Wenjiang, Yang Yanhua. Impact of Climate Change on Runoff in Source Region of Yellow River[J]. Journal of Arid Land Resources and Environment, 2012, 26(4): 97 – 101.

Hu Xinglin, Li Xiaodong, et al. Analysis on Reason of Runoff Decrease in the Source Regions of the Yellow River[R]. Zhengzhou: Yellow River Conservancy Commission, 2004.

Kang Lingling, Wang Jinhua, et al. Impact of Climate Change on Yellow River Upstream and Runoff and Ecological Environment [R]. Zhengzhou: Yellow River Conservancy Commission, 2004.

Runoff Forecasting in Future Climate Change Scenarios of the Yellow River Basin Based on the Water Balance Model

Li Xiaoyu*, *Li Zhuo* and *Qian Yunping

Hydrology Bureau of YRCC, Zhengzhou, 450004, China

Abstract: The Yellow River Basin (YRB) is located in the arid region, water resources are great sensitive to the climate changes, because of the water scarcity and fragile environment. A2 and B2 greenhouse gas emission scenarios which are defined by IPCC, and HadCM3, CSIRO30, INM and MRI climate models had been selected in this study to form the future climatic scenarios of the Yellow River Basin. The Yellow River Water Balance Model (YRWBM) hydrological model was built to forecast the future natural runoff and its temporal-spatial change, with the inputs of the future climatic scenarios. Forecasting results indicate that the runoff of the Yellow River shows decreases trend in the future. It will change -20.6% ~ -1.29% in 2050, and -19.7% ~8.06% in 2100 comparing with present situation. In spatial distribution, the precipitation and runoff are greatly reduced in main water yielding regions upper Lanzhou, and increased in other regions. In annual distribution, the runoff increase in winter and spring, and decrease in summer and autumn.

Key words: climate change scenarios, natural runoff, the Yellow River Water Balance Model, the Yellow River Basin

Hydrology and water resources change affected by climate change is the international issues of common concern. Global climate change causes temperature and evaporation increasing, precipitation changing, accelerates the hydrological cycle, impact natural ecosystems and human environment (Qin Dahe, Ding Yihui, Su Jilan, et al. 2005; Zhang Jianyun, Wang Guoqing, et al. 2007). The Yellow River Basin is located in the semi-arid and arid regions, water resources are great sensitivity to the climate changes, because of the water scarcity and fragile environment. River water resources management must prepare in advance to face the possible future changes in YRB. The study establishes the water-balanced model, with the model input of future climate change scenarios formed by four climate models and two greenhouse gas emission scenarios, and predicts the discharge of main sections in the Yellow River and analyzes the temporal and spatial distribution.

1 Water balance model building

1.1 Model method and framework

Water Balance Model had been built by Thornthwaite (Thornthwaite, C. W. 1948). Thereafter, researchers had improved the model by different structure and assumptions, due to different research purposes and requirements, for example, Van-dewiele had developed the VWBM applied in Belgium River Basin (Van-dewiele, G. L., Xu, C. Y., and Win, N. L. 1992), Boughton had developed the AWBM applied in Australia (Boughton, W. C. 1995), et.. The Yellow River has complex runoff-production sources, at the same time, the study has large research scope and large time and spatial scales. Thus, the Yellow River Water Balance Model (YRWBM) is selected for this project. The model has good adaptability, comprehensive factors, less parameters, and easier calculation. The model had been applied to simulate the runoff in the middle reaches of the Yellow River, and the simulating effect was satisfactory (Zhang Jianyun, Wang Guoqing, et al. 2007).

The YRWBM is based on the principle of water balance in a close basin. The model focuses on the soil moisture system. Its time scale is monthly, spatial scale is river basin cells. The input of this system includes current precipitation and evaporation. The discharge, infiltration and actual evaporation can be calculated according to the initial soil moisture. The output of the system is runoff and the new soil moisture that will be used as initial soil moisture of the next time period. The model equation is shown as Eq. 1, and the model framework is shown as Fig. 1.

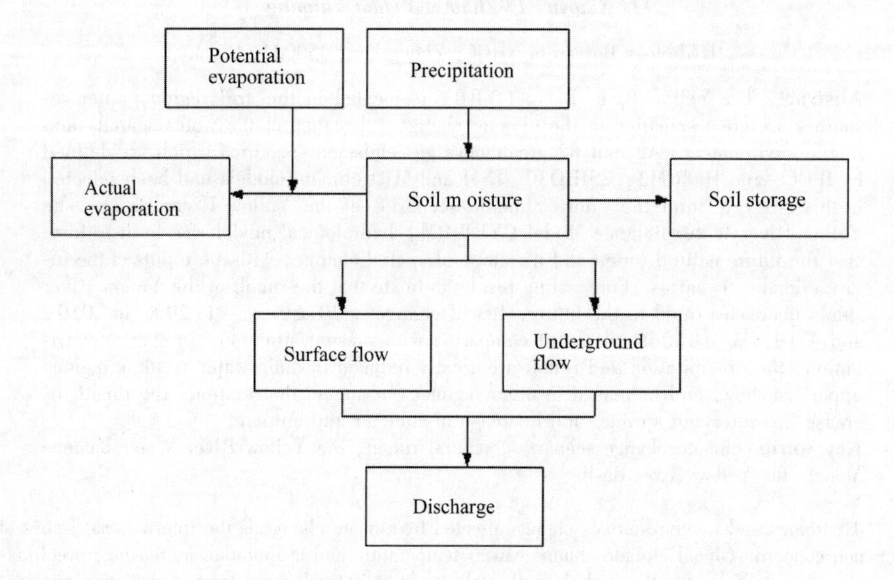

Fig. 1 Process of grid-based monthly water balance model

$$\frac{\partial S}{\partial t} = P - E(S, E_t) - R(P, S) \tag{1}$$

where, S denotes the soil moisture changes within a period; P denotes the precipitation; E denotes the actual evaporation, which is a function of S and the evaporation capacity (E_t) in the catchment; R denotes the runoff, which is a function of P and S.

1.2 Model factors calculation

The runoff should be divided into surface runoff R_s and underground runoff R_g. Owing to the monthly time-scale, division of surface and underground runoff can not be obtained by calculating rainfall intensity and infiltration. So, the model needs to be generalized as follows:

(1) Omit concentration processes of the surface flow.

(2) Omit considering surface runoff generation and outflow at the same time.

(3) Generalize the outflow time of underground runoffs at a certain time period.

(4) Define the outflow of the underground runoff at a time period later than that of the surface runoff.

The generalization can be described as Eq. 2 (in Tab. 1).

Surface runoff and precipitation and soil moisture are proportional relation, the coefficient "Ks" should be quoted to calculate the surface runoff (Eq. 3) (in Tab. 1).

Underground runoff and soil moisture are linear relationship, the coefficient "Kg" should be quoted to calculate the underground runoff (Eq. 4) (in Tab. 1).

Actual evaporation and evaporation capacity and soil moisture are proportional relation, the actual evaporation could be calculated by the ratio of actual moisture and saturation moisture (Eq. 5) (in Tab. 1).

Based on the water balance, the actual soil moisture could be calculated (Eq. 6) (in Tab. 1).

Tab. 1 Model factors formulas

Eq No.	Factor	Eq.	The meaning of variables
2	Runoff	$R_i = R_{si} + R_{gi-1}$	E_i——Actual evaporation E_t——Evaporation capacity
3	Surface runoff	$R_{si} = K_s \cdot \frac{S_{i-1}}{S_{max}} \cdot P_i$	K_s——Surface runoff coefficient K_g——Underground runoff coefficient
4	Underground runoff	$R_{gi} = K_g \cdot S_{i-1}$	P_i—— Precipitation R_i—— Runoff R_{si}——Surface runoff
5	Actual evaporation	$E_i = E_t \cdot \frac{S_i - 1}{S_{max}}$	R_{gi}——Underground runoff R_{gi-1}——Previous time runoff S_i—— Soil moisture
6	Soil moisture	$S_i = S_{i-1} + P_i - R_{si} - R_{gi} - E_i$	S_{i-1}——Previous time soil moisture S_{max}——Saturation soil moisture

1.3 Subarea and calibration of the model

The Yellow River has a large scope for research and has complex precipitation characteristics and diverse runoff-producing sources. The simulation would be distorted if the model uses the same parameters in the whole of YRB. Thus, the basin must be divided into several regions, and the YRWBM model will be calibrated against natural average monthly runoff in each region. The method of the division applies two principles:

(1) It should be based on the geographical distribution taking into account similar precipitation and runoff-producing types.

(2) To avoid an excessive complexity, the division of the region should apply a number of generalizations.

Finally, the study get 6 regions, they are shown as Tab. 2.

Tab. 2 Hydrological characteristics of water resources regions

Regions number	Regions name	Control station	Area ($\times 10^3$ km^2)	Annual runoff ($\times 10^9$ m^3)	Rate of runoff (%)	Runoff - generation
1	Upper Longyangxia	Longyangxia	122	21.04	37.1	saturation excess runoff
2	Longyangxia—Lanzhou	Lanzhou	101	33.81	59.6	saturation excess runoff
3	Lanzhou—Toudaoguai	Toudaoguai	163	34.06	60.1	infiltration excess runoff
4	Toudaoguai—Longmen	Longmen	112	39.15	69.0	infiltration excess runoff
5	Longmen—Sanmenxia	Sanmenxia	184	50.37	88.8	mixed runoff
6	Sanmenxia—Huayuankou	Huayuankou	40	56.23	99.1	mixed runoff
	Whole Basin	Huayuankou	722	56.72	100.0	-

The excessive exploitation and utilization of the Yellow River water resources, the observed river runoff are all factors that are significantly affected by human activities. So it is difficult to reflect the impact of climate change on water resources, therefore natural runoff is analyzed in this report by the Eq. 7. The inputs of the model calibration are the monthly precipitation and evaporation data of the region, which are accumulated by each 50km × 50km grid. The model is calibrated against natural average monthly runoff of the control station of every region from 1961 to 1970, and is verified by the data from 1971 to 2000 (Hydrological Bureau of YRCC, 2005).

$$W_{\text{natural run off}} = W_{\text{observed run off}} + W_{\text{waterusing}} \pm W_{\text{reservoir storing}} \tag{7}$$

The results of the model calibration and verification shows that, the trends of the calculated series are similar to the actual series, and their correlation coefficient are satisfactory (Fig. 2).

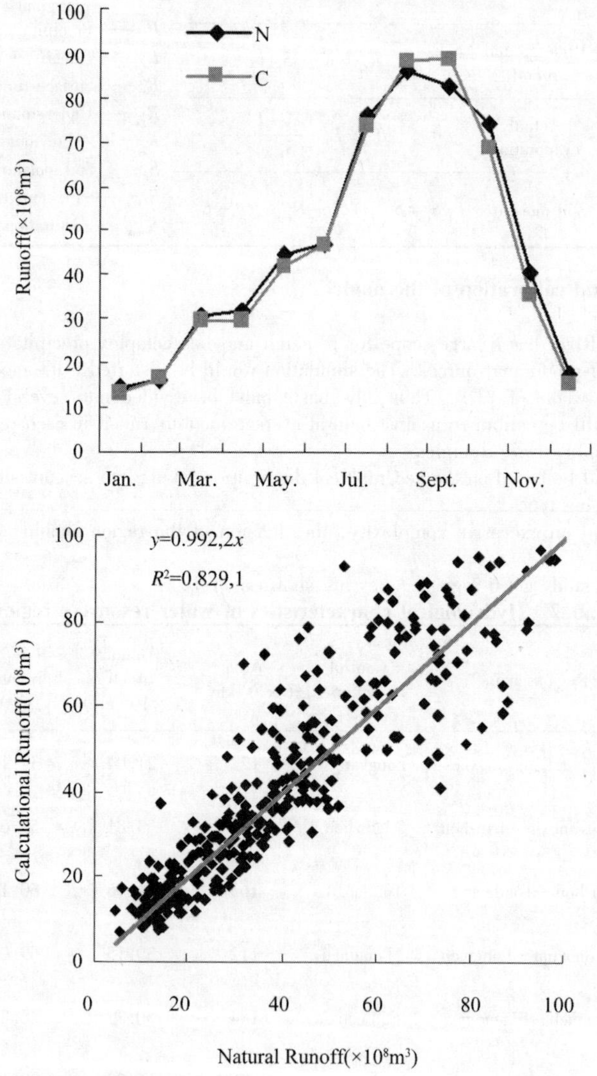

Fig. 2 Monthly runoff verification at the Huayuankou Station

2 Climate model selection

Four future greenhouse gas emission scenarios A1, A2, B1 and B2, had been defined by

IPCC. Usually thinking the A2 and B2 emission scenarios should be proximal to the future development of China (Xu Yinlong, Huang Xiaoying, Zhang Yong, et al. 2005). General Circulation Models (GCMs) are known as an important tool in the assessment of climate change. The GCM should be downscaling when it is applied to a river basin because of the large scale, and the downscaling result is the Regional Climate Model (RCM). In order to avoid the uncertainty of GCM application in YRB, four models has selected in the study, they are HadCM3 of Britain, CSIRO:MK 30 of Australia, INM:CM30 of Russia, and MIROC 3.2 – medres of Japan (Tab. 3). The four models had been applied in China, and the effect was satisfactory. The RCMs matched YRB has been attained by downscaling analysis of the GCMs (Fig. 3), and applied in the hydrological study, in which HadCM3 model had been downscaled by Professor Chen Xiaoqiu of PKU, and other three models had been downscaled by Professor Xu Zhongxue of BNU. The four climate models and two emission scenarios generate eight future climate change scenarios of YRB.

Tab. 3 Introduction of climate model

Climate model	Country	Resolution	Data series
CSIRO:MK30	Australia	1.9°×1.9°	1961 ~ 2000
HadCM3	Britain	2.5°×3.75°	1961 ~ 1999
INM:CM30	Russia	4°×5°	1961 ~ 2000
MIROC3.2_medres	Japan	2.8°×2.8°	1961 ~ 2000

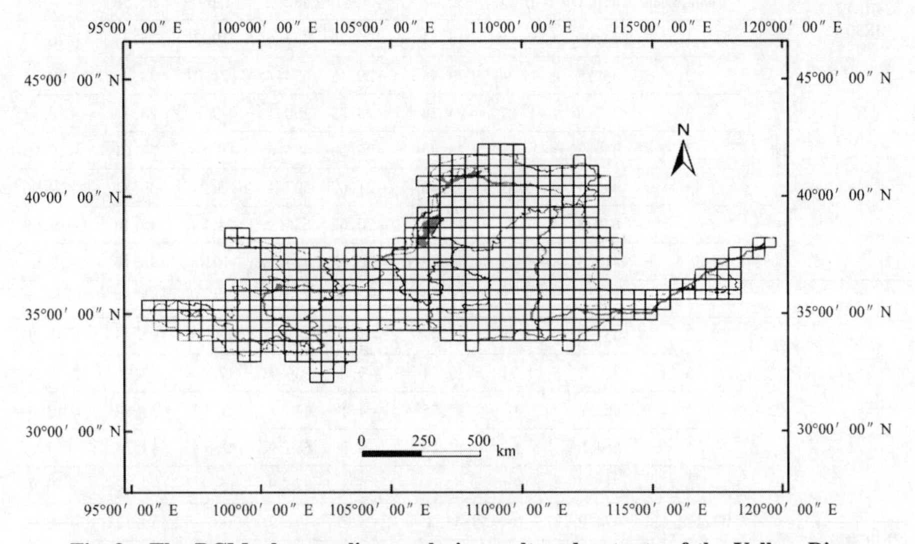

Fig. 3 The RCMs downscaling analysis result and outputs of the Yellow River

3 Future runoff forecasting of YRB

The monthly precipitation, temperature, and evaporation data in every basin cell in 2050 and 2100 are provided by the future climatic scenarios. The data was inserted into the hydrological model to calculate the future runoff of every cell of the Yellow River in the climate change scenarios. Then, the runoff of every region could be forecasted by the cells' runoff cumulating. The runoff forecasted results of the control stations of six regions in 2050 and 2100 are as shown at Tab. 4.

The annual average precipitation is 460.6 mm, and the annual average natural runoff is 0.562,3 $\times 10^8$ m^3 of YRB in present situation. By the models forecasting, in the A2 scenario, in

2050, the annual average precipitation will be from 460.1 mm to 486.9 mm, will change from −1.11% to 5.71% comparing with the present situation, and the annual average natural runoff of the Huayuankou Station will be from 0.526,3 × 10^8 m^3 to 0.555,1 × 10^8 m^3, will change from −6.41% to −1.29% comparing with the present situation. In 2100, the annual average precipitation will be from 482.2 mm to 517.0 mm, will change from 4.69% to 12.27% comparing with the present situation, and the annual average natural runoff of the Huayuankou Station will be from 50.12 billion m^3 to 60.76 billion m^3, will change from −10.9% to 8.06% comparing with the present situation.

Tab. 4 Runoff forecasting results in future climate change scenarios

Climatic model	Scenario	Region number	2050				2100			
			P (mm)	Rate (%)	*R* ($\times 10^9$ m^3)	Rate (%)	*P* (mm)	Rate (%)	*R* ($\times 10^9$ m^3)	Rate (%)
CSIRO: MK30	A2	1	479.6	−11.24	17.65	−16.1	475.6	−11.98	17.43	−17.17
		2	374.3	−14.1	26.69	−21.07	349.9	−19.72	25.71	−23.97
		3	288.5	0.04	27.18	−20.2	301.9	4.67	26.49	−22.22
		4	446	−0.86	32.72	−16.42	487.7	8.39	32.77	−16.28
		5	577.9	7.95	46.31	−8.06	621.6	16.13	48.03	−4.65
		6	734.7	9.85	52.63	−6.41	831.2	24.28	55.56	−1.18
		Whole basin	460.1	−0.11	52.63	−6.41	482.2	4.69	55.56	−1.18
	B1	1	439	−18.75	15.51	−26.29	437.6	−19.01	15.44	−26.63
		2	357.5	−17.98	23.94	−29.18	344	−21.07	23.43	−30.71
		3	262.8	−8.87	24.24	−28.83	260.1	−9.8	23.74	−30.31
		4	384.3	−14.58	28.75	−26.56	406.6	−9.63	28.61	−26.93
		5	498.7	−6.83	39.48	−21.62	507.1	−5.27	39.81	−20.96
		6	640.8	−4.19	44.63	−20.62	656.9	−1.79	45.16	−19.69
		Whole basin	410.4	−10.9	44.63	−20.62	414.2	−10.06	45.16	−19.69
HadCM3	A2	1	506.6	−6.2	18.14	−13.8	519.0	−3.9	17.49	−16.9
		2	431.9	−0.9	29.97	−11.3	431.1	−1.1	27.43	−18.9
		3	320.7	11.2	30.74	−9.8	339.4	17.7	27.96	−17.9
		4	465.3	3.4	37.55	−4.1	478.1	6.3	33.84	−13.6
		5	564.8	5.5	47.18	−6.3	586.9	9.6	43.74	−13.2
		6	715.2	6.9	53.3	−5.2	773.9	15.7	50.12	−10.9
		Whole basin	474.2	3.0	53.3	−5.2	491.2	6.7	50.12	−10.9
	B1	1	506.8	−6.2	17.82	−15.3	514.6	−4.7	17.21	−18.2
		2	435.5	−0.1	29.75	−12.0	436.7	0.2	27.4	−18.9
		3	316.0	9.6	30.48	−10.5	324.5	12.5	27.89	−18.1
		4	465.9	3.6	37.17	−5.1	474.3	5.4	33.63	−14.1
		5	555.8	3.8	46.64	−7.4	564.8	5.5	43.58	−13.5
		6	670.5	0.3	52.01	−7.5	685.9	2.6	49.11	−12.7
		Whole basin	469.0	1.8	52.01	−7.5	476.8	3.5	49.11	−12.7

Continued Tab. 4

Climatic model	Scenario	Region number	2050 P (mm)	2050 Rate (%)	2050 R ($\times 10^9$ m^3)	2050 Rate (%)	2100 P (mm)	2100 Rate (%)	2100 R ($\times 10^9$ m^3)	2100 Rate (%)
INM: CM30	A2	1	471.5	-12.74	17.18	-18.36	439.2	-18.71	15.67	-25.51
		2	367.4	-15.69	26.33	-22.11	299.9	-31.18	22.75	-32.71
		3	303.4	5.2	26.78	-21.37	315.1	9.27	23.52	-30.95
		4	497.4	10.57	33.08	-15.5	543.9	20.89	30.47	-22.16
		5	599.9	12.06	47.61	-5.49	647.6	20.99	47.48	-5.74
		6	793.5	18.65	54.26	-3.51	900.6	34.66	55.4	-1.47
		Whole basin	480	4.21	54.26	-3.51	495.9	7.67	55.4	-1.47
	B1	1	458.4	-15.16	16.44	-21.87	466.8	-13.61	16.99	-19.24
		2	369.5	-15.2	25.59	-24.32	361.5	-17.04	25.98	-23.16
		3	280.3	-2.81	25.7	-24.54	302.1	4.77	26.37	-22.57
		4	464.8	3.31	31.41	-19.78	500.8	11.31	32.73	-16.41
		5	545.6	1.93	44.07	-12.51	599.1	11.92	47.31	-6.08
		6	712.2	6.49	49.69	-11.63	788.5	17.9	53.87	-4.2
		Whole basin	449.7	-2.37	49.69	-11.63	478.4	3.87	53.87	-4.2
MRIOC3.2-medres	A2	1	487.3	-9.81	18	-14.47	498.3	-7.78	18.75	-10.87
		2	428.8	-1.61	29.22	-13.58	422.1	-3.15	29.86	-11.69
		3	307.1	6.48	29.29	-14.01	331.1	14.8	30.28	-11.1
		4	524.4	16.57	35.9	-8.29	569.5	26.59	37.86	-3.29
		5	576.3	7.66	49.38	-1.96	615	14.89	53.6	6.41
		6	751.6	12.38	55.51	-1.29	838.4	25.36	60.76	8.06
		Whole basin	486.9	5.71	55.51	-1.29	517.1	12.27	60.76	8.06
	B1	1	473.5	-12.37	17.36	-17.49	497.7	-7.89	17.05	-18.96
		2	423	-2.94	28.56	-15.52	435.5	-0.06	28.15	-16.76
		3	304.9	5.72	28.72	-15.67	311.1	7.87	28.45	-16.48
		4	507.1	12.72	35.24	-9.99	523.2	16.3	35.24	-9.99
		5	551.1	2.96	48	-4.7	588.5	9.94	49.61	-1.51
		6	710	6.16	53.76	-4.4	773.9	15.71	55.77	-0.81
		Whole basin	471.1	2.27	53.76	-4.4	494.6	7.38	55.77	-0.81

In the B2 scenario, in 2050, the annual average precipitation will be from 410.4mm to 471.1mm, will change from −10.9% to 2.27% comparing with the present situation, and the annual average natural runoff of the Huayuankou Station will be from 0.446,3 $\times 10^8$ m^3 to 0.537,6 $\times 10^8$ m^3, will change from −20.6% to −4.40% comparing with the present situation. In 2100, the annual average precipitation will be from 414.2mm to 494.6 mm, will change from −10.1% to 7.38% comparing with the present situation, and the annual average natural runoff of the Huayuankou Station will be from 0.451.6 $\times 10^9$ m^3 to 0.5577 $\times 10^9$ m^3, will change from −19.7% to −0.81% comparing with the present situation.

Comprehensive analysis the forecasting results of 8 climate scenarios in 2050 and 2100, only

MRI climatic model A2 scenario prediction shows the runoff increasing in 2100, other 15 forecasting results all show the runoff decreasing in the future 50 to 100 years of YRB. For the mean value of the 4 climatic models forecasting results, in A2 scenarios, the annual average runoff of the Huayuankou Station will be 0. 539,3 $\times 10^9$ m^3, less than the historical average runoff of 4. 10% in 2050, and will be 0. 554,5 $\times 10^9$ m^3, less than the historical of 1. 37% in 2100. In B2 scenarios, the annual average runoff of the Huayuankou Station will be 0. 500, 2 $\times 10^9$ m^3, less than the historical average runoff of 11. 04% in 2050, and will be 0. 509, 8 $\times 10^9$ m^3, less than the historical of 9. 35% in 2100.

In spatial distribution, the precipitation and runoff are greatly reduced in main water yielding regions upper Lanzhou, and increased in other regions. In upper Lanzhou region, the precipitation averaged decrease of 10. 7%, coupled with the increased evaporation due to the climate warming, the runoff averaged decrease of 19. 5% in the future. In annual distribution, the runoff averaged increase of 14. 3% in winter and spring, and average decrease of 14. 8% in summer and autumn.

4 Conclusions

The forecasting results based on 4 climatic models indicate that, the runoff will decrease about 4×10^9 m^3 comparing with present situation. Water resources reduction will have severe effect on the administration and development of YRB. Adaptation measures of water management are submitted as the follow:

(1) To produce comprehensive regulations and policies and strengthen integrated management of water resources.

(2) To encourage the inter-basin water transfer.

(3) To establish water-saving culture.

(4) To strengthen construction of water conservancy infrastructure.

(5) To improve water use efficiency.

References

Qin Dahe, Ding Yihui, Su Jilan, et al.. Climate and Environment Evolvement of China [M]. Beijing: Science and Technology Press, 2005: 514 – 535.

Zhang Jianyun, Wang Guoqing, et al.. Climate Change Impacts on Hydrology and Water Resources Research. [M]. Beijing: Science Press, 2007.

Thornthwaite, C W.. An Approach Toward a Rational Classification of Climate [J]. Geogr. Rev. 1948, 38(1): 55 – 94.

Vandewiele, G L., Xu, C Y., Win, N L. Methodology and Comparative Study of Monthly Water Balance Models in Belgium, China and Burma [J]. Journal of Hydrology, 1992(134): 315 – 347.

Boughton, W C.. An Australian Water Balance Model for Semiarid Water-sheds [J]. Journal. Soil and Water Conservation., 1995, 50(5): 454 – 457.

Hydrological Bureau of YRCC. Water Resources Investigation and Assessment of the Yellow River Basin [R]. 2005:58 – 66.

Xu Yinlong, Huang Xiaoying, Zhang Yong, et al.. The 21th Century Climate Change Scenario Statistical Analysis in China [J]. Advances in Climate Change Research, 2005, 1(2):80 – 84.

A Study on Concept of Water Resource Carrying Capacity under Climate Change and its Computing Methods

Zhang Xiuyu[1,2] and *Zuo Qiting*[1]

1. Center for Water Science Research, Zhengzhou University, Zhengzhou, 450001, China
2. North China Institute of Water Conservancy and Electric Power, Zhengzhou,450011,China

Abstract:Water is the source of life, a key factor of production and ecological base. China's basic water conditions can be described as uneven spatial and temporal distribution of water resources as well as its large population with little water. Currently, climate change is regarded as one of the most important global environmental problems at home and abroad, and future climate change may further exacerbate the contradiction between supply and demand of water resources in China.

Research on water resources carrying capacity boasts of an extensive literature at home and abroad. Its definitions can be summarized as the following two viewpoints: one is the water resources development capacity or water resources development scale; another is water resources supporting sustainable development capacity. In practical applications, the calculation of water carrying capacity is generally based on the volume of available water resources with different guarantee rate in the future planning year, which is computed in accordance with historical precipitation data using the probability and statistic method, thus not objectively showing the volume of available water resources in the future years.

The calculation of water resources dynamic carrying capacity is different from that of the traditional water resources carrying capacity, which is based on the volume of available water resources with different guarantee rate in the future planning year. And its volume is calculated in combination with historical rainfall and runoff data through probability statistic method, therefore, it is not a true reflection of water resources utilization in the coming years. The premise of calculation of water resources dynamic carrying capacity is to obtain the volume of available water resources in the coming years with the help of the climate model and hydrology model. Therefore, water dynamic carrying capacity can be summarized as the maximization of the social and economic development water resources system in some basin or region under the climate change, when maintaining the virtuous circle of ecological system, can achieve in the foreseeable period of time.

This paper,reviewing and summarizing the study water resources carrying capacity, systematically expounds on the concept and connotation of water resources dynamic carrying capacity under the climate change, puts forward its calculation frame and main technical method, and analyses its advantages and application range, with an aim to provide the reference for its further research .

Key words: climate change, water resource, dynamic carrying capacity, quantization method

1 Introduction

Water is the source of life, a key factor of production and ecological base. With the raising population and economic and social development, there is a sharp increase in a demand for water. Issues such as wateriness, lack of water, dirty water and the muddy water have become the main

foundation:this study was conducted under the financial of the support of the Natural Science Foundation of china(51279183 和 51079132).

restraining factors for the global economic and social development. China, the largest developing countries suffering from the shortage of water resources, owns only one fourth of the world's average water resources per capita, which shows uneven spatial and temporal distribution of water resources in time and space distribution. Coupled with the rapid economic and social development and deterioration of water environment, China, especially the northwest and north inland areas, displays a prominent shortage of water resources. Drought, floods, water, environmental degradation and soil erosion are the main problems facing water resources in China.

Currently, climate change is regarded as one of the most important global environmental problems at home and abroad, and the global warming has become the most important environmental issue in the world. Climate change will have a significant impact on ecological environment, water resources, socio - economic fields or sectors in the world. The impact of climate change on water resources mainly concentrated on the change of water volume. Due to its complexity of climate change, there are relatively few quantitative researches on river ecology, water quality and extreme hydrological events. The profound changes of the atmospheric system are bound to affect the changes of such hydrological factors as the global hydrological cycle, the amount of precipitation and evaporation, greatly increasing the occurrence frequency of hydrological extreme events, changing the water balance of river basins and regions, and thereby affecting the distribution of water resources. Because of China's large population and the unfavorable national conditions of water resources, coupled with the impact of future climate change, the contradiction between water supply and demand is bound to become more prominent.

Research on water resources carrying capacity boasts of an extensive literature at home and abroad. Its definitions can be summarized as the following two viewpoints: one is the water resources development capacity or water resources development scale; another is water resources supporting sustainable development capacity. In practical applications, the calculation of water carrying capacity is generally based on the volume of available water resources with different guarantee rate in the future planning year, which is computed in accordance with historical precipitation data using the probability and statistic method, thus not objectively showing the volume of available water resources in the future years.

Making a summary of a number of research achievements, this paper puts forwards the basic concepts and framework of water resources dynamic carrying capacity under climate change. Based on the connotation analysis of dynamic carrying capacity of water resources, the paper puts water resources and water environment carrying capacity within a framework.

2 The concept and connotation of the water resources dynamic carrying capacity under climate change

2.1 The introduction of the concept of water resources carrying capacity

The term carrying capacity was originally a physical conception, meaning the maximum load that an object can withstand without producing any damage. Water resources carrying capacity is the concrete application of carrying capacity concept in the field of water resources. With the increasing seriousness of water issue, water resources carrying capacity, part of the carrying capacity of natural resources, was put forward by Chinese scholars in the late 1980s. In recent years, many scholars have done in - depth research on the concepts and computing methods of water resources carrying capacity. Despite the different definitions of water resources carrying capacity, there are no essential differences in the basic concepts and thinking, the definitions that all lay emphasis on "water resources development scale" or "water resources supporting capacity".

2.2 The concept of water resources dynamic carrying capacity

Implementing the most stringent water management system is China's national policy in the current and future period, and there is a higher demand for meticulous management of water resources, therefore, it is particularly urgent to do research on the water resources dynamic carrying capacity. The calculation of water resources dynamic carrying capacity is different from that of the traditional water resources carrying capacity, which is based on the volume of available water resources with different guarantee rate in the future planning year. And its volume is calculated in combination with historical rainfall and runoff data through probability statistic method, therefore, it is not a true reflection of water resources utilization in the coming years. The premise of calculation of water resources dynamic carrying capacity is to obtain the volume of available water resources in the coming years with the help of the climate model and hydrology model. Therefore, water dynamic carrying capacity can be summarized as the maximization of the social and economic development water resources system in some basin or region under the climate change, when maintaining the virtuous circle of ecological system, can achieve in the foreseeable period of time.

2.3 The connotation of water resources dynamic carrying capacity

Adopting the probability analysis, the traditional evaluation of water resources carrying capacity makes a study on water resources in the normal process, failing to fully consider water resources dynamic carrying capacity in face of droughts, floods and other extreme situations. With the frequent occurrence of extreme events and its increasing influence, it is a must to combine the normal process and extreme process to make an integrated analysis. The definition of water resources dynamic carrying capacity contains the following information:

(1) The time scale of analyzing water resources carrying capacity should be defined in the predictable period, and it should reflect the dynamic and relative limit nature of water resources carrying capacity.

(2) The spatial scale of water resources carrying capacity should be defined in the river basin or region as the basic unit. From the perspective of system boundary demarcation and data access, river basin analysis, on one hand, can better reflect the characteristics of the water circulation system, and also have easier access to relevant data; region analysis, on the other hand, can better reflect the characteristics of human social systems, and enjoy more easy access to relevant data.

(3) The two basic boundaries of water resources carrying capacity should be determined: to maintain the realistic demand for the local social and economic sustainable development, to maintain ecosystem stability.

(4) The purpose of water resources carrying capacity should be highlighted, that is, the maximization of the social and economic development that water resources system can achieve.

(5) It should be stressed that water, the most active factor in the natural world and the most important part of the global atmospheric circulation and the global hydrological cycle, is the first to be affected by the climate change.

3 Quantitative research methods of water resources dynamic carrying capacity under climate change

3.1 Introduction of the existing methods of water resources carrying capacity

Evaluation of water resources carrying capacity is a hot issue in the field of water resources research, and professor Xia Jun argues that evaluation methods of water resources carrying capacity can be divided into two categories. As for the first, the index system is established by means of firstly analyzing the phenomena in the water resources carrying capacity system, and water resources carrying capacity, after adopting some kind of evaluation method and evaluation standard comparison, is evaluated, that is, "the greatest supporting ability ". And some main evaluation methods are used, methods that consist of fuzzy evaluation method, gray relational evaluation method and principal component analysis and so on. This type of methods is often limited to the representation of the various factors in water resources carrying capacity system, and has no in – depth study of the correlativity. Besides, because there exists more subjectivity in the choice of index system, evaluation criteria and evaluation methods, evaluation results vary from one to another, as a result, the final result can only be used for qualitative judgment. As for the second one, based on the interaction between various factors in the system of water resources carrying capacity, mathematical equations are constructed to model the development of various factors, and then are transformed through the different variables into a quantitative model for the calculation of water resource carrying capacity, namely, "the largest supporting scale". Some main methods such as conventional method of trend, the system dynamics method and multi – objective analysis and so on are applied. This type of method has made some breakthrough in exploring the interaction between various factors, but the theoretical basis should be further expanded without the support of water cycle theory. Therefore, the circulation and transformation rule of water resources in nature and society should be further analyzed and water resources should also be regarded as a link throughout the research on water resources carrying capacity, and then the interaction between water resources, socio – economic and the ecological and environmental factors in the complex systems of water resources carrying capacity can be ultimately ascertained in essence.

To solve the above problems, Professor Zuo Qiting proposed a simulation – and optimization – based control object inversion model (referred to as the COIM model method) to make up for the drawbacks of the above two methods. Taking water resource systems, social economic systems, ecosystems, mutual constraints (stimulation) model as base models, maintaining a healthy ecological system as control constraints, supporting the largest socio – economic scale as optimization goal, the method establishes the optimization model. "The maximum socio – economic scale" which is achieved through optimizing models solution (or control object inversion model) is water resources carrying capacity. In the COIM model, the mutual restraint relations have been reflected in the water resources system, socio – economic system, ecosystem complexity. And a healthy ecosystem that water resources carrying capacity requires is regarded as a constraint condition in the model. The calculation results of the water resources carrying capacity can be obtained through optimizing the model solution or control object inversion model.

3.2 The calculation framework and techniques of water resources dynamic carrying capacity

The calculation of water resources dynamic carrying capacity starts from analyzing response mechanism of surface water resources, groundwater resources in the river basins or regions and then stimulates and predicts the available water supply in the future years. Meanwhile, the volume of water demand in the coming years is predicted through hierarchical decomposition from river basins to water users by means of the methods of trend extrapolation, regression analysis and quotas, then using system dynamics method, water rationing and use relationship between various water supply projects and water users are constructed to study the future satisfaction degree between water supply and demand and reveal the relations of water resources in terms of "supply – usc – consumption – dewatering". The mechanism of interaction between climate change, water cycle process, water conservancy construction, socio – economic development and utilization of water resources is explored and the composite and interaction model of "water resources – social economy – ecological environment" under the climate change is constructed. Finally, based on the large scale system theory, the intrinsic link and the interaction between climate systems, water systems, socio – economic system and the ecological environment system in the watershed or regions are constructed, and then the quantitative mode of water resources dynamic carrying capacity is therefore built with multi – objective optimization ideas, "social and economic sustainable development virtuous circle and the ecological environment" as the optimization target, the composite interaction model of "water resources – social economy – ecological environment" under the climate change as the base model, while taking other constraints into consideration.

The calculation framework diagram of water resources dynamic carrying capacity under climate change (see Fig. 1) displays the basic ideas of quantitative research methods of water resources dynamic carrying capacity, and its main contents include:

(1) With the maximum socio – economic scale as the objective function, an optimization model is established, with the following equations as constrain conditions, that is, the atmospheric circulation relationship equation, the relationship equation of water resources circulation and transformation, the relationship equation of the pollutant circulation and transformation, the mutual con-

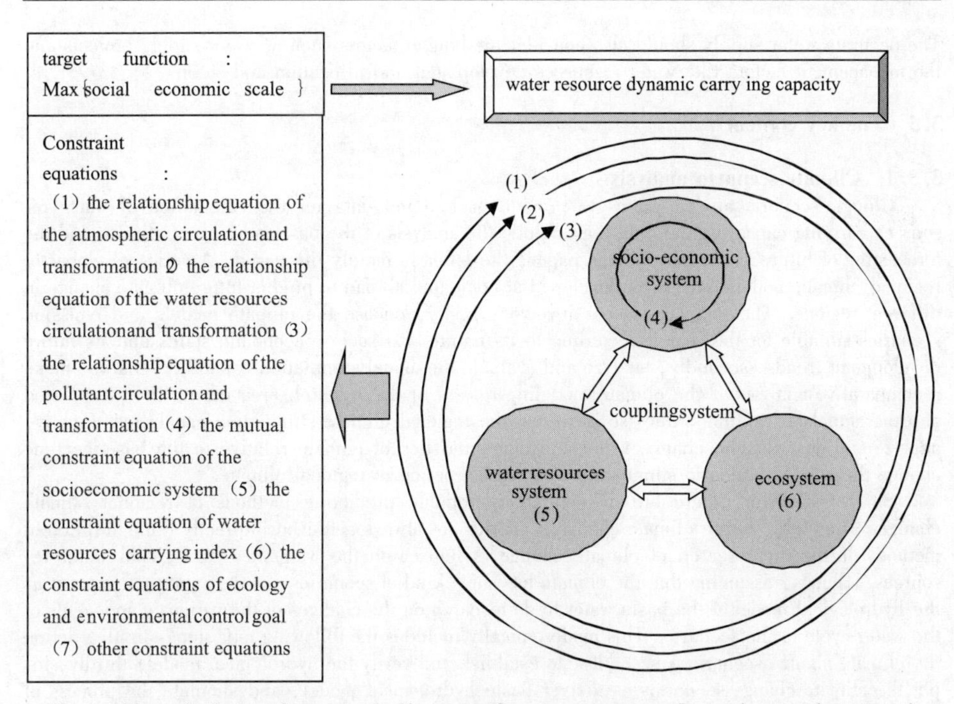

Fig. 1 The calculation framework diagram of water resources dynamic carrying capacity under climate change (remoulded according to the calculation framework diagram of water resources dynamic carrying capacity)

straint equation of the socioeconomic system, the constraint equation of water resources carrying index, the constraint equations of ecology and environmental control goal. Through the solution of the optimization model, the objective function value is water resources carrying capacity.

(2) Climate change has an impact on water resources system, socio – economic system, and ecosystem, making the composite system of "water resources—social economy—ecological environment" experience the dynamic changes, meanwhile, making the complexity and mutual restrictive relationship between water resources system, socio – economic systems and ecosystem fully embody.

(3) The water resources carrying capacity and water environment carrying capacity respectively reflect carrying capacity of water circulation system from human activities in both water quantity and quality. Overloading on either aspect will cause the unsustainable results of the whole system such as ecological water scarcity, groundwater super – mining or water and environmental degradation. Therefore, both of them are put into the same theoretical framework for sake of the sustainability of water resources.

(4) The climate system consists of atmosphere, oceans, land surface, snow – covered layer and biosphere. Solar radiation is the main energy of the system. Influenced by the solar radiation, a series of complex process will produce within the climate system. Through the exchange of material and energy exchange, the various components are closely linked to form an open system. Research on the temporal and spatial variation process of water resources under climate change is the basis for the calculation of water resources dynamic carrying capacity. These changes mainly include the quantitative and qualitative changes of components, structure, function and nature in terms of water resources system. Water resources balance model is often usually used to predict the dynamic changes of water resources in the climate change. The model reflects the impact of climate change on river runoff, mainly through changes in temperature and rainfall. The common water balance model is year water balance model. However, these models only reflect the natural cycle of water resources, not fully reflecting the true situation of the water supply as an economic resource.

The accurate water supply should also consider the human factors such as water supply projects and the management factors like water resources development and utilization and so on.

3.3 The key content

3.3.1 Climate scenario analysis

Climate scenario analysis aims to provide background data for the calculation of water resources carrying capacity, the data that include the analysis of the past and current climate and the forecasting of future scenarios. In this paper, the latter is mainly discussed. A series of global or regional climate models have been developed at home and abroad to predict future climate change in different regions. The general process involves: firstly, choose the climate models and emission scenarios suitable for the region according to its nature and socio – economic status and its future development trends、secondly, analyze and evaluate the model stimulation capability, that is, make a comparative analysis of the climate modeling results in the research area during the past period and the actual observation values so as to test the regional climate stimulation capability. Finally, analyze and describe the change value and characteristics of climate relative to the baseline time change through the scenario stimulation of a certain period of regional climate.

So far, however, there are no completely reliable predicting methods of regional climate change. Therefore, future climate change scenarios are always achieved indirectly through different methods during the research of climate change's impact on the watershed or regional water resources. That is, assuming that the climate has some kind of scenarios change, and then input it as the hydrological model of the basin water to do research on the changes of the various components of the water cycle in the scenario. This mode typically includes the following four steps: firstly, define the climate change scenarios; secondly, to establish and verify the hydrological model; thirdly, input the climate change scenarios as a river basin hydrological model, and simulate the process of change of the regional water cycle; finally, evaluate the impact of climate change on hydrology and water resources through stimulation results of the river basin hydrological model.

3.3.2 Stimulation prediction of change process of water resources system under climate change

The model stimulation, an effective means to predict the impact of climate change on water resources systems, can be divided into empirical regression model stimulation and process model stimulation, both providing a powerful tool support for the process stimulation of water resources system. The empirical regression model predicts the change process of the future water system based on the experience regression relationship of "runoff – temperature – precipitation". Under the simplified conditions of climate change, this model thus has some limitations in the practical applications, not considering that water resources are continuously updated natural resources and water resources systems have the natural – artificial dualistic water cycle mode, and ignoring the function change of water resources system and the lagging impact of runoff on climate change. Process model are involved in the transformation process of the water cycle, and space and time characteristics determined by these processes, therefore, it has a wider range of applications. However, due to the complexity of the process model structure and required larger parameter and other reasons, the model is often subject to the constraints of the lack of background information in practical applications.

Because socio – economic process is influenced by human factors, stimulation prediction of water resources systems changes displays a lot of randomness and uncertainty, currently there are relatively fewer studies at home and abroad of socio – economic changes under climate change. Only the agricultural sector in the socio – economic sectors is affected greatest by natural environmental factors, especially climatic factors, and it is also the most sensitive to climate change, and thus the stimulation of the process of agriculture under climate change is relatively easy. Therefore, the objective stimulation of change progress of socio – economic system under the climate change will be the next focus of research.

4 Conclusions and outlook

Although there is a lot of research topics and discussions on water resources carrying capacity in China, the study of water resources dynamic carrying capacity under climate change, on the whole, is still a new task. Due to global climate change and increasing interferences of human activities, the response of hydrographic elements in the basin water to human activities and climate change are becoming increasingly prominent.

A series of changes such as rainfall, evaporation, infiltration, soil moisture, river runoff, underground runoff, which are caused by the climate changes in atmospheric circulation, increased evaporation, snow and ice conditions, and then come the change of the status quo of the global hydrological cycle and redistribution of water resources in time and space. Therefore, water resources system has become one of the most important areas affected by the global impacts of climate change. Subject to the impact of global climate change and human activities, and future water resources in China will face greater uncertainty, thus the sustainable use and management of water resources is confronted with new challenges. Water resources dynamic carrying capacity under climate change will inevitably involve many aspects of atmospheric circulation, water resources, and socio – economic, ecological environment. So it can be directly seen that the basic theory of water resources dynamic carrying capacity under climate change necessarily are connected with the respective related theories, and interacts with climate system, water system, socio – economic systems and ecological environment system. Therefore, the research on water resources dynamic carrying capacity under climate change is of great significance to the positive response to global climate change, scientifically working out development and utilization program of river basin and regional water resources, laying down plans and policies of watershed and regional industrial structure and industrial development, safeguarding sustainable use of water resources and sustainable development of the national economy.

References

Qian Zhengying, Zhang Guangdou. The Sustainable Development of Water Resources Strategy Research in China[M]. Beijing: China Water Power Press, 2001.

Zhang Jianyun, Wang Guoqing, Liu Jiufu, et al. Review on Worldwide Studies for impact of climate change on water[J]. Yangtze River, 2009, 40(8): 39 – 41.

Xia Jun, Zhang Yongyong, Wang Zhonggen. Water carrying capacity of urbanized area[J]. Journal of Hydraulic Engineering, 2006, 37(12): 1482 – 1488.

Zuo Qiting. City Water Resources Carrying Capacity—Theory · Methods and Applications[M]. Beijing: Chemical Industry Press, 2005.

Zuo Qiting. Relationship between Carrying Capacity and Optimal Deployment of Water Resources [J]. Journal of Hydraulic Engineering, 2005, 36(11): 1286 – 1291.

Zuo Qiting, Ma Junxia, Gao Chuanchang. Study on carrying capacity of urban water environmen [J]. Advances in Water Science, 2005, 16(1): 103 – 108.

Li Jianqiang, Yang Xiaohua, Lu Guihua, et al. Fuzzy Matter – element Model Based on Improved Membership Degree for Comprehensive Assessment of Water Resources Carrying Capacity in River Basins[J]. Journal of Hydroelectric Engineering, 2009, 28(1): 78 – 83.

Cheng Guodong. Evolution of the Concept of Carrying Capacity and the Analysis Framework of Water Resources Carrying Capacity in Northwest of China[J]. Journal of Glaciology and Geocryology, 2002, 24(4): 361 – 367.

Zhang Jianyun, Wang Guoqing. Climate Change Impacts on Hydrology and Water Resources research[M], Beijing: Science Press, 2007.

Drought Trend Assessment for the Middle Reaches of Yellow River Based on Standard Precipitation Index under Climate Change

Liu Yanli[1,2], ***Wang Guoqing***[1,2], ***Gu Ying***[1,2], ***Zhang Jianyun***[1,2], ***Liu Jiufu***[1,2], ***Jin Junliang***[1,2] and ***Liu Jingnan***[1]

1. Nanjing Hydraulic Research Institute, State Key Laboratory of Hydrology – Water Resources and Hydraulic Engineering, Nanjing, 210098, China
2. Research Center for Climate Change of Ministry of Water Resources, Nanjing, 210098, China

Abstract: The middle reaches of Yellow River is one of districts liable to severe drought in China, while the drought trend has become increasing recently as global climate change intensifying. In this study, standard precipitation index (SPI) was employed as drought index, and an improved SPI approach was proposed for better drought representation. The SPI degrees of ten – day series were calculated for Shanxi province located in the middle Yellow River for 40 years from 1970 to 2009. The results show the improved SPI could represent the historical drought situation of this district in a better manner. Allowing for the future drought trends impacted by climate change, the current widely used climate scenarios all over the world—four GCM (CSIRO, NCAR, MPI, PRECIS) projections during the 21st century under four IPCC SRES emission scenarios (A1B, A2, B1, B2)—were imported. The Delta change approach was employed to correct the systematic biases in GCM projections. The drought situations of Shanxi province during 2021 to 2050 were analyzed under future climate scenarios. Thus they were compared with the historical period from 1971 to 1999 in terms of inter – annual variation of drought level. The historical drought formation, development regularity and possible future situation were deduced. The results show under future climate change the drought would show the possibility of more critical against water supply and agriculture production. In view of the uncertainty of future climate change, though not very high possibility for more serious drought, it still should be paid more attention by water resources management.

Key words: drought, climate change, standard precipitation index (SPI), uncertainty, delta change method, the middle reaches of Yellow River

1 Introduction

The middle reaches of Yellow River locates in the arid and semiarid district in China, where is subject to drought disaster. Since 1969, it has shown obvious increasing temperature and decreasing precipitation. Moreover the water consumption increasing, the observed runoff has decreased by 40% in the middle reaches of Yellow River. Most of Shanxi Province lies in the region of the middle reaches of Yellow River. The annual precipitation is obviously lower than the average of China. Aggravated by uneven distribution, the drought frequently occurred. Drought is one of the main meteorological disasters restricting agriculture production of Shanxi province. According to the statistics over the last 40 yrs, the drought of Shanxi province showed long – period, more – influenced, large – area, heavy – disaster, especially affected agriculture critically.

Global environment change charactering by global warming has been more concerned by public recently, which has influenced drought trend of the middle reaches of Yellow River. Under global warming, drought level and distribution could change in future, which may bring new challenges to local water resources sustainable management. Thus it is necessary and urgent to study the possible responses of local drought tendency to future climate change. So far it mainly focuses on the historical drought and characteristics evolution, and less on possible drought trend in future climate change.

Meteorological drought earlier happens than agriculture drought, hydrological drought, socio-economic drought and et al. It is the direct incentive to agriculture drought. Standard precipitation index (SPI) is one of the widely used indexes, which could flexibly reflect precipitation reduction and the source of drought in terms of any time scale. Moreover with the advantages of high space-time suitability, easiness to gather data and simple calculation, it has been widely used all over the world. But the traditional SPI did not allow for the potential effect of antecedent precipitation, which reflects current drought situation and lacks the representational capacity of drought condition for successive time scale.

This study chose the region of the middle reaches of Yellow River located in Shanxi province, which focused on drought issues. Taking antecedent precipitation into account, it proposed a novel approach of monthly SPI, in terms of ten days accumulated scale. It would calculate the SPI for study area and analyze the validity of proposed SPI, thus discuss the evolution characteristic of historical drought and future trend.

2 Data and methodology

2.1 A study case

Shanxi Province locates in the west of north China, the middle reaches of Yellow River, east of Loess Plateau. The territory of Shanxi province is around 156,300 km^2, whereas the range is from 110°14′ E to 114°33′ E, and 34°34′ N ~ 40°43′ N. In this study the study area includes Lvliang City, Taiyuan City and some counties (cities) of Jinzhong City. It consists of 24 counties, and is around 35, 600 km^2. There are 106 rain gauges addressed situated in the study area (Fig. 1).

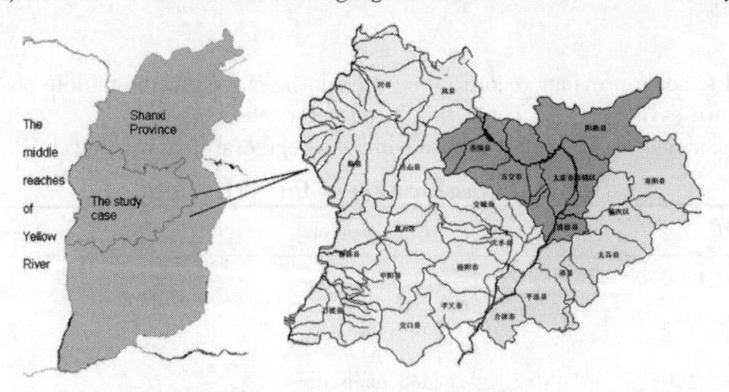

Fig. 1 Schematic of study area

2.2 Ten-day accumulated SPI and calculation methodology

SPI is designed for measuring precipitation scarcity for multiple time scales, which is one of the indexes reflecting the probability of precipitation happened in some period. SPI calculation is based on the long-period rainfall record series for any district. The long period rainfall series is usually fitted by a specific probability distribution, and then is transformed into normal distribution on condition of average zero for a specific period. Positive value of SPI indicates the current precipitation is much more than average (wet), while negative value indicates less than average (drought).

SPI addresses gamma distribution to describe rainfall variation. Rainfall of partial probability distribution is normalized, thus standardized cumulative distribution function is employed to rank drought severity degree. Since gamma distribution is complicated for SPI calculation, this study utilizes a simplified algorithm that empirical cumulative probability of rainfall series from calculation interval to current period is adjusted into standard normal distribution. Thus SPI is standard deviation of current rainfall lower or higher than average. The detailed steps are as follows:

Step 1—basic scheme: collecting ten days rainfall series from beginning of calculation interval to end as basic data.

Step 2—Empirical probability calculation:

$$H = q + (1 + q) \times m/(L + 1) \tag{1}$$

Where, P ($P > 0$) is current rainfall (current interval rainfall for current year); n is sample size; q is the ratio of zero annual rainfall; the non - zero values are sorted ascending, L is the total number; m is the sequence of current rainfall.

Step 3—t calculation:

If $0 < H \leqslant 0.5$

$$t = \sqrt{-\ln H^2} \tag{2}$$

If $0.5 < H < 1$

$$t = \sqrt{-\ln(1 - H)^2} \tag{3}$$

Step 4—SPI calculation:

$$\text{SPI} = \text{sign} \times [t - F/G] \tag{4}$$

where

$$F = c_0 + c_1 \times t + c_2 \times t^2 \tag{5}$$

$$G = 1 + d_1 \times t + d_2 \times t^2 + d_2 \times t^3 \tag{6}$$

If $0 < H \leqslant 0.5$, sign $= -1$; If $0.5 < H < 1$, sign $= 1$.

The constants of c_0, c_1, c_2, d_1, d_2, d_3 take values as follows: $c_0 = 2.515,517$, $c_1 = 0.802,853$, $c_2 = 0.010,328$, $d_1 = 1.432,788$, $d_2 = 0.189,269$, $d_3 = 0.001,308$.

If the current rainfall is equal to zero, then it could be put into the middle stage of all zero values:

$$H = q/2 \tag{7}$$

If P is equal to some previous record, then it could be placed in the middle position. If the sequence is up - down asymmetry, it could be placed before the center.

In this study, negative values of SPI and drought degree are shown in Tab. 1.

Tab. 1 Classification of the SPI values

SPI	Classification	Degree
< -2.0	Extremely dry	4
-2.0 ~ -1.5	Severely dry	3
-1.5 ~ -1.0	Moderately dry	2
-1.0 ~ -0.5	Slightly dry	1
> -0.5	Normal	0

In the traditional SPI calculation, it just addressed the rainfall of current time scale, without considering antecedent precipitation. In the processing of drought analysis, the authors found that the antecedent precipitation may affect the current drought situation to some extent. In order to assess historical and current drought, even predict future drought development, this study introduces an approach of accumulated ten days for monthly SPI computation:

(1) Current monthly SPI: forward from current ten - day to two previous ten - day.

(2) Recording the above monthly SPI into ten - day series as current ten - day SPI.

Taking the first ten - day of February, 1970, forwarding two ten - day (the third ten - day of January and 1970. 1. 2), and the three ten - day is considered as the artificial month. SPI of this artificial month is taken as the SPI of current ten - day (the first ten - day of February, 1970), which is recognized allowing for the antecedent precipitation.

2.3 Drought assessing index

Drought frequency (D_p) and drought station ratio (D_s) were introduced to reflect the drought degree of study area, and to explore the drought space – time distribution.

2.3.1 Drought frequency (D_p)

It was utilized to assess historical drought frequent degree (for time distribution), and could be expressed as:

$$D_P = (n/N) \times 100\% \tag{8}$$

where, N is the total year number of meteorological data; n is the drought year.

2.3.2 Drought station ratio (D_s)

It was addressed to show the drought of space distribution, and could be calculated as:

$$D_s = (m/M) \times 100\% \tag{9}$$

where, M is the total of meteorological stations; m is the drought stations; s is the order of stations.

2.4 Future climate scenarios

This study allows for the widely used ten climate scenarios: four climate models, CSIRO—Commonwealth Scientific and Industrial Research Organization, NCAR—National Center for Atmospheric Research, MPI_ECHAM5—Max Planck Institute of Germany, PRECIS—Hadley Center of UK; four emission styles, low emission of B1 and B2, moderate emission of A1B and high emission of A2. While A scenario emphasizes on economic growth, and B scenario focuses on global environment when allowing for economic growth. 1 places emphasis on global scale, and 2 on regional scale of economy, society, environmental sustainable schemes. The applied scenarios in this study have been downscaled aiming at regions of China, which were listed in Tab. 2.

Tab. 2 Future climate scenarios

Emission	MPI_ECHAM5	PRECIS	NCAR	CSIRO
B1			√	√
B2		√		
A1B	√	√	√	√
A2		√	√	√

3 Results and discussion

3.1 Inter – annual variation analysis of historical drought

3.1.1 SPI performance for observed drought

According to drought data statistics of Shanxi province in 1990 ~ 2007, the year of 1999 and 2006 showed severe drought. In 1999 drought happened in the whole study area over the year, while in 2006 periodical drought occurred in the whole study area. The drought situation of each counties and corresponding SPI degree were listed in Tab. 3 (the SPI degree was the average of whole year). It could be seen in 1999, all SPI are bigger than 0.6, and basically increasing with drought aggravating (e.g., for Lanxian, the drought ratio was 100.00%, while the SPI reached 1.14). But it is not the strict direct proportion between SPI and drought ratio. For example, Taigu, drought ratio was 88.80%, and SPI was 0.67; in contrast with Yangqu, drought ratio was 29.19%, but SPI was higher as 0.69. Tab. 3 also listed the drought situation, period (months) and SPI to analyze the performance of SPI to describe drought in 2006, which indicated that the SPI basically reflect the seasonal drought. In general, the proposed SPI could be employed to study the drought of study area.

Tab. 3 Observed drought and SPI

City& County	Drought period (drought ratio)_1999	SPI degree _1999	Drought period (drought ratio)_2006	SPI degree _2006
Taiyuan	Whole year(52.09)	0.89	Mar. - Sep. (57.48)	0.38
Gujiao	Whole year(100.00)	0.83	Mar. - Sep. (100.00)	0.52
Qingxu	Whole year(100.00)	0.92	Mar. - Sep. (89.78)	0.33
Yangqu	Whole year(29.19)	0.69	Mar. - Sep. (32.18)	0.48
Loufan	Whole year(84.43)	1.06	Mar. - Sep. (100.00)	0.62
Yuci	Whole year(45.17)	0.69	Apr. - Jul. (35.04)	0.58
Jiexiu	Whole year(53.38)	0.92	Apr. - Jul. (15.97)	0.17
Shouyang	Whole year(98.58)	0.83	Apr. - Jul. (39.12)	0.58
Taigu	Whole year(88.80)	0.67	Apr. - Jul. (46.36)	0.42
Qixian	Whole year(14.15)	0.78	Apr. - Jul. (76.56)	0.25
Pingyao	Whole year(57.64)	0.83	Apr. - Jul. (10.96)	0.42
Lishi	Whole year(87.67)	0.94	Apr. - Sep. (64.87)	0.22
Xiaoyi	Whole year(78.32)	0.78	Apr. - Sep. (56.75)	0.28
Fenyang	Whole year(87.73)	0.97	Apr. - Sep. (68.89)	0.33
Wenshui	Whole year(96.99)	0.81	Apr. - Sep. (55.67)	0.39
Zhongyang	Whole year(61.60)	0.92	Apr. - Sep. (56.77)	0.22
Xingxian	Whole year(100.00)	1.03	Apr. - Sep. (71.89)	0.67
Linxian	Whole year(91.54)	0.97	Apr. - Sep. (58.23)	0.33
Fangshan	Whole year(100.00)	1.06	Apr. - Sep. (95.02)	0.28
Liulin	Whole year(67.79)	0.81	Apr. - Sep. (58.16)	0.33
Lanxian	Whole year(100.00)	1.14	Apr. - Sep. (78.44)	0.39
Jiaokou	Whole year(48.89)	0.78	Apr. - Sep. (70.51)	0.50
Jiaocheng	Whole year(82.77)	0.92	Apr. - Sep. (68.68)	0.44
Shilou	Whole year(100.00)	0.75	Apr. - Sep. (99.85)	0.44

3.1.2 Drought and SPI of study area in last 40yrs

Ten - day scale was used in SPI analysis, while the highest degree SPI of all ten - day in one year was considered as the drought level in this year (same as below). The SPI degrees of 40 years from 1970 to 2009 for whole study area were described in Fig. 2. All SPI of 1970 to 2007 were equal or higher than level 1 (slightly dry). If level 1 was defined as drought, thus the drought probability reached up to 100%. While severely dry probability was 60% and 2.5% (one time for level 4) for extremely dry. For the drought occurring period statistics, in 1999, the probability of more than level 1(including level 1) was around 44.4% , and more than level 3 was 11.1%. In 2006, the above two types of probability were separately 33.3% and 2.8%.

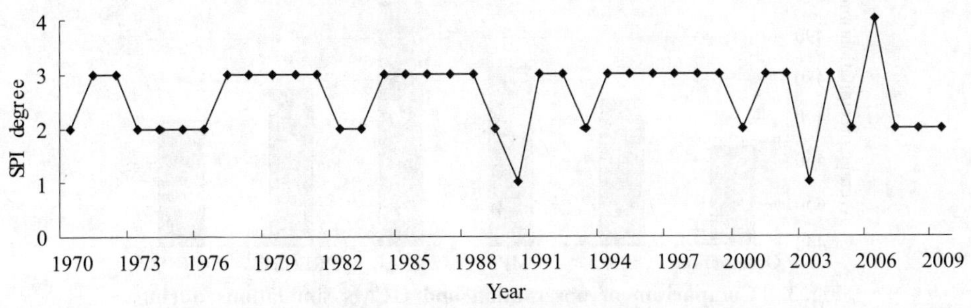

Fig. 2 SPI degree of study area from 1970 to 2009

In view of the general drought in last 40 years, Xingxian experienced the severest drought, while the average annual drought degree reached 0.52. The extremely dry happened 4 times, and the probability was 0.3%; the severely dry above happened 86 times, and the probability was 6.0%; the slightly dry above happened 436 times, and the probability was 30.3%. For the more serious drought ten - day, in the third ten - day of February in 1999, there were 9 counties occurred extremely dry, and the probability came up to 25%. The SPI degree average of all 24 counties were more than 3, thus severely dry probability reached up to 100%. In the third ten - day of March in 2006, the whole study area showed extremely dry, among 24 counties there were 18 counties' SPI experienced extremely dry, and the probability was 75%. The same as the above ten-day in 1999, the SPI degree average of all 24 counties were more than 3, and the probability reached up to 100%.

3.2 Future climate

3.2.1 Delta change method

It is recognized that great uncertainties exist in GCMs and downscaling processes, leading to inevitable biases for regional climate. The widely used method for dealing with these biases is delta change method, which, to some extent, could eliminate simulation biases of future climate because of global climate models' systematic deviation. The algorithm is described as follows. First, the deviation series (average annual series is in common use) between future and baseline period of simulated climate is derived, which then is added to observed series of baseline period to produce the modified future climate series. It could be expressed as Eq. (10):

$$X_{future,TD} = X_{baseline,TD} + (\overline{XS}_{future,TD} - \overline{XS}_{baseline,TD}) \quad (10)$$

where, Eq. (1) shows the algorithm in terms of ten-day series; $X_{future,TD}$ is the demanding modified future climate series; $X_{baseline,TD}$ is the observed series of baseline period; $\overline{XS}_{future,TD}$ is the average annual of projected future climate; $\overline{XS}_{baseline,TD}$ is the average annual of observed baseline climate.

3.2.2 Future climate processing

Fig. 3 shows climate comparisons between four GCMs (CSIRO, MPI, and NCAR) in baseline period and observation. The precipitations of all three GCMs are projected greater than observation, and the projections of PRECIS are relatively close to observation for average annual value. It could be seen that gaps exist between projected and observed data, which may indicate systematic residuals in future projections. For different GCM, the residuals are different. However, for the same GCM, the residual may keep consistent in baseline period and future climate, which demonstrates that it could be processed by some approach. Therefore, the delta change method was employed to tackle the systematic residuals. For the future period of 2021 ~ 2050, Fig. 4 indicates the ten-day rainfall variation for each scenario in contrast with that of observation. These variations were introduced in the processing to generate the modified future climate scenarios by means of delta change method.

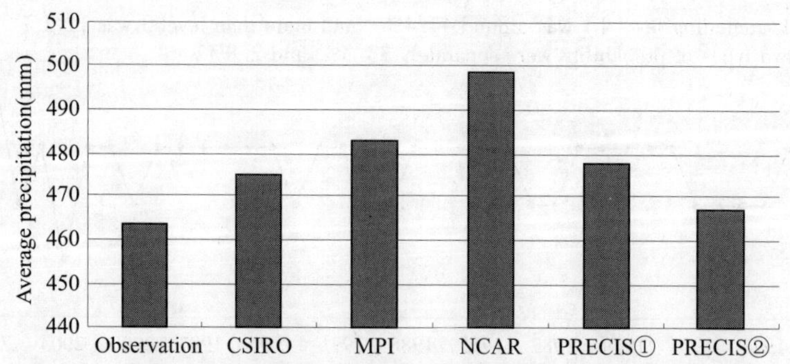

Fig. 3 Comparison of observation and GCMs simulations during the baseline period (1970 ~ 1999)

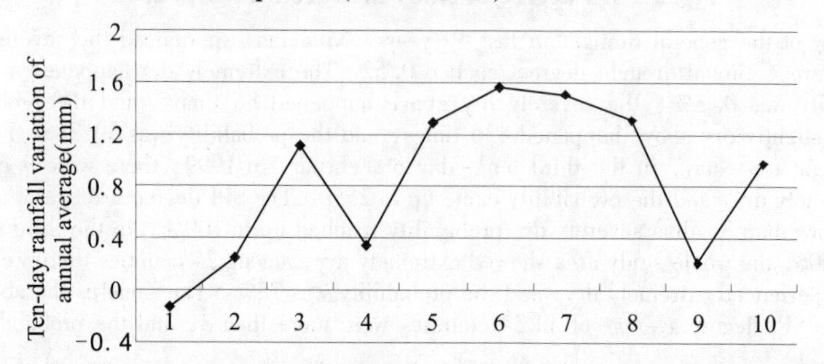

Fig. 4 The precipitation comparison between observation and GCMs in future period (2021 ~ 2050)

3.3 Future drought trend analysis

It could be seen from Fig. 3 that the two deviations of PRECIS are relatively small compared with observation in baseline period. Allowing for the errors of GCMs projections for the baseline period and similarly for the future climate, the three scenarios (A1B, A2, B2) of PRECIS were chosen to discussion the future drought situation to eliminate the potential big projected errors. Fig. 4 exhibits that only the scenario of MPI – A1B shows decreasing precipitation, which is subject to drought. Therefore the MPI – A1B was employed for future drought trend prediction as well.

Fig. 5 describes the comparisons between drought under four future scenarios during the period of 2021 ~ 2050 and the one of 1970 ~ 1999. It could conclude that the general of drought situation in the future period would remain almost the same, with the probability of more than level 1 reaches up to 100%. According to statistics, during the period from 1970 ~ 1999, there were 20 times for severe drought (level 3) with 66.7% of probability. While during the period from 2021 ~ 2050, under the scenario of MPI – A1B severe drought happened 26 times with 86.7% of probability; under PRECIS – A1B it happened only 9 times and 30% probability; under PRECIS – A2 and PRECIS – B2 they were the same 10 times and 33.3%. Compared with the baseline period of 1970 ~ 1999, the severe drought under MPI – A1B possess the risk of increasing trend. For moderate drought, during the period from 1970 ~ 1999, it appeared for 29 times and the probability was as high as 96.7%. During the period of 2021 ~ 2050, under MPI – A1B, it reached up to 100% with all 30 times; under PRECIS – A1B they were 23 times and 76.7%; under PRECIS – A2 they were 25 times and 83.3%; under PRECIS – B2 they were 26 times and 86.7%. It could be concluded that the moderate drought probability of occurrence in 2021 ~ 2050 would correspond largely to the observed drought in baseline period.

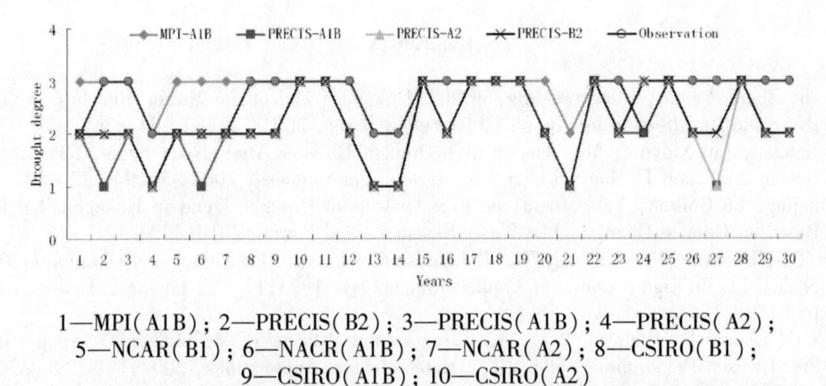

1—MPI(A1B); 2—PRECIS(B2); 3—PRECIS(A1B); 4—PRECIS(A2);
5—NCAR(B1); 6—NACR(A1B); 7—NCAR(A2); 8—CSIRO(B1);
9—CSIRO(A1B); 10—CSIRO(A2)

Fig. 5 Comparison analysis between drought under four future scenarios during the period of 2021 ~ 2050 and the one of 1970 ~ 1999

Given all that, it may be inferred that the future drought condition would be comparable with the one of baseline period, whereas the one under MPI – A1B would be more serious. Though in Fig. 4, in the total of ten scenarios only one (MPI – A1B) shows precipitation declining and might bring more drought. Due to the huge uncertainty of future projections from GCMs, the possible worse drought should be paid enough importance.

4 Conclusions

It is found that the current drought condition was not only controlled by the rainfall of current period, but affected by antecedent precipitation. An approach allowing for effect of antecedent precipitation was presented in this study, in terms of a month scale SPI based on accumulated ten – day. The SPI degree was executed on the relatively dry region of Shanxi province located in the middle reaches of Yellow River for last 40 years (1971 ~ 2009). The results show the proposed SPI could perform well in representing the observed drought situation of study area.

Global climate change has caused certain variation of water cycle, which could affect the future drought condition. This study absorbed the widely used four GCMs and four possible emission styles to investigate the future climate scenarios of study area. MPI – A1B, PRECIS – A1B, PRECIS – A2 and PRECIS – B2 were chosen to analyze the future drought trend of 2021 ~ 2050. The results indicate under MPI-A1B the drought may increasing while basically unchanged under other nine scenarios, compared with that of baseline period (1971 ~ 1999). But it is recognized that big deviations and large uncertainty of GCMs, therefore the projected drought trends from most scenarios may not very reliable. The potential worse drought predict should be aware especially for future water security and agriculture management.

It is worth noting that it focuses on the improved SPI and its application. Since only future projected precipitation was considered, the drought tendency prediction is somehow limited. The temperature may play a comparable role in drought processing, which should be addressed into the integrated analysis of future drought in the further study.

Acknowledgements

The study is financially supported by the Major State Basic Research Development Program of China (973 Program) (No. 2010CB951103), the National Natural Science Foundation of China (No. 51009094), the Nonprofit Research Project of Ministry of Water Resources (No. 201001042), the International Science & Technology Cooperation Program of China (No. 2010DFA24330), the ACCC project funded by DFID, SDC and DECC. The scenario data were provided by Dr. LvLiu LIU (China Meteorological Administration) and Prof. Xu Yin – long (Chinese Academy of Agricultural Sciences). They are gratefully acknowledged.

References

Wang Qi, Zhang Yamin, Kang Lingling, et al.. Drought Trend of the Middle Reaches of Yellow River and its Affection for Runoff [J]. Yellow River, 2004, 26(8): 34-36.

Wang Jichuan, Liu Xiuying, Mao Yu, et al.. Drought Disaster Analysis for Shan Xi Province in Recent 23 Years[J]. Journal of Shanxi Agricultural Sciences, 2008, 36(9): 12-14.

Hao Jingjing, Lu Guihua, Yan Guixia, et al.. Analysis of Drought Trend in Huanghuaihai Plain Based on Climate Change [J]. Water Resources and Power. 2010(11): 12-14.

McKee T B, Doesken N J, Kleist J. The Relationship of Drought Frequency and Duration to Times Scales[J]. In Eighth Conf. on Applied Climatology, 1993, 17-22 January, Anaheim, CA, pp. 179-184.

Patel N R, Chopra P, Dadhwal V K. Analyzing Spatial Patterns of Meteorological Drought using Standardized Precipitation Index [J]. Meteorological Applications, 2007, 14: 329-336.

Vicente-Serrano M. Spatial and Temporal Analysis of Droughts in the Iberian Peninsula (1910-2000) [J]. Hydrological Sciences Journal, 2006, 51(1): 83-97.

Min S K, Kwon W T, Park E H, et al.. Spatial and Temporal Comparisons of Droughts over Korea with East Asia [J]. International Journal of Climatology, 2003, 23: 223-233.

Li W, Fu R, Juárez R, et al.. Observed Change of the Standardized Precipitation Index, Its Potential Cause and Implications to Future Climate Change in the Amazon Region, Phil. Trans. R. Soc. B. 2008, 363: 1767-1772.

Yuan Yun, Li Dongliang, An Di. Winter Aridity Division in China Based on Standardized Precipitation Index and Circulation Characteristics[J]. Journal of Desert Research, 2010, 30(4): 917-925.

Ma Guofei, Zhang Xiaoyu, Duan Xiaofeng, et al.. Analysis of Drought Evolvement Characteristics Based on Standardized Precipitation Index (SPI) in the Mountain Area of Ningxia[J]. Acta Agriculturae Boreali-Occidentalis Sinica, 2010, 19(10): 101-106.

The people's kepublic of China State Administration of Quality Supervision, Inspeltion and Quarantine, China National Strandardization Management Committee. GB/T 20481—2006 Classification of Meteorological Drought[S]. Beijing: China Standard Press, 2006.

Zhang Xueqin, Peng Lili, Lin Zhaohui. Progress on the Projections of Future Climate Change with Various Emission Scenarios[J]. Advances in Earth Science, 2008, 23(2): 174-185.

IPCC. Climate change 2001: The Scientific Basis, Summary for Policy Maker[M]. Cambridge: Cambridge University Press, 2001.

Liu Lvliu, Ren Guoyu. Percentile Statistical Downscaling Method and Its Application in the Correction of GCMs Daily Precipitation in China [J]. Plateau Meteorology, 2012, 31(3): 715-722.

Hay L E, Wilby R L, Leavesley G H. A Comparison of Delta Change and Downscaled GCM Scenarios for three Mountainous Basins in the United States [J]. Journal of the American Water Resources Association, 2000, 36(2): 387-397.

Gao Ge, Chen Deliang, Xu Ying. Impact of Climate Change on Runoff in the Huaihe River Basins [J]. Journal of Applied Meteorological Science, 2008, 19(6): 741-748.

Analysis on Variation of Runoff and Sediment Load of the Comparison Small Watersheds in the Third Sub-Region of Hilly and Gully Loess Region of the Loess Plateau

Zhang Linling, *Zhang Manliang*, *Wang Hong*, *Zhang Haiqiang* and *Liu Xiao*

Tianshui Test Station of Water and Soil Conservation, YRCC, Tianshui, 741000, China

Abstract: Water loss and soil erosion on the Loess Plateau has been a major concern for eco-environment conservation in China; so much attention has been paid to reducing water and soil losses. In order to understand the cause, development, and effectiveness of water and soil conservation measurements in the third sub-region of the hilly Loess Plateau, Tianshui Test Station of Water and Soil Conservation deployed a comprehensive soil erosion measurement system at East Qiaozigou Small Watershed (with comprehensive control) and West Qiaozigou Small Watershed (without comprehensive control) that were subject to the third sub-region of hills and gully area of the Loess Plateau. The measurement system was carried out following the idea of parallel contrast between the two comparison small watersheds. This system is the first comparison measurement system in China aiming at evaluating the effects of rainfall, runoff interception, water conservation and soil sediment measurement. More than 20 years' soil and water conservation observation data has been collected since 1985. By analyzing the characteristics of annual soil erosion with statistical methods, this paper developed the relationships between rainfall and runoff, rainfall and soil erosion, as well as runoff and sediment yield, in addition, researched the runoff-sediment transport process in the single flood event. This paper also gives systematic analysis on the effects of water and soil conservation practices based on Water Conservation Regulation. The paper also discussed overall effects of the measurements on the small watershed areas and soil erosion improvement caused by different water conservation methods. The results of this study will be of great value and will provide important reference to the systematic investigation on soil erosion pattern, and eco-environment effects of water conservation measurements for small watersheds.

Key words: hilly and gully loess region, comparison watersheds, variation of runoff and sediment load

1 Survey of the Qiaozigou Small Watershed

Qiaozigou Small Watershed, which is the integration of the East Qiaozigou Small Watershed and the West Qiaozigou Small Watershed, is located in the north suburb of Tianshui City. It is a pair of adjacent gullies at the left side of the lower reaches of Luoyugou Watershed which stands at the north bank of the Xihe River and belongs to Weihe River drainage system. Qiaozigou Small Watershed belongs to Luoyugou Watershed and consists of two nested small watersheds. Having the area of 2.45 km^2, Qiaozigou Small Watershed was selected as the site of comparison test for study of local water loss and soil erosion by Tianshui Test Station in 1985. Qiaozigou Small Watershed consists of the East Qiaozigou Small Watershed and the West Qiaozigou Small Watershed. The east one is fan-shaped with the area of 1.36 km^2, the gully length of 2.04 km and the average gradient of 16.6%; whereas, the west one is feather-shaped with the area of 1.09 km^2, the gully length of 2.12 km and the average gradient of 16.7%. For comparison study of the water loss and soil erosion in these two small watersheds, 44.8% of the East Qiaozigou Small Watershed was under comprehensive control while only 10.1% of the West Qiaozigou Small Watershed was controlled. Following the idea of parallel comparison between the two small watersheds, in 1987, observation and experimental investigation began to be made at the East Qiaozigou Small Watershed (with comprehensive control) and the West Qiaozigou Small Watershed (without comprehensive control), and four rainfall observation spots, two runoff and sediment observation stations (one in the East and the other in the West) and one meteorological station were built.

2 Analysis of the characteristics of rainfall, runoff and sediment of the comparison watersheds

2.1 Rainfall with uneven annual distribution and great inter-annual variation

The East Qiaozigou Small Watershed has the mean annual rainfall of 527.9 mm (1987 ~ 2010), and the West Qiaozigou Small Watershed, 538.3 mm (1987 ~ 2010). Local rainfall shows no obvious spatial misdistribution because of the small watershed area, but it shows extremely uneven time distribution. Taking the West Qiaozigou as example, the rainfall in the flood season (May ~ October) amounts to 81.89% of the annual rainfall, and the rainfall coming in the period from July to September contributes 47.62% of the annual rainfall. The rainfall in each month of a year differs greatly from each other. The maximum monthly rainfall comes in July in the flood season, amounting to 16.86% of the annual rainfall, followed by August, September and July in descending order, amounting to 16.25%, 14.51%, and 13.64% of the annual rainfall respectively. See Tab. 1 and Fig. 1 for details.

Tab. 1 Monthly rainfall of the East and the West Qiaozigou Small Watersheds

Place	Monthly rainfall (mm)											
	1	2	3	4	5	6	7	8	9	10	11	12
East Qiaozigou Small Watershed	8.7	10.9	26.2	33.0	58.6	72.0	89.0	85.8	76.6	50.3	12.0	4.8
West Qiaozigou Small Watershed	8.3	10.8	24.7	33.3	59.9	75.4	89.5	89.6	76.9	52.5	12.5	4.9

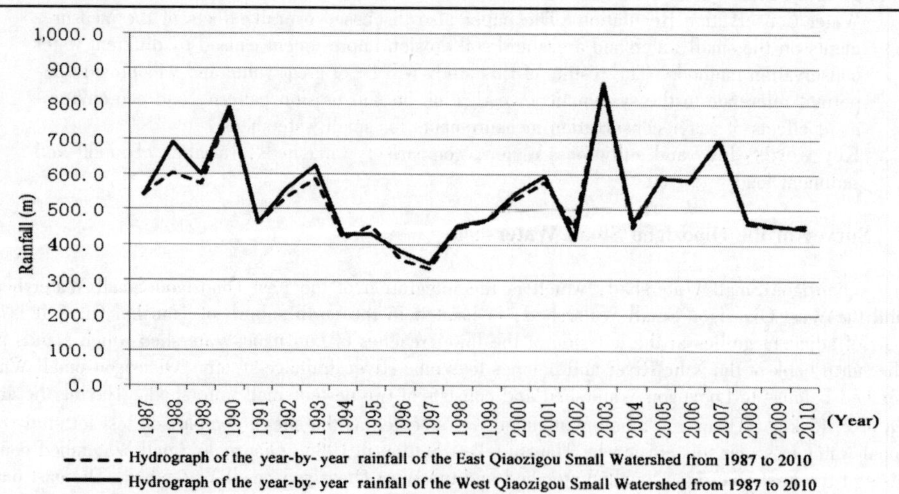

Fig. 1 Hydrograph of the year-by-year rainfall of the East and the West Qiaozigou Small Watersheds

2.2 Study on relation between runoff and rainfall

(1) The East Qiaozigou Small Watershed has the mean annual runoff of 8,900 m^3 and the runoff modulus of 6,471 m^3/km^2. Local runoff mainly comes from rainfall and has the time distribution similar to that of local rainfall. The distribution of annual runoff is extremely uneven, mostly coming the flood season in form of flood. Local mean annual runoff in the flood season is 8,800

m^3, amounting to 98.88% of total annual runoff. The maximum monthly runoff occurs in August, amounting to 37.08% of the annual runoff, followed by July and June in descending order, amounting to 19.10% and 17.98% of the annual runoff respectively. Local runoff also shows great inter-annual variation with coefficient *Cv* value of 1.76. The measured data presented by the gully head station indicated that local maximum runoff was 57,300 m^3 (1988).

(2) The West Qiaozigou Small Watershed has the mean annual runoff of 19,200 m^3 and the runoff modulus of 17,615 m^3/km^2. The distribution of annual runoff is extremely uneven. Local mean annual runoff in the flood season is 19,100 m^3, amounting to 99.48% of total annual runoff. The maximum monthly runoff occurs in August, amounting to 33.33% of the annual runoff, followed by September and June in descending order, amounting to 19.27% and 17.71% of the annual runoff respectively. Local runoff also shows great inter-annual variation with coefficient Cv value of 1.26. The measured data presented by the gully head station indicated that local maximum runoff was 85,500 m^3 (1988).

See Tab. 2 for detailed monthly runoff of the East and the West Qiaozigou Small Watersheds, and see Fig. 2 for the hydrograph of the year-by-year runoff.

Tab. 2 Monthly runoff of the East and the West Qiaozigou Small Watersheds

Place	Monthly runoff ($\times 10^4$ m^3)											
	1	2	3	4	5	6	7	8	9	10	11	12
East Qiaozigou Small Watershed	0	0	0	0.01	0.04	0.16	0.17	0.33	0.08	0.10	0	0
West Qiaozigou Small Watershed	0	0	0	0.01	0.15	0.34	0.29	0.64	0.37	0.12	0	0

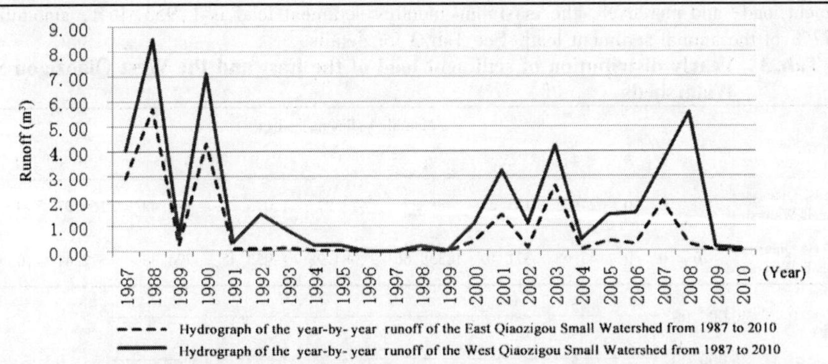

Fig. 2 Hydrograph of the year-by-year runoff of the East and the West Qiaozigou Small Watersheds

It can be seen from Fig. 2 that the later period of the 1980s was a high-water period, the 1990s was a low-water period, and 2000 ~ 2010 is a normal-water period, and moreover, the runoff of the West Qiaozigou Small Watershed is obviously greater than that of the East Qiaozigou Small Watershed.

2.3 Analysis of the influence of rainfall on sediment

2.3.1 Sources of sediment

The sediment of the Qiaozigou Small Watershed is mainly produced by surface soil erosion on slopes and in gullies. The mean annual sediment load of the East Qiaozigou Small Watershed is 2,512.76 t, and the West Qiaozigou Small Watershed, 5,961.46 t. According to the investigation of rainstorm and soil erosion happened on April 19, 1987, the soil erosion on slope surface and in gullies was in the proportion of 74.0% and 26.0%, where the East Qiaozigou Small Watershed held the proportion of 91.5% and 8.5%, and the West Qiaozigou Small Watershed, 56.4% and 43.6%.

The soil erosion modulus of the East Qiaozigou Small Watershed was $1.009,0 \times 10^8$ t/km^2, and the West Qiaozigou Small Watershed, $1.243,0 \times 10^8$ t/km^2.

2.3.2 Characteristics of sediment yield

The underlaying surfaces composed of different substances show great difference in sediment yield, no matter the mode or the course. Gravity erosion on valley-side slopes and channel erosion in gullies feature largely in the regions covered by the loess, and slope surface erosion is common in the inter-rill land. The study area belongs to the third sub-region of the loess hilly and gullied area where the complex soil is dominant, featured by the thick and loose soil layer with vertical development. Being highly water-disintegrable, such soil layer shows poor erosion resistance. Therefore, local gully density is high, accompanied with fragmented surface and steep slopes and high erosion modulus. The rainstorm investigation and the measured data show that the sediment runoff modulus of the Qiaozigou Small Watershed is generally higher than 10,000 t/km^2.

2.3.3 Characteristics of sediment load

The sediment load of the Qiaozigou Small Watershed show centralized yearly distribution but great inter-annual variation. Soil erosion and sediment yield manly happen in the six-month flood season mostly in form of several rainstorms. The sediment load of the East Qiaozigou Small Watershed in the flood season is 2,485.13 t, amounting to 98.90% of the annual sediment load; and the sediment load in the period from July to September is 1,723.66 t, amounting to 68.60% of the annual sediment load; and moreover, the maximum monthly sediment load is 1,075.27 t, amounting to 42.79% of the annual sediment load. The sediment load of the West Qiaozigou Small Watershed in the flood season is 5,915.78 t, amounting to 99.23% of the annual sediment load; and the sediment load in the period from July to September is 3,833.31 t, amounting to 64.30% of the annual sediment load; and moreover, the maximum monthly sediment load is 1,953.45 t, amounting to 32.77% of the annual sediment load. See Tab. 3 for details.

Tab. 3 Yearly distribution of sediment load of the East and the West Qiaozigou Small Watersheds

Place	Monthly sediment load (t)											
	1	2	3	4	5	6	7	8	9	10	11	12
East Qiaozigou Small Watershed	0	0	0	27.65	103.23	488.27	473.97	1,075.27	174.42	169.97	0	0
West Qiaozigou Small Watershed	0	0	0	45.68	371.40	1,338.66	1,084.98	1,953.45	794.88	372.41	0	0

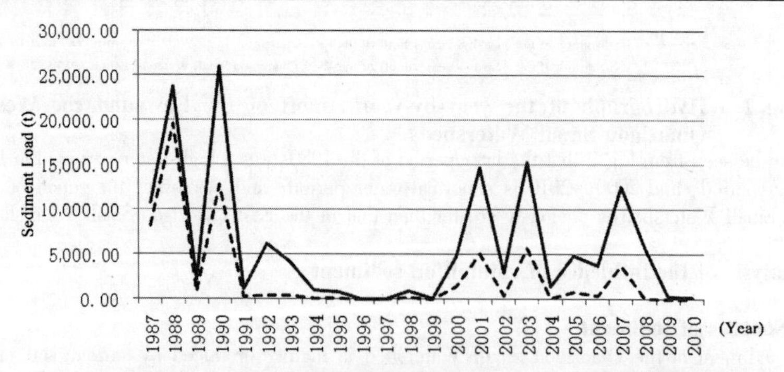

Fig. 3 Hydrograph of the year-by-year sediment load of the East and the West Qiaozigou Small Watersheds

It can be seen from Tab. 3 the maximum monthly sediment load of the East and the West Qiaozigou Small Watersheds appears in August, followed by June and July in descending order, and the minimum monthly sediment load appears in January.

Local sediment load also shows great inter-annual variation. The measured data presented by the gully head station at the East Qiaozigou Small Watershed indicated that local maximum sediment load was 19,602.09 t (1988), and the measured data presented by the gully head station at the West Qiaozigou Small Watershed indicated that local maximum sediment load was 25,969.94 t (1990). See Fig. 3 for the hydrograph of the year-by-year sediment load.

It can be seen from Fig. 3 that the later period of the 1980 s was a period with high sediment yield, the 1990s was a period with low sediment yield, and 2000 ~ 2010 is a period basically with the average value, and moreover, the sediment load of the West Qiaozigou Small Watershed is obviously greater than that of the East Qiaozigou Small Watershed.

2.3.4 Sediment concentration

The mean annual sediment concentration of the East Qiaozigou Small Watershed is 239.5 kg/m^3 and its measured maximum cross-sectional sediment concentration comes to 1,290.0 kg/m^3 (April 20, 1989). The mean annual sediment concentration of the West Qiaozigou Small Watershed is 286.7 kg/m^3 and its measured maximum cross-sectional sediment concentration comes to 1,260.0 kg/m^3. Local flood basically belongs to the water flow with high sediment concentration, which indicates that local sediment is highly centralized. See Tab. 4 for the hydrograph of sediment concentration of the East and the West Qiaozigou Small Watersheds. See Fig. 4 for details.

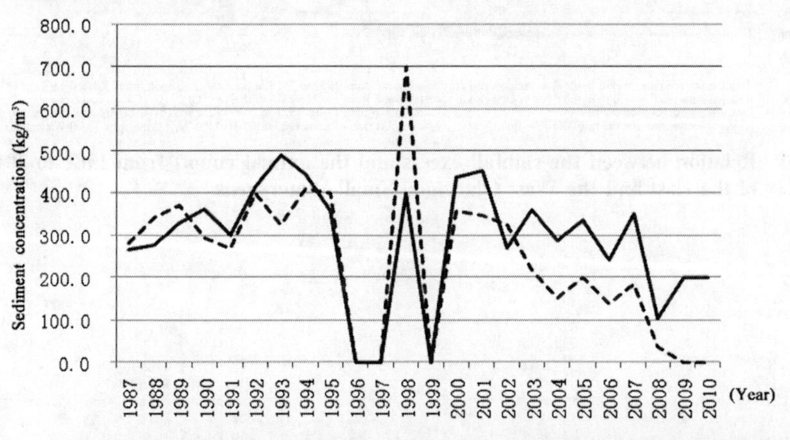

Fig. 4 Hydrograph of the mean annual sediment concentration of the East and the West Qiaozigou Small Watersheds

2.3.5 Relation between rainfall and runoff and between rainfall and sediment yield

Through analyzing the measured data of local rainfall, runoff and sediment from 1987 to 2010, it is shown that the linear correlation coefficients of the rainfall in the flood season and the annual runoff of the East and the West Qiaozigou Small Watersheds are 0.28 and 0.47 respectively, and the linear correlation coefficients of the rainfall excess and the annual runoff are 0.31 and 0.44 respectively, and the linear correlation coefficients of the rainfall excess and the annual sediment load are 0.22 and 0.60 respectively, and moreover, the linear correlation coefficients of the rainfall in the flood season and the annual sediment load are 0.17 and 0.66 respectively. See Fig. 5 ~ Fig. 8 for details.

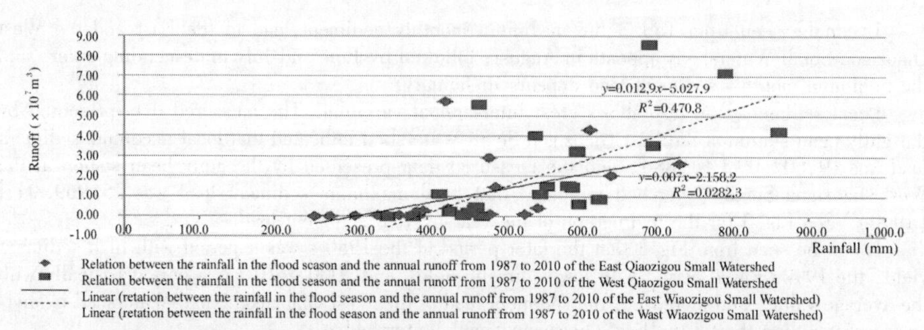

Fig. 5 Relation between the rainfall in the flood season and the annual runoff from 1987 to 2010 of the East and the West Qiaozigou Small Watersheds

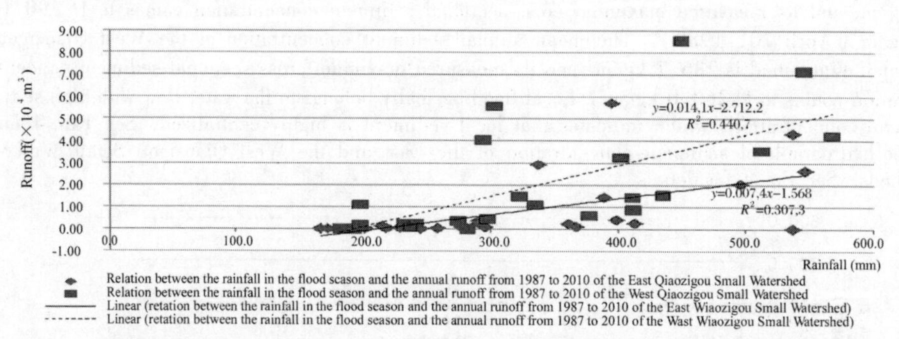

Fig. 6 Relation between the rainfall excess and the annual runoff from 1987 to 2010 of the East and the West Qiaozigou Small Watersheds

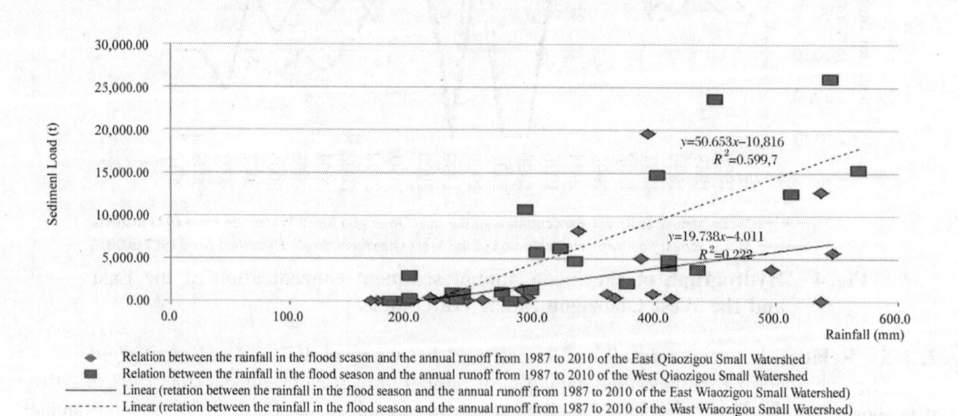

Fig. 7 Relation between the rainfall excess and the annual sediment load from 1987 to 2010 of the East and the West Qiaozigou Small Watersheds

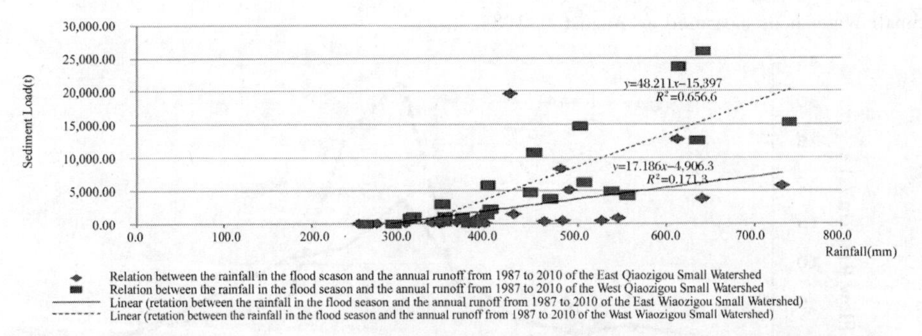

Fig. 8 Relation between the rainfall in the flood season and the annual sediment load from 1987 to 2010 of the East and the West Qiaozigou Small Watersheds

2.3.6 Relation between runoff and sediment load

The relational expression of the annual runoff and the annual sediment load is worked out by using the data of the runoff and the sediment from 1987 to 2010 measured at the gully head stations of the East and the West Qiaozigou Small Watersheds. See Fig. 9 for details.

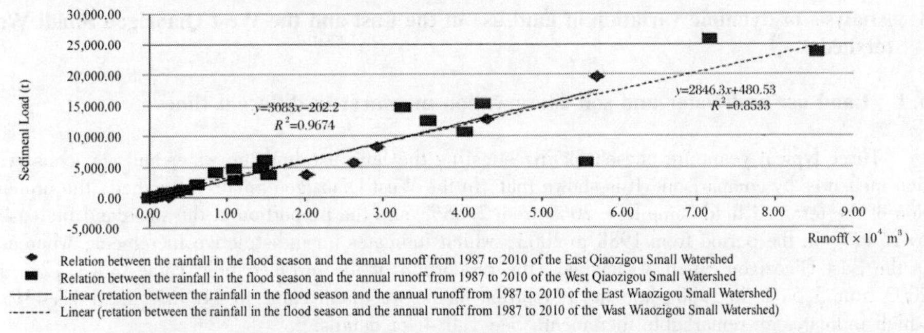

Fig. 9 Relation between the annual runoff and the annual sediment load from 1987 to 2010 of the East and the West Qiaozigou Small Watersheds

It can be known from the figure that the linear correlation coefficients of the runoff and the sediment load of the East and the West Qiaozigou Small Watersheds are 0.97 and 0.85 respectively.

2.4 Typical characteristics of rainstorm, runoff and sediment

2.4.1 Rainstorm with centralized rainfall, short duration and high intensity

The study area is a place subject to frequent rainstorms with centralized rainfall, short duration, and high intensity, mostly occurring in the period from May to September. On April 19, 1987, the two rainfall stations in the Qiaozigou Small Watershed obtained the rainfall data of 52.8 mm and 49.1 mm respectively, averagely 51.0 mm, and the maximum hourly rainfall of 31.3 mm. On August 7, 1988, a single rainstorm of 98.0 mm happened in the Qiaozigou Small Watershed, and the maximum hourly rainfall came to 56.8 mm. The investigation of the rainstorm happened on June 15, 2010, showed that the rainfall at the rainstorm center came to 68.8 mm in 55 min.

2.4.2 Abrupt rise and fall of flood peak

The East and the West Qiaozigou Small Watersheds are characterized by runoff yield under excess infiltration. Because the rainstorm has centralized rainfall, high intensity and short duration, the volume of runoff under excess infiltration is great and the flood generally rises and falls abruptly, having the characteristics of high flood peak, short duration, high peak discharge, and high sediment concentration. See Fig. 10 for the flood hydrograph of the East and the West Qiaozigou

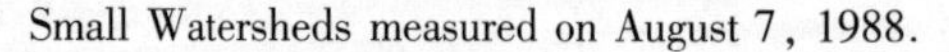
Small Watersheds measured on August 7, 1988.

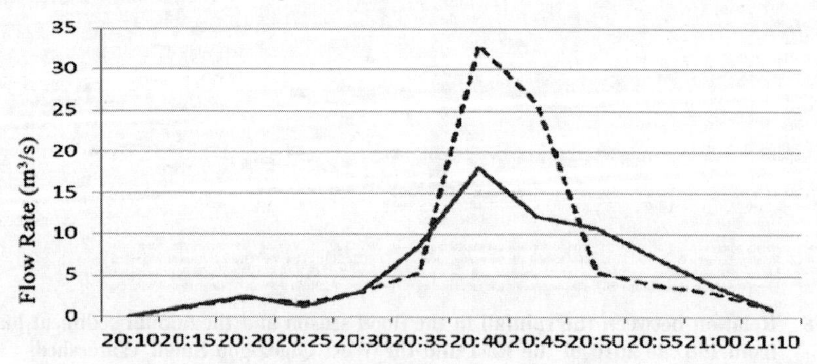

Fig. 10 Flood hydrograph of the East and the West Qiaozigou Small Watersheds measured on August 7, 1988

3 Analysis of dynamic variation of land use in the East and the West Qiaozigou Small Watersheds

3.1 Land use and water and soil conservation measures in different times

Three typical years are chosen for investigating the land use and the water and soil conservation measures by comparison. It is shown that, in the West Qiaozigou Small Watershed, the proportion of the terraced field came to 4.20% from 2.83% and the proportion of the woodland increased by 2.49% in the period from 1988 to 2003, which indicates an indistinctive increment; whereas, in the East Qiaozigou Small Watershed, the proportion of the terraced field came to 43.12% in 2003 from 3.54% in 1988, and the proportion of the woodland came to 44.96% from 41.43%, which indicates an remarkable increment. See Tab. 4 for details.

Tab. 4 Comparison of land use in the East and the West Qiaozigou Small Watersheds in three typical years

Land type	West Qiaozigou Small Watershed						East Qiaozigou Small Watershed					
	Year 1988		Year 1991		Year 2003		Year 1988		Year 1991		Year 2003	
	hm^2	%	hm^2	%	hm^2	%	hm^2	%	hm^2	%	hm^2	%
Terraced field	3.09	2.83	3.18	2.92	4.58	4.20	4.81	3.54	31.85	23.42	58.64	43.12
Hillside farmland	84.40	77.40	83.20	76.40	80.12	73.50	66.79	49.11	54.53	40.10	11.30	8.31
Forest land	0.96	0.88	1.39	1.28	2.57	2.36	8.39	6.17	17.28	12.71	29.94	22.01
Young forest land	1.20	1.10	1.30	1.19	1.52	1.39	2.01	1.48		0.00	13.75	10.11
Open woodland		0.00	0.49	0.45	0.55	0.50	41.49	30.51	10.02	7.37	0	0.00
Economic forest	0.62	0.57	0.68	0.62	0.77	0.71	0.68	0.50	16.68	12.26	16.73	12.30
Wild wood	0.52	0.48	0.52	0.48	0.61	0.56	3.77	2.77	0.73	0.54	0.73	0.54

Continued Tab. 4

Land type	West Qiaozigou Small Watershed						East Qiaozigou Small Watershed					
	Year 1988		Year 1991		Year 2003		Year 1988		Year 1991		Year 2003	
	hm²	%	hm²	%	hm²	%	hm²	%	hm²	%	hm²	%
Artificial pasture		0.00		0.00		0.00	0.37	0.27	3.29	2.42	3.29	2.42
Unused land	15.80	14.50	15.80	14.54	15.44	14.17	7.38	5.43	1.59	1.17	1.59	1.17
Village	2.33	2.14	2.31	2.12	2.84	2.61	0.31	0.23	0.03	0.02	0.03	0.02

3.2 Comparative study of soil erosion in the East and the West Qiaozigou Small Watersheds

The East Qiaozigou Small Watershed has the mean annual sediment load of 3,662.7 t and the sediment runoff modulus of 2,692.9 t/(km² · a), and the West Qiaozigou Small Watershed has the mean annual sediment load of 7,514.7 t and the sediment runoff modulus of 6,893.8 t/(km² · a) From the view of the sediment runoff modulus, the East is 60.94% less than the West. See Tab. 5 for the comparative analysis of 3 typical years.

Tab. 5 Comparison of soil erosion and sediment yield in the East and the West Qiaozigou Small Watersheds in three typical years

Year			June ~ September						Whole year	Sediment runoff modulus (t/(km² · a)
			June	July	August	Sepstember	Subtotal	Proportion		
1988	East	Rainfall (mm)	29.2	78.2	220.4	23.1	350.9	55.03	637.6	
		Runoff (m³)	259.20	2,946.24	51,693.12	1,555.20	56,453.76	98.60	57,257.28	
		Sediment load (t)	98.50	779.41	18,159.55	417.31	19,454.77	99.25	19,602.09	14,413.30
	West	Runoff (m³)	1,814.40	5,356.80	70,709.76	4,665.60	82,546.56	96.56	85,484.16	
		Sediment load (t)	552.10	1,633.82	19,605.89	1,218.24	23,010.05	97.11	23,695.37	21,738.87
	East less than West (%)									33.70
1991	East	Rainfall (mm)	38.3	48	48.2	95.9	230.4	47.74	482.6	
		Runoff (m³)	259.20	160.70	0	259.20	679.10	86.37	786.24	
		Sediment load (t)	139.97	8.03	0	72.58	220.58	87.28	252.72	231.85
	West	Runoff (m³)	518.40	1,071.36	0	777.60	2,367.36	68.84	3,438.72	
		Sediment load (t)	191.81	361.58	0	181.44	734.83	70.83	1037.49	951.83
	East less than West (%)									75.64

Continued Tab. 5

Year			June ~ September						Whole year	Sediment runoff modulus (t/(km² · a)
			June	July	August	Sepstember	Subtotal	Proportion		
2003	Rainfall (mm)		72.1	146.4	196.7	188.9	604.1	69.66	867.2	
	East	Runoff (m³)	0	1,607.04	3,214.08	5,443.20	10,264.32	38.58	26,602.56	
		Sediment load (t)	7.78	463.36	916.01	1,544.83	2,931.98	51.74	5,666.63	4,166.64
	West	Runoff (m³)	777.60	4,553.28	11,517.12	8,294.40	25,142.40	59.09	42,552.00	
		Sediment load (t)	158.11	1,215.99	3,669.41	4,510.08	9,553.59	62.25	15,346.97	14,079.79
	East less than West (%)									70.41

3.3 Analysis of the benefits from the water and soil conservation measures

3.3.1 Analysis of the benefits of flood and sediment reduction by water and soil conservation measures

The 'hydrological method' is taken for analyzing the benefits of flood and sediment reduction by water and soil conservation measures. The hydrological method is to analyze the variation of runoff and sediment load depending on the hydrologic data measured at the watersheds and the general theory of statistics.

3.3.1.1 Analysis of the benefits of flood reduction

The benefits of flood reduction are investigated by comparative analysis of the runoff and peak discharge in each month, in each year and in the flood season.

(1) Comparative analysis of monthly runoff.

See Tab. 6 for the results of comparative analysis of monthly runoff in the East and the West Qiaozigou Small Watersheds.

Tab. 6 Comparative analysis of monthly runoff in the East and the West Qiaozigou Small Watersheds

Watershed	Period (Year)	Average monthly runoff depth (mm)											
		1	2	3	4	5	6	7	8	9	10	11	12
East Qiaozigou Small Watershed	1987 ~ 2010	0	0	0	0.048	0.292	1.140	1.219	2.414	0.603	0.763	0	0
West Qiaozigou Small Watershed	1987 ~ 2010	0	0	0	0.119	1.351	3.101	2.693	5.862	3.428	1.096	0	0
Absolute value	1987 ~ 2010	0	0	0	0.071	1.059	1.961	1.474	3.448	2.825	0.333	0	0
Relative value	1987 ~ 2010	0	0	0	59.66	78.39	63.24	54.73	58.82	82.41	30.38	0	0

Notes: 1. The absolute value = East Qiaozigou − West Qiaozigou

2. The relative value (%) = The absolute value/ West Qiaozigou × 100%.

It can be seen from Tab. 6 that the monthly runoff depth of both the East and the West Qiaozigou Small Watersheds in the period from January to March and from November to December is zero, and the monthly runoff depth of the West Qiaozigou Small Watershed in the period from April to October is greater than that of the East Qiaozigou Small Watershed, and moreover, the absolute value is 0.071 mm ~ 3.448 mm and the relative value is 30.38% ~ 82.41%. The effect of monthly runoff reduction is obvious.

(2) Comparative analysis of the runoff in each year, in the flood season and in the period from July to September.

See Tab. 7 for the results of comparative analysis of the runoff in each year, in the flood season, and in the period from July to September of the East and the West Qiaozigou Small Watersheds.

Tab. 7 Comparative analysis of the runoff in each year, in the flood season and in the period from July to September of the East and the West Qiaozigou Small Watersheds

Watershed	Period (Year)	Annual average runoff depth (mm)	Average runoff depth in the flood season (mm)	Average runoff depth in the period from July to September (mm)
East Qiaozigou Small Watershed	1987 ~ 1989	21.69	21.43	18.83
	1990 ~ 1999	3.48	3.46	1.47
	2000 ~ 2010	5.06	5.04	2.77
	1987 ~ 2010	6.48	6.43	4.24
West Qiaozigou Small Watershed	1987 ~ 1989	40.29	39.89	27.88
	1990 ~ 1999	9.63	9.49	5.04
	2000 ~ 2010	18.77	18.75	13.96
	1987 ~ 2010	17.65	17.53	11.98
Absolute value	1987 ~ 1989	18.60	18.46	9.05
	1990 ~ 1999	6.15	6.03	3.57
	2000 ~ 2010	13.71	13.71	11.19
	1987 ~ 2010	11.17	11.10	7.74
Relative value	1987 ~ 1989	46.17	46.28	32.46
	1990 ~ 1999	63.86	63.54	70.83
	2000 ~ 2010	73.04	73.12	80.16
	1987 ~ 2010	63.29	63.32	64.61

Notes: 1. The absolute value = East Qiaozigou − West Qiaozigou.
2. The relative value (%) = The absolute value/ West Qiaozigou × 100%.

It can be known from Tab. 7 that the annual average runoff depth of the West Qiaozigou Small

Watershed in the period from 1987 to 2010 is 17.65 mm, whereas the East Qiaozigou Small Watershed, 6.48 mm, having the difference of 11.17 mm; and the annual runoff depth of the East Qiaozigou Small Watershed is 63.29% less than that of the West Qiaozigou Small Watershed. In the three different periods from 1987 to 1989, from 1990 to 1999, and from 2000 to 2010, the annual runoff depth of the East Qiaozigou Small Watershed is 46.17%, 63.86%, and 73.04% less than that of the West Qiaozigou Small Watershed respectively.

(3) Comparative analysis of the maximum peak discharge.

The measured data show that the mean annual maximum peak discharge of the West Qiaozigou Small Watershed is 2.85 m^3/s, and the East Qiaozigou Small Watershed, 2.33 m^3/s. Water and soil conservation measures make the peak discharge reduced by 18.20%.

(4) Comparison of runoff coefficient.

Analysis shows that the mean annual runoff coefficient of the West Qiaozigou Small Watershed is 0.033, and the East Qiaozigou Small Watershed, 0.012. The runoff coefficient of the West is greater than that of the East by the ratio of 2.75, which indicates that impoundment of surface runoff performs better in the East than in the West.

3.3.1.2 Analysis of the benefits of sediment reduction

The benefits of sediment reduction are investigated by comparative analysis of the sediment load and cross-sectional maximum sediment concentration in each month, in each year, in the flood season, and in the period from July to September.

(1) Comparative analysis of monthly sediment load.

See Tab. 8 for the results of comparative analysis of monthly sediment load in the East and the West Qiaozigou Small Watersheds.

Tab. 8 Comparative analysis of monthly sediment load in the East and the West Qiaozigou Small Watersheds

Watershed	Period (Year)	Monthly sediment runoff modulus (t/km^2)											
		1	2	3	4	5	6	7	8	9	10	11	12
East Qiaozigou Small Watershed	1987 ~ 2010	0	0	0	20.3	75.9	359.0	348.5	790.6	128.3	125.0	0	0
West Qiaozigou Small Watershed	1987 ~ 2010	0	0	0	41.9	340.7	1,228.1	995.4	1,792.0	729.2	341.7	0	0
Absolute value	1987 ~ 2010	0	0	0	21.6	264.8	869.1	646.9	1,001.4	600.9	216.7	0	0
Relative value	1987 ~ 2010	0	0	0	51.6	77.7	70.8	65.0	55.9	82.4	63.4	0	0

Notes: 1. The absolute value = East Qiaozigou - West Qiaozigou.
2. The relative value (%) = The absolute value/ West Qiaozigou × 100%.

It can be seen from Tab. 8 that the monthly sediment load of both the East and the West Qiaozigou Small Watersheds in the period from January to March and from November to December is zero, and the monthly sediment load of the West Qiaozigou Small Watershed in the period from April to October is greater than that of the East Qiaozigou Small Watershed, and moreover, the relative value is 51.6% ~ 82.4%. The effect of sediment reduction in September is the greatest.

(2) Comparative analysis of the sediment load in each year, in the flood season, and in the period from July to September.

See Tab. 9 for the results of comparative analysis of the sediment load in each year, in the flood season and in the period from July to September of the East and the West Qiaozigou Small Watersheds.

Tab. 9 Comparative analysis of the sediment load in each year, in the flood season and in the period from July to September of the East and the West Qiaozigou Small Watersheds

Watershed	Period (Year)	Annual sediment runoff modulus (t/km^2)	Sediment runoff modulus in the flood season (t/km^2)	Sediment runoff modulus in the period from July to September (t/km^2)
East Qiaozigou Small Watershed	1987 ~ 1989	7,026.01	6,874.18	6,181.03
	1990 ~ 1999	1,041.90	1,039.42	445.30
	2000 ~ 2010	1,167.81	1,167.12	674.67
	1987 ~ 2010	1,847.62	1,827.29	1,267.39
West Qiaozigou Small Watershed	1987 ~ 1989	11160.26	10987.46	7811.46
	1990 ~ 1999	3,673.97	3,627.37	2,048.70
	2000 ~ 2010	5,549.18	5,547.23	3,680.14
	1987 ~ 2010	5,469.23	5,427.32	3,516.79
Absolute value	1987 ~ 1989	4,134.25	4,113.28	1,630.43
	1990 ~ 1999	2,632.07	2,587.95	1,603.40
	2000 ~ 2010	4,381.37	4,380.11	3,005.47
	1987 ~ 2010	3,621.61	3,600.03	2,249.40
Relative value	1987 ~ 1989	37.04	37.44	20.87
	1990 ~ 1999	71.64	71.35	78.26
	2000 ~ 2010	78.96	78.96	81.67
	1987 ~ 2010	66.22	66.33	63.96

Notes: 1. The absolute value = East Qiaozigou - West Qiaozigou.
2. The relative value (%) = The absolute value/ West Qiaozigou × 100%.

It can be known from Tab. 9 that, from 1987 to 2010, the mean annual sediment load of the East Qiaozigou Small Watershed in each year, in the flood season and in the period from July to September is 66.22%, 66.33%, and 63.96% less than that of the West Qiaozigou Small Watershed respectively. In the three different periods from 1987 to 1989, from 1990 to 1999, and from 2000 to 2010, the annual sediment load of the East Qiaozigou Small Watershed is 37.04%, 71.64%, and 78.96% less than that of the West Qiaozigou Small Watershed respectively.

(3) Comparative analysis of the cross-sectional maximum sediment concentration.

The measured data in the period from 1987 to 2010 show that the mean annual cross-sectional maximum sediment concentration of the West Qiaozigou Small Watershed is 802.38 kg/m^3, and the East Qiaozigou Small Watershed, 537.20 kg/m^3, indicating that the East Qiaozigou Small Watershed has a reduction of 33.05%.

3.3.2 Investigation and analysis of the benefits of controlling rainfall erosion by water and soil conservation measures

According to the investigation of the soil erosion in Qiaozigou Small Watershed caused by the rainstorm on April 19, 1987, the erosion-reducing performance of artificial pasture averagely reduced soil erosion by 43.0% compared with the hillside farmland, i.e. the sediment yield of each hectare of artificial pasture was 51.12 t less than that of each hectare of hillside farmland; and the sediment yield of each hectare of artificial woodland was 41.00 t less than that of each hectare of hillside farmland, reducing the sediment yield by 34.4% ; and similarly, the sediment yield of each hectare of bench terrace was 119.00 t less than that of each hectare of hillside farmland. See Tab. 10 for the results.

Tab. 10 Calculation of sediment reduction benefits from water and soil conservation measures in the Qiaozigou Small Watershed

Place	Terraced field			Artificial pasture			Artificial woodland			Sediment reduction benefits	
	Area (hm^2)	Sediment reduction index (t)	Quantity of sediment reduction (t)	Area (hm^2)	Sediment reduction index (t)	Quantity of sediment reduction (t)	Area (hm^2)	Sediment reduction index (t)	Quantity of sediment reduction (t)	Quantity of sediment reduction (t)	Benefits (%)
East Qiaozigou Small Watershed	3.60	119	428.40	0.39	51.12	19.94	8.07	41	330.87	779.21	22.60
West Qiaozigou Small Watershed	0	119	0	0.20	51.12	10.22	5.80	41	237.80	248.02	6.00

Note: Sediment reduction benefits (%) = Quantity of sediment reduction / (Measured soil erosion + Quantity of sediment reduction) ×100%

It can be known from Tab. 10 that various measures for water and soil conservation had taken a certain effect of reducing the slope surface erosion caused by this rainstorm, i. e. 779.21 t and 248.02 t of sediment had been retained in the East and the West Qiaozigou Small watersheds, having the sediment reduction benefits of 22.6% and 6.0% respectively.

Investigation and analysis of this rainstorm show that the slope surface erosion caused by rainstorm has the following characteristics. Firstly, the slope surface erosion becomes greater along with the increase of slope gradient. As to hillside farmland, the erosion on the slope surface with the gradient of 10° ~20° is 61.4% higher than that on the slope surface with the gradient of <10°, and similarly, the erosion of 20° ~30° is 78.3% higher than that of <10°, and the erosion of 20° ~30° is 43.7% higher than that of 10° ~20°. As to artificial pasture, the erosion on the slope surface with the gradient of 20° ~30° is 53.1% higher than that of 10° ~20°. Secondly, the slope surface erosion varies with different vegetation and different water and soil conservation measures applied. Averagely the erosion on artificial woodland is 34.4% less than that on cropland and the erosion on artificial pasture is 43.0% less than that on cropland. The erosion-reducing performance of terraced field almost comes to 100%.

3.4 Analysis of economic benefits

Comprehensive control of soil erosion and water loss brings about remarkable economic benefits. A piece of powerful evidence is that the income of local people has obviously improved through further developing agriculture, forestry, livestock husbandry, and side occupation. Investigations show that the annual income of each hectare of fruit trees in the fruiting period can come to 15,000 ~150,000 yuan, which is much higher than the income of cropland. What is more, control of water loss and soil erosion can not only alleviates or eliminates the damages caused by serious water loss and soil erosion, but also reduces economic loss, thereby bringing about great indirect economic benefits.

3.5 Analysis of social benefits

Application of engineering measures of soil and water conservation has not only improved local living environment, production conditions, land use efficiency, and farmland quality, but also enhanced local environmental carrying capacity and environmental resistance against disasters. Meanwhile, reducing sediment concentration of stream flow enables less sediment to go to the lower reaches, which can alleviate silt deposition and improve the flood regulating and flood storage performance of river channels, and thus flood disasters can be reduced in the lower reaches.

4 Conclusions

(1) The East and the West Qiaozigou Small Watersheds are two small watersheds arranged for observation of water loss and soil erosion by comparison, i. e. with comprehensive control and without comprehensive control, in the loess hilly and gullied area. The two small watersheds having outstanding comparability are equipped with perfect observation facilities by which detailed and reliable observation data are acquired. More than twenty years' measured data have become important fundamental material for studying the characteristics of water loss and soil erosion in small watersheds and for investigating the benefits of water and soil conservation measures.

(2) The rainfall in the loess hilly and gullied area shows great inter-annual variation and obvious difference in the high-flow years and the low-flow years. The annual variation of local rainfall is also great while most rainfall comes in the period from June to September in the form of frequent rainstorm. The runoff, sediment and rainfall show the positive correlation. Typical rainstorm is the main cause of water loss and soil erosion, and the gully erosion caused by the typical rainstorm becomes the main source of sediment.

(3) Water and soil conservation measures have made remarkable influence on runoff and sediment. By taking water and soil conservation measures, under the same rainfall amount, the runoff of unit area can be reduced by 63.26%. Comparing with that of the West Qiaozigou Small Watershed, the peak discharge of the East Qiaozigou Small Watershed is reduced by 18.20%. Comparison of the sediment load in each year, in the flood season, and in the period from July to September shows the difference of 66.22%, 66.33%, and 63.96% between the small watershed with comprehensive control and the small watershed without comprehensive control.

References

Zhang Shengli, Yu Yiming, et al., Hydrologic calculation method of benefits on reduction of water and sediment runoff through soil conservations measures [M]. Beijing: China Environment Press, 1995.

Zhang Xiaoming, Yu Xinxiao, Wu Sihong, et al., Effects of land use/cover change on runoff sediment discharge in typical watershed in the loess Gullied-Jilly region of China [J]. Journal of Beijing Forestry University, 2007, 29 (6).

Yuan Cuiping, Li Shuqin, Lei Qixiang, et al., Study on water erosion from two small watersheds with and without management on Loess Plateau [J]. Journal of China Agricultural University, 2010 (6).

Wang Hong, Kang Xuelin. Investigation and analysis on "1987. 4. 19" storm, flood and tts coursed soil erosion in Qiaozigou watershed [J]. Yellow River, 1995, (10).

Comparing Hydrological Effects of Treated and Untreated Watersheds at Experimental Scale

Wang Guoqing[1,2], *Zhang Jianyun*[1,2], *Xuan Yunqin*[3], *Jin Junliang*[1,2], *Shang Manting*[1,2] and **Bao Zhenxin**[1,2]

1. State Key Laboratory of Hydrology – Water Resources and Hydraulic Engineering, Nanjing Hydraulic Research Institute, Nanjing, 210029, China
2. Research Center for Climate Change, Ministry of Water Resources, Nanjing, 210029, China
3. UNESCO – IHE Institute for Water Education, Delft, Netherlands

Abstract: Taking two typical tributaries in the Loess Plateau as case, hydrological effects of treated and untreated watersheds were studied with parallel comparison method. Results show that soil and water conservation measures implemented can reduce annual runoff and sediment module, low down flood peak discharge and sediment concentration, lag occurring time of flood peak, and shorten flood duration. Although treated measures can change the relationships between runoff, sediment and rainfall, it has no significant effects on correlativity.

Key words: treated watershed, the Loess Plateau, parallel comparison method, soil and water conservation measures, hydrological effects

1 Introduction

The Loess Plateau is well known in the world with serious losses of soil and water (Xu et al., 2006; Zhang et al., 1998). Thereby, the Loess Plateau has been becoming a key area of China national ecological environment recovery and implementation. Better understanding hydrological effects of soil and water conservation measures in the Loess Plateau is essential for benefit assessment of soil and water conservation measures as well as ecological environment construction planning. There are many literatures available for hydrological effect of forest vegetation cover and soil conservation measures adopting parallel comparative analysis method or hydrological simulation method (Liu et al., 2005; Mu et al., 2002; Wang et al., 2006), but the previous studies focus on water, less on sediment. Taking two adjacent experimental gullies (Yangdao and Chacaizhu) in the Loess Plateau as case, hydrological comprehensive effects of Soil and Water Conservation Measures (SWCMs) are analyzed.

2 Study watersheds

Yangdao gully and Chacaizhu gully are two adjacent gullies with same flow direction in Lishi County of Shanxi Province, and both are located at left side of Wangjiagou tributary in the Loess Plateau. Soil type, geomorphology and topography in the two watersheds are similar. Drainage areas of the gullies are 0.206 km^2 and 0.193 km^2 respectively. In order to study benefits of soil and water conservation measures, The Chacaizhu gully was treated with reforestation, terrace land construction, et al. from 1954 while Yangdao gully has been keeping in natural. As of 1956, the harnessing percentage of the Chacaizhu gully was up to 78.3%.

Because the two gullies are adjacent and their area is very small, the climatic condition in the watershed can be taken as same. According to the data from 1956 to 1970, statistical results showed that mean annual precipitation was about 544 mm, and precipitation in flood period from June to September accounted for 72% of annual precipitation. Soil and water losses were mainly caused by several high-intensity storms in the period. Thereby, runoff and sediment were concentrated in the flood period as well. Precipitation in dry period was little with low density, resulting in less runoff and sediment yield in this period. The mean annual runoff modules of the two gullies were 27,740 m^3/km^2 and 14,115 m^3/km^2, and the corresponding sediment modules were 20,811 t/km^2 and 8,504 t/km^2 respectively.

3 Results and discussion

3.1 Effect of SWCMs on annual runoff and sediment yield

As rainfall condition is very similar in the two gullies, the differences of the recorded runoffs and sediment yields between the two gullies reflected the benefits of soil and water conservation measures. Taking 1963, 1967 and 1957 as typical rainy year, normal year and dry year, the effects of soil and water conservation measures on annual runoff and sediment under different rainfall condition were compared (Tab. 1).

Tab. 1 Statistical values of hydrological characteristics of Yangdao gully and Chacaizhu gully

Items	Annual precipitation (mm)	Annual runoff module(m^3/km^2)		Water reduction benefits		Annual sediment module (t/km^2)		Sediment reduction benefits	
		Yangdao gully	Chacaizhu gully	(m^3/km^2)	(%)	Yangdao gully	Chacaizhu gully	(t/km^2)	(%)
Rainy year (1963)	732.7	35,600	19,900	15,700	44.1	31,562	14,023	17,539	55.6
Normal year (1967)	492.8	21,600	10,000	11,600	53.7	18,730	6,902	11,829	63.2
Dry year (1957)	376.5	7,300	2,400	4,900	67.1	3,674	645	3,030	82.5

Tab. 1 shows that annual runoff modules and sediment modules of the two watersheds are quite different even under the same year, indicating significant effect of soil and water conservation measures on runoff and sediment yield. In the rainy year (1963), runoff module and sediment module of treated watershed, Chacaizhu gully, were reduced by 15,700 m^3/km^2 and 17,539 t/km^2 respectively as comparing to untreated watershed, Yangdaogou gully, the corresponding reduction percentages are 44.1% and 55.6%. In the normal year (1967), the reduced modules of runoff and sediment are less than that in the rainy year with higher reduction percentages of 53.7% and 63.2%. In the dry year (1957), although the reduced modules of runoff and sediment yield are the least, 4,900 m^3/km^2 and 3,030 t/km^2 respectively, with highest reduction percentages of 67.1% and 83.2%, possibly resulting from low runoff and sediment modules in the dry year.

3.2 Effects of SWCMs on flood and sediment yield

Flood and sediment processes are dominated by ground cover and precipitation conditions. Under the similar groundcover and rainfall condition, there should be no big difference in hydrological characteristics for different watershed. Intensive human disturbance to watershed with soil and water conservation measures will change groundcover to some extent, as a result, hydrological characteristics, such as flood duration, peak flood discharge, sediment yield process, et al, will be changed in responding to land cover change. Processes of flow and sediment content of Chacaizhu gully and Yangdaogou gully under a rainstorm event numbered 19660813 were shown in Fig. 1 and Fig. 2.

Fig. 1 shows flow discharge from natural watershed Yangdao gully rises quickly, and reaches to peak at 28 min late after rising. The occurring time of peak discharge for treated watershed, Chacaizhu gully, is 5 min late in comparison to natural watershed. The recession process of flood process for natural watershed is much quicker than that of treated watershed. The peak discharge of Yangdao gully is 4.1 m^3/s, which is approximately 3.3 times of that from treated watershed. The

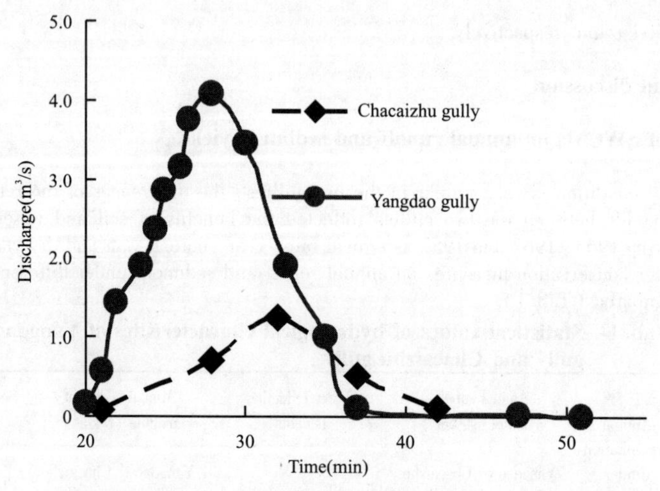

Fig. 1 Discharge processes of Yangdao gully and Chacaizhu gully under a rainfall event (No. 19660813)

sediment contents of the two gullies both decrease slightly after it reached to peak value. But sediment content in natural gully is obviously higher than that in harnessed gully, and meanwhile, sediment content of Chacaizhu gully gets to the peak much earlier than that of Yangdao gully (Fig. 2).

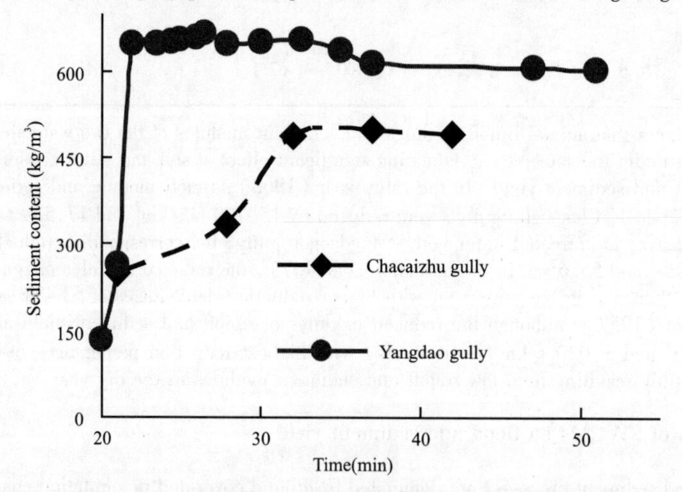

Fig. 2 Silt content processes of Yangdao gully and Chacaizhu gully under a rainfall event (No. 19660813)

Analysis indicates that soil and water conservation measures increased vegetation cover area on one hand; on the other hand, it also changed soil structure and increased soil porosity. Thereby, soil infiltration capacity and soil water storage both increased as well. The changed soil property could positively low down flood peak discharge, and make flow recession process long. High vegetation cover, low flow discharge will decrease soil erosion to some extent.

In order to compare the effects of soil and water conservation measures under the different rainfall amount and intensity, 3 typical rainfall events were selected, considering similar rainfall amount or rainfall intensity. The characteristics of runoff and sediment under the typical rainfall conditions in Yangdao gully and Chacaizhu gully are given in Tab. 2.

Tab. 2 Runoff and sediment characteristics under the typical rainfall events

Flood events	Watershed	Rainfall amount (mm)	Rainfall intensity (mm/h)	Flood duration (h:min)	Runoff depth (mm)	Sediment module (t/km^2)	Maximum discharge (m^3/s)	Flood peak module ($m^3/s/km^2$)
	Chacaizhu gully			00:37	0.09	9.3	0.017	0.088
19640820	Yangdao gully	9.2	5.51	00:40	0.32	118.2	0.11	0.534
	Benefits (%)				71.9	92.1	84.5	83.5
	Chacaizhu gully			00:32	0.8	625.9	0.475	2.461
19660728	Yangdao gully	9.5	11.31	00:38	2.3	2,970	2.181	10.587
	Benefits (%)				65.2	78.9	78.2	76.8
	Chacaizhu gully	-		07:31	12.9	12,190.2	2.228	11.544
19580729	Yangdao gully	48.7	11.46	08:29	18.2	13,521.1	3.325	16.141
	Benefits (%)				29.1	9.8	33.0	28.5

Tab. 2 shows that the ability of retaining runoff and sediment in harnessed basin has been improved significantly compared to the natural basin. Runoff are decreased by 29% ~ 72% under the 3 typical rainfall events, meanwhile, the sediment are reduced by 10% ~ 93%. The peak discharge reduction rate is about 25% ~ 85%, and flood duration is shortened to some extent. The effects of soil and water conversation measures are different under the different rainfall conditions. The benefit percent is most evident under the less rainfall amount and lower intensity.

3.3 Effects of SWCMs on relationships between runoff, sediment and rainfall

The correlative analysis shows that annual runoff and sediment has very close relationship, and correlative coefficient exceeds 0.95. The annual runoff, sediment has higher correlation coefficient with short interval rainfall amount than longer interval rainfall. And correlative coefficients between annual runoff, sediment and runoff-yielding rainfall amount are higher than 0.80. These results show that rainfall intensity could affect runoff and sediment significantly.

According to the data from 1956 to 1970, the linear relationship formula between runoff, sediment and runoff-yielding precipitation are established and correlative diagram are given in Fig. 3.

$$W_{YDG} = 0.019P_f - 0.338,4 \quad (r = 0.76) \quad (1)$$

$$W_{CCZG} = 0.008,9P_f - 0.136,8 \quad (r = 0.74) \quad (2)$$

$$S_{YDG} = 0.013,7P_f - 0.227,8 \quad (r = 0.69) \quad (3)$$

$$S_{CCZG} = 0.006,2P_f - 0.155,6 \quad (r = 0.70) \quad (4)$$

Where, W_{YDG} and S_{YDG} are runoff and sediment module in Yangdao gully respectively; W_{CCZG} and S_{CCZG} are runoff and sediment module in Chacaizhu gully respectively; P_f is runoff – yielding precipitation.

The harnessed basin and natural basin get similar correlative coefficients between runoff, sediment and rainfall, but different relationships, which indicate that soil and water conservation measures have great impact on relationship of rainfall and runoff, but weak effect on correlation.

The comparison in Fig. 3 shows that runoff and sediment module in natural gully is obviously greater than that in harnessed gully at the same rainfall condition, and difference between these modules is more obvious with precipitation increasing. Runoff and sediment reduction module caused by soil and water conservation measures is greater under the greater rainfall condition.

Runoff and sediment calculation formula in harnessed basin is different from that in natural basin. Slop in natural liner calculation formula is obviously greater than that in the harnessed watershed. This shows that unit precipitation increasing would make runoff and sediment amount in natural basin larger than that in harnessed basin. Soil and water conservation measures changed relationship between runoff, sediment and precipitation, and reduced water and sediment amount.

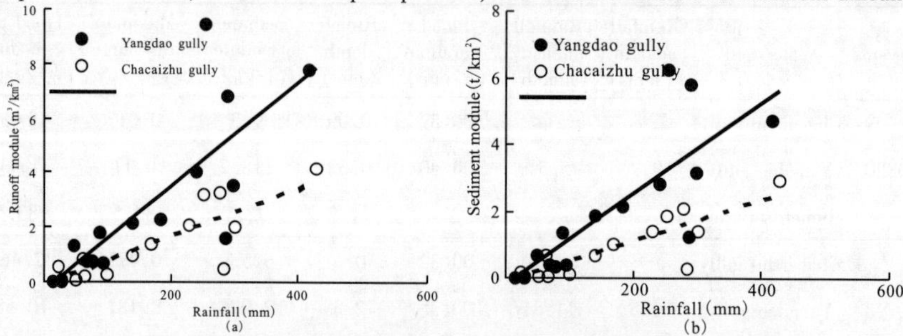

Fig. 3 Correlative diagram between runoff module, sediment module and precipitation of runoff yielding

4 Conclusions

Hydrological effects of soil and water conservation measures in small - scale watershed in the middle reaches of the Yellow River Basin have been studied with parallel compared analysis method. Results show that soil and water conservation measures can reduce annual runoff and sediment module, cut down flood peak discharge and reduce sediment concentration efficiently. And meanwhile, it also can delay the occurred time of flood peakand shorten flood duration. Although the measures can change the relationships between runoff, sediment and rainfall, it has no significant effects on correlation of precipitation and runoff - sediment. At the same time, parallel compared analysis method can only derive comprehensive effects of many measures, how to separate hydrological response of different measure in the middle and large-scale basin should be further analyzed.

Acknowledgments

This study was jointly supported by the National Basic Research Program of China (grant 2010CB951103), the International Science & Technology Cooperation Program of China (grant 2010DFA24330), funded by Ministry of Science and Technology, and the Adapting to Climate Change in China (ACCC) project, funded by the UK Department for International Development (DFID), the Swiss Agency for Development & Co-operation (SDC), and the UK Department for Energy and Climate Change (DECC). Thanks also to the anonymous reviewers and editors.

References

Mu Xingmin, Wang Wenlong, Xu Xuexuan. Effect of Soil and Water Conservation on Surface Flow in Small Basin of Loess Plateau[J]. Journal of Hydraulics, 1999 (3).

Hu Xianglai. Effect of Forest Vegetation on Water Resources in Loess Rolling Region in Gansu Province[J]. Journal of Hydraulics Advance, 2000 (2).

Liu Changming, Zhong Junxiang. Primary Analysis on Effect of Forest on Annual Runoff in Loess Plateau[J]. Journal of Geography, 1978, 33(2): 112 - 127.

Pomiplium Mita. The Effect of Forest on Runoff Generation[R]. Proceedings of Hilsimli Symposium, 1980.

Wang Guoqing, Zhang Jianyun, He Ruimin, et al. Impact of Environmental Change on Runoff in the Fenhe River Basin[J], Advances in Water Science, 2006(5).

Strategies of Water Resources Management for Response to Climate Change in Pollution Induced Water Shortage Areas

Wang Lihui[1] and *Li Jie*[2]

1. College of Civil Engineering of Fuzhou University, Fuzhou, 350108, China
2. Yinzhou District Water Conservancy Bureau, Ningbo, 315199, China

Abstract: There is growing concern for water shortage due to climate change. With the socio – economic development, particularly in pollution induced water shortage areas, the water crisis is worsened. Xiaoshan – Shaoxing – Ningbo (XSN) Plain is typical pollution induced water shortage district, which lied on the south bank of Hangzhou Bay, Zhejiang Province. In the present study, XSN Plain is chosen as research areas. Firstly, the paper statistic XSN's meteorological, hydrological and environmental data in the past 50 years, with comprehensive summarization of former domestic and foreign researches. Secondly, modern mathematical statistical methods were applied with the trend of water resource by climate change. Study showed that, climate change impacted on the hydrological cycle, the redistribution of water resources on spatial – temporal and furtherly influenced on water quality. It also put forward to adapting strategies to solve the water crisis in pollution induced water shortage district by climate change, according to Zhejiang practical situation.

Key words: climate change, water resources management, strategies, pollution, water shortage

The XSN Plain belongs to water sufficient area with complicated river network. The present situation: the gross amount of water resource is rich, but it is scare if based on the water own by per person, misdistribution on spatial and temporal, and the pollution is serious, which is a typical one of the pollution induced water shortage areas.

Climate change had great influence on water resources in Zhejiang Province. Although precipitation slightly increases in the past 50 years, but Zhejiang Province entered continuing low water and water total capacity below the annual water amount, with temperatures rising, more evaporation, and the excessive centralized precipitation since the 21st century (Mao Minjuan, et al, 2009). On the other hand, the demand of water is increasingly increased with the social development, which results in water shortage. There are serious land subsidence and seawater intrusion because of excessive exploitation and unreasonable use of groundwater in coastal plain region. Currently, with the development of industrial economy and the urbanization, the internal contradiction between water shortage and demand is apparent, and pollution induced water shortage is an obvious restrictive factor of the regional economic and social development. At the same time, the delaying of researching climate change impact on water management, also affect water resource management and protection on water shortage region.

In this paper, the XSN Plain is chosen as research case, studied comprehensive management of water resources on a typical pollution induced water shortage areas. The countermeasures of water resources management be discussed to adapt climate change. The study can provide typical demonstration project with the implement of the most strict water resources management in Zhejiang Provice.

1 The general situation of water resources in the XSN Plain

The XSN Plain (Fig. 1) is located in the northeast of Zhejiang Province, on the south bank of Hangzhou Bay, which formed with the Qiantang, Cao'e and Yong River alluvial and Hangzhou Bay current sedimentation. The total topography characteristic along the route is that the southwest is higher than the northeast; the total area is about 8,000 km^2. The river system are mainly Yong, Cao'e, and Xiaoshao; mainly river way which are Yong, Yao, Fenghua, Cao'e, Hangzhou –

Ningbo (Xiaoshao) canal, little East and West river; the upstream is mountain areas and downstream are coastal plain in Yong and Cao' e River Basin. The region is divided into Xiaoshao and Yao river plain with Cao' e River.

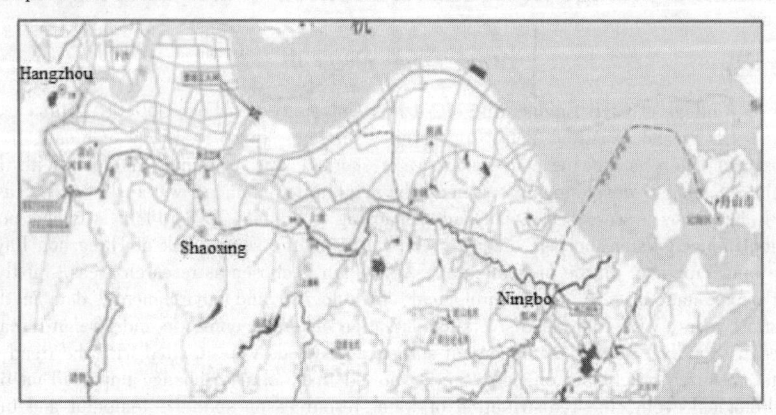

Fig. 1 Schematic of the Xiaoshan – Shaoxing – Ningbo (XSN) Plain

The XSN Plain is one of the economic developed districts in Zhejiang Province, but also the typical pollution induced water shortage areas. There are significant characteristics being different from the other rivers, due to the complicated natural and physical conditions. According to the survey, the Xiaoshao main canal (Xiaoshan section) average water quality is Class IV in XSN Plain; little West River is III, III ~ IV in Jian lake; the water quality of the river Shaoxing City portion is inferior to Class V; Yao River is III ~ IV in east of Cao' e and Ningfeng plain; network river and town water is Class V or inferior to Class V. The current water resources become dissatisfy with economic and social development.

The XSN Plain is located in subtropical monsoon climate zone. According to the precipitation, there is wet zone with plenty of rainfall, but there are dense population, per capita water resources quantity is little. The XSN Plain have inhomogeneity water resources problems. For Temporal distribution, plum rain (every year in mid June to early July) and typhoon period is the peak rainfall, but there is the least precipitation during summer drought. The rainfall is above 1,800 mm in abundant year, but only 900 mm in special low water year. The maximum annual precipitation are nearly two times than the minimum one (Zhejiang Province climate bulletin,2010). In case of continuous high temperature and drought summer, The problem of water resources shortage become more obvious, because of the water level drops quickly in rivers.

There are large discharge capacities of pollutant in the region. The content of pollutant in drained waste water gets high in the river and species are becoming more complicated, with life and industrial sewage discharge into the river are continuously increasing. In particular recent years, with industries booming, some enterprises cause serious pollution but a few prevention measures. At the same time, with non – point pollution such as pesticide, chemical fertilizer and agricultural and river vessel oil pollution are continuously increasing, which formed combined pollution between urban and rural, static water quality be worsen in river network, and become black and odorous because of hypoxia and eutrophication. With economic development and the improvement of the living standard, industrial sewage and domestic wastewater are rapidly increasing, but pollution management is backward, urban infrastructure construction and environment improvement obvious lag. All above are attributed to pollution and eutrophication be worsen in the XSN region.

In spatial distribution, limited by geographic and natural conditions, the XSN Plain water resource loss rapidly and the retaining and storage capacity is poor. Impeded drainage often subject to tidal backwater in downstream estuarine region, where water flow blocked and seawater intrusion, the water can not be used as a drinking water source. There are also considerably affected by industrial and domestic wastewater pollutions, and long time stranded in the middle reach of river network. In addition, the lack of river regulation, sedimentation and riverbed rising in the past years, caused the weak function of water storage and water self – purification ability. With industrial and economic development, the pollution has trend of upstream and reservoir upstream extend, the situation is worsen.

The basic characteristics of climate, hydrology and water quality, determined the sensitivity and vulnerability of climate change impact on water resources in the XSN Plain. Especially in the last thirty years, there are the increasing water demands by industrial and economic development, but the decreasing water resource capacity by climate change in developed region. Water supply is hard to meet human consumption needs in Zhejiang Province, water management has become one of the severest problems.

2 Impacts of climate change on water resources in the XSN region

Water resources in Zhejiang Province are deeply affected by climatic change, for example, the disequilibrium of its distribution, because of the typhoon. Response for the global warming, extreme weather and climate events shows enhanced and increased. With the climate warming, in the future, the probability of extremes climate will continuously increase; the tropical cyclone will be more frequent and violent. The ultra – strong typhoon will be more active; rainfall rate will increase, and its temporal and spatial distribution will be more disequilibrium; the total drought and rainstorm will be increase, which will be more easily induced secondary geological disaster such as landslides and debris flow. Sea – level rise is also caused by Higher temperatures, impeded drainage will be aggravated on coastal area. Economic loss will be increase significantly by extreme events and the weather disaster.

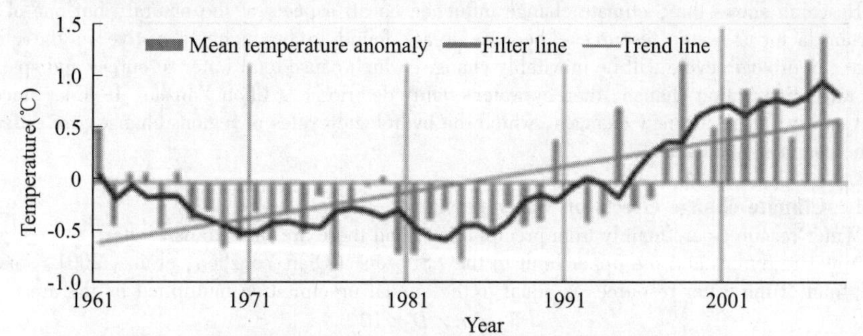

Fig. 2 Zhejiang Province annual mean temperature changes (Zhejiang Province climate bulletin, 2010)

2.1 Climate change characteristics in the XSN Plain

Using statistics of meteorological and hydrological data since the 1950s in the Zhejiang Province and the XSN zone (Fig. 2) (Zhejiang Province climate bulletin, 2010), it is showed that, the average temperature gradually increased since 1961, and the tendency is increases 0.25 ℃/10a with the linear regression calculation, in which the most significant one is after the 1980s, the rate is increases 0.6 ℃/10a. The general trend of climate change in Zhejiang Province is basically the same with that in China and even in the world, but seasonal average temperature change is not the same, the largest change of average temperature in winter, followed by the spring, The summer and autumn rising trend only appear after the 1990s. The lowest temperature increased more than the highest temperature increases, and diurnal change become smaller.

Using statistical Zhejiang Province and the XSN zone's precipitation and hydrological data since the 1950s (Fig. 3) (Zhejiang Province climate bulletin, 2010). The Zhejiang Province annual precipitation is slightly increased since 1961. The linear increased rate is 39.2 mm/10a during 1961 ~ 2008. In the past 50 years, the annual precipitation are the total increasing, but there

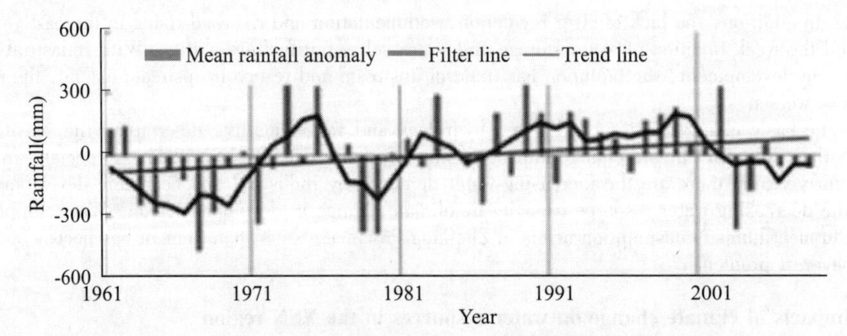

Fig. 3 Zhejiang Province annual mean precipitation changes (Zhejiang Province climate bulletin, 2010)

are some oscillation, such as two times suddenly increase in 1972 ~ 1973 and 1979 ~ 1980, and suddenly decrease in 2002 ~ 2003, which are lowerer than the standard value. The perennial precipitation in 2008 is slightly less 22 mm than in 2007, and less 10% than perennial. Precipitation is more concentrated, and the mean rainstorm days are increased. There is disequilibrium of seasonal precipitation, increase in summer, autumn and winter, but decrease in spring.

2.2 Climate change impact on regional water resource

Research shows that, climate change influence on all aspects of the natural, but one of most important factor is water resources, because it are found to be highly sensitive in the climate change. Hydrologic cycle will be inevitably change, which cause total water resources and spatial-temporal redistribution change, thereby water quality deteriorate (Chen Yanzhu, 1990). Since the 21th Century, there are new changes, which the hydrologic cycles of regions changes are different, but the common is decrease.

2.2.1 Climate change effects on water reserves

Water resources are mainly from precipitation, and there are litter transit water, so precipitation is water resources and the upper limit in the XSN zone (Qian Yanzhen, et al., 2001), namely upper limit of the water resources is equal to the annual precipitation multiplied by the area:

$$W = r \times D \times 10^3 \tag{1}$$

Where, W is the total amount of water resources limit, m^3/a; R is the regional annual precipitation, mm; D is the size of the area, km^2.

No all water resources usable, quite part of it will be loss, including evaporation and leakage. Evaporation is far greater than leakage, so the loss is main evaporation. Studies show that, comparing with rainfall and runoff depth, in general the smaller annual evaporation variation rate, which approximately take as a constant.

Currently available water is main the runoff, namely as runoff depth in hydrologic. The runoff is average precipitation minus average regional evaporation in the XSN zone. Runoff depth is multi-years annual runoff multiplied by the area:

$$W = f \times D \times 10^3 \tag{2}$$

Where, f is the regional mean runoff depth, $f = r - e$, e is the regional average evaporation.

In hydrological, the multi-years annual runoff is water reserves, The regional available water mainly depends on water reserves. Some scholars study (Chen Yanzhu, 1990) shows, with climate change, when the precipitation change in 46.8%, changes 88% on runoff, the ratio is 1∶2 in the XSN zone. So the annual precipitation of small variety will cause large changes on water resources.

So, with climate change, the water reserves relative amplitude is nearly 2 times than annual rainfall in the XSN zone, and the impact of climate change on water resources is very obvious.

2.2.2 Temporal disequilibrium of precipitation reduces the available water

The main precipitations are rainy and tropical cyclone rainstorm in the XSN zone, mainly concentrated in June and September. Plum rain period, there are often appear continuous rainstorm, such as the onset and out plum rains, flood control ability is limited in the XSN zone. In general it need release flood waters at earlier stage, only be stored in reservoir on the out plum rains, so much the available water resource is no well utilized.

Tropical cyclone (typhoon) is the same. Typhoon precipitation accounted for the total 10% ~ 20% in the XSN zone. Using statistical meteorological data in recent 50 years show that, there are annual 2 ~ 3 events of tropical cyclones influencing on the XSN zone, and a tremendous storm in per 2 years, causing serious consequences (Chen Yanzhu, 1990). With the change of climate, tropical cyclone become more frequent and violent, and the strong typhoon is more active. Tropical cyclone (typhoon) can bring out a lot of water resources, on the other hand, but the limit of flood control capacity, often causing regional flood disaster and the waste of water resources. As the tropical cyclone are suddenly bring out continuity and heavy rainstorm, often make release waters form reservoir, and available water is losses.

The rainfall becomes more concentrated by climate change. According to statistic, in recent 50 years, annual days of heavy rain and rainstorm are increased. Temporal disequilibrium of precipitation greatly reduces the available water rate.

2.2.3 Water resources more serious with the global warming

Climate changes lead to global warming. Since the 1980s, the annual temperature continually rise by human activities, environment and other reasons. The trend is very obvious in the 21th Century. it increases the evaporation, in generally, with the high temperature, the water demand is more, especially agricultural and living water (National assessment report on climate change, 2007). So water resources are more serious with the global warming.

2.2.4 The extreme winter weather to increase evaporation

Using statistical weather data in recent 50 years, it shows that, there are cold - air activities nearly every 5 ~ 7 d from December to February, which bring the 4 ~ 6 or up to 6 ~ 8 level southward strong wind (Qian Yanzhen, et al., 2001). Climate change aggravate it, the probability of warmer winters and extreme cold weather are increasing, with frequently cold - air activities, the evaporation will be greatly improve, which affect water reserves and other physical quantities. The cold - air activities are worsening, so, we should pay attention to protect water resources in winter.

2.3 Climate change impact on water quality

Study on the relationship between climate change and water quality is no comprehensive, because water quality study is more complex than water study. Many factors impact on water quality, such as climate, hydrology, land use and the interaction between natural and human activities, and the interaction multi - scales on the spatial and temporal impact scales.

The worsen of water quality and the serious consequences by climate change, included inferior water amount, the non - point source pollution load increased (runoff and permeability) by the strong precipitation, hydraulic facilities failure in flood, and waste water treatment measurements overload running during extreme precipitation (National assessment report on climate change, 2007).

Climate change impacts on water quality are multi - scales on spatial, mainly temperature. With water temperature increasing, the activation of molecular ratio increases, accelerating chemical reaction rate (kinetic) and nutrient cycling (Xia Jun, et al, 2008). In addition, it will reduce the concentration of dissolved oxygen, river self - purification ability and biological degradation available oxygen reducing.

According to the analysis, with climate warming, extreme events such as flood and drought will become more frequent and serious, and also influence to water quality. Currently, the probability of extreme precipitation is increasing in the XSN zone, and affecting pollutants ratio (nutrients, heavy metals, pathogenic microorganisms and toxic substances) with overland flow concentration into the river, non – point source pollutants go into the river with strong precipitation. During the flood, the city storm water drainage system and sewage pipe network may also be sources of pollution.

The drought also affects the water quality, but it is different with the flood in the mechanism. Because of drought, decreasing water have disadvantage of nutrient dilution and pollutant loads. In fact, dilution is not possible to fundamentally solve the water pollution, but for water pollution degree is very important. If less flow, water quality will be serious with the same sewage.

3 Strategies

Climate change is the undisputed fact, the climate changes bring many negative effects on water resources, climate change will aggravate water crisis on the future, especially in pollution induced water shortage regions, and the problem is worsen. It is necessary to evaluate climate change impact on water and put forward adaptation strategies (Bates B C, et al, 2008).

For pollution induced water shortage, water saving is the key to the solution of regional water problem. At the same time, it need strengthen comprehensive management of the water resources and environment, control the pollution, optimization allocation and the market mechanism, and establish water price policy for water saving and management system of quato. The aim is improving the water resources use efficiency, and realize the maximization benefit of economic, social and environmental. For comprehensive adapting to climate change, it need change the water management thought and concepts (Environment Agency, 2007), scientific assessment the influence of climate change on water evolution, optimizing basin and regional water optimal allocation, improving the spatial and temporal regulation capability (Brekke L D, et al,2009).

3.1 Pollution control

For adapting to climate change, and solving the pollution induced water shortage, it is the key to uninstall pollution load, namely is decrease the pollutant. It is purposes that reclaimed water reuse after unloading pollution. The water pollution control concept must change in practice. It need take the breakthrough point from pollutant reduction extended to the real unloading of the water environment and renovation of wastewater, and improving the social benefit of water pollution control.

Currently, using the transition period of Zhejiang Province' s industry economy, government should put forward to reasonable industry distribution, guide to priority developing green industry for reducing pollution, improve urban sewage treatment system, and strengthen pollution control efforts. At the same time, scientific fertilizers and pesticides will be implemented in agriculture, controlling continuous deterioration of pollution. Using the Zhejiang Province' s "Purification of River Pollution" Project, rivers will be scientific planning, overall arrangement and comprehensive management by stages. Adopting the dredge and water system connection, the aim is that increasing the storage capacity of river network, and improving the self – purification. According to the XSN zone located on coastal and relatively abundant precipitation, the measurements of diverting water to flush out pollutants and water exchange will be implemented, utilizing the advantage of tidal power, the flow – head of natural river ,and short duration extraordinary storm, which is beneficial to improve the water quality.

3.2 Water

Climate change aggressive spatial and temporal water disequilibrium in the XSN zone. So, the actions of built, building and under construction reservoirs will be fully developed, improved the water resources infrastructure construction, in order to as possible utilization of water resources. The allocation capacity will be enhanced by building water transfer projects. Flood – water resources utilization will improved with researching the utilization of unconventional water resources, fully using the regulation condition of rivers and lakes, and strengthening flood forecast and dispatching optimization. Seawater desalination technology will be developed with the advantage of coastal district. In order to solve the water resources disequilibrium, the distribution of industry must fully consider the accessibility of water.

In the upstream basin region of Cao'e and the Yongjiang River, water storage and water diversion project will be constructed, giving full play to the regional water resource. The water resources of Cao'e and Fuchun River can meet to regional water demand, using the Cao'e River sluice, which make river networks to interconnect in the XSN Plain, and improves the networks of supply and transport water. At the same time, a portion of water supplement from Fuchun River will divert into the XSN Plain. With above measures, the basic life, production and environment and ecological water demand should be ensured.

3.3 Existing water conservancy project operation plan adjustment

The water management of adapting to climate change need compile flexible operation project, for giving full play to the available water resources. The operation project based on the historical climate data, for example, the reservoir flood dispatching rule mainly using statistical hydrological and meteorological data, but it is the shortcoming of steady analysis. With the climate changing, the basis of meteorological and hydrological data will be changed. Considering the available hydrological conditions, weather forecast and the different scenarios of runoff, the modified water regulation rules can achieve satisfactory result. Only forcasting drought year, reservoir regulation will reduce the control capacity, and short – term of torrential rainfall, the flood vacated capacity should be arranged. Using statistical climate change, hydrology and socio – economic data, the more reasonable measure of storage water will be adopted under flood security, which will be achieved more socio – economic value than the present. Flood storage will be reassessment with the updated hydrologic records and future forecast. Joint operation of the river basin engineering will be safety and effective, if the project is taken as a part of whole instead of separately itself.

3.4 The conversion mechanism based on market of water resources

Referencing and development "Yiwu – Dongyang" water right trade model, the multi – stage water rights trading market will be established with cultivating incentive mechanism, which the water right and water resources will be free transaction in different users. Facing increasing trade stress by climate and demand pattern change, it should increase the opportunity of water right transfer between owner and user, with innovation application of water leasing, bank and market. Water transfer should be long – term, as example purchasing water rights; also be temporary, such as signing water contract in drought. The measure should save water in incentive mechanism with water market and the high price. Especially in drought, limited supply water resource should be more reasonable utilization.

3.5 Strengthen management and water saving

Water management need strengthen the water management of the demand and use. In order to comprehensively build a water – saving society, the key is decreasing the consumption of water with improving water saving and efficiency. The water authority will encourage people water – saving with the market means such as water measurement and price. On agriculture, it need select drought – enduring crops, improve irrigation method and renew farming technique. All above, the consumption of water will be reduced; efficiency and benefit will be continuous increased. The most strict water resources management will be implemented. At the same time, reclaimed water will reuse with strengthening reclaimed water reuse construction, giving full play to water resources. Using the measurement of water – saving and management, the water crisis will be reasonable solved in pollution induced water shortage areas.

4 Summary

The impact of climate change on water resources is the no - dispute fact. Especially in pollution induced water shortage areas, it is more significant. The paper shows that:

(1) With climate change, the precipitation worsening spatial - temporal disequilibrium and decreasing effective utilization of water resources led to the water shortage will be more serious in the XSN zone.

(2) Water quality will be worsening in pollution induced water shortage areas such as the XSN zone, because the original distributions of meteorological and hydrological data were changed by climate.

(3) For adapting to climate change, water saving is the key to solve pollution induced water shortage areas. It need control the pollution, optimizing of water resources allocation, improve the water resources infrastructure construction, strengthen management and water - saving. At the same time, water authority and society need establish water resources transfer market on market mechanism, and changing the idea of use and management. Above all, the reasonable utilized water and the most strict water resources management will be implemented.

Acknowledgements

This research was sponsored by the National Natural Science Foundation of China under Grant No. 51079130.

References

Bates B C, Kundzewicz Z W, Wu S, et al. Climate Change and Water. Technical Paper of the Intergovernmental Panel on Climate Change [R]. Geneva: IPCC Secretariat, 2008.

Environment Agency. Water Resources Planning Guideline [M/OL]. www. environmentagency. gov. uk, 2007.

Brekke L D, Kiang J E, Olsen J R, et al. Climate Change and Water Resources Management—A Federal Perspective [M]. U. S. Geological Survey Circular 1331, 2009.

Chen Yanzhu. Climate Change and Water Resources in Ningbo City [J]. Zhejiang Meteorological Science and Technology, 1990, 1: 5 - 6.

Wu Shaohong, Zhao Zongci. Climate Change and Water the Latest Scientific Cognition [J]. Advances in Climate Change Research, 2009, 5(3).

Xia Jun, Tanner Thomas, Ren Guoyu, et al. Climate Change on Water Resources in China Affect the Suitability Assessment and , Management Framework of [J]. Advances in Climate Change Research, 2008, 4(4).

"National Assessment Rport on Climate Change" the Compile Committee. National Assessment Report on Climate Change [M]. Beijing: Science Press, 2007.

Mao Minjuan, Chen Baode, Fang Gaofeng, et al. Global Cimate Cange of Zhejiang Province under the Regional Response of [C] //; Twenty - sixth session of Chinese Meteorological Society Annual Meeting on Climate Change of Field Collection, 2009.

Zhejiang Province Meteorological Bureau. Zhejiang Province Climate Bulletin [EB/OL]. http: // zj. weather. com. cn/qhbh/qhbhgb/12/1217649. shtml, 2010.

Qian Yanzhen, Zhang Jianxun, Hu Yadan, et al. Ningbo City Impact of Climate Change on Water Resources [J]. Meteorology, 2001, 27(6): 51 - 54.

Study on Runoff Trends of Key Hydrological Station in Typical Basins in China over the Past 60 Years

Zhang Liru[1,2], *Zhang Jianyun*[1,2], *Wang Guoqing*[1,2] and *Liu Jiufu*[1,2]

1. Nanjing Hydraulic Research Institute, Nanjing, 210029, China
2. Research Center for Climate Change, MWR, Nanjing, 210029, China

Abstract: The decreasing of stream runoff, directly affects water resources planning, development, utilization and the development of industrial and agricultural and soci – economic. In order to study the trends of runoff in china, this study selected several typical basins in Yellow River, Huaihe River, Haihe River and Liaohe River, which are seriously affected by climate change and human activity in recent years. Based on the observed streamflow of 12 key hydrological control stations in the four selected basins for the past 60 years, this study employed the Spearman test, the Kendall test and the Mann – Kendall test to detected trends of annual runoff respectively. Results showed that annual runoff in the four selected basins has generally decreased during the periods, especially in Yellow River, Haihe River and Liaohe River, there are remarkable decreasing trends.

Key words: runoff, inter – annual variation, climate change, trends

1 Runoff trends of key hydrological station in typical basins in china over the past 60 years

Results show that hydrological cycle in China has changed greatly by the effective of climate change and human activity in recent years. It was found that annual runoff in the six larger basins has generally decreased during the periods, with significant decreases detected in northern China. Especially in Yellow River, Huaihe River, Haihe River and Liaohe River, there are remarkable decreasing trends. In order to further study the trends of runoff in China, this study selected 12 key hydrological control stations from Yellow River, Huaihe River, Haihe River and Liaohe River, which are seriously affected by climate change and human activity in recent years. Based on the data of observed streamflow of 12 key hydrological control stations for nearly 60 years, this study analyzed the characters of variation of hydrological variables. The spatial distributions of 12 key hydrological control stations are showen in Fig. 1.

There are two indexes which can show the non – uniformity of the special distribution of rainfall (Jiao et al., 2001), here can be used to analyze the character of annual runoff.

$$C_v = \sqrt{\frac{\sum_{i=1}^{n}(Q_i/\sqrt{\overline{Q}} - 1)^2}{n-1}}, \qquad K_m = \frac{Q_{max}}{Q_{min}}$$

Where C_v is the coefficient of variation of annual runoff; K_m is the ratio between maximum and minimum annual runoff; Q_i is the annual runoff(m^3/s); $\overline{Q}$ is the average runoff of many years(m^3/s), Q_{max} is the maximal annual runoff(m^3/s), Q_{min} is the minimum annual runoff(m^3/s).

There are 12 key hydrological control stations selected from Yellow River Basin, Huaihe River Basin, Haihe River Basin and Liaohe River Basin. The ratio of extreme runoff K_m and coefficient of variation C_v of every key hydrological control station are shown in Tab. 1. From the Tab. 1, it is easy to find that the ratio of extreme runoff K_m and coefficient of variation C_v of every key hydrological control station is different from one to another, especially for the ratio of extreme runoff K_m of every key hydrological control station, the changes of some key hydrological control stations are larger, like Guantai hydrological control station, the value of K_m is 153.57, and some changes are smaller, like Daning hydrological control station, the value of K_m is only under 10.

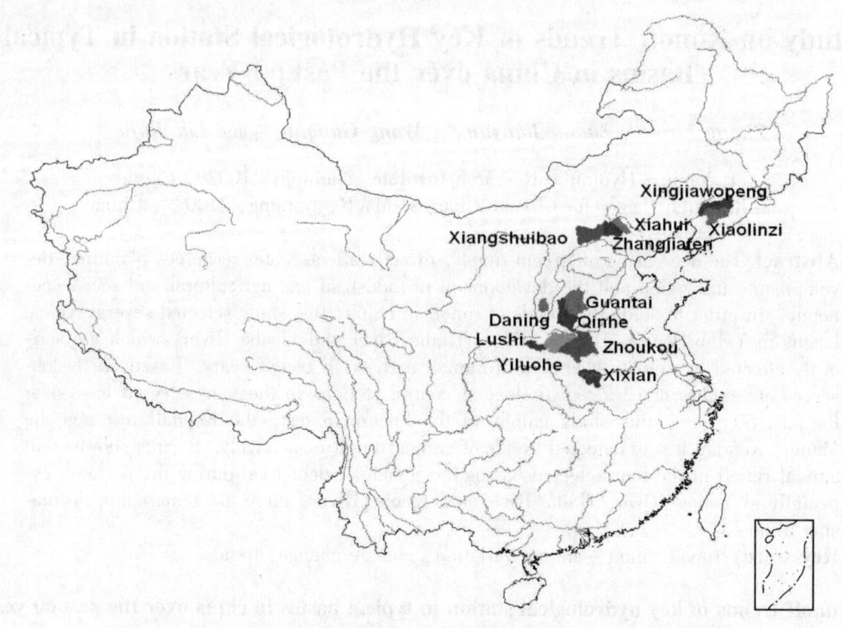

Fig. 1 Spatial distribution of key hydrological control stations

Tab. 1 Characters of annual runoff of typical basins

Basins	Station	Annual runoff (m^3/s)	C_v	The maximal year Runoff (m^3/s)	The maximal year Year	The minimal year Runoff (m^3/s)	The minimal year Year	K_m
Yellow River	Lushi	26.74	0.74	104.0	1958	3.27	1997	31.80
	Heishiguan	84.82	0.65	301.7	1964	17.6	1995	17.14
	Wulongkou	29.73	0.59	89.46	1963	4.83	1991	18.52
	Daning	4.69	0.58	13.13	1958	1.37	1997	9.58
Huaihe River	Zhoukou	107.65	0.65	376.19	1964	21.88	1966	17.19
	Xixian	120.55	0.53	294.42	1956	21.72	1966	13.56
Haihe River	Guantai	29.10	1.03	147.43	1963	0.96	1999	153.57
	Zhangjiafen	16.32	0.79	63.57	1956	2.02	2002	31.47
	Xiahui	8.17	0.70	28.14	1973	1.49	2002	18.89
	Xiangshuibao	10.78	0.67	31.27	1959	0.74	2001	42.26
Liaohe River	Xiaolinzi	72.18	0.55	176.9	1985	17.04	1980	10.38
	Xingjiawopeng	60.76	0.65	212.65	1995	13.15	1989	16.17

In order to further study the trend of streamflow, streamflow of years is analyzed as followed. Different streamflow variation of years can be analyzed by the mean and departure of streamflow in different years.

Tab. 2 The departure and mean runoff in different years

Station	1950 s		1960 s		1970 s		1980 s		1990 s		2000 s	
	Mean (m^3/s)	Departure (%)	Mean (m^3/s)	Departure (%)	Mean (m^3/s)	Departure (%)	Mean (m^3/s)	Departure (%)	Mean (m^3/s)	Departure (%)	Mean (m^3/s)	Departure (%)
Lushi	51.7	97.9	35.7	36.6	20.0	-23.5	28.2	7.9	12.6	-51.8	23.1	-11.6
Heishiguan	132.2	47.9	112.3	25.6	64.9	-27.5	95.5	6.8	45.9	-48.7	78.7	-12.0
Wulongkou	58.1	90.9	46.8	53.8	28.0	-8.0	20.7	-32.0	15.7	-48.4	19.1	-37.2
Daning	9.4	115.5	6.5	49.0	4.6	5.5	3.4	-22.1	3	-31.2	1.7	-61.0
Zhoukou	137.0	29.5	128.1	21.1	80.4	-24.0	105.8	-0.01	89.3	-15.6	94.3	-10.9
Xixian	130.1	8.6	121.3	1.3	109.5	-8.6	128.3	7.1	110.7	-7.6	118.7	-0.9
Guantai	61.3	110.2	56.5	93.7	30.9	6.0	11.2	-61.6	10.6	-63.7	9.6	-67.1
Zhangjiafen	39.1	139	18.5	16.4	20.5	22.6	9.4	-44.9	11.2	-32.6	6.0	-63.2
Xiahui	-	-	9.67	22.4	12.2	46.9	6.5	-14.3	8.8	10.2	2.8	-63.3
Xiangshuibao	19.3	78.6	15.4	42.5	14.3	32.4	7.4	-31.5	6.5	-39.8	1.8	-83.3
Xiaolinzi	95.2	32.7	83.3	16.2	69.6	-3.0	72.7	1.4	61.3	-14.5	56.7	-20.9
Xingjiawopeng	79.3	26.6	67.2	7.3	55.2	-11.9	59.9	-4.4	64.1	2.3	57.5	-8.3

From Tab. 2, it is interesting to note that there is a various decreasing trend in one degree or another for annual runoff in typical hydrological control stations of Yellow River, Huaihe River, Haihe River and Liaohe River basins after 1970 s. The largest amount of decrease for 1970 s is Xiahui hydrological control station which is selected in Haihe River basin, with a decrease of more than 45%. The largest amount of decrease for 1980 s is Guantai hydrological control station which is selected in Haihe River Basin, with a decrease of more than 60%. The larger amount of decrease for 1990s are Lushi hydrological control station in Yellow River Basin and Guantai hydrological control station in Haihe River Basin, with a decrease of more than 50% and more than 60% percent respectively. After 2000 years, the largest amount of decrease occurred in Haihe River Basin, with 60% of decrease for all the four typical hydrological control staions.

2 Methodology

2.1 Spearman test

Spearman test is one non - parametric rank - order test. Given a sample data set $\{X_i, i=1,2,\cdots,n\}$, the null hypothesis H_0 of the test against trend tests is that all the X_i are independent and identically distributed; the alternative hypothesis is that X_i increases or decreases with i, that is, trend exists. The test statistic is given by (Sneyers, 1990)

$$r = 1 - \frac{6 \cdot \sum_{i=1}^{n} d_i^2}{n^3 - n}$$

And

$$d_i = R_i - i$$

where, R_i is the rank of ith observation X_i in the sample of size n.
If R_i is similar to i, the value of d_i will be little, and the value of r will be big, and the trend of series will be significant.
In general, T test is used to detect the significance of series trend; T test statistic is given by

$$T = r\sqrt{(n-4)(1-r^2)}$$

Under the null hypothesis of no trend, the distribution of T is asymptotically T with a degree of freedom of $n-2$, the test statistic can be calculated. And then select the significance level of α. At the significance level of α, finding the value of $t_{\alpha/2}$ in the table of t distribution, if $|T| \geqslant t_{\alpha/2}$, then existing trend is considered to be statistically significant.
Statistical T can be made as levels of measurement of significance of hydro – series, with the value of $|T|$ large enough, the trend significance of series can be got to some extent.

2.2 Kendall test

The Kendall test is based on the test statistic U defined as follows:

$$U = \frac{\tau}{[V_{ar}(\tau)]^{1/2}}$$

And

$$\tau = \frac{4p}{n(n-1)} - 1$$

$$V_{ar}(\tau) = \frac{2(2n+5)}{9n(n-1)}$$

where, n is the lenth of data set. p is the number of allelomorph if $x_i < x_j$. When n increases, the statistic U is approximately standard normal distribution.

Suppose the trend of time series is no significance, and then select the significance level of α. At the significance level of α, finding the value of $U_{\alpha/2}$ in the table of normal distribution, if $|U| \geqslant U_{\alpha/2}$, then existing trend is considered to be statistically significant.

2.3 Mann – kendall test

The time series of all the hydrologic variables were analyzed using the MK non – parametric test for trend. Mann(1945) originally used this test and Kendall(1975) subsequently derived the test statistic distribution. This test was found to be an excellent tool for trend detection by other researchers in similar applications.

The MK test is based on the correlation between the ranks of a time series and their time order. For a time series $X = \{x_1, x_2, x_3, \cdots, x_n\}$, the test statistic S_k defined as follows:

$$S_k = \sum_{i=1}^{k} r_i (k = 2,3,4,\cdots,n)$$

where

$$r_i = \begin{cases} +1 & x_i > x_j \\ 0 & x_i \leqslant x_j \end{cases} \quad (j = 1,2,\cdots,n)$$

And n is the sample size, the x_i and x_j are the sequential data values.

If no ties between the observations are presented and no trend is present in the time series, the standardized test statistic UF_k is computed by

$$UF_k = \frac{[S_k - E(S_k)]}{\sqrt{var(S_k)}} \quad (k = 1,2,\cdots,n)$$

where, $UF_k = 0$, $E(S_k)$, and $\mathrm{var}(S_k)$ are the mean and the variation of the time series.

If these time series $x_1, x_2, \cdots, x_n$ are independent of each other, and have the same distribution, then they can be calculated as follows:

$$E(S_k) = \frac{k(k+1)}{4}$$

$$\mathrm{var}(S_k) = \frac{k(k-1)(2k+5)}{72}$$

UF_i is standard normal distribution, they are statistic series which are calculated by sequential data values of $x_1, x_2, \cdots, x_n$, At the significance level of α, find the value of U_α in the table of normal distribution, if $|UF_i| \geqslant U_\alpha$, then existing trend is considered to be statistically significant.

3 Significance of trend results

The results of the trend tests can be used to determine whether or not the observed collection of time series for a hydrologic variable exhibit a number of trends that is greater than the number that is expected to occur by change.

Results obtained from the trend tests were analyzed using a local significance level of 5%. Results caculated by Spearman test and Kendall test for 12 typical stations were showed in Tab. 3.

Tab. 3 Test of runoff change for typical stations

Basins	Station	Spearman $\|T\|$	critical value	Kendall $\|U\|$	critical value	Slope of runoff (m^3/s)	Trend
Yellow River	Lushi	4.04	2.01	3.61	1.96	-0.82	significance
	Heishiguan	4.16	2.01	3.66	1.96	-1.83	significance
	Wulongkou	6.34	2.01	4.96	1.96	-1.17	significance
	Daning	4.68	2.01	4.03	1.96	-0.14	significance
Huaihe River	Zhoukou	1.72	2.01	1.62	1.96	-0.95	no significance
	Xixian	0.16	2.01	0.19	1.96	-0.12	no significance
Haihe River	Guantai	7.66	2.01	5.66	1.96	-1.16	significance
	Zhangjiafen	8.40	2.01	6.04	1.96	-0.52	significance
	Xiahui	4.23	2.01	3.62	1.96	-0.17	significance
	Xiangshuibao	11.83	2.01	7.39	1.96	-0.34	significance
Liaohe River	Xiaolinzi	2.96	2.01	2.98	1.96	-0.89	significance
	Xingjiawopeng	2.36	2.01	2.32	1.96	-0.46	significance

From the statistical results (Tab. 3), it is easy to find that the statistic $|T|$ and $|U|$ for every key hydrological control stations selected from Yellow River Basin, Haihe River Basin and Liaohe River Basin are significant bigger than the value of significance level of 5%, and shows remarkable decreasees trend. Especially for Heishiguan station, Wulongkou station and Guantai station, with slope of runoff of 1.83 m^3/s, 1.17 m^3/s and 1.16 m^3/s respectively.

In order to further study the character of streamflow, the Mann – Kendall non – parametric test were used to detecte the trend of runoff for every control station. Fig. 2, Fig. 3 and Fig. 4 show the results of Mann – Kendall test for annual Mean runoff trend for key stations in Yellow River Basin,

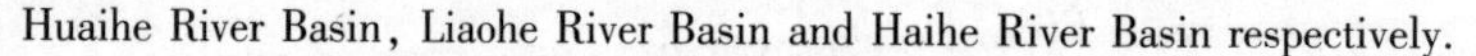
Huaihe River Basin, Liaohe River Basin and Haihe River Basin respectively.

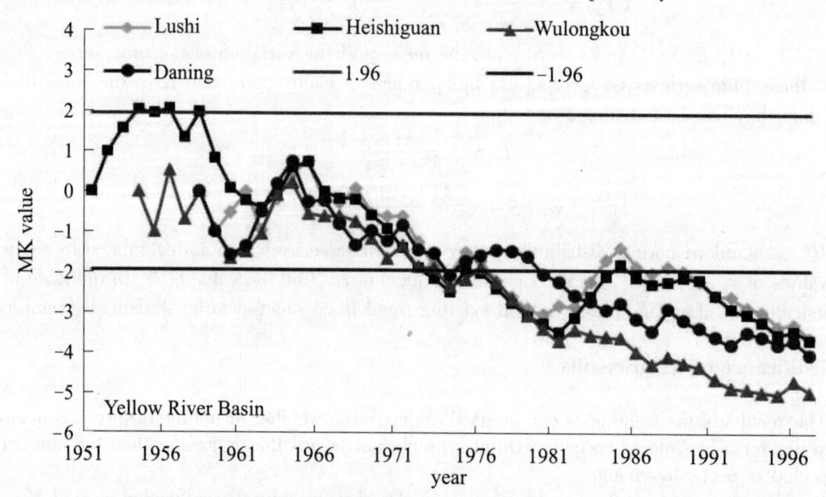

Fig. 2 Mann-Kendall test for annual mean runoff trend for key stations in Yellow River Basin

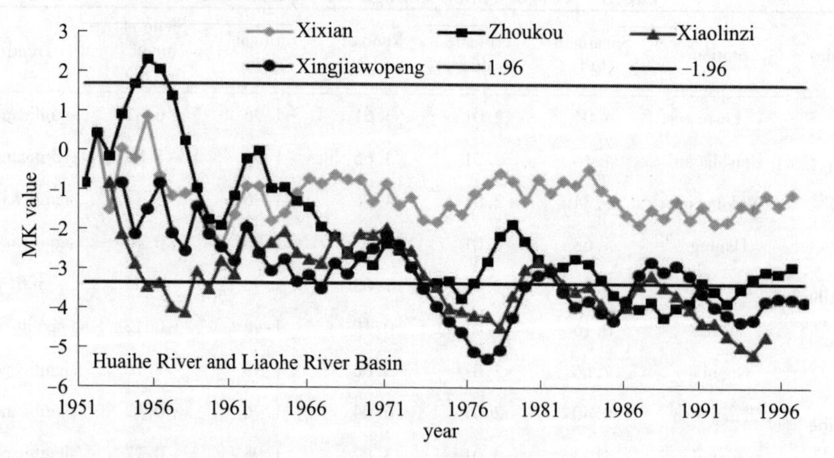

Fig. 3 Mann - Kendall test for annual mean runoff trend for key stations in Huaihe River and Liaohe River Basins

Fig. 2 shows Mann-Kendall test for annual mean runoff trend for key stations in Yellow River Basin. It is easy to find that the trend of runoff decreasing for all four key hydrological control stations are remarkable since the year of 1965, especially for Lushi station between 1974 and 1983, with average MK value of -2. 39, and far exceed the value of significance level of 5%. For Heishiguan station, runoff decreasing remarkably occurred the year between1973 and 1984, with average MK value of -2. 62, and also far exceed the value of significance level of 5%. For Wulongkou station, runoff decreasing remarkably occurred the year of 1974, with average MK value of -3. 76, and also far exceed the value of significance level of 5%. For Daning station, runoff decreasing remarkably occurred the year of 1980, with average MK value of -3. 18.

Fig. 3 shows Mann - Kendall test for annual mean runoff trend for key stations in Huaihe River Basin and Liaohe River basin. For Xixian station, runoff decreasing trend was not remarkable, with all the MK value within the confine of significance level of 5%. For Zhoukou station, runoff de-

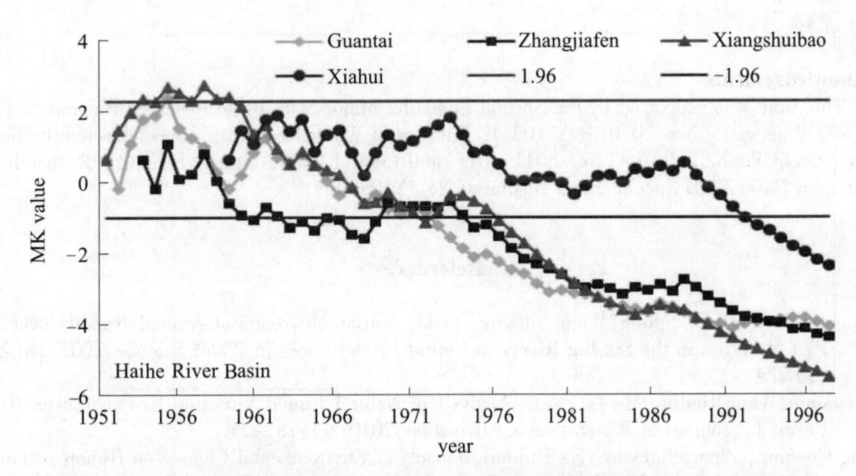

Fig. 4 Mann – Kendall test for annual mean runoff trend for key stations in Haihe River Basin

creasing trend was not remarkable as a whole, but remarkable between the year of 1993 and 1999, with average MK value of -2.39, and far exceed the value of significance level of 5%. For Xiaolinzi station and Xingjiawopeng station, runoff decreasing occurred since the year of 1950. For Xiaolinzi station, runoff decreasing remarkably occurred the year of 1979, with average MK value of -2.36, before the year of 1979, runoff decreasing and runoff increasing occurred alternately. For Xingjiawopeng station, runoff decreasing trend was not remarkable before the year of 1978, with slight decreasing trend but within the confine of significance level of 5%; after the year of 1978, runoff decreasing are remarkable, with average MK value of -2.34.

Fig. 4 shows Mann – Kendall test for annual mean runoff trend for key stations in Haihe River Basin. It is also easy to find that the trends of runoff decreasing for all four key hydrological control stations are remarkable in the whole. For Guantai station, runoff decreasing occurred since the year of 1964, but remarkable after the year of 1978, with average MK value of -4.67. For Zhangjiafen station, runoff decreasing occurred since the year of 1960, but remarkable after the year of 1981, with average MK value of -4.41. For Xiangshuibao station, runoff decreasing occurred since the year of 1963, but remarkable after the year of 1983, with average MK value of -5.15. For Xiahui station, runoff decreasing occurred a little later, since the year of 1979, but remarkable after the year of 2003, with average MK value of -2.82.

4 Conclusion

This study demonstrates that the annual of every typical station becoming decreasing since 1960s, especially from later of 1970s. It was found that annual runoff was remarkable decreasing in Yellow River Basin and Haihe River basin, with 40% ~ 50% decreasing in Yellow River and 30% ~ 70% Bdecreasing in Haihe River basin after the year of 1970. From the value of Mann-Kendall test for annual mean runoff trend in the four basins, it is easy to find that the trend of runoff decreasing for key hydrological control stations in Yellow River Basin, Haihe River Basin and Liaohe River Basin are remarkable. with average MK value of -2.39 for lushi station, with average MK value of -2.62 for heishiguan station, with average MK value of -3.76 for wulongkou station, with average MK value of -3.18 for daning station, with average MK value of -2.36 for xiaolinzi station, with average MK value of -2.34 for xingjiawopeng, with average MK value of -4.67 for guantai station, with average MK value of -4.41 for zhangjiafen station, with average MK value of -5.15 for xiangshuibao station and with average MK value of -2.82 for xiahui station, which are far exceed the value of significance level of 5%. For Xixian station and Zhoukou station, runoff decreasing trend was not remarkable, with all the MK value within the confine of significance level of 5%.

Acknowledgements

This work was supported by the Special Funds for Major State Basic Research Program of China (973 Program) (No. 2010CB951103), Ministry of Water Resources' Special Scientific funds for non-profit Public industry (No. 201101015) and Central Public-interest Scientific Research Institutes for Basic R&D Special Fund Business (No. Y509004).

Referance

Zhang Jianyun, Zhang Silong, Wang Jinxing, et al. Variation Trends of Annual Runoffs over the Past 50 Years in the Six Big Rivers in China[J]. Advances in Water Science, 2007, 18(2): 230-234.

Liu Guojun, Wang Hailin, Ma Li, et al. Analysis of Natural Runoff Variation Law in Dawen River Basin[J]. Journal of Water - saving Irrigation, 2010(6): 18 - 25.

Wang Guoqing, Zhang Jianyun, He Ruimin. Impacts of Environmental Change on Runoff in Fenhe River Basin of the Middle Yellow River[J]. Advances in Water Science, 2006, 17(6): 853 - 858.

Zhang Jianyun, Wang Guoqing, He Ruimin, et al. Variation Trends of Runoff in the Middle Yellow River Basin and Its Response to Climate Change[J]. Advances in Water Science, 2009, 20(2): 153 - 158.

Liu Chunzhen, Liu Zhiyu. Study of Trends in Runoff for the Haihe River Basin in Recent 50 Years [J]. Journal of Applied Meteorological Sciench, 2004, 15(4): 385 - 393.

Wang Jinxing, Zhang Jianyun, Li Yan, et al. Variation Trends of Runoffs Seasonal Distribution of the Six Larger Basins in China over the Past 50 Years[J]. Advances in Water Science, 2008, 19(5): 656 - 661.

Zhang Jianyun, Wang Guoqing. Impact of Climate Change on Hydrology and Water Resources [M]. Beijing: Science Press, 2007.

Hirsch RM, Slack JR. A Nonparametric Trend Test for Seasonal data with Serialdependence[J]. Water Resources Research, 1984, 20(6): 727 - 732.

Hirsch RM, Slack JR, Smith RA. Techniques of Trend Analysis for Monthly Water Quality Data [J]. Water Resources Research, 1982, 18(1): 107 - 121.

The Impact of Climate Change on Water Resources in the Upper Yellow River Basin in the 21st Century

***Bao Zhenxin*[1,2], *Wang Guoqing*[1,2], *Yan Xiaolin*[1,2], *Song Xuan*[3], *Zhang Aijing*[4] and *Shang Manting*[1,2]**

1. State Key Laboratory of Hydrology - Water Resources and Hydraulic Engineering, Nanjing Hydraulic Research Institute, Nanjing, 210029, China
2. Research Center for Climate Change, MWR, Nanjing, 210029, China
3. Nanjing Water Planning and Designing Institute, Ltd., Nanjing, 210006, China
4. Faculty of Infrastructure Engineering, Dalian University of Technology, Dalian, 116024, China

Abstract: Climate change would lead to a significant impact on hydrological cycle and water resources system. The water resources in the Upper Yellow River Basin (UYRB) in the 21st century was assessed by the Variable Infiltration Capacity (VIC) model and PRECIS under three scenarios. With the VIC model being calibrated in four sub - catchments of the UYRB, the model parameters in other regions were estimated by spatial proximity and physical similarity. As a result, the hydrological regime could be simulated for the whole UYRB, and that was used to investigate the relative change of water resources in future scenarios (2010 ~ 2099) compared to it in baseline period (1961 ~ 1990), using the output from PRECIS. The results indicated that: 1) There might be a sustaining decreasing trend for water resources in the 21st century under the three scenarios in UYRB, with a lowest decreasing trend in 2020s, followed by 2050s, and then 2080s. 2) The highest decreasing trend was under A2 scenario for every period. In 2020s and 2080s, the lowest decreasing trend was under B2 scenario, but A1B scenario had the lowest decreasing trend in 2050s. 3) There would be a high variability of the decreasing trend for water resources in different grids, and the decreasing trend in west part was higher than that in east part of UYRB. The decrease of future water resources might lead to more serious water shortage and related economic and environmental problems in the Yellow River basin during the 21st century.

Key words: water resources, climate change, Upper Yellow River, impact

1 Introduction

The Yellow River is the second longest river and regarded as the "Mother River" of China, but its water shortage problem is very severe. Although the drainage basin area accounts for 8.3% of the total area of China, the water resources in the watershed accounts for only 2.6% of the total water resources. The amount of water resources per capita in the Yellow River Basin is only 1/3 of the average in China. The area and water resources in the upper reaches (above the Lanzhou station) are 28.0% and 47.3% of the whole Yellow River Basin, respectively. The water crisis in the Yellow River Basin is becoming more critical now. For example, there is a statistically significant decreasing trend for observed streamflow in the Yellow River Basin during the last several decades (Zhang, et al., 2007; Fu, et al., 2004). Future climate change might lead to a significant impact on hydrological cycle and water resources system (Bao, et al., 2011). Thus it is essential and important to assess the impact of climate change on future water resources in the Upper Yellow River Basin (UYRB) for regional water resources programming and management, especially under future global warming scenarios.

Generally, the impact of climate change on future water resources is assessed by the following stages: ①projecting future climatic scenarios from GCMs simulations; ②downscaling the global - scale climatic variables to regional - scale; ③calibrating the hydrological model with observed hydro - climatic data; ④assessing water resources using calibrated hydrological model and climatic scenarios (Xu, 1999). Based on this methodology, the impact of climate change on water resources was widely studied in many basins over the world. Arnell (1999) studied the effects of climate change by the 2050s on hydrological regimes at the continental scale in Europe, at a spatial resolution of 0.5° ×0.5°, using a macro - scale hydrological model and four climate change scenarios, and pointed out that there would be a general reduction in annual runoff in southern Europe, and an increase in the north. Vaze, et al. (2011) using two hydrological models: SIMHYD and Sacramento model, investigated the impact of future climate on runoff generation in the Macquarie - Castlereagh River basin in Australia, based on outputs from 15 of the 23 IPCC AR4 GCMs for the A1B global warming scenario. Setegn, et al. (2011) investigated the impact of climate change on the hydroclimatology of Lake Tana basin in Ethiopia, using 15 GCMs and SWAT model. Wang, et al., (2012) assessed the impact of climate change on water resources during 2021 ~ 2050 in China, using the Variable Infiltration Capacity model and PRECIS.

2 Study area and dataset

2.1 Study area

The Upper Yellow River Basin (above the Lanzhou station) is located in the northwest China (95.8° ~ 104.3° E, 32.7o ~ 38.3° N), with the elevation varying from 1,532 m to 6,253 m (Fig. 1). Covering 222,551 km^2 drainage areas, most of the UYRB are located in Qinghai province, and others are in Gansu Province and Sichuan province. UYRB has a semi - humid plateau climate. The annual mean temperature and precipitation of UYRB are -0.83 ℃ and 471.81 mm respectively (1951 ~ 2008). The precipitation in flood season (June - September) generally contributes most parts of the annual precipitation. The length of the main stream in UYRM is 2,119 km, and there are some major sub - rivers following into it, such as Daxiahe River, Taohe River, Huangshui River, etc.

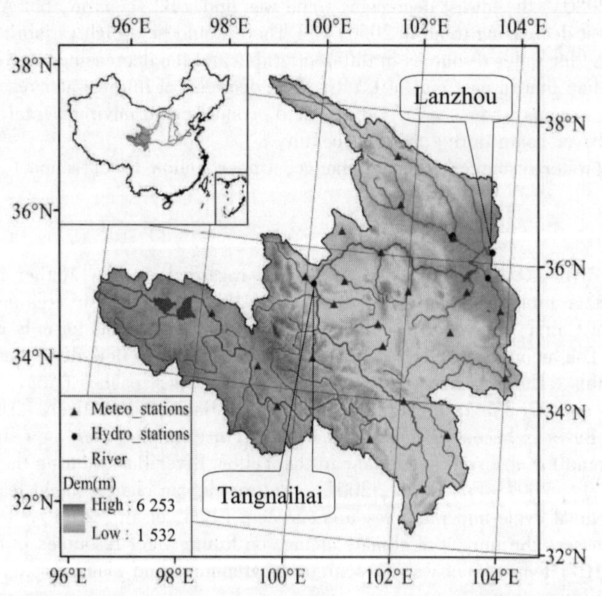

Fig. 1 Upper Yellow River Basin and its location in China

2.2 Observed and GCM dataset

The meteorological data in 23 stations with daily precipitation, mean temperature, maximum temperature and minimum temperature from 1951 to 2008 were collected from National Meteorological Administration of China, which applied data quality control before releasing these data. Using the inverse distance weighting method, the station data were interpolated into grid data as the input for hydrological model. Five hydrological stations were used in this study (Tab. 1). The monthly streamflow data at the five hydrological stations were extracted from the "China's Hydrological Year Book" which was published by the Hydrological Bureau of the Ministry of Water Resources, China.

Tab. 1 The basic information of the five hydrological stations

Catchment	River	Lon. (°E)	Lat. (°N)	A(km^2)	P (mm)	T(℃)	Available data
Tangnaihai	Yellow River	100.15	35.50	121,972	500.8	-2.06	1956 ~ 1976
Minhe	Huangshui River	102.80	36.33	15,342	443.3	1.80	1968 ~ 1988
Xiangtang	Datonghe River	102.83	36.35	15,126	410.0	-2.37	1953 ~ 1973
Hongqi	Taohe River	103.57	35.80	24,973	561.2	2.18	1955 ~ 1975
Lanzhou	Yellow River	103.82	36.07	222,551	471.8	-0.83	1968 ~ 1986

Note: A: Area; P: annual precipitation; T: annual mean temperature.

The future precipitation and temperature scenarios were obtained from the PRECIS, which was based on the Hadley Centre's regional climate modeling system. PRECIS was developed in order to help generate high - resolution climate change information for as many regions of the world as possible. The spatial resolution of the PRECIS used in this study was 0.25° in latitude and in longitude. In order to cover different social - economic development scenarios, three scenarios: SRES A1B, A2, and B2, were used in this study. The changes compared to the baseline period (1961 ~ 1999) from the 20C3M, were calculated for three periods: 2020s (2010 ~ 2039), 2050s (2040 ~ 2069), and 2080s (2070 ~ 2099).

3 Methodology

3.1 Assessing the impact of climate change on water resources

The relative runoff change (ΔR_{Re}) in future scenarios compared to it in baseline period is expressed as:

$$\Delta R_{Re} = \frac{R_{FG} - R_{BG}}{R_{BG}} \times 100\% \tag{1}$$

where, R_{FG} is simulated average annual runoff using data in future scenarios from GCMs, and R_{BG} is simulated average annual runoff using data in baseline period from GCMs.

3.2 A brief introduction of the VIC model

The Variable Infiltration Capacity (VIC) model is used for streamflow simulation in the UYRB. VIC is a semi - distributed macro - scale hydrological model, and could balance both the water and surface energy budgets within the grid cell (Liang, et al., 1994, 1996). The key characters of the VIC model are the representation of multiple land cover types, spatial variability of soil moisture capacity, soil water moving between the three soil layers, surface flow considering both infiltration excess and saturation excess, and non - linear base flow. With refined describing of hydrologic process on land surface and finer performance of streamflow simulation, VIC model was applied in a number of catchments over the world. For more information about the VIC model, the reader is referred to the official VIC web - site: http://www.hydro.washington.edu/Lettenmaier/Models/VIC.

There are six parameters in the VIC model needing to be calibrated: b, D_m, D_s, W_s, d_2, and d_3. The parameter d_2 and d_3 are the thicknesses of the second and third soil layer, respectively. The six parameters are optimized by two objectives: Nash - Sutcliffe coefficient (Nsc) and relative error (Re), which are defined as:

$$Nsc = 1 - \frac{\sum (Q_{obs} - Q_{sim})^2}{\sum (Q_{obs} - \overline{Q}_{obs})^2}$$

$$Re = \frac{R_{sim} - R_{obs}}{R_{obs}} \times 100\% \quad (2)$$

where, Q_{obs} and Q_{sim} are the observed and simulated streamflow, respectively; $\overline{Q}_{obs}$ is the mean value of Q_{obs}; R_{obs} and R_{sim} are the observed and simulated average annual runoff respectively.

The VIC model was firstly calibrated in four hydrological stations: Tangnaihai, Minhe, Xiangtang, and Hongqi. The model parameters for the ungauged regions were estimated by a parameter regionalization methodology: spatial proximity and physical similarity (Oudin, et al., 2008). Then the hydrological regime for the UYRB could be simulated.

4 Results and discussion

4.1 Hydrological simulation

Based on parameters calibration, the VIC model was used for streamflow simulation in four catchments, at a 0.25° spatial and daily temporal resolution. The monthly Nsc and Re values denoted good performance of VIC model (Tab. 2). For example, the Nsc and Re values were 0.91 and −2.14%, respectively, at Tangnaihai station. The model parameters in other regions were estimated by spatial proximity and physical similarity. Then streamflow at the Lanzhou station could be simulated (Fig. 2). The simulations fitted the observations well, although there were some obvious differences at the peak and vale values. Between the Tangnaihai station and Lanzhou station, there were three reservoirs: Longyangxia, Lijiaxia, and Liujiaxia reservoir, in the main stream of the Yellow River. Because of the regulation of reservoir, the simulated streamflow was higher than observed streamflow at the flood season, vice versa. Overall, the accuracy of VIC model was acceptable for monthly streamflow simulation in the UYRB.

Tab. 2 The performance of VIC model in the four catchment

Station	Tangnaihai	Minhe	Xiangtang	Hongqi
Nsc	0.91	0.73	0.78	0.89
Re (%)	−2.14	−2.02	−0.79	−2.69

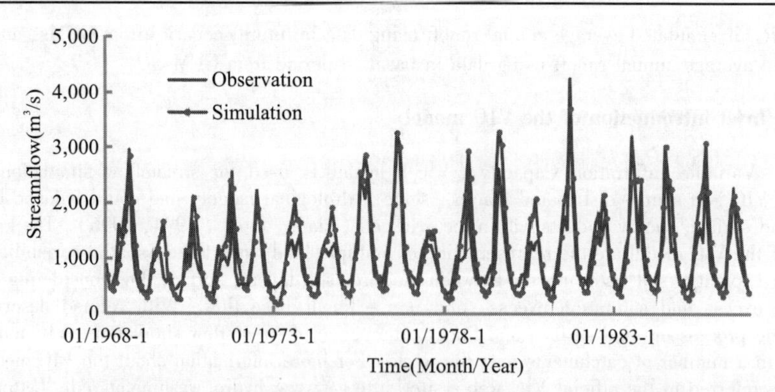

Fig. 2 Observed and simulated monthly streamflow at the Lanzhou station

4.2 Future scenarios for water resources

In UYRB, the relative change of water resources in the future scenarios compared to it in baseline period was expressed in Tab. 3. Generally, there might be a sustaining decreasing trend for water resources in the 21st century under the three scenarios, with a lowest decreasing trend in 2020s, followed by 2050s, and then 2080s. For example, under A1B scenario, the water resources in UYRB would decreased by 8.52%, 16.67%, and 27.89% in 2020s, 2050s, and 2080s, respectively, compared to it in the baseline period. The highest decreasing trend was under A2 scenario, and was -9.77%, -21.93%, and -31.57%, in 2020s, 2050s, and 2080s, respectively. In 2020s and 2080s, the lowest decreasing trend was -7.87% and -23.31%, respectively, under B2 scenario, but A1B scenario had the lowest decreasing trend (-16.67%) in 2050s.

Tab. 3 The relative change of water resources in future scenarios compared to it in baseline period for the whole region

Periods	A1B	A2	B2
2020s	-8.52%	-9.77%	-7.87%
2050s	-16.67%	-21.93%	-17.07%
2080s	-27.89%	-31.57%	-23.31%

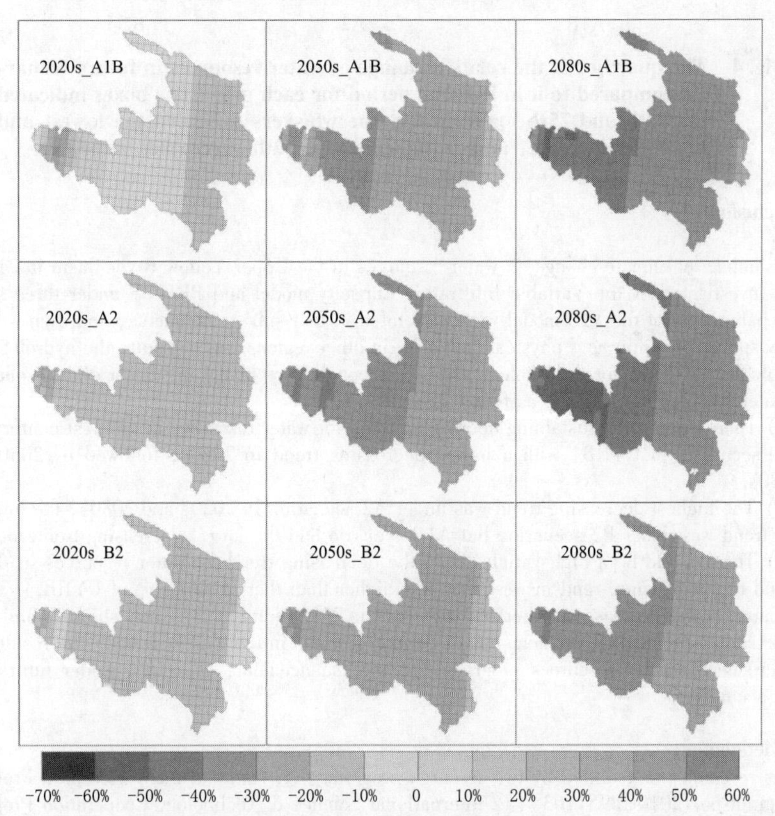

Fig. 3 The spatial distribution of the relative change of water resources in future scenarios compared to it in baseline period for each grid

Spatially, there was a specific distribution of the relative change of water resources in future scenarios compared to it in baseline period (Fig. 3). The water resources would decrease at each grid in 21st century under the three scenarios, except for it in 2020s under A1B scenario, in which the water resources in west UYRB would increase. Generally, there was a high variability of the decreasing trend for water resources in different grids, and the decreasing trend in west part was higher than that in east part of UYRB (Fig. 3, Fig. 4). For example, in 2050s under A2 scenario, the highest decreasing trend for water resources was −51%, but the lowest one was only −23%.

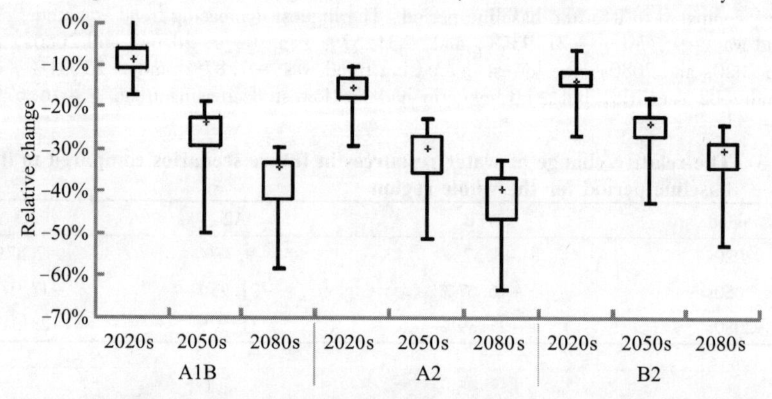

Fig. 4 The quartiles of the relative change of water resources in future scenarios compared to it in baseline period for each grid (The boxes indicated the 25th and 75th, percentiles; the whiskers indicated the lowest and highest data value; and + indicated the 50th percentiles value)

5 Conclusion

The impact of climate change on water resources in the Upper Yellow River Basin in 21st century was investigated by the Variable Infiltration Capacity model and PRECIS under three scenarios. The parameters of the VIC model were calibrated in four sub − catchments, and then were estimated by spatial proximity and physical similarity in other regions. As a result, the hydrological regime could be simulated for the whole UYRB. The conclusions of the impact of climate change on future water resources were made as following:

(1) There might be a sustaining decreasing trend for water resources in the 21st century under the three scenarios in UYRB, with a lowest decreasing trend in 2020s, followed by 2050s, and then 2080s.

(2) The highest decreasing trend was under A2 scenario. In 2020s and 2080s, the lowest decreasing trend was under B2 scenario, but A1B scenario had the lowest decreasing trend in 2050s.

(3) There would be a high variability of the decreasing trend for water resources in different grids, and the decreasing trend in west part was higher than that in east part of UYRB.

Because of the decrease of water resources in the 21st century, the water shortage and related economic and environmental problems might be more serious in the Yellow River. The results could be a reference for water resources programming and management, especially under future global warming scenarios.

Acknowledgement

This research was founded by two research programs: ① National Basic Research Program of China (grant no. 2010CB951103); ②International Science & Technology Cooperation Program of China (grand no. 2010DFA24330).

References

Arnell N W. The Effect of Climate Change on Hydrological Regimes in Europe: A Continental Perspective[J], Global Environmental Change, 1999, 9: 5 − 23.

Bao Z, Zhang J, Liu J, et al. Sensitivity of Hydrological Variables to Climate Change in the Haihe River Basin[J], China, Hydrological Processes, 2012,26(15):2294 – 2306.

Fu G, Chen S, Liu C, et al. Hydro – climatic Trends of the Yellow River Basin for the Last 50 Years[J], Climatic Change, 2004, 65: 149 – 178.

Liang X, Lettenmaier D P, Wood E F, et al. A Simple Hydrologically Based Model of Land Surface Water and Energy Fluxes for General Circulation Models[J], Journal of Geophysical Research, 1994, 99: 14415 – 14428.

Liang X, Wood E F, Lettenmaier D P. Surface Soil Moisture Parameterization of the VIC – 2L Model: Evaluation and Modification[J], Global and Planetary Change, 1996, 13: 195 – 206.

Oudin L, Andréassian V, Perrin C, et al. Spatial Proximity, Physical Similarity, Regression and Ungaged Catchments: A Comparison of Regionalization Approaches Based on 913 French Catchments [J], Water Resources Research, 2008, 44, W03413, doi: 10. 1029/2007WR006240.

Setegn S G, Rayner D, Melesse A M, et al. Impact of Climate Change on the Hydroclimatology of Lake Tana Basin, Ethiopia[J]. Water Resources Research, 2011, 47, W04511, doi: 10. 1029/2010WR009248.

Vaze J, Davidson A, Teng J, et al. Impact of Climate Change on Water Availability in the Macquarie – Castlereagh River Basin in Australia[J]. Hydrological Processes, 2011, doi: 10. 1002/hyp. 8030.

Wang G, Zhang J, Jin J, et al. Assessing Water Resources in China Using PRECIS Projections and a VIC Model[J]. Hydrology and Earth System Sciences, 2012, 16: 231 – 240.

Xu C. From GCMs to River Flow: A Review of Downscaling Methods and Hydrologic Modeling Approaches[J]. Progress in Physical Geography, 1999, 23(2): 229 – 249.

Zhang J, Zhang S, Wang J, et al. Study on Runoff Trends of the Six Larger Basins in China over the Past 50 Years[J]. Advances in Water Science, 2007, 18(2): 230 – 234 (in Chinese).

Runoff Simulation in the Heiyukou Sub - basin of Weihe River Basin and its Responses to Climate Change Based on HIMS

Liang Kang[1,2], *Liu Changming*[1,3], *Wang Zhonggen*[1] and *Liu Xiaowei*[4]

1. Institute of Geographic Sciences and Natural Resources Research, CAS, Beijing, 100101, China
2. Graduate University of Chinese Academy of Sciences, Beijing, 100049, China
3. College of Water Science, Beijing Normal University, Beijing, 100875, China
4. Hydrological Bureau of Yellow River Water Conservancy Commission, Zhengzhou, 450004, China

Abstract: Weihe River, as the first largest branches of the the Yellow River, is intensely affected by human activities. We took the HeiYuKou Sub - basin (HYK), located in the middle areas of Weihe River Basin, as the study area and constructed a lumped hydrological model to simulate daily runoff and its responses to climate change. Based on the hydrological data, we choose the span from 1991 to 1996 as calibration period, and the period from 1997 to 2000 as validation period. In addition, we established eight climate change scenarios to contribute to analyze the responses of runoff to climate change based on the validated hydrological model. Main results are as follows: ① Nash - Sutcliffe Coefficient (*NSC*) and Correlation Coefficient (*CC*) are relatively high both in the calibration and validation period. In the calibration period, NSC and CC are 0.64 and 0.65 respectively, while in the validation span they are 0.63 and 0.66 respectively. ②Water balance Error (*WE*) in the validation is 10%, which is unsatisfactory. The reason is the poor performance of HIMS in the simulation of peak floods, which need to be improved. ③the daily runoff is sensitive to the changes of precipitation and potential evaporation, and the sensitivity value to precipitation is higher than to evaporation. the relationship between runoff change and precipitation change is positive, ④while that between runoff change and potential evaporation change is inverse. Both of the two responses are nonlinear.

Key words: simulation of runoff, hydro - informatic modeling system, climate change, Weihe River basin

1 Introduction

Influence of climate change on the water cycle and water resources security is the global issue at present. Since 1980's, lots of plans and researches about climate change and water cycle have been carried out in the world. For example, WCRP - GEWEX, IGBP - BAHP, UNESCO - IHP, GWSP and IPCC technical reports. Observational records and climate projections in the world provide abundant evidence that freshwater resources are vulnerable and have the potential to be strongly impacted by climate change, with wide - ranging consequences for human societies and ecosystems. In the background of global worming, drought in the north of China in the past 30 years, especially in the North China Plain, has been intensified, and the hydro - ecological environments have also been deteriorated, which have imposed restrictions on the sustainable development of economy. Climate change in the future is most likely to have more remarkable influence on the existing pattern of "southern flood and northern drought" in China in recent 30 years and the regional distributions of water resources in the future. And climate change will also has bad effect on the food security, and major national hydraulic projects such as the South - to - North Water Transfer Project, flood control planning on the main rivers in China.

Hydrological models, as approximate simulating to the complicated hydrologic phenomenon in

foundation: Nation Natural Science Foundation of China, No. 40971023; National Basic Research Program of China, No. 210CB428406 nature, are one kind of tools and methods to help to analyze and solve some issues in hydrological science. At present, hydrological models are regular tools to do quantitative evaluation on relationship and interaction between climate change and water resources. Modern hydrological models appeared in the 1930's when the applied hydrology just became to rise. After decades of development, the scholars in the world have put forward some famous hydrological models, in which the representative models are Stanford Watershed Model, Sacramento Model, TANK Model, API Model, SCS Model, SHE Model, TOPMODEL, SWAT Model, VIC Model, Time Variant Gain Model, et al. These models have been widely applied in the main rivers in the world.

The Yellow River is the second longest river in China, there are many domestic scholars have applied many hydrological models to the Yellow River and its tributaries, in order to study relationship and interaction between climate change and water resources. Some representative researches are as follows: SWAT Model applied to Luohe River Basin and the source regions of the Yellow River, WEP – L Model applied to the Yellow River Basin, Monthly Water Balance Model applied to the upper and the middle reaches of the Yellow River, Distributed Time Variant Gain Model (DTVGM) applied to the Jing River Basin, VIC Model applied to the Wei River Basin and the source regions of the Yellow River, and HIMS applied to the Lushi Sub – basin in Luohe River Basin.

We used Hydro – Informatic Modeling System (HIMS) to construct a lumped hydrological model to simulate the daily runoff in the HeiYuKou Sub – basin, and then eight kinds of climate change scenarios were set to simulate the runoff's response to climate change. The research can help to predict runoff in the middle reaches of Weihe River Basin, provide services for the water resources management.

2 Brief introduction of HIMS

HIMS was developed to facilitate water resources management and water environment protection. To realize the purposes of the system, the framework of HIMS was designed to be open with modular based incorporating knowledge from water related disciplines such as hydrology, climatology, ecology, geography, hydrogeology, and geomorphology, etc. HIMS includes a Hydrologic Information System (HIS) and a Hydrologic Model Library (HML). Integrated with Global Information System(GIS) and Remote Sensing (RS), the hydrologic information system of HIMS was well organized to provide functions to deal data with different sources and obtain geographical characteristics from DEM. The current version of HML consists of over 115 individual component models offering the alternation for hydrologic modeling in different spatial and temporal scales. The framework enables HIMS to be a platform of cross – or multi – disciplinary researches in relation to water. The other merit of HIMS is that the advanced users can use the existing modules to build a new model fitting with their study areas and research interests, which greatly enhances the flexibility of the system.

2.1 Structure of HIMS and its main controlling equations

Elements of model structures of HIMS, shown in Fig. 1, include precipitation, evaporation, infiltration, runoff generation, routing, and base flow, etc. In the calculation of runoff generation and routing, HIMS integrates the current many kinds of mature hydrological simulation methods. The main controlling equations used in HIMS are shown in references of NO. 42 and NO. 45.

2.2 Parameters calibration and evaluation of simulation

In HIMS, parameters calibration is done by automatic optimization method provided by HIMS, as well as combining automatic optimization method with artificial adjustment. There are three coefficients to evaluate the results of simulation, namely Water balance Error (*WE*), Nash – Suttcliffe Coefficient (*NSC*) and Correlation Coefficient (*CC*):

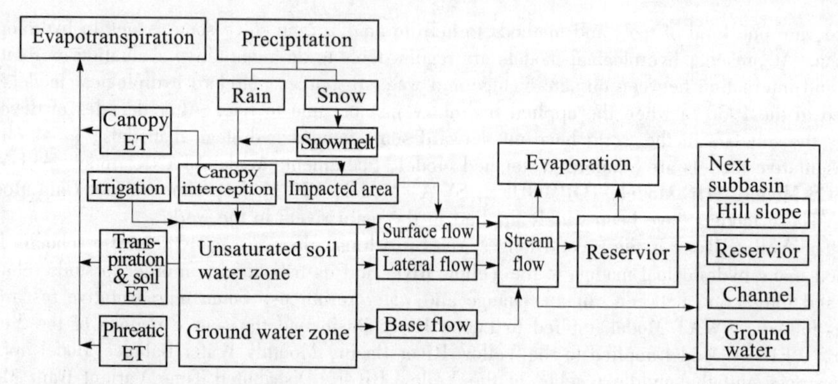

Fig.1 Structure of runoff generation model for a hydrological unit in HIMS

2.2.1 Water balance Error (*WE*)

When *WE* is in the range of ±10%, we think water is in balance. *WE* is defined as:

$$WE = \overline{Q_{sim}} / \overline{Q_{obs}} - 1$$

2.2.2 Nash – Suttcliffe Coefficient (*NSC*)

Nash – Suttcliffe Coefficient (*NSC*) is defined as:

$$NSE = 1 - \frac{\sum (Q_{obs,s} - Q_{sim,i})^2}{\sum (Q_{obs,i} - \overline{Q_{obs,i}})^2}$$

2.2.3 Correlation Coefficient (*CC*)

Correlation Coefficient (*CC*) is defined as:

$$R^2 = \frac{[\sum (Q_{sim,i} - \overline{Q_{sim}})(Q_{obs,i} - \overline{Q_{obs}})]^2}{\sum (Q_{sim,i} - \overline{Q_{sim}})^2 \cdot \sum (Q_{obs,i} - \overline{Q_{obs}})^2}$$

3 Model application

3.1 Study area

Weihe River is the biggest tributary of the Yellow River. It heads up in Wushu Mountains in Weiyuan county, Gansu Province. From west to east, it flows through Tianshui city of Gansu, Baoji city, Xianyang, Xi'an and Weinan et al, and flows into the Yellow River in Tongguan county, Shaanxi province. The length of main stream is 818 km, drainage area is 134,766 km^2, however, the geographic coordinate are between 104°00′E ~ 110°20′E and 33°50′N ~ 37°18′N. Weihe River Basin belongs to arid zone and semiarid zone. The annual average temperature is 6 ~ 14 ℃. The annual average rainfall is 500 ~ 800 mm. The ranges of evaporation capacity are from 1,000 mm to 2,000 mm.

The Heiyukou Sub – basin is located in the middle reaches of Weihe River Basin. The area is 1,476 km^2. The specific location was shown in Fig. 2.

3.2 Runoff simulation in the Heiyukou Sub – basin

According to the data of three weather stations in study area, the Thiessen Polygon method was used to calculate the area rainfall. The parameters of runoff yield and concentration were determined on the basis of land use and soil types. The model parameters were calibrated automatically according to observed runoff data. The simulation results of daily runoff in calibration period (1991 ~ 1996) and validation period (1997 ~ 2000) were showed in Tab. 1 and Fig. 3 to Fig. 6.

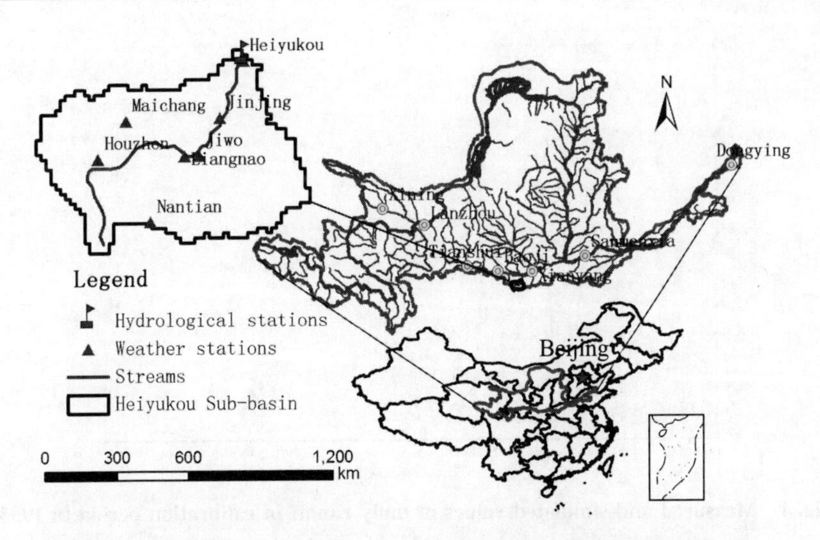

Fig. 2 Sketch map of Heiyukou Sub - basin in Weihe River Basin

Tab. 1 The results of simulation and verification of HIMS

Period	Calibration(1991 ~ 1996)			Validation(1997 ~ 2000)		
Objective functions	*NSE*	*WE*(%)	*R*	*NSE*	*WE*(%)	*R*
Results	0.64	-5	0.65	0.63	-10	0.66

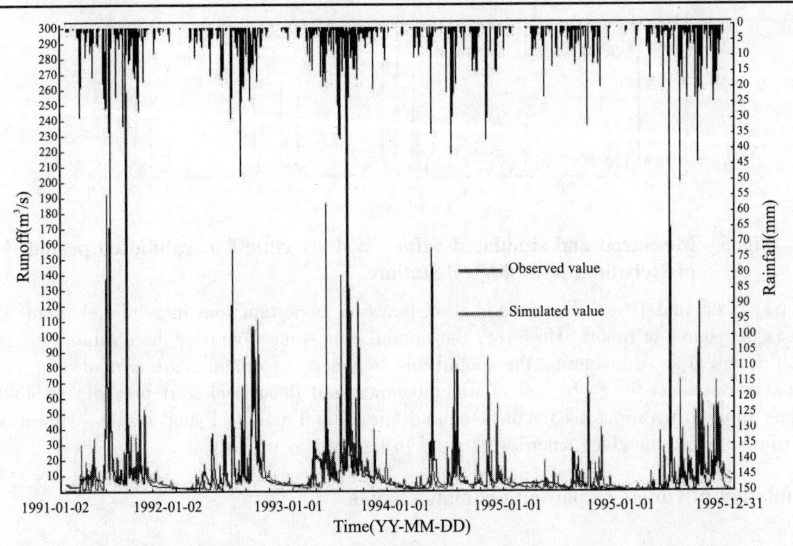

Fig. 3 Measured and simulated values of daily runoff in calibration period of Heiyukou hydrological station

The model had a good performance in daily runoff simulation. The calibration period was from 1991 to 1996 and the validation period was from 1997 to 2000. The *NSC*, *CC*, *WE* were 0.64, 0.65 and 5% in calibration period, respectively. In validation period, the three evaluation indexes

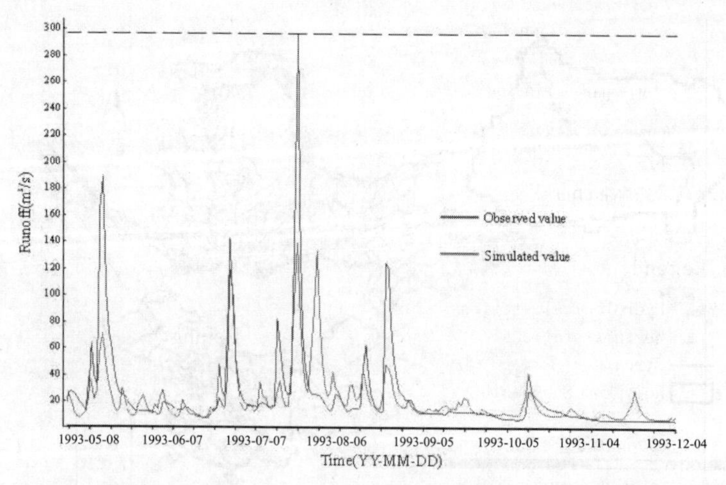

Fig. 4 Measured and simulated values of daily runoff in calibration period in 1993

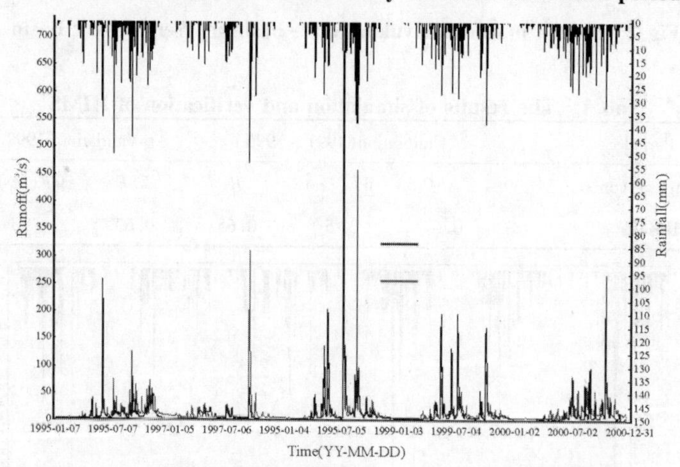

Fig. 5 Measured and simulated values of daily runoff in validation period of Heiyukou hydrological station

were 0.63, 0.66 and 10%. The precipitation plays an important role in rainfall - runoff simulation. It is the source of runoff. However, the rainfall data come from only three rainfall gauged stations in this research. Considering the spatial inhomogeneity of rainfall, the rainfall data had a related poor performance in coefficient of water balance and flood peak. At present, HIMS system model has applied in various basins at home and abroad and achieved good results. However, it is still in trial version, therefore, more work need to be done to improve it.

4 Simulation of runoff response to climate change

Based on the precipitation and evaporation data from 1991 to 1996 in Heiyukou Sub - basin, eight scenarios was employed to simulate the runoff responses to climate change in future which are shown in Tab. 2. All of the runoff responses to different climate change are given in Tab. 3. Fig. 7 gives the simulation of averaged monthly runoff under evaporation increased 10% or decreased 10% scenarios, while Fig. 8 gives the simulation under precipitation increased 10% or decreased 10% scenarios.

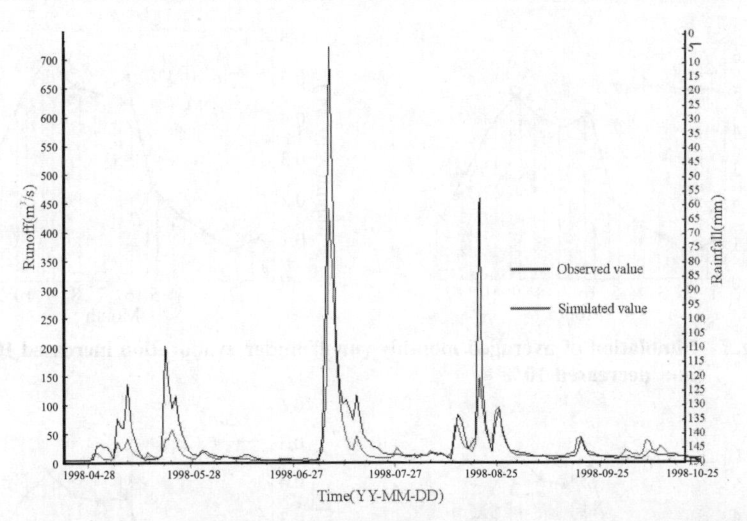

Fig. 6 Measured and simulated values of daily runoff in validation period in 1998

Tab. 2 The climate scenarios

		Rainfall change		
		$P\times(1+10\%)$	P	$P\times(1-10\%)$
Evaporation change	$E\times(1+10\%)$	S11	S21	S31
	E	S12	S22	S32
	$E\times(1-10\%)$	S13	S23	S33

Tab. 3 The results of runoff response to climate change

			Rainfall change		
			$P\times(1+10\%)$	P	$P\times(1-10\%)$
Evaporation change	Annual runoff ($\times10^8$ m^3)	$E\times(1+10\%)$	3.9	3.3	2.8
		E	4.1	3.5	2.9
		$E\times(1-10\%)$	4.2	3.6	3.1
	Runoff change ($\times10^8$ m^3)	$E\times(1+10\%)$	0.4	−0.2	−0.7
		E	0.6	0.0	−0.6
		$E\times(1-10\%)$	0.8	0.2	−0.4
	Change rate(%)	$E\times(1+10\%)$	12.0	−4.6	−19.8
		E	17.1	0.0	−16.0
		$E\times(1-10\%)$	22.7	5.0	−11.7

Based on the simulations, several conclusions could be obtained:

(1) The impact of climate change to runoff was conspicuous. The annual runoff increased conspicuous with the evaporation decreasing and precipitation increasing. There was a negative relationship between annual runoff and evaporation. When evaporation increased 10%, annual runoff would decrease 0.2×10^8 m^3. When precipitation increased 10%, annual runoff would increase

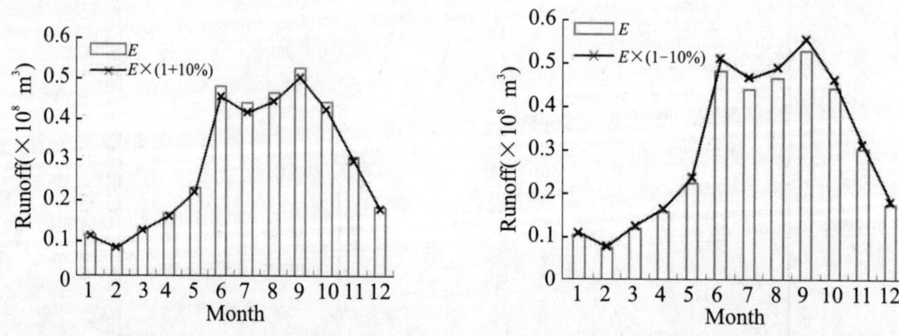

Fig. 7 Simulation of averaged monthly runoff under evaporation increased 10% and decreased 10%

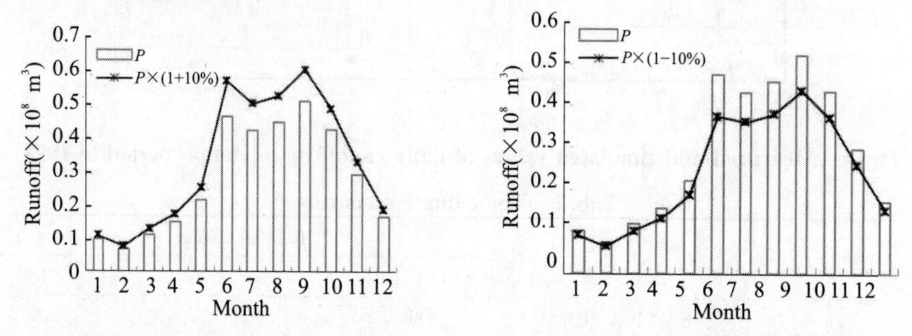

Fig. 8 Simulation under precipitation increased 10% and decreased 10%

0.6×10^8 m^3.

(2) The impact of precipitation change to runoff was more remarkable than evaporation change. The change rate of annual runoff was 17.1% and −16.0% under precipitation changed 10% and −10% respectively. The change rate of annual runoff was −4.6% and 5.0% under evaporation changed 10% and −10% respectively. It could be inferred that the change rate of annual off by precipitation change was 3 ~ 4 times to evaporation change which might be that precipitation was the direct resource of runoff.

(3) Change in evaporation and change in precipitation had the similar impact on the inter−annual distribution of runoff. Increase in evaporation led to the decrease in monthly runoff, and vice verse, while the amplitude was not obvious. The runoff was more sensitive to the change in precipitation, and the increase in precipitation led to the increase in runoff. In addition, the change in runoff was mainly concentrated in the flood season (June to September), because the changes in precipitation were more significant in flood season. The 10% increase in precipitation led to 19.5% increase in runoff, while 10% decrease in precipitation led to 17.6% decrease in runoff in flood season. The change in precipitation and runoff indicated that the change in precipitation was the main reason of change in runoff both in the inter − annual and annual scale at the Heiyukou station.

(4) The change in runoff fluctuated significantly between different extreme climate scenarios. The analysis of different conditions of changes in evaporation and precipitation showed that the largest decrease in runoff would occur when the evaporation increases by 10% and the precipitation decrease by 10%, which would lead to the decrease in runoff by 19.8%. The largest increase in runoff would occur when the evaporation decreases by 10% and the precipitation increase by 10%, which would lead to the increase in runoff by 22.7%.

5 Results

In this study, the HIMS hydrological model was built in the Heiyukou watershed of Weihe midstream basin. The daily process of runoff at the Heiyukou station was simulated based on the HIMS model. In addition, the response of runoff to the climate change was simulated in 8 scenarios. The following results were concluded:

(1) HIMS model had a good performance in the daily simulation of runoff process during 1991 ~ 2000 at the Heiyukou station. However, the limited number of metrological station had a negative impact on the simulation efficiency.

(2) Water balance Error was a little bit large because of the error in the simulation of flood peak, which should be improved in the future.

(3) The runoff had a positive relationship with precipitation, while it had a negative relationship with evaporation. Both of the relationships were non – linear.

(4) The impact of climate change on runoff was significant, and runoff was more sensitive to the change in precipitation than to the change in evaporation, which means the impact of change in precipitation on runoff was more significant than the impact of change in evaporation.

References

UN. The 3rd United Nations World Water Development Report[R]. Water in a Changing World (WWDR – 3),2009.

International GEWEX Project Office (IGPO). About GEWEX [EB/OL]. http://www. gewex. org/gewex_overview. html, 2004 – 12 – 01.

Bates B C, Kundzewicz Z W, Wu S, et al. Climate Change and Water[C]// Technical Paper of the Intergovernmental Panel on Climate Change. IPCC Secretariat, Geneva, 2008.

GWSP. The Global Water System Project: Science Framework and Implementation Activities[R]. Earth System Science Partnership (DIVERSITAS, IGBP, IHDP, WCRP) Report No. 3, 2005.

IPCC. Climate Change 2007—The Physical Science Basis[C]// Contribution of Working Group I to the Third Assessment Report of the IPCC. Cambridge: Cambridge University Press,2007.

IPCC. Climate Change 1990: The IPCC Impacts Assessment[M]. Canberra: Australian Government Publishing Service, 1990.

IPCC. Climate Change 1995: Impacts, Adaptations and Mitigation of Climate Change: Scientific – Technical Analyses[C]// Contribution of Working Group II to the Second Assessment Report of the Intergovernmental Panelon Climate Change. Cambridge, UKand NewYork: Cambridge University Press, 1996.

IPCC. Climate Change 1995: The Science of Climate Change[C]//. Contribution of Working Group I to the Second Assessment Report of the Intergovernmental Panel on Climate Change. Cambridge, UK and New York,: Cambridge University Press, 1996.

IPCC. Climate Change 2001: Impacts, Adaptation, and Vulnerability[C]//. Contribution of Working Group II to the Third Assessment Report of the Intergovernmental Panel on Climate Change. Cambridge, UK and New York: Cambridge University Press, 2001.

IPCC. Climate Change 2001: The Scientific Basis[C]// Contribution of Working Group Ito the Third Assessment Report of the Intergovernmental Panel on Climate Change. Cambridge, UK and New York:Cambridge University Press, 2001.

IPCC. Climate Change 2007: Impacts, Adaptation, and Vulnerability[C]// Contribution of Working Group II to the Forth Assessment Report of the Intergovernmental Panel on Climate Change. Cambridge, UK and New York: Cambridge University Press, 2007.

Bates B C, Kundzewicz Z K, Wu S, et al. Climate Change and Water[C]//. Technical Paper of the Intergovernmental Panel on Climate Change. IPCC Secretariat, Geneva, 2008.

Xia Jun, Liu Changming, Ding Yongjian, et al. Water Issues Vision in China (Volume 1)[M],

Beijing, 2011 (in Chinese).

Anderson M G, Burt T P. Process studies in hill slope hydrology: an overview[C]// Process studies in hill slope hydrology: an overview. Chichester: John Wley & Sons Ltd. , 1990.

Singh V P. Computer Models of Watershed Hydrology [M]. USA: Water Resources Publications, 1995.

Singh V P. Hydrologic Systems WATERSHED MODELING[M]. Translation by Zhao Weimin, et al. Zhengzhou: Yellow River Conservancy Press,2000 (in Chinese).

Craw for d N H, Linsley R K. Digital Simulation in Hydrology: Stanford Watershed Modelö [R]. Technical Report, Dept of Civil Engineering , Stanford University, 1966.

Kachroo R K. River Flow Forecasting. Part 5. Applications of a Conceptual Model[J]. Journal of Hydrology, 1992, 133: 141 - 178.

Nash J E, Sutcliffe J. River Flow Forecasting Through Conceptual models, Part 1, A Discussion of Principles [J]. Journal of Hydrology, 1970(10) : 282 - 290.

Zhao Renjun. Basin Hydrological simulation—Xinanjiang Model and Northern Shaanxi Model[M]. Beijing: China WaterPower Press,1984 (in Chinese).

Richard H McCuen. A Guide to Hydrologic Analysis Using SCS Methods[M]. Prentice - H all, Englewood Cliffs, 1982.

Jonch - Clausen T. System Hydrologique European: A Short Descript ion [C]//. SHE Report. Horsholm, Denmark, 1979.

Beven K J, Kirkby M J. A Physically Based Variable Contributing Area Model of Basin Hydrology [J], Hydrol. Sci. Bull. ,1979, 24(1) : 43 - 69.

Luzio M D, Srinivasab R, Arnold J . ArcView Interface for SWAT 2000, Users' guide[R]. Blach Land Research Center Texas Agricultural Experiment Station , 2001.

Liang X, Dennis P L , Wood E. F , et al. A Simple Hydrologically Based Model of Land Surface Water and Energy Fluxes for Heneral Circulation Models[J]. Geophys. Res. , 1994, 99 (D7): 14415 - 14428.

Xia Jun. Real - time Rainfall - runoff Forecasting by Time Variant Gain Models and Updating Approaches[C]// Research Report of the 6th International Workshop on River Flow Forecasting. UCG, Ireland, 1995.

Xia Jun, O Connor K M, Kachroo R K, et al . A Non - linear Perturbation Model Considering Catchment Wetness and Its Application in River Flow Forecasting[J]. Hydrologic Journal, 1997, 200: 164 - 178.

Hao Fanghua, Zhang Xuesong, Yang Zhifeng. A Distributed Non - point Source Pollution model: calibration and validation in the Yellow River Basin[J]. Journal of Environmental Sciences. 2004,16(4): 646 - 650.

Zhang Xuesong, Srinivasan R, Hao Fanghua. Predicting Hydrologic Response to Climate change in the Luohe River Basin Using the SWAT Model[J]. American Society of Agricultural and Biological Engineers, 2007, 50(3):901 - 910.

Yang Guilian, Hao Fanghua, Liu Changming, et al. The Study on Baseflow Estimation and Assessment in SWAT—Luohe Basin as An Example[J]. Progress in Geography. 2003,22(5): 463 - 471(in Chinese).

Xu Z X, Zhao F F, Li J Y. Response of Streamflow to Climate Change in the Headwater Catchment of the Yellow River Basin[J]. Quaternary International, 2009 (208),62 - 75.

Zhang Xuesong, Raghavan Srinivasan, Bekele Debele, et al. Runoff Simulation of the Headwaters of the Yellow River Using The SWAT Model With Three Snowmelt Algorithms[J]. Journal of the American Water Resources Association, 2008,44(1):48 - 61.

Chen Liqun, Liu Changming. Influence of Climate and Land - cover Change on Runoff of the Source Regions of Yellow River[J]. China Enviromental Science. 2007,27(4):559 - 565 (in Chinese).

Li Daofeng, Tian Ying, Liu Changming. Distributed Hydrological Simulation of the Source Regions of the Yellow River under Environmental Changes[J]. Acta Geographica Sinica. 2004, 59 (4): 565 - 573 (in Chinese).

Liu Jin, Li Zhengjia. Distributed Hydrological Modeling in the Source of the Yellow River[J].

Yellow River. 2007,29(9):30 – 32 (in Chinese).

Yangwen Jia, Hao Wang, Zuhao Zhou, et al. Development of the WEP – L Distributed Hydrological Model and Dynamic Assessment of Water Resources in the Yellow River Basin[J], Journal of Hydrology, 2006, 331:606 – 629.

Yangwen Jia, Cunwen Niu, Hao Wang, Integrated Modeling and Assessment of Water Resources and Water Environment in the Yellow River Basin[J], Journal of Hydro – environment Research, 2007(1):12 – 19.

Jia Yangwen, Wang Hao, Wang Jianhua, et al.. Development and Verification of a Distributed Hydrologic Model for the Yellow River Basin[J]. Journal of Natural Resources. 2005,20(2):300 – 308 (in Chinese).

Wang Guoqing, Wang Yunzhang, Kang Lingling. Analysis on the Sensitivity of Runoff in Yellow River to Climate Change[J]. Journal of Applied Meteorological Science. 2002,13(1):117 – 121 (in Chinese).

Wang Gangsheng, Xia Jun, Zhu Zhong, et al. Distributed Hydrological Modeling Based on Nonlinear System Approach[J]. Advances in Water Science. 2004,15(4):521 – 525 (in Chinese).

Xie Zhenghui, Liu Qian, Yuan Fei, et al. Macro – scale Land Hydrological Model Bbased on 50 km × 50 km Grids System[J]. Jaurnal of Hydraulic Engineering. 2004,(5):76 – 82 (in Chinese).

Liu Changming, Wang Zhonggen, Zheng Hongxing, et al. Development of Hydro – Informatic Modelling System and its Application[J]. Sci. China Ser E – Tech Sci., 2008, 51(4):456 – 466.

Liu Changming, Wang Zhonggen, Zheng Hongxing, et al. Multi – scale Integrated Simulation of Hydrological Processes Using HIMS with Verified Case Studies[J]. Journal of Beijing Normal University (Natural Science), 2010(3): 268 – 273 (in Chinese).

Wang Zhonggen, Zheng Hongxing, Liu Changming. A Modular Framework of Distributed Hydrological Modeling System: HydroInformatic Modeling System, HIMS[J]. Progress in Geography. 2005(6): 109 – 115 (in Chinese).

Liu Changming, Wang Zhonggen, Zheng Hongxing, et al. Development of Hydro – Informatic Modelling System and its Application[J]. Sci. China Ser E – Tech. Sci., 2008 (3): 350 – 360 (in Chinese).

Liu Changming, Xia Jun, Guo Shenglian, et al. Advances in Distributed Hydrological Modeling in the Yellow River Basin[J]. Advances in Water Science. 2004(04): 495 – 500 (in Chinese).

Liu Changming, Zheng Hongxing, Wang Zhonggen. Water Cycle Distributed Simulation on Basin Hydrological Cycle[M]. Beijing: Yellow River Hydraulic Press, 2006 (in Chinese).

Recent Changes of Runoff and Their Causes in Upper Yangtze River Basin

Wang Miaolin[1,3] and *Xia Jun*[2]

1. Upper Yangtze River Survey Bureau of Hydrology & Water Resources, Bureau of Hydrology, Changjiang Water Resources Commission, Chongqing, 400014, China
2. State Key Laboratory of Water Resources & Hydropower Engineering Sciences, Wuhan University, Wuhan, China
3. Chongqing Jiaotong University, Chongqing, 400074, China

Abstract: The Yangtze River from the source regions to the Yichang station is called as the upper Yangtze River. Firstly, the runoff changes at sixteen main control stations in the upper reaches of Yangtze River Basin were analyzed. The results show that except in the Jinsha River Basin, the runoff in the other basin decreased more or less. Among them, the runoff in the Minjiang River, Hengjiang River, Tuojiang River and Jialing River decreased significantly. The runoff at the Cuntan station increased before 1968, but decreased significantly after 1993. A large scale distributed monthly water balance model was developed and applied in the upper Yangtze River. The model was used to evaluate the impacts of climate change and human activities on the changes of runoff generation in the upper Yangtze River Basin and to explore the causes of runoff change. The runoff at the Cuntan station of Yangtze River decreased significantly since 1993 and climate change was the main factor that contributed 71.43% of the runoff change. Meanwhile, it was needed to pay attention that the runoff in the Minjiang River and Jialing River Basin decreased obviously and the contribution of human activities to runoff change reaches to about 50%. It was related to the intense human activities in those areas. Human activities will most probably play a dominant role in influencing the discharge of the Minjiang River and Jialing River Basin.

Key words: runoff change, climate change, human activities, distributed hydrologic model, the upper Yangtze River

1 Introduction

The Yangtze River (Changjiang) is the largest river in China and often called the equator of China. The river from the source regions to the Yichang station is called as the upper Yangtze River. Its area is about 1×10^6 km^2. The main tributaries of the upper Yangtze River include Jinsha River, Minjiang River, Tuojiang River, Jialing River and Wujiang River etc. The runoff amount of the upper Yangtze River occupies 47% of the total amount of the Yangtze River Basin.

The previous studies on runoff changes of the Yangtze River Basin mainly focused on the Yichang station in the upper Yangtze River and Datong station in the lower Yangtze River. The coincident conclusion is that the runoff at Yichang station decreased and runoff at Datong station increased (Chen et al, 2001; Xiong & Guo, 2004; Wang et al, 2005; Jiang et al, 2005; Yang et al, 2005). Other conclusion was that the runoff at the Zhimenda station in the original area of Yangtze River decreased (Xie et al, 2003; Li et al, 2004) and the runoff at the Pingshan station of the upper Yangtze River increased (Wang et al, 2005).

The objectives of this paper are to analyze the runoff changes of the upper Yangtze River basin and their causes: ① to detect the changing trend of runoff of major stations along the main stream and the major tributaries in the upper Yangtze River basin; ② to develop a large scale distributed monthly water balance model; ③ to explore and discuss the possible influences of human activities and climatic variability on runoff in the upper Yangtze River Basin.

2 Runoff changes in the upper Yangtze River Basin

According to water resources engineering and hydrologic stations distribution, 16 main control stations in the upper reaches of Yangtze River Basin were selected (see Tab. 1).

Tab. 1 Runoff changes

Station	Basin	Area(km^2)	Range	M	Changing trend
Shigu		232,651	1956 ~ 2006	1.172	↑
Panzhihua	JinSha River	284,540	1965 ~ 2006	1.864	↑
Pingshan		465,099	1950 ~ 2006	1.077	↑
Gaochang	Minjiang River	135,378	1950 ~ 2006	-2.629	↓ *
Hengjiang	Hengjiang River	147,81	1964 ~ 2006	-3.717	↓ *
Lijiawan	Tuojiang River	19,613	1952 ~ 2000	-2.400	↓ *
Xiaoheba		29,420	1952 ~ 2006	-3.260	↓ *
Luoduxi		38,064	1954 ~ 2006	-0.884	↓
Wusheng	Jialing River	79,714	1944 ~ 2006	-3.432	↓ *
Beibei		156,142	1943 ~ 2006	-2.367	↓ *
Zhutuo	Yangtze River	694,725	1954 ~ 2006	-0.343	↓
Cuntan		866,559	1939 ~ 2006	-0.936	↓
Gongtan	Wujiang River	64,200	1940 ~ 2006	-0.710	↓
Wulong		83,035	1952 ~ 2006	-0.007	↓
Wuxi	Daning River	13,721	1972 ~ 2006	-1.150	↓
Yichang	Yangtze River	1,005,501	1952 ~ 2004	-1.081	↓

Note: M: Mann-Kendall test statistic. The sign "↑" means an increasing trend, while "↓" means a decreasing trend. The sign * means the trend is statistically significant at 10% level.

The runoff changes were analyzed by the Mann-Kendall statistical test methods. The rank-based non-parametric Mann-Kendall statistical test has been commonly used for trend detection (Yue et al., 2002) due to its robustness for non-normally distributed and censored data, which are frequently encountered in hydro-climatic time series. This method defines the test statistic M as:

$$M = S/\sigma \tag{1}$$

with
$$S = \sum_{i=1}^{n-1}\sum_{j=i+1}^{n} \mathrm{sgn}(x_j - x_i) \tag{2}$$

and
$$\sigma_s^2 = \frac{n(n-1)(2n+5) - \sum_{i=1}^{n} t_i i(i-1)(2i+5)}{18} \tag{3}$$

where n is the data record length, x_j and x_i are the sequential data values. The function sgn(x) is defined as:

$$\mathrm{sgn}(x) = \begin{cases} 1 & \text{if } x > 0 \\ 0 & \text{if } x = 0 \\ -1 & \text{if } x < 0 \end{cases} \tag{4}$$

Eq. (3) gives the standard deviation of S with the correction for ties in data with t_i denoting the number of ties of extent i. The null hypothesis of an upward or downward trend in the data cannot be rejected at the α significance level if $|M| > u_{1-\alpha/2}$, where $u_{1-\alpha/2}$ is the 1-α/2 quantile of the standard normal distribution (Kendall, 1975). A positive M indicates an increasing trend in the time series, while a negative M indicates a decreasing trend.

Tab. 1 show the analysis results of runoff changing trend of 16 stations. The results show that except the runoff in the Jinsha River Basin increased slightly, the runoff in the other river decreased more or less. Among them, the runoff at the Gaochang station on the MinJiang River; Hengjiang station on the Hengjiang River; Lijiawan station on the Tuojiang River; Xiaoheba, Wusheng and Beibei station on the Jialing River decreased significantly. And the runoff at the Cuntan station, one of the control stations of the upper Yangtze River decreased slightly.

Fig. 1 show the annual mean flow series at Beibei station on the Jialing River. It was found that the annual mean flow at Beibei station show a significant decreasing trend. Further, through the analysis on the accumulated deviation from the mean runoff, it was found that the runoff at Gaochang, Beibei and Cuntan station increased before 1968, but decreased significantly after 1993 (see Fig. 2).

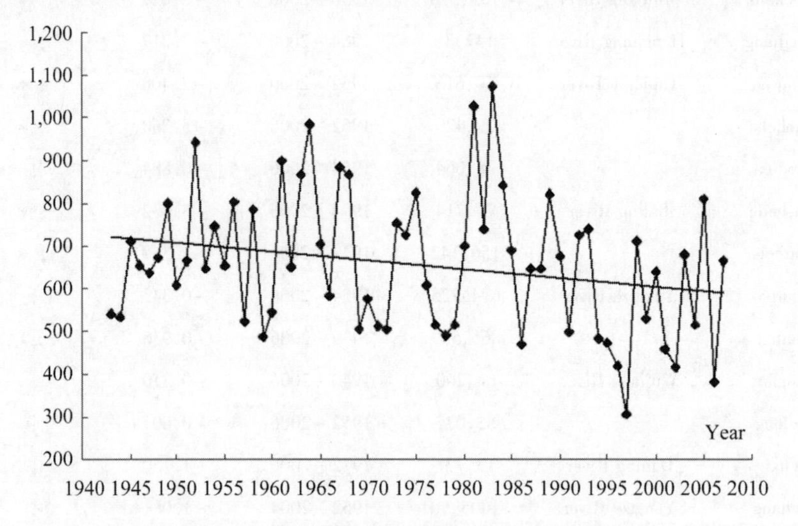

Fig. 1 Runoff hydrograph at Beibei station

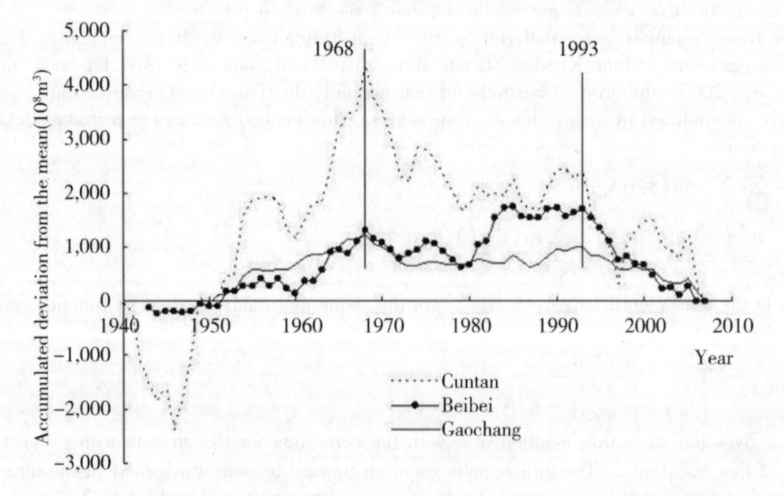

Fig. 2 Residual mass curves of Gaochag, Beibei and Cuntan station

3 Monthly water balance model

Climate change and human activities are expected to alter the timing and magnitude of runoff and soil moisture etc. Quantitative estimates of the hydrological effects of the climate change and human activities are essential for understanding and solving potential water resources problems. The question is, which of the two crucial reasons for runoff changes—climate change and human activities—is the leading factor and how much are their contribution? In order to probe this issue, a monthly water balance model was developed. Such a distributed monthly hydrological model, called the Distributed Time Variant Gain Model (DTVGM), was adopted to analyze the impact of climate change and human activities on the runoff. The model structure and procedure are briefly described below.

In the DTVGM, the whole basin is divided into sub areas according to DEM, and the land use and cover information is obtained from the RS images. The water balance procedure can be expressed as:

$$\Delta S_t = S_{t+1} = S_t = P_t - ETa_t - RS_t - RSS_t - WU_t \tag{5}$$

where ΔSt is the change of soil moisture storage in the units of mm; St and St + 1 are the soil moisture storage at month t and (t + 1) respectively; Pt is the precipitation; ETat is the actual evapotranspiration; RSt、RSSt are the surface runoff and subsurface runoff; WUt is the net water consumption, including water use, sink filling, ineffective evapotranspiration and seepage loss.

Normally, due to the difficulty of evaluating the value of WUt, in the study, this term is included in the computation of ETa, RS and RSS through adjusting the parameter values related to human activities.

3.1 Evapotranspiration model

The actual monthly evapotranspiration, ETa_t, can be calculated by (Xiong & Guo, 2004):

$$ETa_t = Kaw \times EP_t \times tanh(P_t / EP_t) \tag{6}$$

where EP_t is the monthly potential evapotranspiration; Kaw is the evaporation coefficient.

3.2 Runoff generation model

In the DTVGM, the rainfall and antecedent soil-moisture content are used to model runoff generation. Analyzing the observations of runoff and soil moisture over several river basins, XIA Jun et al. (2005) found that the surface runoff generation coefficient is time-variant, and is a function of the antecedent soil-moisture content. Therefore, the surface runoff (RS_t) generated in a subbasin can be described as follows:

$$RS_t = g_t \left(\frac{S_t}{SM}\right)^{g2} \cdot P_t \tag{7}$$

where SM is the saturated soil moisture content; g_1 and g_2 are the parameters in the time-variant gain function.

In the DTVGM, the RSS_t was calculated by a linear storage-outflow relationship (Thompson, 1999):

$$RSS_t = Kr \cdot S_t \tag{8}$$

where Kr is the subsurface runoff generation coefficient.

Then, the total runoff (R_t) generated during the month t is the sum of surface and subsurface runoff in the subbasin:

$$R_t = RS_t + RSS_t \tag{9}$$

3.3 Parameters related to the influence of human activities

Intuitively, the impacts of human activities on the terrestrial hydrologic processes can be observed through various aspects of runoff generation, evapotranspiration and soil moisture movements. Consequently, some parameters in the DTVGM will be influenced when such the impacts are considered in the model.

The parameters *SM*, g_1 and g_2 used in Eq. (7) are related to surface runoff generation, and their values would be adjusted during modelling the influences of human activities. The other two parameters, *Kr* in Eq. (8) and *KAW* in the Eq. (6) can represent the influences of human activities through manipulating the soil moisture simulation in the model.

4 Analysis on driving forces of runoff changes

According to water resources engineering and hydrologic station distribution, the upper Yangtze River basin was divided into 6 areas: ① Jinsha River Basin (Above Pingshan station); ②Minjiang River Basin(Above Gaochang station); ③Tuojiang River Basin(Above Lijiawan station); ④ Jialing River Basin(Above Beibei station); ⑤Cuntan area(from Pingshan, Gaochang, Lijiawan and Beibei stations to Cuntan station); ⑥Wujiang River Basin(Above Wulong station).

The proposed model was used for quantitative estimates of the hydrological effects of the climate change and human activities. Here climate changes mainly include the rainfall and temperature. Human activities are decided by model parameters. According Fig. 2, the runoff process of the upper Yangtze River can be divided into three stages: ① from 1955 to 1968; ②from 1969 to 1992; ③from 1993 to 2001. In this paper, we main compare the runoff from 1955 to 1968 and from 1993 to 2001 (see Tab. 2). The first stage (from 1955 to 1968) was considered as the reference period which human activities were not obvious.

Tab. 2 Runoff simulation stages

Stage	Range	Number of years	Climate data set	Human activities-related parameter set (HAPS)
I	1955 ~ 1968	14	Data1	Para1
II	1993 ~ 2001	9	Data2	Para2

The proposed model was used for assessing the effects of climate change and human activities as follows: ①Firstly, the runoff depth, R_0, was calculated by using the human activities-related parameter set (HAPS) Para1 and the climate data set (CDS) Data1. ②Secondly, the runoff depth, R_1, was calculated by using the human activities Para1 and the climate data Data2. So the impact of climate change on runoff can be evaluated by comparing R_1 with R_0. ③Thirdly, the runoff depth, R_2, was calculated by using the human activities Para2 and the climate data Data1. So the impact of human activities on runoff can be evaluated by comparing R_2 with R_0. ④Finally, the runoff depth, R_3, was calculated by using the climate data Data2 and the human activities Para2. So the impact of climate change and human activities on runoff can be evaluated by comparing R_3 with R_0. And we can divide their contributions to runoff change.

Tab. 3 Optimized parameters at different stages for Jialing River Basin

HAPS	Parameters:					Model efficiency	
	Wum	g_2	g_1	*Kaw*	K_r	R^2(%)	*RE*(%)
Para1	255	0.161	0.475	1.129	0.290	84.72	2.33
Para2	271	0.114	0.460	1.230	0.129	88.38	1.82

For example, in the Jialing River Basin, the human activities-related parameter set in two periods were calibrated and obtained respectively (See Tab. 3). Through the model calculation, the simulated runoff in stage I was 454 mm and it was defined as the reference value. If keep the human activities not change (Para1) and climate data changed to stage II (Data2), the simulated runoff was 375 mm. It decreased 79 mm compared to the reference value. It occupied 57.66% of total change (137 mm) and so the contribution rate of climate change is 57.66%. If keep the climate data changed not change (Data1) and human activities changed to Para2, the simulated runoff was 390 mm. It decreased 64mm compared to the reference runoff and occupied 46.72% of total change (137 mm). So the contribution rate of human activities was 46.72 % (see Tab. 4).

Tab. 4 Contribution of climate change and human activities to runoff decrease in Jialing River Basin

Stage	I	Climate impact	Human influence	II	
Variable and parameter	Data1 Par1	Data2 Par1	Data1 Par2	Data2 Par2	Change
Annual observed runoff (mm)	456			319	-30.04%
Annual simulated runoff (mm)	454	375	390	317	-30.18%
Diff = Sim - Sim(I) (mm)		-79	-64	-137	
Contribution ratio (%)		57.66	46.72	100	

In turn, the contribution of the climate change and human activities to runoff change in 7 areas of the upper Yangtze River Basin were calculated respectively (see Tab. 5). It was found that the total contribution rates of the climate change and human activities may not equal to 100%. It would be caused by simultaneous changes of the climate change and human activities and the nonlinear features of the DTVGM in simulating runoff (Wang et al., 2006).

Tab. 5 Contribution of climate change and human activities to runoff change in 7 basins of upper Yangtze River Basin

Basin	Jinsha	MinJiang	TuoJiang	Jialing	Wujiang	Cuntan
Runoff change (mm)	13	-60	-165	-137	77	-28
Change percent (%)	4.25	-9.04	-23.84	-30.18	13.32	-6.75
Climate impact (%)	62.90	-55.00	-73.33	-57.66	58.44	-71.43
Human influence (%)	32.77	-48.33	-28.48	-46.72	40.26	-25.00

In short, the runoff at the Cuntan station decreased significantly since 1993 and climate change is the main factor. However, it needs to pay attention that the runoff in the Minjiang River and Jialing River Basin decreased obviously and the contribution of human activities to runoff change reaches to about a half. It was related to the intense human activities in those areas. The middle and lower reaches of the Minjiang and River Jialing River is mostly located in Sichuan Province of China. Sichuan Province is one of the agricultural and industrial centres in China. The population of Sichuan Province in 1952 was about 4.6×10^7 and in 2000 was about 1.1×10^8 (including the population of Chongqing City). The population boom and the rapid development of agriculture and industry in those areas greatly increased water consumption. Furthermore, more than 5,000 reservoirs had been built from the 1950s to the 1980s, with total volume capacity of 3,400 106 m3 in the Jialing River Basin (Chen, et al, 2001). These human activities have influenced the water cycle, with strong implications for water resources use. Changes in the water cycle are also linked to changes in biogeochemical cycles. In the future decades, the increasing trend in population and development of economy will continue, and many new reservoirs will be built in the catchments, which will inevitably lead to further decrease in discharge. Human activities will most probably play a dominant role in influencing the discharge of the Minjiang River and Jialing River Basin.

5 Conclusion

In this study, the runoff changes of 16 main control stations at the upper reaches of Yangtze

River Basin were analyzed by the Mann-Kendall statistical test methods. The results show that except the runoff in the Jinsha River basin increased slightly, the runoff in the other river decreased more or less. Among them, the runoff of the MinJiang River, Hengjiang River, Tuojiang River, and Jialing River decreased significantly. And the runoff at the Cuntan station, the control station of the upper Yangtze River decreased slightly. Furthermore, it was found that the runoff at the Cuntan station increased before 1968, but decreased significantly after 1993.

Then a large scale distributed monthly water balance model (DTVGM) was applied in the 7 basins of the upper Yangtze River Basin. It shows that the distributed monthly water balance model can obtain high precision. In the calibration period, the average value of Nash-Sutcliffe efficiency is 85.00% and that in the verification period is 81.69% and the value of the relative error of the volumetric fit in the calibration and verification period are close to zero. So the model can simulate monthly runoff well in the upper Yangtze River basin.

Finally, the DTVGM was used to evaluate the impacts of climate change and human activities on the changes of runoff generation in upper Yangtze River Basin. Using the fixing-changing factor technique, the contribution rates of climate change and human activities on runoff change were evaluated. The runoff at the Cuntan station of Yangtze River decreased significantly since 1993 and climate change is the main factor. But in the non-flood period the reservoirs operation in the upper Yangtze River Basin increased runoff more or less. Meanwhile, it is need paid attention to that the runoff in the Minjiang River and Jialing River Basin decreased obviously and the contribution of human activities to runoff change reaches to about a half. It is related to the human activities in those areas. Human activities will most probably play a dominant role in influencing the discharge of the Minjiang and River Jialing River Basin.

Acknowledgement

This research was supported by the National Basic Research Program of P. R. China (973 Project, No. 2012CB417001).

References

Chen Xiqing, Zong Yongqiang, Zhang Erfeng, et al. Human impacts on the Changjiang (Yangtze) River basin, China, with special reference to the impacts on the dry season water discharges into the sea[J]. Geomorphology, 2001(41): 111-123.

Jiang Tong, Su Buda, Wang Yanjun, et al. Trends of temperature, precipitation and runoff in the Yangtze River Basin from 1961 to 2000[J]. Advances in Climate Change Research, 2005, 1 (2): 65-68.

Li Lin, Wang Zhenyu, Qin Ningsheng et al. Analysis of the relationship between runoff amount and its impacting factor in the upper Yangtze River[J]. Journal of Natural Resources, 2004, 19 (6): 694-700.

Thompson, S. A. Hydrology for water management, 1999, 362, A. A. Balkema, Rotterdam.

Wang Gangsheng, Xia Jun, Wan Donghui, et al. A distributed monthly water balance model for identifying hydrological response to climate changes and human activities[J]. Journal of Natural Resources, 2006, 21(1): 86-91.

Wang Yanjun, Jiang Tong, Shi Yafeng. trends of climate and runoff over the upper reaches of the Yangtze River from 1961 to 2000[J]. Journal of Glaciology and Geocryology, 2005, 27(5): 709-714.

Xia Jun, Wang Gangsheng, Ye Aizhong, et al. A distributed monthly water balance model for analysing impacts of land cover change on flow regimes[J]. Pedosphere, 2005, 15(6): 761-767.

Xie Changwei, Ding Yongjian, Liu Shiyin , etc. Comparison analysis of runoff change in the source regions of the Yangtze and Yellow Rivers[J]. Journal of Glaciology and Geocryology, 2003, 25(4): 414-422.

Xiong L. , Guo S. Trend test and change-point detection for the annual discharge series of the Yangtze River at the Yichang hydrological station[J]. Hydrological Sciences Journal, 2004, 49 (1): 99-112.

Yang S L, Gao A, Hotz Helenmary M, et al. Trends in annual discharge from the Yangtze River to the sea (1865 ~2004)[J]. Hydrological Sciences Journal, . 2005, 50(5): 825-836.

Yue S, Pilon P, G Cavadias. Power of the Mann-Kendall and Spearman's rho tests for detecting monotonic trends in hydrological series[J]. Journal of Hydrology, 2002, (259): 254-271.

Study of Spatial and Temporal Variation Characteristics of Drought in the Yellow River Basin in Recent 50 Years

Yang Zhiyong[1,2] *Yuan Zhe*[2,3], *Zhang Peng*[1,2] and **Yu Dongying**[1,2]

1. State Key Laboratory of Simulation and Regulation of Water Cycle in River Basin, China Institute of Water Resources and Hydropower Research (IWHR), Beijing, 100038, China
2. Department of Water Resources, IWHR, Beijing, 100038, China
3. College of Soil and Water Conservation, Beijing Forestry University, Beijing, 100083, China

Abstract: With reference to daily rainfall data recorded by 112 meteorological stations in reaches of the Yellow River Basin and its surrounding regions in the past 50 years (1961 ~ 2010), applying data processing platform of Geo – information system (GIS) technology and in combination with Mann – Kendall Method and Moving t – test method, the article tried to analyze spatial and temporal variation characteristics of drought in the Yellow River Basin. The analysis showed that ①the situation of drought in the Yellow River Basin was becoming more and more serious due to the decreasing of precipitation; ②the spatial and temporal patterns of drought had been changed which was mainly represented by the increasing of drought area and expanding of areas prone to drought.
Key words: the Yellow River Basin, drought area, areas prone to drought, variation characteristics

1 Introduction

As the negative impacts brought about by global warming have become increasingly substantial, the environment problems induced by climate changes have drawn people's attention extensively. The IPCC Report has shown that the average temperature of global surface increased by 0.56 ~ 0.92 ℃ (IPCC, 2007). The stability of water circulation system is declining and extreme climate processes like drought and flood have happened frequently due to acceleration of water cycle which is the consequence of temperature increasing. Not only does the climate change adversely affect industrial and agriculture productions, it also threatens rural and urban water safety, and even significantly damages ecological environment, thus restrains the sustainable development of economy and society (Qin Dahe, Ding Yihui, Su Jilan, et al., 2005). The Yellow River Basin is one of top 10 river basins in China and it plays an important role in the development of social and economic. But the precipitation is decreasing while the temperature is increasing since 1990s and the basin has suffered from drought (Liu Jifeng, Xu Zhuoshou, Wang Lin, et al., 2009; Xu Zongxue and He Wanlin, 2005). So analyzing the spatial and temporal distribution characteristics of drought events is beneficial to the rationalization of industrial structure, optimization of water resources allocation and aversion of disaster risk.

2 Study site

The Yellow River Basin across the arid and semi – arid zone. The west of basin is dry while the east is wet. The Yellow River originates from the Tibetan Plateau, wandering eastward through 9 provinces such as Qinghai, Sichuan, Gansu, Ningxia, Inner Mongolia, Shanxi, Shaanxi, Heibei and Shandong. With a length of 5,464 km and a area of 794,712 km^2, the basin is located on the east longitudes 95°53′ ~ 119°05′ and north latitudes 32°10′ ~ 41°50′N (Fig. 1). The average of annual precipitation is 466 mm and the average of annual evaporation ranges from 700 mm to 1,800 mm. It has been reported that the drought is frequent and widespread extreme hydrological event causing great economic losses in the Yellow River Basin.

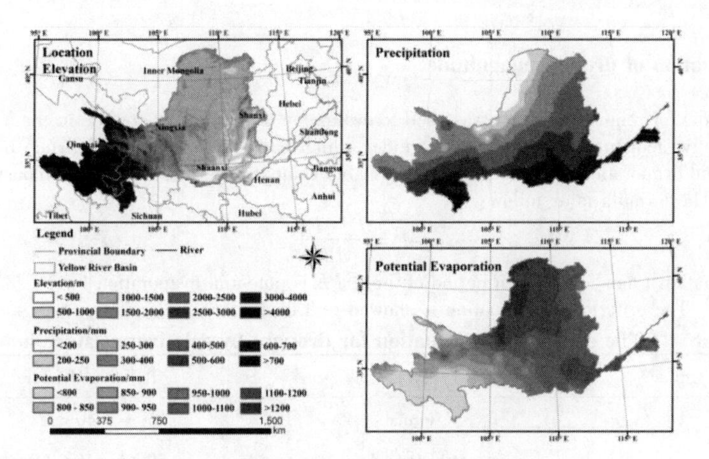

Fig. 1 Map of the Yellow River Basin

3 Materials and methods

3.1 Data source

The data including precipitation, maximum temperature, minimum temperature, average temperature, sunshine hours, wind speed, relative humidity, latitude, longitude and elevation were obtained from the National Meteorological Information Center, China Meteorological Administration for the year 1961 to 2010. The study period of 1961 to 2010 was chosen because it was well represented in the station data. Pre – 1961 records of precipitation were less consistent than that during the 1961 to 2010 period. In some circumstances that the individual stations contained partial or no data, they were deleted from the analysis of precipitation. In order to improve the precision of spatial analysis, station with less than 80% temporal coverage during the 50 year period were omitted. A total of 122 meteorological stations were used for the analysis after selected. Fig. 2 shows the distribution of meteorological stations which are in reaches of the Yellow River and its surrounding regions.

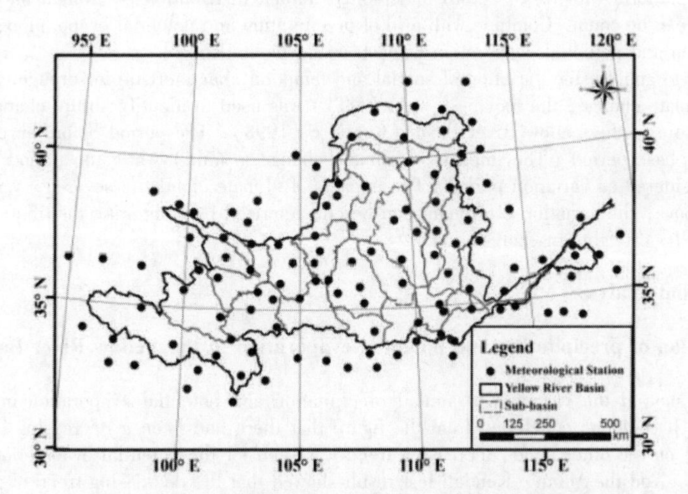

Fig. 2 The distribution of meteorological stations

3.2 Evaluation of drought magnitude

The index of relative moisture was used to evaluation of drought magnitude in the Yellow River Basin. Relative moisture was an important index which can characterize the balance between precipitation and evaporation and it was more reasonable compared with the index only considering precipitation. The formula is as follow:

$$M = \frac{P - PE}{PE} \tag{1}$$

P is precipitation (mm) in a given period of time; PE is potential evaporation (mm) in a given period of time. The criterion of evaluation is showed in Tab. 1.

Tab. 1 The criterion of evaluation for drought by relative moisture index

Magnitude	Type	Relative Moisture
1	Normal	$-0.40 < M$
2	Mild Drought	$-0.65 < M \leqslant -0.40$
3	Moderate Drought	$-0.80 < M \leqslant -0.65$
4	Severe Drought	$-0.95 < M \leqslant -0.80$
5	Extreme Drought	$M \leqslant -0.95$

The formula of Penman – Monteith was chosen to estimate the potential evaporation. The detail can be found in the reference (Gong L, Xu C, Chen D, et al., 2006).

3.3 Analytical method of spatial and temporal characteristic of drought

Based on the estimate of drought area and frequency of drought, the study analyzed the characteristic of drought in spatial and temporal term. Before the study of drought area, the continuous grid surface with the size of 10 km × 10 km for annual precipitation and potential evaporation was created with the IDW (Inversed Distance Weighted) method. Then the data of relative moisture index of each year for each grid can be obtained. Taking -0.40 as criterion, the research selected cells on GIS platform with $M < -0.40$ and plotted annual distribution of drought and the annual drought area can be count. Combine with grid of precipitation and potential evaporation, time series of precipitation and potential evaporation in each would be statistics.

In order to analyze the variation of spatial and temporal characteristic for drought in the background of climate change, the moving t – test (MTT) was used to identify abrupt change of annual relative moisture in the Yellow River Basin (Cao M S, 1998). The period before abrupt year was considered as base period (The impacts of climate change is mild) while the period after abrupt year was considered as variation period (The impacts of climate change is severe). With the same process as above, the variation of drought can be characterized by comparing distribution of spatial and temporal for droughts in each period.

4 Results and analysis

4.1 Variation of precipitation and potential evaporation in the Yellow River Basin

Fig. 3 showed the variation of annual precipitation and potential evaporation in the Yellow River Basin. It could be concluded from the figure that there had been a decreasing trend for the precipitation; on the other hand, increasing trend was seen for the potential evaporation during recent 50 years. And the Mann – Kendall test result showed that the decreasing trend of precipitation had reached the significant level of $\alpha = 0.1$ ($M_P = -1.380$) while the increasing trend of potential evaporation had not reached the significant level ($M_{PE} = 0.644$). The dotted line in Fig. 3 represented the variation of relative moisture index which was calculated by annual precipitation and potential evaporation. As illustrated in Fig. 4 which is the result of MTT, the abrupt year of relative moisture index appeared in 1992. So we can conclude that base period was from 1961 ~ 1992 and variation period was from 1993 ~ 2010. The impact of climate change during later period was more serious than that of the former period.

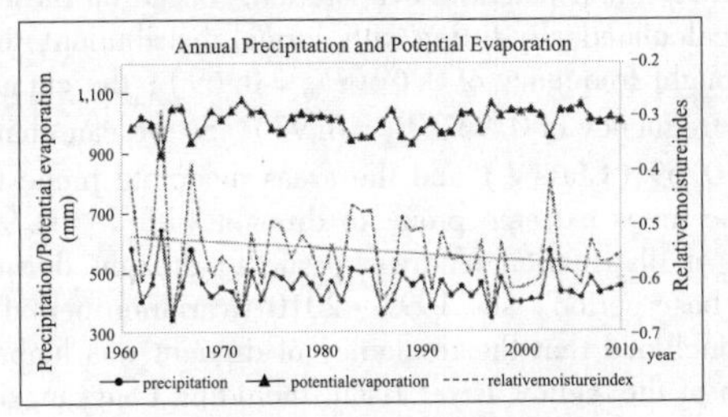

Fig. 3 The v of P and PE

Fig. 4 The result of MTT

4.2 Variation of drought area in the Yellow River Basin

The research analyzed annual drought magnitude of each gridding cell and counted the number of different magnitudes based on the relative moisture index and the criterion of evaluation for drought, figured out the proportion of cell numbers to entire basin cell numbers, and plotted the diagram of change of drought area along with time change (Fig. 5). Tab. 2 showed the comparison results of drought areas before and after 1992 (abrupt change). Comprehensively analyzing Fig. 4 and Tab. 2 we can conclude that: ① the drought happened frequently and impacted extensively. The area of drought was about 75.3% of whole basin from 1961 to 2010 and the main magnitude was moderate drought which was about 65.2% of whole basin; ②there had been a high increase in areas with drought during recent 50 years and the trend passed significance of $\alpha = 0.1$ ($M_{\text{drought}} = 1.564 > 1.282$) by the method of Mann - kendall text. Since the late of 1990s, the drought had appeared with the characteristics of wide spread and long duration; ③ the drought area in variation period (1993 ~ 2010) increased by 10.4% than the base period (1661 ~ 1992) including that moderate drought area and severe drought increased by 11.6% and 3.1% respectively.

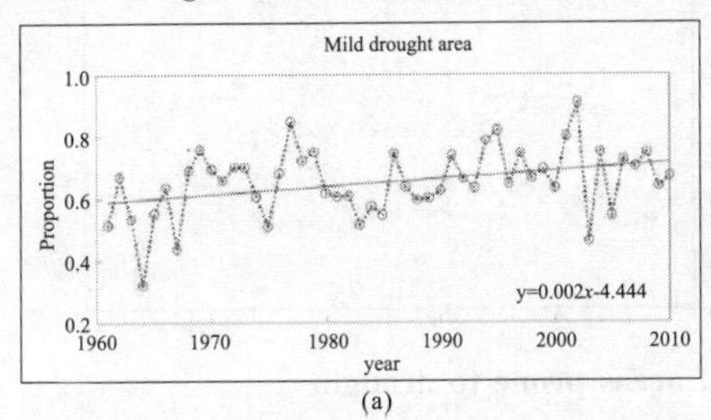

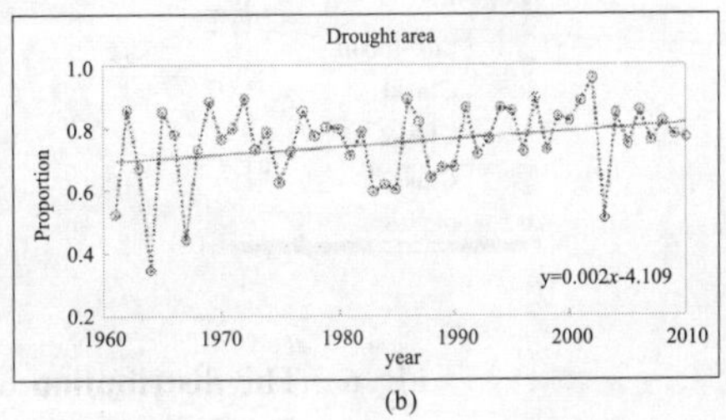

Fig. 5 Time sequences of drought areas (proportion to the entlre basin)

Tab. 2 Drought areas during different period

Magnitude	1961 ~ 2010	1961 ~ 1992	1993 ~ 2010	Variation/%
Normal	0.000	0.000	0.000	—
Mild Drought	0.652	0.626	0.698	11.6
Moderate Drought	0.101	0.100	0.103	3.1
Severe Drought	0.000	0.000	0.000	—
Extreme Drought	0.753	0.726	0.801	10.4

4.3 Areas prone to drought

The annual drought magnitude in 29 sub – basins of the Yellow River Basin can be obtained with the GIS technology and the annual of precipitation and potential evaporation. Based on these, the frequency of drought in each sub – basin was calculated. By fitting with normal distribution, the cumulative frequency of 30% corresponded to drought frequency of 0.69 (P_{30} = 0.69); the cumulative frequency of 60% corresponded to drought frequency of 0.93 (P_{60} = 0.93). So we can characterize the areas mild prone to drought with $P < 0.69$ (Class 1) and the areas moderate prone to drought with $0.69 < P < 0.93$ (Class 2) and the areas extreme prone to drought with $P > 0.93$ (Class 3). Based on this criterion, the figures of distribution of areas prone to drought during 1961 ~ 2010 (whole period), and 1961 ~ 1992 (base period) and 1993 ~ 2010 (variation period) were plotted. Comprehensive analysis of Fig. 6 concluded that the frequency of drought was higher in the north than the south. The area in the north of the Yellow River Basin including Longyangxia – Lanzhou – Hekouzhen and the Internal flow area were extreme prone to drought (Class 3). It also showed by the figure that the area extreme prone to drought during the variation period (1993 ~ 2010) was expand compared with base period (1961 ~ 1992). The areas such as the upper of Baojixia in the Wei River, upper of Zhangjiashan in the Jing River, upper of Zhuangtou in the Beiluo River and the area from Hekouzhen to the left of Longmen were transformed from Class1 or Class2 to Class 3.

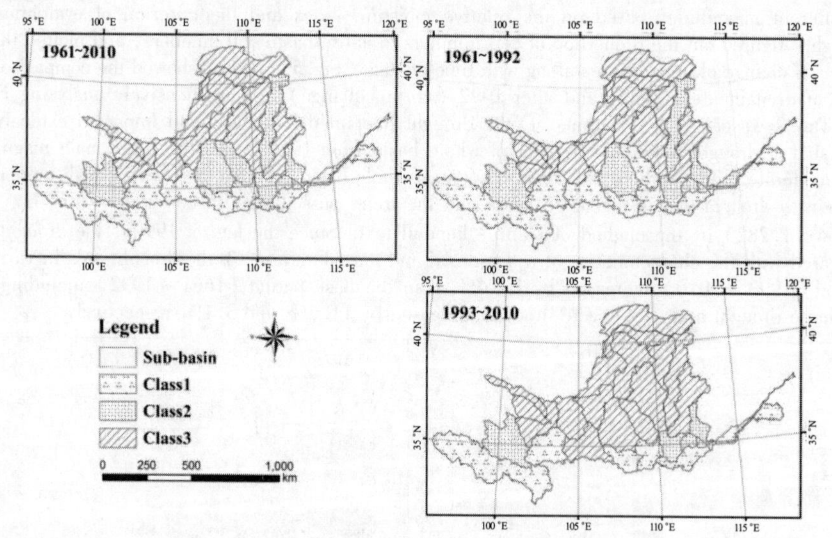

Fig. 6 The distribution of areas prone to drought

5 Conclusions

The precipitation has been decreasing while the potential evaporation has been increasing during recent 50 years. The former passed significance of $\alpha = 0.1$ while the later did not. The abrupt year of relative moisture index appeared in 1992. The period before abrupt year (1961 ~ 1992) was considered as base period while the period after abrupt year (1993 ~ 2010) was considered as variation period. The impact of climate change during later period was more serious than former.

The area of drought was about 75.3% of whole basin from 1961 to 2010 and the main magnitude was moderate drought. There had been a high increase of areas with drought during recent 50 years and the trend passed significance of $\alpha = 0.1$ and The drought area in variation period increased by 10.4% than the base period including that moderate drought area and severe drought increased by 11.6% and 3.1% respectively.

The frequency of drought was higher in the north than the south. The areas in the north such as Longyangxia—Lanzhou—Hekouzhen and the Internal flow area were extremely prone to drought. The areas of extreme proneness have expanded to south since 1992. It may increase drought risk in the Yellow River Basin.

Acknowledgement

This study is jointly financed by National Program on Key Basic Research Project of China (973 Program) (No.: 2010CB951102); Project supported by the National Natural Science Foundation of China (No.: 51021066;51009148)

References

IPCC. Summary for Policymakers of Climate Change 2007: The Physical Science Basis. Contribution of Working Group I to the Fourth Assessment Report of the Intergovernmental Panel on Climate Change [M]. Cambridge: Cambridge University Press, 2007.

Qin Dahe, Ding Yihui, Su Jilan, et al.. Assessment of Climate and Environment Changes in China (I): Climate and Environment Changes in China and Their Projection [J]. Advances in Climate Change Research, 2005,1(1):4 –9.

Liu Jifeng, Xu Zhuoshou, Wang Lin, et al.. Features of Climate and Water Resources Evolution in the Yellow River Basin [J]. China Water Resources, 2009, (13):23 –25.

Xu Zongxue, He Wanlin. Analysis on the Long – Term Trend of Pan Evaporation in the Yellow River Basin over the Past 40 Years [J]. Hydrology, 2005, 25(6):6 – 11.

Wang Hao, Yan Denghua, Qin Dayong, et al.. A Study of the Spatial Shift of 400 mm – rainfall Contours in the Yellow River Basin during Recent 50 Years [J]. Advance in Earth Science, 2005,20(6):649 –655.

Shao Xiaomei, Xu Yueqing, Yan Changrong. Analysis on the Spatial and Temporal Structure of Climatic Water Deficit in the Yellow River Basin [J]. Climatic and Environmental Research, 2007,12(1): 74 –80.

Zhang Qiang, Zou Xukai, Xiao Fengjin, et al.. GB/T20481—2006 Classification of Meteorological drought[S]. Beijing: Standards Press of China, 2006.

Gong L, Xu C, Chen D, et al.. Sensitivity of the Penman – Monteith Reference Evapotranspiration to Key Climatic Variables in the Changjiang (Yangtze River) Basin [J]. Journal of Hydrology. 2006, 329(3): 620 –629.

Cao M S. Detection of Abrupt Changes in Glacier Mass Balance in the Tianshan Mountains [J]. Journal of Glaciology. 1998, 44(147): 352 –358.

Sensitivities of Flood Events to Climate and Land Surface Change in the Qingjianhe Catchment

Zhu Ruirui[1], ***Zheng Hongxing***[1], ***Liu Changming***[1], ***Huo Shiqing***[2] and ***Xu Keyan***[2]

1. Institute of Geographic Sciences and Natural Resources Research, Chinese Academy of Sciences, Beijing, 100101, China
2. Hydrological Bureau, Yellow River Conservancy Commission of the Ministry of Water Resources, Zhengzhou, 450004, China

Abstract: Climate and land surface change of a catchment could lead to changes in the hydrological processe consequently, especially the changes of extreme hydrologic events like flood. The Qingjianhe Catchment (QJH) is one of the most important tributaries of the Yellow River. Since 1950s, large scale water and soil conservation countermeasures have greatly changed the land surface property of the region. More recently, mining industries have further intensified the process on land use change. In this paper, on the basis of a well calibrated hydrological model GR4J, the sensitivity of flood events to precipitation and land surface changes have been investigated. The results show that the peak flow of flood event is positively related to precipitation intensity with elasticity coefficient varying at 1.67 ~ 4.52. The coefficient decreases with the increase of recurrence period of the flood events. Peak flow tends to be negative with parameters representing production store and routing time, but be positive with the routing store. Moreover, the sensitivity of the flood event to climate and land surface change varies with the recurrence period of the flood. The longer the recurrence period of the flood event, the lower sensitivity the peak flow to precipitation and production store is.

Key words: flood events, hydrological model, climate change, land use change, Qingjianhe Catchment

1 Introduction

It has been reported that the frequency of extreme hydrologic events such as floods and droughts have increased significantly owing to global climate change (Prudhomme et al., 2003; Meehl and Tebaldi, 2004). The changes of extreme hydrologic events have become challenges for water resources management, and resulted in serious environmental and ecological problems as well (Huntington, 2006; Oki and Kanae, 2006; Labat et al., 2004; Min et al., 2011; Pall et al, 2011; Plummer et al., 1999; IPCC, 2007). In 2010, the total economic loss due to floods reached to $ 84.7 billion in China (MWR, 2010). Therefore, there are increasing concerns on assessing the impacts of climate change on extreme hydrologic events. The extreme hydrologic events, however, depend not only on climate conditions, but also on the land surface conditions and human activities. For a certain storm, whether an extreme flood event occurs or not could be determined by the properties of the land surface (e. g. the ratio of forest and soil moisture). It is therefore important for us to detect the sensitivity of the extreme hydrologic event to changes in climate and land surface separately.

Among the researches on the sensitivities of hydrologic events to climate changes and land use/land cover changes, hydrological model (lumped or distributed) is one of the most widely used tools (Bahremand and Smedt, 2008). The advantage of the hydrological model is that we can assess the hydrologic impacts of a driving factor individually while keep the other driving factors to be invariant. The sensitivity analysis based on hydrological models could be of rather high uncertainty, which may inherit from uncertainties in model structure, model parameters, and moreover from non-stationary of the hydrologic system. The uncertainties have attracted a lot of researches in recent decade, and different frameworks of uncertainty analysis have been developed (Dooge, 1992; Dooge et al., 1999; Kuhnel et al., 1991; Sankarasubramanian et al., 2001; Chiew, 2006; Zheng et al., 2009). However, in the sense of a reliable sensitivity analysis through hydrological model, the fundamental issue is to set up a robust modeling system.

The scope of this paper is to assess the sensitivity of flood event to climate and land surface changes for river in the Loess Plateau with high flood risk (Zhang and Hui, 1994). The performance of GR4J model (Perrin et al., 2003) will firstly be calibrated and evaluated for flood events with different recurrence period. The sensitivity of the peak flow to rainfall and land surface parameters will then be investigated basing on the concept of elasticity coefficient.

2 Study area and data

The Qingjianhe Catchments (QJH) is located at the middle reaches of the Yellow River in China. The length of river channel is 167.8 km with area of 4,080 km^2 (Fig. 1). Climatologically, the catchment is in the arid to semi – arid region with annual average temperature between 9 ℃ to 10 ℃ and annual precipitation amount of 477.6 mm. The temporal – spatial distribution of precipitation is of highly uneven. More than 75% of annual precipitation falls during the period from June to September (Zhang and Hui, 1994). The rainfall in the rainy season used to be heavy storm showing high density, low duration and small spatial cover. In terms of geomorphology, the QJH belongs to the Loess Hilly region with numerous of gullies. The density of gulley is about 4 ~ 7 km/km^2. The vegetation of the catchments mostly consists of grass and shrubs. The low vegetation cover rate has been considered as an important factor for serious soil erosion in the region. The area with high erosion rate amounts to 4,006 km^2, with erosion coefficient larger than 10,000 t/km^2.

There are 16 rainfall stations located in the basin. The hydrological station, Yanchuan, at the outlet controls a drainage area of 3,468 km^2. In this paper, the hourly precipitation and runoff data are from Hydrological Bureau of Yellow River Conservancy Commission. Daily potential evapotranspiration is estimated based on routine meteorological observations from three stations near the catchment by using Penman-Monteith method recommended by FAO (1998). Daily potential evapotranspiration is then downscaled to hourly data by a simple line function. All the precipitation, potential evapotranspiration and runoff data are in the period from 1980 to 2007 with missing data in 1987 ~ 1991. In order to investigate the difference in the response of flood events to climate and land surface changes, flood events with different recurrence period or accumulated probability (α) are selected for further sensitivity analysis. The flood events selected are $\alpha = 1\%$, 5%, 10%, 25%, 50%, 75% and 90% respectively, where $\alpha = 1\%$ represents the flood event with the largest peak flow.

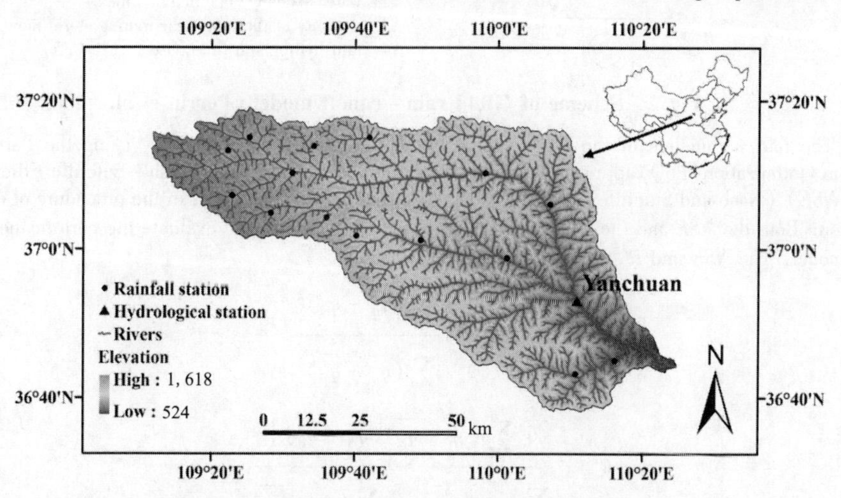

Fig. 1 Sketch map of the Qingjianhe Watershed

3 Methodology

3.1 Rainfall – runoff model

The GR4J model proposed by Perrin et al. (2003) is used in this paper. A schematic of the model is shown in Fig. 2. The inputs of the model are rainfall and potential evapotranspiration, and the four free parameters to be calibrated are *X*1, *X*2, *X*3 and *X*4. The production store (*X*1) represents storage in soil surface and is dependent on catchment soil type. It is assumed that the lower the soil porosity, the larger the production store. Groundwater exchange coefficient (*X*2) is a function of groundwater storage. A negative value of *X*2 indicates water seepage into the depth aquifer, while a positive value indicates water flow from aquifer to routing storage. The routing storage (*X*3) is the maximum amount of water that can be stored in soil porous, depending on soil type and soil water content. The time base of the unit hydrograph (*X*4) represents the time from rainfall to peak flow.

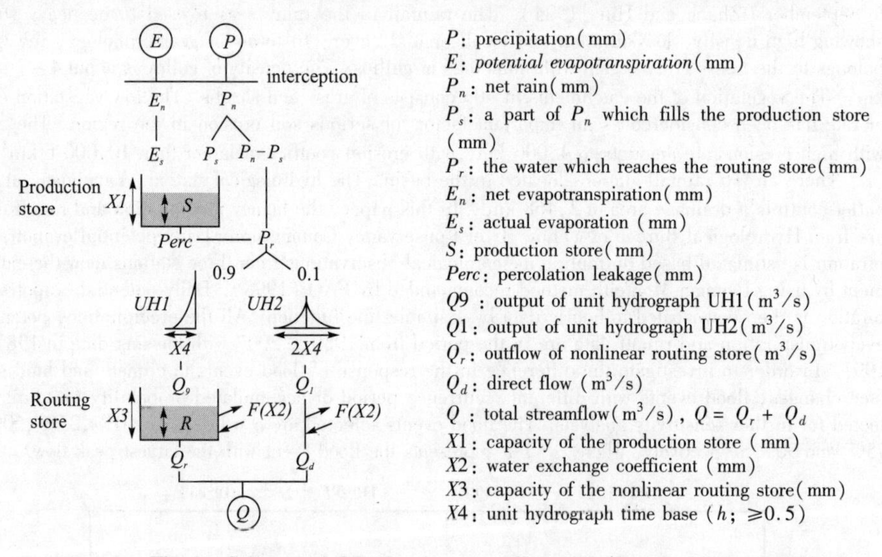

Fig. 2 Scheme of GR4J rain – runoff model (Perrin et al., 2003)

The four parameters of the GR4J model are optimized simultaneously by using the Particle Swarm Optimization(PSO) approach (Eberhart and Kennedy, 1995). The Nash – Sutcliffe Efficiency (*NSE*) (Nash and Sutcliffe, 1970) is adopted as the objective function in the procedure of calibration. Both the *NSE* and the determined coefficient (R^2) are used to evaluate the performance of the model. The *NSE* and R^2 are calculated as:

$$NSE = 1 - \frac{\sum_{i=1}^{n}(q_{ri} - q_{ci})^2}{\sum_{i=1}^{n}(q_{ri} - \overline{q_{ci}})^2} \tag{1}$$

$$R^2 = \frac{\left[\sum_{i=1}^{n}(q_{ri} - \overline{q_{ri}})(q_{ci} - \overline{q_{ci}})\right]^2}{\sum_{i=1}^{n}(q_{ri} - \overline{q_{ri}})^2 \cdot \sum_{i=1}^{n}(q_{ci} - \overline{q_{ci}})^2} \tag{2}$$

Where, q_{ri} and q_{ci} are observed and simulated streamflow respectively; $\overline{q_{ri}}$ and $\overline{q_{ci}}$ are the means of q_{ri} and q_{ci}.

3.2 Sensitivity analysis

When a model is well calibrated, the response of the flood event to climate and land surface changes can be investigated through sensitivity analysis, where a specific driving factor is assumed to have a proportional change while the other driving factors keep constant. In this paper, the driving factors under consideration are precipitation and three model parameters which are highly related to land surface properties, i. e., *X*1, *X*3 and *X*4 in the GR4J model. The sensitivity of the flood event to changes in the driving factors can be measured by the corresponding proportional change in the peak flood, or the elasticity coefficient proposed by Schaake (1990).

The elasticity coefficient of a hydrological variable (i. e peak flow in this paper) to changes of the driving factor is defined as the ratio between the proportional change of the hydrological variable (Q) and the driving factor (V_i) (Schaake, 1990), expressed as:

$$\varepsilon = \frac{\Delta Q/Q}{\Delta V_i/V_i} = \frac{H(V_i + \Delta V_i + \Delta V_i, C)/H(V_i, C)}{\Delta V_i/V_i} \tag{3}$$

Where, H represents a specific hydrological model; C means other variables in the hydrological system; ΔQ and ΔV_i are changes of the hydrological variable and the driving factor respectively.

The advantage of the elasticity coefficient is that it is a dimensionless value, which enables the comparability among different driving factors. According to the concept described above, if the elasticity coefficient of peak flow to precipitation equals to 2.0, it indicates that if precipitation is 10% higher, the peak flow will be 20% larger. Because of the nonlinearity of the hydrologic system, the sensitivity of the hydrological variable to the driving factor could vary with the change rate of the factor. Therefore, the proportional change and the elasticity coefficient of the peak flow are both investigated by assuming that the change rates of the driving factors are in the range [-15%, 15%].

4 Results

4.1 Performance of the model

Fig. 3 shows the simulation results of the 59 flood events for the period 1980 ~ 2007. The proportion of events with *NSE* higher than 0.5 is about 68%, where 19% of simulations has *NSE* higher than 0.8. The average *NSE* of the 59 cases is about 0.60. When calibrated simultaneously, the *NSE* and R^2 of the selected flood events with accumulated probability equal to 1%, 5%, 10%, 25%, 50%, 75% and 90% are 0.66 and 0.67 respectively, with the four parameters *X*1, *X*2, *X*3 and *X*4 optimized to be 8.84 mm, -0.817 mm, 1.385 mm and 7.17 h respectively. The rather high consistency between the simulation and observation indicates that the GR4J model can be adopted to simulate hydrological process in the QJH Basin.

4.2 Sensitivity of flood event to precipitation

Fig. 4 shows the proportional changes of the peak flow owing to changes in the precipitation. It is obvious that peak flow increases (decreases) with the increasing (decreasing) of precipitation. However, it is worth noting that the proportional change of peak flow is not the same to that of the precipitation. For instance, when precipitation is 15% higher, the peak flow of the flood event can be 63% larger, which implies that the increasing rate of peak flow could be larger than that of precipitation. One may also notice that the proportional changes are different among the flood events when the change rates of precipitation keep the same. The small flood (e. g. F7 in Fig. 4) tends to have higher proportional change than big flood (e. g. F1 in Fig. 4).

The elasticity coefficient further depicts the different responses of the flood events to precipitation. As shown in Fig. 5, the elasticity coefficient of peak flow to precipitation varies in the range 1.67 ~ 4.52, which means that when precipitation is 10% higher, the peak flow could increase at the rate of 16.7% ~ 45.2%. The elasticity coefficient generally increases with the accumulated

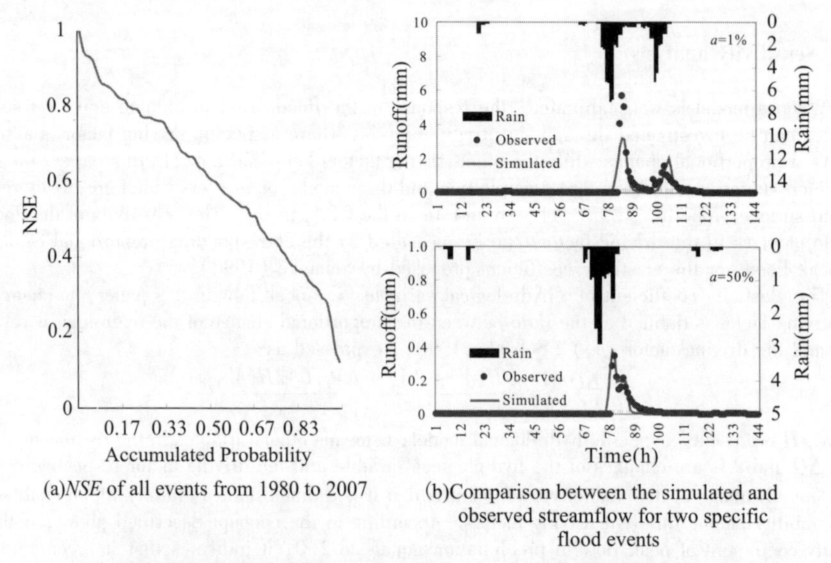

(a)*NSE* of all events from 1980 to 2007

(b)Comparison between the simulated and observed streamflow for two specific flood events

Fig. 3 Performance of GR4J model in flood event simulation at QJH Basin

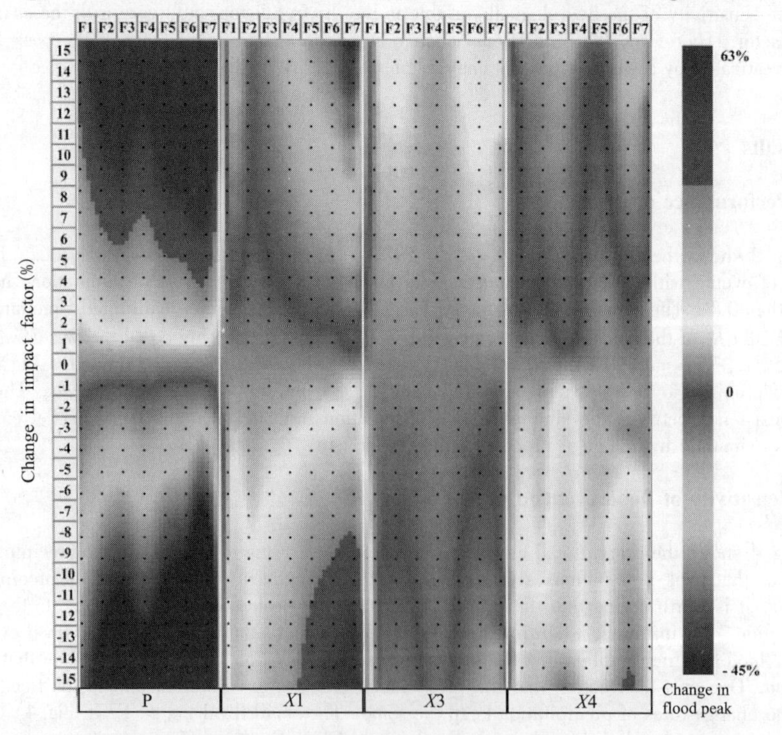

(F1 ~ F7 correspond to flood events with accumulated probability of 1% , 5% , 10% , 25% , 50% , 75% and 90% respectively)

Fig. 4 Proportional changes of peak flow with the precipitation and land surface changes

probability of the flood event. In other words, the smaller the flood (e. g. F7 in Fig. 5), the higher the elasticity coefficient. The results indicate that smaller flood is more sensitive to precipitation change.

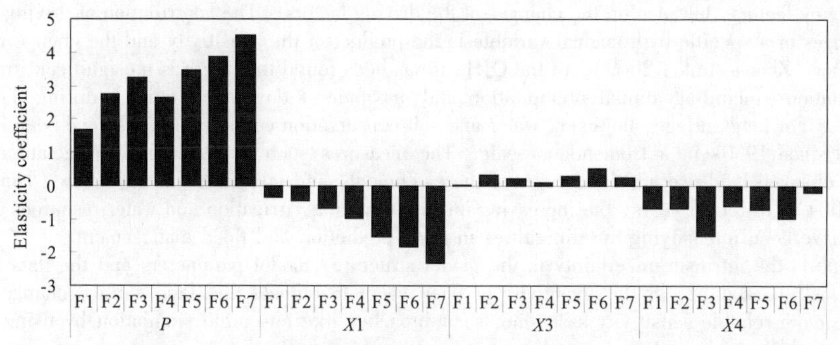

(F1 ~ F7 correspond to flood events with accumulated probability of 1%, 5%, 10%, 25%, 50%, 75% and 90% respectively)

Fig. 5 Sensitivity of peak flow to precipitation and the parameters change

4.3 Sensitivity of flood event to land surface

Fig. 4 also shows the proportional changes of peak flow with the changes of the parameters related to land surface properties. For all flood events, peak flow tends to have negative relation with the parameters $X1$ and $X4$, but have positive relation with parameter $X3$. The decrease of $X1$ or $X4$, or increase of $X3$ could all result at increase of the peak flow. However, it is also noticed that the proportional change of peak flow is not the same to that of the parameters. For instance, for a mild flood ($\alpha = 50\%$), when $X1$, $X3$ or $X4$ increases 15%, the proportional change of the peak flood could be −15.5%, 5.4% and −2.1% respectively; but if $X1$, $X3$ or $X4$ reduced at a rate of 15%, the peak flow of the mild flood could change at the rate of 18.7%, −6.7% and 2.7%. Among the three parameters concerned, it is quite clear that peak flow is most sensitive to $X1$. Moreover, as shown in Fig. 4, the proportional change of peak flow is different for flood events with different recurrence period. In case of $X1$, the proportional change of peak flow tends to be higher for small flood, which implies that small flood (e. g. F7) is more sensitive to changes of $X1$ than bigger flood (e. g. F1).

The results are further confirmed by the elasticity coefficient shown in Fig. 5. It is noticed that the elasticity coefficients of peak flow to the parameters $X1$ and $X4$ (ε_{X1} and ε_{X4}) are negative, but is positive to parameter $X3$ (ε_{X3}). Among the three parameters, in general, we have the relation as: $|\varepsilon_{X1}| > |\varepsilon_{X4}| > |\varepsilon_{X3}|$. With respect to $X1$, the elasticity coefficient varies at the range −0.33 ~ −2.34, and tends to decrease with the recurrence period. The ranges of elasticity coefficients related to $X3$ and $X4$ are 0.01 ~ 0.52 and −0.21 ~ −1.49 respectively, but there are no any significant trend between the elasticity coefficient and the recurrence period of the flood event.

5 Discussions

The occurrence and development of flood event is a systemic response involving series of complicated procedures. Climate and land surface can be two vital driving factors changing the intensity, duration and frequency of the flood events (Deng et al., 2003; Jing et al., 2005; Li et al., 2007), and result at higher flood risk. Precipitation is the source of flood. The changes of intensity, duration, tempo-spatial pattern in precipitation could directly lead to the change of the flood risk. On the other hand, land use and land cover changes could alter the store capacity of a catchment, and may change the flow direction as well. According to the results in the QJH, as shown above, the increasing of store capacity (increasing in $X1$) could efficiently reduce the flood risk.

However, it is worth noting that peak flow of a flood event is more sensitive to precipitation than the land surface parameters.

It should be pointed out that the changes of flood event depend not only on the sensitivity to the driving factors, but also on the changes of the driving factors . The contribution of driving factor to changes of a specific hydrological variable is the product of the sensitivity and the change rate of the factor (Zheng et al. , 2009). In the QJH, it has been found that there is no significant trend in precipitation (including annual precipitation and maximum 1-day precipitation) during the past decades. For land surface, however, water and soil conservation countermeasures have been implemented since 1950s in a tremendous scale. The measures such as terracing, revegetation have deeply changed land use and land cover of the catchment, and attenuate the flood flow (Ding and Mu, 2004). In recent years, the increasing impact of mining, irrigation and water resources developing have been intensifying the difficulties in flood prediction and flood management.

Due to the intrinsic uncertainty in the model structure, model parameters and the data used, there exist uncertainties in the sensitivity of flood event to climate and land surface changes. To provide more reliable sensitivity assessments, approaches like ensemble simulation by using multi models could be of great help.

6 Conclusions

The intensity and frequency of extreme hydrologic events such as flood and drought are changing with global climate and land surface change. Assessing the response of the extreme event is believed to be helpful for human community to adjust the design and operational rules of hydro-engineering, and develop adaptive measures to mitigate the threats of the extreme events. On the basis of a well calibrated hydrological model, in this paper, the sensitivity of flood events to precipitation and land surface changes have been investigated. The main findings of the paper are: ①The peak flow of flood event positively related to precipitation intensity with elasticity coefficient ε_P varies at 1.67 ~ 4.52. The coefficient decreases with the increase of recurrence period of the flood events; ②Peak flow tends to be negative with parameters representing production store (*X*1) and routing time (*X*4), but positive with the routing store (*X*3); ③ Peak flow is more sensitive to precipitation and *X*1, which holds $|\varepsilon_P| > |\varepsilon_{X_1}| > |\varepsilon_{X4}| > |\varepsilon_{X3}|$; ④The sensitivity of the flood event varies with the recurrence period. The longer the recurrence period of the flood event, the lower is the sensitivity of the peak flow to precipitation and production store. It is worth pointing out that the GR4J model used in this study is a lumped conceptual model. The uncertainties of the simulation as well as the spatial pattern of precipitation and land surface properties need more investigations in the future study.

Acknowledgements

This study is financially supported by the Public Welfare Projects of China Ministry of Water Resources (No: 200901016) and China National Natural Science Foundation (No: 41101032).

References

Bahremand A, De Smedt F. Distributed Hydrological Modeling and Sensitivity Analysis in Torysa Watershed, Slovakia[J]. Water Resources Manage, 2008, 22:393 - 408.

Chiew F H S. Estimation of Rainfall Elasticity of Streamflow in Australia[J]. Hydrol. Sci. J., 2006, 51(4), 613 - 625.

Deng H, Li X, Chen J, et al. Simulation of Hydrological Response to Land Cover Changes in the Suomo Basin[J]. Acta Geographica Sinica, 2003, 25(1):53 - 62

Ding L, Mu X. The Effects of Soil and Water Conservation on Temporal Change of the Surface Runoff in Watersheds[J]. Journal of Arid Land Resources and Environment, 2004, 18(3): 103 - 106.

Dooge J. Sensitivity of Runoff to Climate Change: A Hortonian Approach[J]. Bull. Amer. Mete-

or. Soc. , 1992,73(12): 2013 -2024.

Dooge J C I, Bruen M, Parmentier B. A Simple Model for Estimating the Sensitivity of Runoff to Long Term Changes in Precipitation without a Change in Vegetation[J]. Advances in Water Resources, 1999,23: 153 -163.

Eberhart R C, Kennedy J. A New Optimizer Using Particle Swarm Theory[C]// Proceeding of the Sixth International Symposium on Micro Machine and Human Science. Nagoya, Japan, 39 - 43. Piscataway, NJ:IEEE Service Center,1995.

Food and Agriculture Organization of the United Nations. Crop Evapotranspiration:Guidelines for Computing Crop Requirements[M]. Italy:Food & Agriculture Org.

Huntington T G. Evidence for Intensification of the Global Water Cycle: Review and Synthesis[J]. Journal of Hydrology, 2006,319: 83 -95.

IPCC. Climate Change 2007: Synthesis Report[C]//. Contribution of Working Groups Ⅰ, Ⅱ and Ⅲ to the Fourth Assessment Report of the Intergovernmental Panel on Climate Change. IPCC, Geneva, Switzerland,2007.

Jing X, Wang G, Lu F, et al. Impact of Soil and Water Conservation Measures on Flood Runoff in Qingjianhe River Basin[J]. Water Resources and Hydropower Engineering, 2005,36(3): 66 -68.

Kennedy J, Eberhart R C. Particle Swarm Optimization[C]// Proc. IEEE Int'l, Conf. on Neural Networks, IV, 1942 -1948. Piscataway, NJ: IEEE Service Center, 1995.

Kuhnel V, Dooge J C I, O'Kane J C I, et al. Partial Analysis Applied to Scale Problems in Surface Moisture Fluxes[J]. Surv. Geophys. , 1991, 12: 221 -247.

Labat D, Godderis Y, Probst J L, et al. Evidence for Global Runoff Increase Related to Climate Warming[J]. Advances in Water Resources, 2004, 27: 631 -642.

Li L, Jiang D, Li J, et al. Advances in Hydrological Response to Land use /Land Cover Change [J]. Journal of Natural Resources, 2007,22(2):211 -224

Meehl G A, Tebaldi C. More Intense, More Frequent, and Longer Lasting Heat Waves in the 21st Century[J]. Science, 2004, 305:994 - 997.

Min S K, Zhang X B, Zwiers E W, et al. Human Contribution to More - intense Precipitation Extremes[J]. Nature, 2001, 470: 378 -381

Nash J E, Sutcliffe J V. River Flow Forecasting Through Conceptual Models. Part I: A Discussion of Principles[J]. Journal of Hydrology, 1970, 27(3): 282 -290.

Oki T, Kanae S. Global Hydrological Cycles and World Water Resources[J]. Sciences. 2006, 313: 1068 -1072.

Pall P, Aina T, Stone D A, et al. Anthropogenic Greenhouse gas Contribution to Flood Risk in England and Wales in Autumn 2000[J]. Nature, 2011, 470: 382 -385.

Perrin C, Michel C, Andre'assian V. Improvement of a ParsiMonious Model for Stream Flow Simulation[J]. Journal of Hydrology, 2003, 279(1 -4):275 -289.

Plummer N, Salinger M J, Nicholis N, et al. Changes in Climate Extremes over the Australian Region and New Zealand During the Twentieth Centuries[J]. Climate Change, 1999, 42:183 -202.

Prudhomme C, Jakob D, Svensson C. Uncertainty and Climate Change Impact on the Flood Regime of Small UK Catchments[J]. Journal of Hydrology, 2003, 277: 1 -23.

Sankarasubramanian A, Vogel R M, Limbrunner J F. Climate Elasticity of Streamflow in the United States[J]. Water Resources Research, 2001, 37(6):1771 -1781.

Schaake J C. From Climate to Flow[C]// Climate Change and U. S. Water Resources. John Wiley & Sons Inc. , New York, USA, 1990.

The Ministry of Water Resources of the People's Republic of China (MWR). Communiqué on china flood and drought disaster [EB/OL]. http://www. mwr. gov. cn/zwzc/hygb/zgshzhgb/.

Zhang J, Hui P. Analysis of Runoff and Sediment Variations in Qingjianhe River[J]. Soil and Water Conservation in China, 1994,11(152):21 -26.

Zheng H, Zhang L, Zhu R, et al. Responses of Streamflow to Climate and Land Surface Change in the Headwaters of the Yellow River Basin [J]. Water Resources Research, 2009, 45: W00A19.

Temperature Variation Characteristics Analysis in the Ice Flood Season of the Yellow River Basin

Fan Minhao, ***Zhang Ronggang*** and ***Li Mingzhe***

Hydrology Bureau of the Yellow River Conservancy Commission,
Zhengzhou, 450004, China

Abstract: This article is based on the daily mean temperature data of the Yellow River basin to analyze variation characteristics of average daily temperature and negative accumulated temperature of four sampling meteorological stations in the ice flood season by using statistical analysis, linear trend, the cumulative anomaly analysis and other methods. It attempts to explore the temperature variation characteristics and laws of the Yellow River basin in the ice flood season since 1950s. The results shows an increasing trend in general temperature change of the Yellow River basin in the ice flood season since 1950s. The warming rate of Baotou, Zhengzhou, Jinan, Beizhen was 0.6 ℃/10 a, 0.34 ℃/10 a, 0.35 ℃/10 a, 0.38 ℃/10 a respectively. In the ice flood season the negative accumulated temperature increasing rate of Baotou and Beizhen was 7.67 ℃/a and 2.36 ℃/a respectively, the total number of days which daily mean temperature were under 0 ℃ decreased by the rate of 3 d/10 a. Negative accumulated temperature variation characteristics and trends of Baotou station and Beizhen station are basically the same. On the premise of overall temperature upward trend, the temperature of the Yellow River basin before 1985 was cold phase, from 1986 to 2010 was warm phase, and abruptly change of temperature occurred around 1986 in which negative accumulated temperature rose and river frozen relatively weakened. Since the mid - 1980s, the temperature and negative accumulated temperature of Ningxia-Inner Mongolia reach and the lower reach of the Yellow River have been increasing obviously that significantly influenced on the Freeze-up Time, the Break-up Time and the degree of river frozen.
Key words: temperature, ice flood, negative accumulated temperature, climatic change, the Yellow River basin

1 Introduction

Affected by northwest wind in winter, the climate in the Yellow River basin is dry and cold and many reaches of Yellow River begin to freeze. But in the early spring every year, ice slush floods, i. e. ice flood, occur in its Ningxia-Inner Mongolia reach and lower Huayuankou Reach flush. As to the Ningxia-Inner Mongolia reach, located the northernmost Yellow River basin, is a stable frozen reach for it is extremely cold in winter. Generally, ice water flowing appears in November; it is frozen in early or middle December, while it is unfrozen in middle or later March with an ice period of about 120 d in average. As to the lower reach, the temperature in Henan is obviously higher than Shandong where Yellow River estuary is located because of the difference of latitude. Thus the frozen period begins in the later December and lasts to the middle of next February with an average ice period of 45 d. The lower reach are frozen in 80% of the years. When the weather changes drastically, the river may be frozen and unfrozen several times in only one ice flood period, causing serious threat. Affected temperature and course of river, the unfrozen period of the Ningxia-Inner Mongolia reach and the lower reach usually begins from the upper to the lower, the releasing amount of tank water storage increases continuously, tending to form ice flood disaster.

It is revealed in the fourth assessment report from IPCC, the average global surface tempera-

Founded by Hydrology Bureau of Yellow River Conservancy Commission (Hydro - 1103) ture increases 0.74 centigrade from 1906 to 2005 according to the data of global surface temperature thermometer. In China, the temperature increases significantly in recent 100 years: since the winter of 1986/1987, China has experienced 21 warm winters in a row . The ice condition forecast is related to the safety of the people and their property along the Yellow River basin. Only when timely prevention to the ice water in time based on the measurement of ice flood is possible can the ice flood disasters be reduced and regimen, ice condition and temperature along the river be mastered, which is a significant insurance for ice flood forecast service. Temperature is one of the important factors of ice condition. Variation of temperature has significant influence to frozen time and unfrozen time, tank storage water release amount and ice flood peak, etc. The seriousness of ice flood is related to the amount of tank storage water and ice in the course, while the negative accumulated temperature has a close linear relation with the thickness of ice. Therefore, in the background of global warming, it is of great significance to research the characteristics and tendency of temperature variation in ice flood period for ice flood prevention and disasters reduction in life and production.

2 Data and approach

Baotou is selected as the representative of Ningxia-Inner Mongolia reach, with the daily average temperature data in November to December in 1956 to 2010 and January to March in 1957 to 2011 (ice flood period during 1956 to 2010, similarly hereinafter). Zhengzhou, Jinan and Beizhen are selected as the representatives of lower reach, with daily average temperature data in December in 1951 to 2010 and January to February in 1952 to 2011 (ice flood period during 1951 to 2010, similarly hereinafter). Arrangement to daily average temperature value of each observation station is applied to get monthly average temperature series and negative accumulated temperature related series. To ensure the continuity of the series, the data of neighbor stations can be interpolated when there is lack of some data. It is to analyze variation characteristics of average daily temperature and negative accumulated temperature (in this article, negative accumulated temperature refers to the daily average temperature below 0, measuring the coldness in winter.) of four sampling meteorological stations in the ice flood season by using statistical analysis, linear trend, the cumulative anomaly analysis and other methods. It attempts to explore the temperature variation characteristics and laws of the Yellow River basin in the ice flood season since 1950s.

3 The temperature variation tendency in Ningxia-Inner Mongolia reach

3.1 Basic temperature characteristics

The average temperature in ice flood period is −5.8 ℃ at Baotou observation station in Ningxia-Inner Mongolia reach (1956 ~ 2010), with the highest −2.7 ℃ in 2011/2012 and the lowest −10.7 ℃ in 1967/1968. Its monthly average temperature in December to March in the next year is −1.9 ℃, −9.4 ℃、−11.5 ℃、−6.9 ℃、and 0.8 ℃, with the highest in March and the lowest in January.

3.2 Temperature variation tendency

The temperature in 1950 to 2010 shows an overall increase during ice flood period with an rate of 0.6 ℃/10 a. It shows in Fig. 1 that the average temperature is lower than year average value only in 1995 and 2009 while higher in other years since 1987. It is thus clear that the temperature increase is faster since the mid of 1980s. The analysis of the monthly temperature statistics indicates that the temperature in December to the next March has an upward tendency with a highest rate in February of 0.9 ℃/10 a, flowing the December of 0.66 ℃/10 a. Temperature increase rates are 0.47 ℃/10 a, 0.57 ℃/10 a and 0.4 ℃/10 a respectively in December, January and March. Generally, it is frozen in early or middle December and unfrozen in middle or later March. The increase of temperature affect the frozen/unfrozen date and the development of ice flood to some extend.

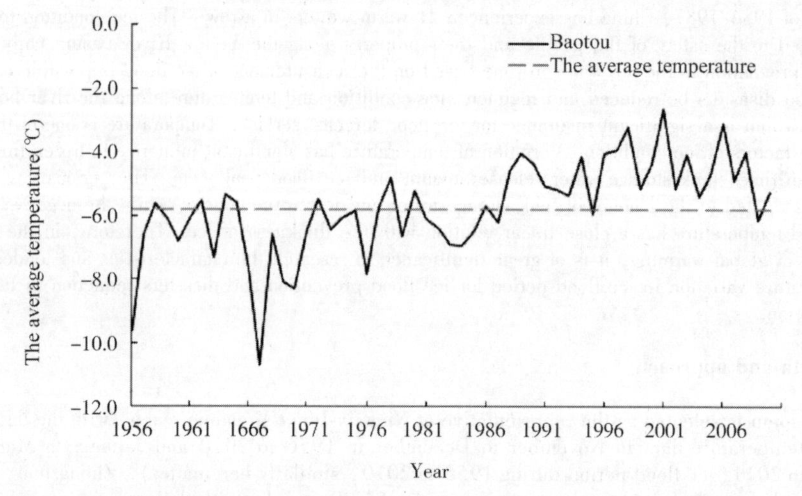

Fig. 1 The average temperature change curve of Baotou in the ice flood season

3.3 Variation characteristics of negative accumulated temperature

Ice condition has some relationship with hydro meteorology factors. On the whole, the main factors affecting the ice condition include water flow, temperature and wind speed, in which the wind speed mainly affects regional temperature while it has little influence to the motive factors of water flow . Negative accumulated temperature is the sum total of daily average temperature below 0 ℃ during a certain period, measuring the coldness in winter. The higher the negative accumulated temperature is, the colder in winter.

The damage of the ice flood is related to the condition of the river course, the tank storage water in the course and the variation of the temperature at the unfrozen period. Ice condition affects the water flowing under the ice. The thicker the ice is, the harder the water flows. Thus it is easier to form a stable ice cover. Therefore, it is of great significance to study the temperature variation characteristics and tendency of negative accumulated temperature for overall mastering the ice condition and preventing ice flood.

3.3.1 The basic characteristics of negative accumulated temperature

During the ice flood period in 1956 to 2010, the average negative accumulated temperature is -976.6 ℃, with a total of 120 d below 0 ℃ in average daily temperature. Tab. 1 shows the average of negative accumulated temperature from Baotou observation station in different period. It is indicated that, the average negative accumulated temperature in the three periods before 1986 is lower than -1,000 ℃, while it increases obviously after 1986, higher than -1,000 ℃. In a word, the temperature increase is more obvious since the middle and later 1980s.

Tab. 1 The average negative accumulated temperature of Baotou in each period

Time interval	1956 ~ 1965	1966 ~ 1975	1976 ~ 1985	1986 ~ 1995	1996 ~ 2005	2006 ~ 2010
The average negativeaccumulated temperature(℃)	-1,069.4	-1,166.6	-1,039.2	-874.1	-800.1	-844.5

3.3.2 The variation tendency of negative accumulated temperature

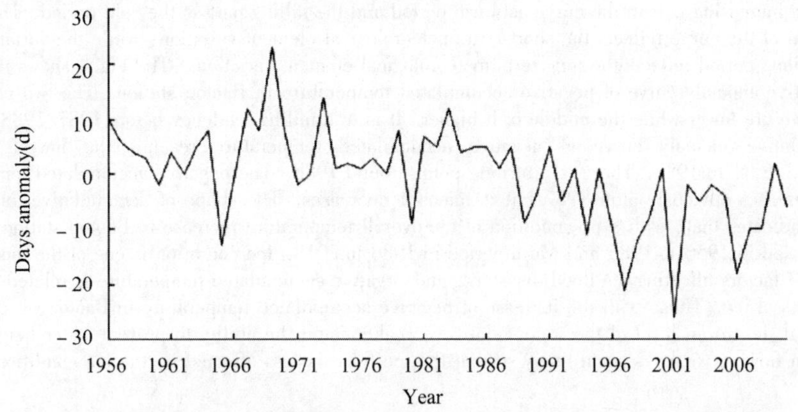

Fig. 2 The total days anomaly with Baotou average daily temperature below 0 ℃ in the ice flood season

The number of the total days with Baotou average daily temperature below 0 ℃ is in an obvious falling tendency in ice flood season in 1956 to 2010, with a falling rate of 3 d/10 a. In 1988, the total days with average daily temperature below 0 ℃ before ice flood period are mainly in positive anomaly, while in 1988 to 1993, the total days with average daily temperature below 0 ℃ in ice flood period are shown in positive and negative anomaly and the total days is mainly in positive anomaly after 1995/1996 (see Fig. 2).

In ice flood season, the negative temperature anomaly of Baotou are in an upward tendency with the change of time and they are in an increasing tendency rate of 7.67 ℃/a (see Fig. 3). In 1988, the negative accumulated temperature before ice flood period is mainly in negative anomaly and it turns into positive in ice flood period and the value of negative accumulated temperature increases and its amount (the absolute value of negative accumulated temperature) decreases, indicating that the days with daily average temperature below 0 ℃ in ice flood period decrease or the value of daily average temperature below 0 ℃ increases. It conforms to the falling tendency of the days with daily average temperature below 0 ℃ and the increase tendency of monthly temperature.

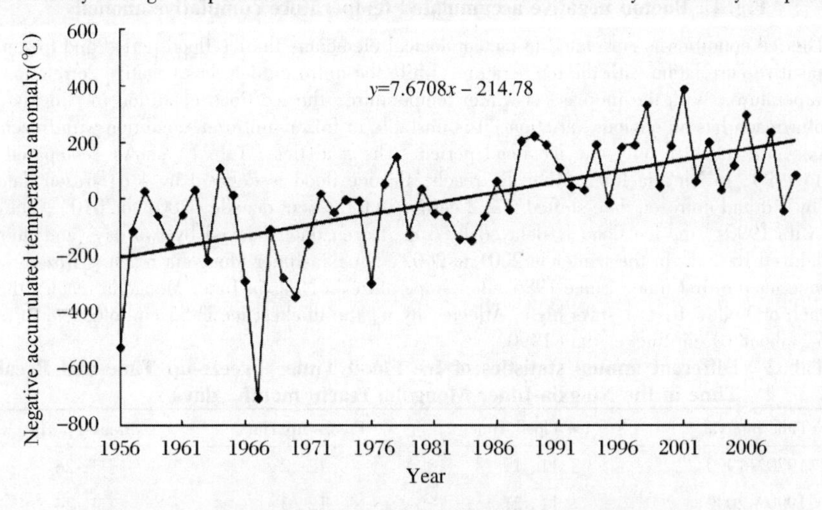

Fig. 3 Baotou negative accumulated temperature anomaly

Accumulative anomaly curve can be used to analyse the tendency of climate variation, determining the increasing part in the curve as warm period and the falling part as the cold period. The tiny variation of the curve reflects the short-term meteorological element variation, while the variation of it in a long period reflects the long-term meteorological element variation . The Fig. 4 shows the accumulative anomaly curve of negative accumulated temperature in Baotou station. The two ends of the curve are lower while the middle of it higher. It is in a falling tendency before 1987/1988. The accumulative anomaly curve of negative accumulated temperature reaches the lowest value −3,452.6 ℃ in 1985. There is a turning point around 1986. The negative accumulated temperature increases obviously after 1987 but its amount decreases. The shape of accumulative anomaly curve indicates that, with a precondition of the overall temperature increase in Baotou station, it is cold period in 1956 to 1985 and warm period in 1986 to 2010. Ice condition is one of the most importance factors affecting ice flood disasters, and negative accumulated temperature is related to the thickness of ice. Thus, with the increase of negative accumulated temperature in Baotou, it can be seen that the frozen level of the water relatively weakens and the ability to prevent water below the ice from flowing weakens accordingly so that the ice flood is less affected by the ice condition.

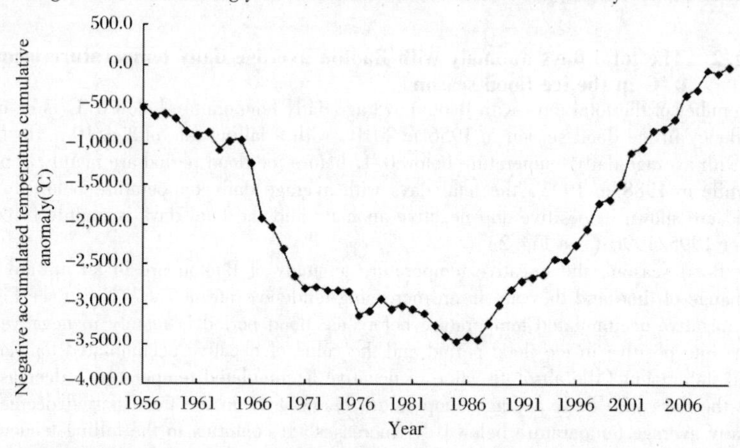

Fig. 4 Baotou negative accumulated temperature cumulative anomaly

The ice condition is correlated to meteorological elements: the ice flood period and frozen date have positive correlation with the temperature, while the unfrozen date has negative correlation with the temperature. With the increase of winter temperature, the ice flood condition in Ningxia-Inner Mongolia reach has an obvious variation: it is unstable in frozen/unfrozen condition with frozen/unfrozen several times in only one ice flood period. The statistics (Tab. 2) shows, compared with 1970 to 1990, in Ningxia-Inner Mongolia reach, the ice flood is delayed by 4 d, frozen date delayed by 2 d and unfrozen date shifted to 2 d earlier in the resent decade (2001 to 2010); but compared with 1990s, the ice flood is delayed by 6 d, frozen date delayed by two days and unfrozen date delayed by 1 d. In the winter in 2001 to 2002, Ningxia-Inner Mongolia reach is frozen/unfrozen twice for the first time. Since 1986, the temperature in Ningxia-Inner Mongolia reach (the upper reach of Yellow River) stays high. Affected by it, the thickest ice is 53 cm to 63 cm in recent decade, about 10 cm thicker than 1990.

Tab. 2 Different annual statistics of Ice Flood Time, Freeze-up Time and Break-up Time in the Ningxia-Inner Mongolia reach(month ,day)

Time interval	Ice Flood Time	Freeze-up Time	Break-up Time
1970 ~ 1990	11, 19	12, 2	3, 26
1990 ~ 2009	11, 21	12, 3	3, 24
2000 ~ 2010	11, 23	12, 3	3, 25

4 The temperature variation tendency in lower reaches

4.1 Basic temperature characteristics

During 1951 to 2010, the average temperatures in observation stations at Zhengzhou, Jinan and Beizhen are 1.7 ℃, 0.8 ℃ and −1.5 ℃ respectively in ice flood period. As to the monthly average temperature (see Tab. 3) in December to next February, it is lowest in January and generally flat in December and February. Because of latitude deference, the temperature in estuary area is obviously lower than upper reaches. River frozen generally begin from the lower reaches to upper, while river unfrozen is the opposite. If it is improperly controlled in upper reaches reservoirs, it is prone to increasing water level, causing ice flood disaster.

Tab. 3 The lower reaches of the Yellow River Station three years average temperature statistics of ice flood period

Average temperature(℃)	Zhengzhou	Jinan	Beizhen
December	2.1	1.4	−0.7
January	0.1	−0.9	−3.1
February	2.8	2.0	−0.4
December ~ February	1.7	0.8	−1.5

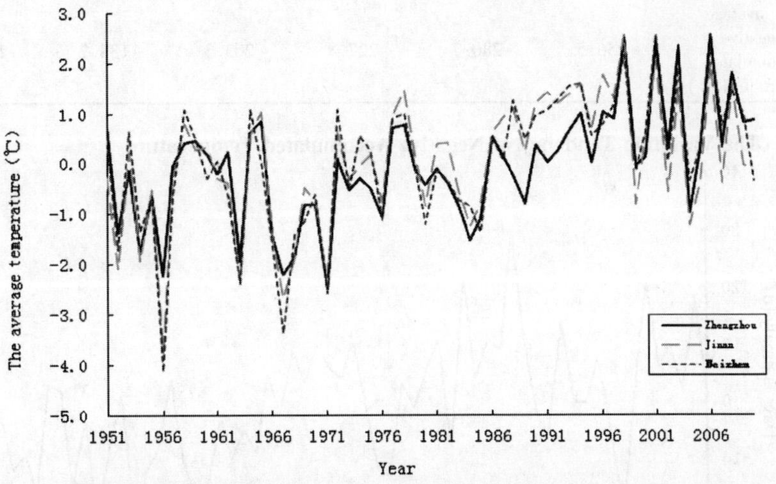

Fig. 5 The average temperature curve of three stations in lower reaches in the ice flood season

4.2 Temperature variation tendency

The temperature in 1950 to 2010 shows an overall increase during ice flood period at Zhenegzhou, Jinan and Beizhen, with rates of 0.34 ℃/10 a, 0.35 ℃/10 a and 0.38 ℃/10 a respectively. The historic average temperature variation curves in these three places move in the same direction during ice flood period. Basically, the average temperature of Jinan in ice flood period is 2 ℃ to 3 ℃ higher than that of Beizhen, while average temperature of Zhengzhou is 3 ℃ to 4 ℃ higher than Beizhen as well. It is indicated in Fig. 5 that, except 1988 and 1989, the variation curves of the average temperature anomaly during ice flood period are of very slight deference, in which the curve of Zhengzhou in 1989 shows 0.3 ℃ lower but the curves of Jinan and Beizhen show 0.2 ℃ and 0.5 ℃ higher respectively. The higher and lower average temperatures in Jinan and Beizhen during ice flood period are very close. In 1990 to 1996, the average temperature anomalies of Jinan and Beizhen in ice flood period are significantly higher than that of Zhengzhou, indicating the increasing range in Beizhen and Jinan is relatively large during that period.

4.3 Variation characteristics of negative accumulated temperature

4.3.1 The basic characteristics of negative accumulated temperature

The frozen period is longest in lower reaches near estuary, river frozen lasting from later December to the middle of next February. The average daily temperature is this area conforms to it in Beizhen. Therefore, the observation station in Beizhen is selected as a representative to analyse the development characteristics of negative accumulated temperature in ice flood period in recent 6 decades. In ice flood period from 1951 to 2010, the average negative accumulated temperature is −213.4 ℃ with 58 d in average daily temperature below 0 ℃. Calculating the average negative accumulated temperature value in a 10-year term (Tab.4), the result shows a relative lower average negative accumulated temperature value in the 3 decades before 1980s. The lowest appears in 1961 to 1970. But the average negative accumulated temperature value increases obviously after 1990s, while the highest appears in 1991 to 2000, with an increase from −280.7 ℃ to −138.2 ℃, more than twice.

Tab.4 The average negative accumulated temperature of Beizhen in each period

Time interval	1951 ~ 1960	1961 ~ 1970	1971 ~ 1980	1981 ~ 1990	1991 ~ 2000	2001 ~ 2010
The average negative accumulated temperature(℃)	−256.5	−280.7	−227.8	−203.3	−138.2	−174.1

4.3.2 The Variation Tendency of Negative Accumulated Temperature

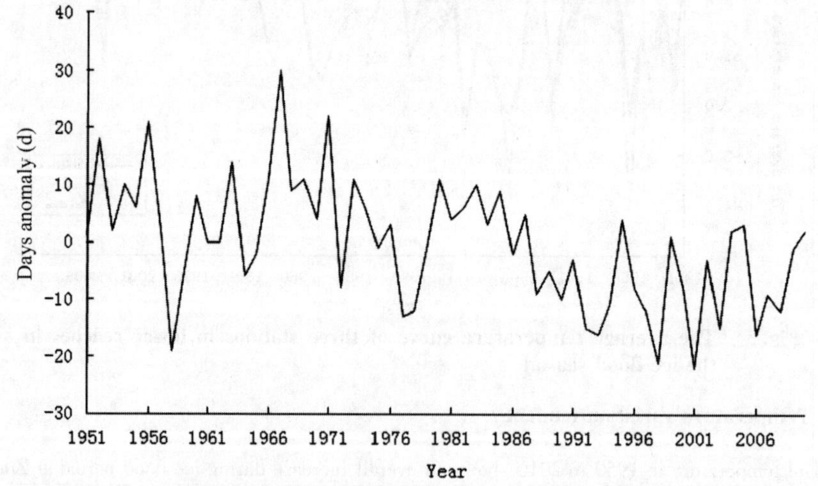

Fig.6 The total days anomaly with Beizhen average daily temperature below 0 ℃ in the ice flood season

The number of the total days with average daily temperature below 0 ℃ is in an obvious falling tendency in ice flood period in 1951 to 2010 in Beizhen, with a falling rate of 3 d/10 a. In 1986, the total days with average daily temperature below 0 ℃ before ice flood period are mainly in positive anomaly, while in 1987, the total days with average daily temperature below 0 ℃ in ice flood period are mainly in negative anomaly (Fig. 6), indicating in 1986 to 1987 it is the turning point of the total days in average daily temperature below 0 ℃ in ice flood period.

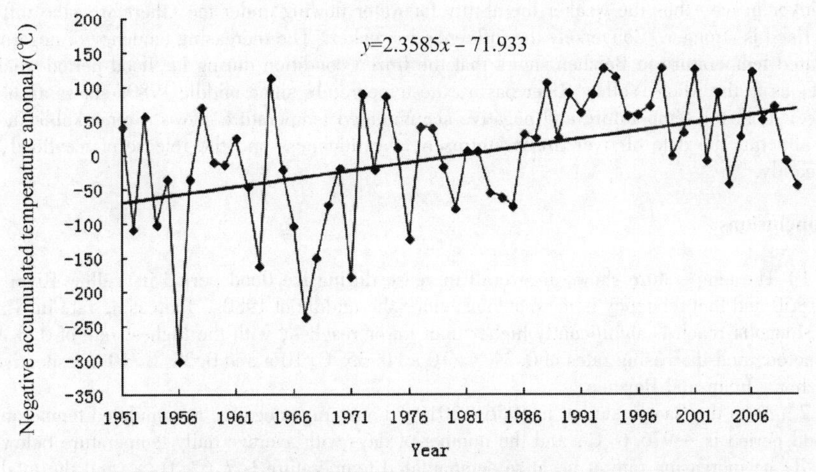

Fig. 7 Beizhen negative accumulated temperature anomaly

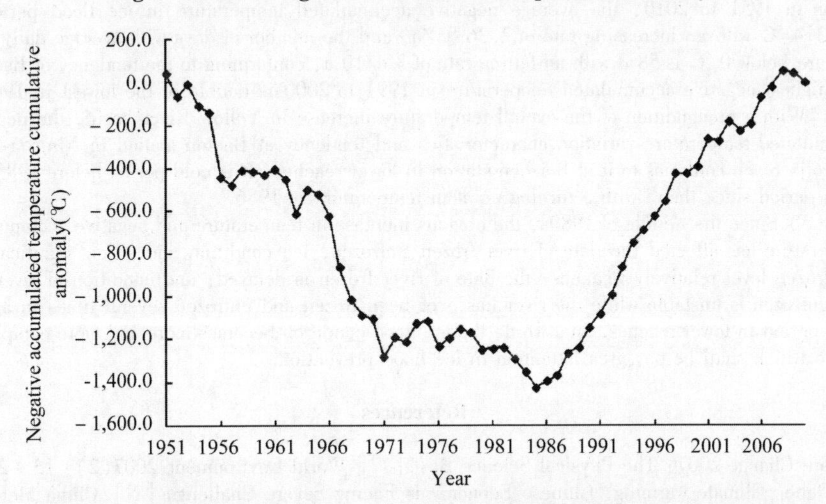

Fig. 8 Beizhen negative accumulated temperature cumulative anomaly

In ice flood period, the negative accumulated temperature values are in an upward tendency with the change of time and they are in a linear increasing tendency rate of 2.36 ℃/a (see Fig. 7) in Beizhen. In 1986, the negative accumulated temperature before ice flood period is of relative the same positive and negative anomaly. Except the negative value in 2004, 2009 and 2010 after 1986, negative accumulated temperature is positive in other years, indicating the temperature increases significantly in Beizhen after 1986. The Fig. 8 shows that, similar with Baotou station, the two ends of the Beizhen curve are lower while the middle of it higher. It is a falling period before the ice flood period in 1985/1986, and the accumulative anomaly of negative accumulated temperature reaches the lowest point of −1,425.6 ℃ in the ice flood period of 1985/1986, but the curve shows a monotonic increase since then, indicating that with a precondition of the overall temperature increase in Beizhen, it is cold period in 1951 to 1985 and warm period in 1986 to 2010. It conforms to the temperature variation in Baotou. Negative accumulated temperature reflects the coldness in winter. The higher the absolute value of negative accumulated temperature, the colder in winter and the thicker in ice, thus the weaker the ability for water flowing under ice, therefore, the influence to ice flood is stronger. Conversely the influence is weaker. The increasing tendency of negative accumulated temperature in Beizhen shows that the frozen condition during ice flood period weakens. In fact, as to the whole Yellow River basin, the temperature since middle 1980s shows an obvious increase. Both the temperature and negative accumulated temperature shows a remarkable increase later, affecting the date of river frozen/unfrozen, the thickness and the release of ice flood, etc. significantly.

5 Conclusions

(1) The temperature shows an overall increase during ice flood period in Yellow River basin since 1950 and that tendency is more obvious since the middle of 1980s. Increasing rate in Ningxia-Inner Mongolia reach is significantly higher than lower reaches, with the highest rate of 0.6 ℃/10 a in Baotou, and increasing rates of 0.34 ℃/10 a, 0.35 ℃/10 a and 0.38 ℃/10 a respectively in Zhengzhou, Jinan and Beizhen.

(2) As to the Baotou station in 1956 ~ 2010, the average negative accumulated temperature in ice flood period is −976.6 ℃, and the number of days with average daily temperature below 0 ℃ is 120 d; he increasing rate of negative accumulated temperature is 7.67 ℃/a, and the total days with average daily temperature below 0 ℃ decrease in a rate of about 3 d/10 a. As to the Beizhen station in 1951 to 2010, the average negative accumulated temperature in ice flood period is −213.4 ℃ with an increasing rate of 2.36 ℃/a, and the number of days with average daily temperature below 0 ℃ is 58 d with an falling rate of 3 d/10 a, conforming to the tendency of Baotou. The highest negative accumulated temperature in 1991 to 2000 is double of the lowest in 1961 to 2010. With a precondition of the overall temperature increase in Yellow River basin, the negative accumulated temperature variation characteristics and tendency at Baotou station in Ningxia-Inner Mongolia reach conforms to it at Beizhen station in lower reaches. It is cold period before 1985 and warm period since then, with a turning point in temperature in 1986.

(3) Since the middle of 1980s, the obvious increase in temperature and negative accumulated temperature has affected the date of river frozen/unfrozen, ice condition and so on significantly. The frozen level relatively weakens; the date of river frozen is delayed; and condition of river frozen/unfrozen is unstable while the river has ever been frozen and unfrozen several times in an ice flood period in lower reaches. In a word, the ice flood condition becomes more and more complicated, to which shall be pay great attention in ice flood prevention.

References

Climate Change 2007: The Physical Science Basis[J]. World Environment,2007(2): 13 − 22

Qin Dahe. Climate warming, Chinese Economy is Facing Severe Challenges[N]. China Meteorological News, 2007 − 3, 20.

Gong Deji, Bai Meilan, Wang Qiuchen. Forecasting Research on Ice Run of the Yellow River [J]. Meteorological Monthly,2001, 27(5): 38 − 42.

Wang Yirong. Evolvement of Temperature at the Bend of the Yellow Rive during Freeze Phase [J]. China Population Resources and Environment, 2006,16(3): 88 − 92

Statistical Analysis on Characteristics of Ice Regime Variation at Inner Mongolia Section of the Yellow River [J]. Yellow River, 2010,32(4): 53 − 58.

Zhao Jinan,Yang Jie. Haihe Water Resources,2000,6. 29 − 31

Wei Fengying. Modern Climatic Statistical Diagnosis and Prediction Technology [M]. Beijing: China Meteorological Press,2007.

Liu Jifeng, Wang Jinhua, Jiao Minhui, et al. Response of Water Resources in the Yellow River Basin to Global Climate Change [J]. Arid Zone Research, 2011, 28(5): 860 − 865.

E. Ecological Protection and Sustainable Water Utilization in River Basins

The Prospect for Application of Ecosystem Approach in Trans-boundary Water Environment Management

Li Pei*, *Zhang Fengchun* and *Zhang Xiaolan

Foreign Economic Cooperation Office, Ministry of Environmental Protection, Beijing, 100035, China

Abstract: Due to administrative divisions and their own economic benefits, responsibilities and authorities, etc., there are frequent contradictions among the provincial, municipal and county governments that are involved in trans-boundary water bodies. It is necessary to introduce some scientific, rational and sustainable techniques for trans-boundary water environment protection and management in order to change the current situation. Given the fact that cross-border water environment management involves complicated cross-regional, interdisciplinary, and cross-sectoral issues, for effective management, it needs to take into account the regional responsibilities, rights and interests, the cooperation and complementarity between disciplines, as well as inter-department coordination and cooperation. The ecosystem approach, as a new tool for integrated natural resource management, has been widely used for trans-boundary water environment management both at home and abroad in recent years, and has been quite successful. In this study, based on the characteristics and principles of the ecosystem approach, by analyzing the domestic cases of its application in trans-boundary water environment management, recommendations were provided on the basis of the current issues and favorable conditions for trans-boundary water management in China. It is suggested to have a comprehensive assessment of the current trans-boundary water environment management from the perspective of ecosystem management; to provide data support for its application; to have pilot and demonstrations for cross-border water environment management by ecosystem approach; and to establish a coordination committee for it, in order to improve the capacity and effectiveness in applying the ecosystem approach to trans-boundary water environment management. An advisory committee of experts should be established to provide scientific and technical support for the relevant decisions in the ecosystem approach to manage trans-boundary water environment. Websites shall be established to serve as a platform for information and technology exchange, as well as experience sharing. A long-term mechanism should be established to ensure timely communication and coordination. Joint monitoring and regulatory mechanisms shall be put into place for trans-boundary water environment management, which will help to improve the trans-boundary water environment management with ecosystem approach.

Key words: ecosystem approach, trans-boundary water management, inter-government management, water resource management

1 The current situation with trans-boundary water environment management

China's rapid economic development, spurs the utilization of regional water resources, causing increasing contradictions between water supply and water demand, deteriorating water quality, and growing pressure on pollution control in China (Shengxian Zhou, 2007). Besides, given the administrative divisions and related regional economic interests, duties and privileges, etc., the protection of water resources and water environment management are lagging behind. The trans-boundary water bodies are facing particularly severe environmental problems among them, conflicts between all levels of governments involved in trans-boundary water bodies are frequent, such as disputes caused by uneven distribution of trans-boundary water resources and water pollution accidents, etc.. Around trans-boundary water bodies, multiple contradictions often come up between provinces, municipalities cities and counties, as well as between upstream and downstream areas. The trans-boundary water environment quality is declining in both water-scarce and water-rich areas, and drinking water security problems can be found frequently (Zhao Ming, Hu Xiquan, 2009; Ming Cheng, 2009; Chen Zujun, et al., 2011). Trans-boundary water resource allocations, pollution incidents and the treatment and disposal schemes have become prominent issues for the relevant parties to take into consideration for water resource conservation and water environment management. It is now of great urgency to improve the water environment quality of trans-boundary waters and to work out reasonable water sharing plans.

Given the existing problems with trans-boundary water environment in China, governments at all levels have taken unremitting efforts over the years in terms of the legislation, policy, institutional setting, planning and so on. For example, inter-departmental cooperation has been achieved at decision-making and implementation levels in Taihu Lake, Songhuajiang River and Liaohe River. At the same time, research agencies are also actively exploring more scientific and effective management approaches. Although these efforts made some impacts, the problems faced with trans-boundary water environment still cannot be fundamentally solved.

Our existing water pollution management policies, institutions and technical methods can not meet the needs for trans-boundary water environmental protection and the sustainable use of resources. For example, there are almost no specific rules and regulations in our constitution and local regulations for inter-governmental cooperation. The law stipulates that governments at all levels are only responsible for the affairs within the scope of their jurisdiction and cross-domain affairs should be under the jurisdiction of superior authorities (or central government) (Yi Zhibin, Ma Xiaoming, 2009). Therefore, there is of a variety of water environment management activities with the current cross-border coordination and cooperation mechanisms. The lack of legitimacy and authority also restrict the roles played by various coordination and cooperation committees and groups. Currently, horizontal inter-departmental cooperation has not been established for trans-boundary water environmental management in most regions in China. Even when inter-sectoral cooperation is established, there are still no communities composed of multiple departments that are of concern.

The cross-border environmental problems essentially reflect the conflicts among human being, and that between human being and nature. They are not only the natural problems, but also economic, social, political, technical and cultural issues, and these factors are intertwined and influenced by each other. Trans-boundary water environment management is faced with such complex cross-regional, cross-disciplinary, and cross-sectoral issues (Sustainable Development Research Group of CAS; Tang Guojian, 2010). For effective management, it is necessary to take into consideration the geographical interests, responsibilities and rights, the cooperation and complementarity between disciplines, as well as inter-sectoral coordination and cooperation.

In addition to the deficiency of legislation, regulations and institutions mentioned above, our current techniques and methods for trans-boundary water environment management can not meet the needs of the new situation. As proved by research and practices, in respect of China's cross-border water environment management, the problems of resource allocation contradictions and pollution control, etc., are caused by not only natural factors such as climate change and water scarcity, not policies, laws and regulations, and institution-building, but also lacking effective integrated management tools. In order to fundamentally solve the trans-boundary water environmental issues, it is highly necessary to introduce scientific, reasonable and sustainable techniques for trans-boundary water environmental protection in addition to improving the legislation, regulations, and institutional settings.

The ecosystem approach, as one of the new integrated means for natural resource management, has been developed and used internationally in recent years to specifically deal with cross-regional, cross-disciplinary and cross-sectoral environmental issues. The ecosystem approach, with its scientific, rigorous, and advanced design, has been widely applied to trans-boundary water environment management both at home and abroad and desired results have been achieved. In recent years, research on the ecosystem approach has also been started in China which lays the foundation for further improvement of the ecosystem approach to be used in the field of trans-boundary water environment management. The studies also include a look at the existing water environmental protection laws and regulations from the perspective of the ecosystem approach, based on which it is proposed that attention should be given to the natural properties of the ecosystem of the trans-boundary water environment in the process of legislation and its enforcement (Chen Xiaojing, Xiao Qiangang; 2007). Most of these studies emphasize that particular trans-boundary waters should be regarded as complete ecosystems; the connections between the sub?? -ecosystems of natural watersheds and social subsystems should be the key factor for the application of ecosystem approach to manage trans-boundary water environment; the overall macro-management strategy of the basin should be balanced with the local interests and the distribution of power, in order to ultimately realize the integration of ecological, social and economic benefits of the water environment (Zhang Lei, 2002; Zhou Jinsong, et al., 2007; Liu Yong, et al., 2007). In addition, some studies have been carried out on specific technologies and methods such as the study to build the water environment pressure partition index system by using the ecosystem approach (Tian Fujiao, etc., 2011). Researcher Hu Zhenpeng (2010) proposed to use the ecosystem approach to remedy the problems in Poyang Lake such as its decreasing water tables in dry season and extended duration of dry seasons, which resulted in deterioration of water quality, destruction of wetland ecosystems, habitat for migratory birds threatened, as well as industrial and domestic water, irrigation, shipping and aquaculture fishing severely affected.

There are also some domestic studies on trans-boundary water environment management, which mainly focus on exploring improvement of effective methods for trans-boundary water environment management (Zhang Chao, 2007). For example, researcher Cheng Ming (2009) studied the trans-boundary water source development model in Beijing in terms of water shortages and water pollution and used examples of Yanqing County in Beijing and Huailai County in Hebei Province that are part of the Guanting Reservoir basin. By using the index system for comparative analysis, study was made about the current management methods, which came out with the conclusion that there is still a long way to go, from the current development models and the development situation in Huailai County to realize the goal of ecological sustainability, and it is mainly the way of administrative division that leads to the different development modes of trans-boundary waters inside and outside Beijing. Researchers such as Chen Zujun (2011) studied the joint pollution control mechanism to control cyanobacteria blooms in lakes that cross the provincial borders, and proposals were made for the design of joint pollution control mechanisms by Jiangsu Province and Shanghai Municipality to control the cyanobacteria in the trans-boundary Dianshan Lake. Zhu Demi (2009), with the example of Taihu Lake, divided the intergovernmental relations involved in trans-boundary water management into three levels: the relationship among central government departments, that between the central government departments and provincial governments, and that among the local government departments. Li Yuansheng (2010) proposed a basic framework for the basin network governance mechanism, and made suggestions to establish democratic consultation mechanism, economic cooperation mechanism, emergency response mechanism and ecological compensation mechanism to meet cross-border water management needs. Yi Zhibin (2009) proposed to have innovative intergovernmental cooperation mechanisms for cross-boundary water pollution control from the political, administrative and legal aspects. Wang Yong (2009) proposed that the intergovernmental governance coordination mechanism for water environmental protection in the basin should take the forms of Watershed Public Energy Field, intergovernmental e-governance in the basin and intergovernmental alliance in the watershed, etc.. Currently, the trans-boundary water environment management methods used in China involve policies, regulations, coordination mechanisms, early warning, emergency response, arbitration, compensation, etc., which make certain contributions in improving trans-boundary water environment management. But, so far, there has not been much exploration about how to apply the ecosystem approach to trans-boundary water environmental management.

Europe, the United States and Canada all started early in the study of joint management of trans-boundary water environment, and have gained a lot of successful experiences. Europe had been seeing more accidents in trans-boundary basins ever since the industrial revolution. But with the emphasis on water environment control, European countries took early measures in trans-boundary water environmental management that were based on the unit of basins. For example, the countries that the Rhine River and Danube River flow through co-developed countermeasures and took joint efforts so that the water environment had been greatly improved in the basin. After a long period of practices and continuous improvement, Europe has formed a more scientific and mature mode for trans-boundary basin management (Wan Wei et al., 2009; Jing Chunyan, et al., 2011).

The multinational governance model in Rhine River is recognized as a successful example. The nine countries that River Rhine flows through believe that basin environmental protection are of common interest. They have developed long-term collaboration mechanism, and have made unified planning in terms of policy, regulation, institutional and technical co-ordination, with the entire river basin taken as a whole. The water environment management in Rhine River has become an international successful example for pollution control in trans-boundary basins (Hong Yu, 2008). In addition, the case with Danube in Europe, the joint governance model of the Great Lakes between the United States and Canada, and the joint governance of cross-boundary rivers such as Tennessee River in US, the Thames River in Britain, and Po across in Italy have all been regarded as successful practices (Huang Dechun et al., 2009). These integrated watershed management of trans-boundary rivers or lakes have in common that they are all integrated water planning and management with basin as unit, focus on the relationship between water environmental capacity in the basin and the economic development, and have established coordination and consultation institutions for water resource management in the basin, as well as a cross-border integrated management mode. They also encourage public participation in the management. In terms of the content and methods of all these practices, they have all been permeated with the concept and technology of the ecosystem approach (Avramoski, 2001).

2 The ecosystem approach

2.1 The definition and application of the ecosystem approach

The ecosystem approach is a comprehensive and coordinated approach for natural resource management that can promote protection and sustainable use of land resources, water resources and biological resources. It attaches importance to the harmonization of economic and social development, and the protection and sustainable use of natural resources. The ecosystem approach believes that all aspects involved in environmental management, including policies, laws and regulations, administrative agencies and public awareness are all integral parts. Therefore, the application of the ecosystem approach will help to achieve the secondary goals under the overall goal, to coordinate and balance the natural resource conservation and sustainable use, and to achieve a win-win situation. For the trans-boundary water environment management, the ecosystem approach is to integrate economic, social and cultural factors into the mechanism and process of trans-boundary water resource management (Fig. 1).

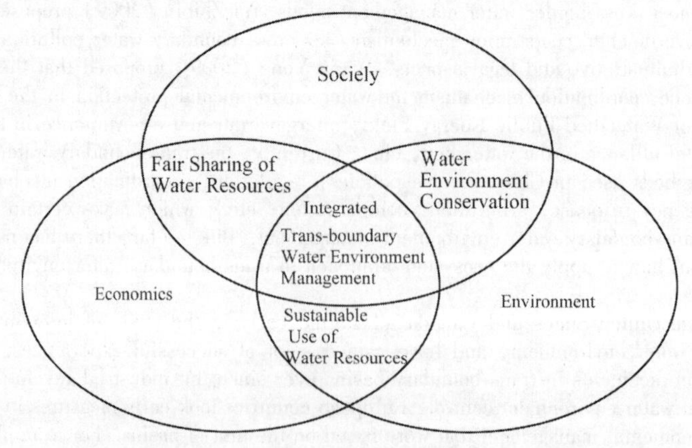

Fig. 1 Diagram of ecosystem approach in trans-boundary water environment management

The most important pioneering application of the ecosystem approach should be " the Convention on Biological Diversity. In the second Conference of the Parties of the Convention on Biological Diversity held in Jakarta in 1995, the ecosystem approach was taken as the basic method under Convention on Biological Diversity to deal with various issues, and it was written in the conference resolutions. At that time, the ecosystem approach was developed and introduced for the cross-regional, cross-disciplinary and cross-sectoral characteristics of biodiversity conservation and sustainable use. Subsequently, the ecosystem approach was used for the management of a variety of multi-disciplinary and cross-sectoral subjects under the Convention on Biological Diversity. Currently, the ecosystem approach has been considered to be one of the most effective tools for sustainable environmental management and has been widely applied in the management of water resources, land resources and biological resources.

2.2 Characteristics of the eco-system approach

2.2.1 Inclusive

The ecosystem approach is above the level, but it does not exclude other effective methods for natural resource management. The ecosystem approach is to integrate various management methods in order to deal more effectively with the complex environmental management issues.

2.2.2 Comprehensive

The eco-system approach is based on integrated management thinking. By the coordination and harmonization among the various natural, social, economic, legal, policy, and technological factors, it's aimed at maximizing the unified benefits in economic, social and environmental aspects.

2.2.3 Systemic

Given that the eco-system approach is cross-sectoral, inter-regional, inter-disciplinary and covers multi levels, it is supposed that all factors involved in the management of natural resources are parts of the system and they affect each other in interactions. The ecosystem approach is to effectively integrate the economic, social and environmental factors into a systematic set of management objectives.

2.2.4 Dynamic

The ecosystem itself is in continual succession or evolution, and keeps changing in time and space. Therefore, the ecosystem approach puts an emphasis on following the laws of nature, and making changes accordingly, in order to continually adjust and improve the management practices.

2.2.5 Persistent

The purpose of the ecosystem approach is to achieve both conservation and sustainable use of natural resources. Therefore, the management strategies and tools used in this approach are supposed to be effective and sustainable for a long time. It focuses on the long-term benefits of the resources, with an emphasis on long-term and sustainable development of the ecosystem, and the balance between the interests of present and future generations.

2.2.6 Scientific

The ecosystem approach provides a platform to apply theories and methods of multiple disciplines to the specific management practices. Therefore, joint multi-disciplinary research is adopted in the ecosystem approach, including environmental, economic, and social discipline, etc., in order to have scientific management of natural ecosystems.

2.2.7 Flexible

The ecosystem approach is not composed of a set of immutable management measures. It emphasizes to manage resources according to the specific situation by following the laws of nature, and flexible measures should be taken according to the different natural, economic and social conditions as well as regional differences.

2.2.8 Harmonious

The ecosystem approach attaches great importance to the relationship between man and nature. It is aimed at maximizing the nature's production capacity with no damage caused, in order to have harmonious development of human being and the nature.

2.3 Principles for the ecosystem approach

The Convention on Biological Diversity has formulated 12 pieces of principles for the ecosystem approach. The main idea is that this approach should be ecosystem-based, and to realize conservation and sustainable use of natural resources by adopting scientific and flexible means, to coordinate the relationship between man and nature, and therefore in order to achieve sustainable economic, social and environmental development and continuous improvement. The 12 pieces of principles are as follows:

(1) The natural resource management by ecosystem approach is a question of social choice to unify the economic, cultural and social needs of different regions and sectors of the community;

(2) The management authority is decentralized to the lowest level, to have all stakeholders involved in the management process and balance the local interests with the overall public interest;

(3) The ecosystem approach should consider the impact of its activities on the surrounding and adjacent ecosystems: appropriate adjustments can be made to the ecosystem approach in order to achieve the synergies with the surrounding ecosystem;

(4) It is necessary to understand and manage the ecosystem approach from the perspective of economic efficiency, to protect natural resources for the rational development and sustainable use of natural resources;

(5) The priority target of the ecosystem approach is to protect the structure and function of the ecosystem, in order to maintain the sustainability of ecosystem services;

(6) The management objective is to maintain the natural functions of the ecosystem;

(7) The application of the ecosystem approach needs appropriate spatial and temporal scales;

(8) It is necessary to develop a long-term goal, in order to adapt to the time lag effect caused by changes in the ecosystem;

(9) The ecosystem approach should be adjusted to changes in the internal and external conditions;

(10) The ecosystem approach seeks the balance and unity between the conservation and sustainable use of natural resources;

(11) The ecosystem approach is based on all the information available, and therefore it is necessary to have effective mechanisms and operational procedures that could help to get access to all information needed;

(12) The ecosystem approach requires the participation of all relevant sectors and disciplines.

3 Suggestions on the application of ecosystem approach to trans-boundary water environment management

3.1 Assessment of trans-boundary water management stuation in china from the perspective of the ecosystem approach

A clear picture has been made about the current problems and situation with the trans-boundary water environment management in China, but there is still a lot of information missing in terms of applying the ecosystem approach to trans-boundary water environment management. It is necessary to carry out extra investigations based on the special needs of the ecosystem approach, to re-evaluate the existing problems in the existing cross-border water environment management as well as the favorable conditions for the establishment of the ecosystem approach, which will provide a basis and foundation for the development of effective cross-border water environment management by ecosystem approach. The information should specifically include: ① with each basin as a complete ecosystem, and in accordance with the natural laws of trans-boundary waters ecosystem, coherence and consistency of the existing policies, laws and regulations, development planning, resource utilization and protection actions shall be re-examined; ②analysis and evaluation should be carried out with respect to the interests of and within governments and government departments involved in trans-boundary waters, in order to have a good understanding of the duplication, deletion, conflict and contradictions among governments or government departments; ③ it is necessary to collate and analyze the existing management modes and methods and to summarize the successful ones. these models and technologies, after being improved and integrated, will serve as important technical parts of the ecosystem approach for trans-boundary water environment management; ④ analysis should be made about the participation of all stakeholders and their roles, including governments, communities, research and teaching institutions, international agencies, non-governmental organizations and enterprises; ⑤based on the results of the assessment, the needs of the ecosystem approach for trans-boundary water environment management should be summarized and management programs should be worked out or improved.

3.2 Pilots and demonstrations for the application of ecosystem approach in trans-boundary water environment management

There have been some attempts for the application of ecosystem approach to cross-border water environment management at home and abroad, but it is still in the exploratory stage with constant improvement. In this case, pilots of the ecosystem approach for trans-boundary water environment management should be based on the real situation in China, by exploring applicable technologies, testing management effectiveness, and summarizing experiences and lessons. Furthermore, with the pilots, new technologies can be tested, new ways integrated, and existing technologies assembled and improved. The successful technologies and management models can be extended to cross-border water environment management on a larger scale.

3.3 Establishment of coordination committee for cross-boundary water environment management

Trans-boundary Water Environment Management Coordination Committees should be set up, composed of all stakeholders (i. e., the parties within the region involved with trans-boundary waters) at different levels in line with the requirements of the ecosystem approach. The Committee is responsible for coordinating the distribution of water resources and water conservation duties among different governments based on the characteristics of the natural ecosystems of trans-boundary water bodies, as well as the local environmental, economic and social conditions. It is supposed to strengthen horizontal cooperation between the governments at the macro policy level, and to set up network nodes and a supporting framework for trans-boundary water management network. By optimizing the internal structure of the basin management institutions, it is also supposed to help to establish a basin management system with unified planning, unified management and vertical leadership, to realize the integration and unification of trans-boundary water environmental management, and therefore to maximize resource utilization efficiency with the premise of water environment protection. The Trans-boundary Water Environmental Management Coordination Committee should be different from similar organizations in the past in terms of member compositions, ways of working, and rights and responsibilities, etc.. The Committee should be composed of governments involved and policy-making departments, research institutes, communities, and business sector. They should be jointly responsible for decisions on environmental protection, economic development, and social progress in trans-boundary waters. Most importantly, the Committee must have decision-making power for a variety of economic development and environmental protection issues, or at least has the right to participate in decision-making process.

3.4 The establishment of a consulting committee

An Advisory Committee shall be established under the Environmental Management Coordination Committee, which is composed of experts and scientists in all areas related to the trans-boundary water environment management, in order to provide scientific and technological support for cross-border water management. For the composition of the Advisory Committee, it is also necessary to take into account the balance between regions, departments and major stakeholders. The Advisory Committee is supposed to provide technical advice and technical support for decision making on cross-border water environment management and play a key role in scientific and technical support by assessing the policies, programs and actions related to trans-boundary water environment management. There are a lot of successful cases about scientific and technical support mechanisms for trans-boundary basin management in Europe, the United States and other developed countries. For example, in the professional organizations, such as the board of directors for water quality and pollution control under the Great Lakes Water Quality Convention, professional and technical personnel are engaged in the day-to-day technical issues to ensure the effectiveness of the mechanism and objectivity for the implementation of the management methods. The advisory committee should also be concerned about the training of professional and technical staff, to enable technical competence for new technologies and fulfill complex horizontal cross-regional, cross-sectoral and interdisciplinary integrations.

3.5 The establishment of a platform for information and technology sharing and communication

The ecosystem approach for trans-boundary water environment management needs some real, systematic and comprehensive information supporting, to avoid decision-making deviation caused by single or one-sidedness information channel. According to successful experiences abroad, it is effective to access and share the information needed by establishing an electronic decision-making support platform based on computer-based network and with the trans-boundary basin as a unit. The platform also helps to put forward information and technical exchanges between various departments and provide a more efficient tool for multi-sectoral joint decision-making and unified actions. The platform should not be constrained to the limitations of individual administrative regions and departments. It should effectively transmit both government information and technical information. It should be an open platform for the basin as a whole, to ensure that all the stakeholders in the basin be able to have instant access to the required information (Yong Wang, 2009). There have been some successful attempts in this respect both at home and abroad. For example, a knowledge management platform for water resources and water environment has been established in Hai River Basin; an inter-departmental Species Data Bank has been established for knowledge management in Norwegian biodiversity construction.

3.6 To strengthen joint monitoring and supervision

The ecosystem approach for trans-boundary water environment management is a dynamic and long-term management method. The management techniques and measures are in continuous improvement, which requires information constantly updated for cross-border water utilization and environmental protection, and necessary adjustments need to be made accordingly. Therefore, it is of necessity to establish a joint inter-basin monitoring and supervision mechanism. The monitoring and supervision should be led by the governmental departments concerned, and under the guidance of the Advisory Committee. Monitoring and supervision mechanisms should be established with participation of the environmental protection and water conservancy departments of the same level. Monitoring and supervision shall cover the entire process from pollutant sources to wastewater treatment, disposal and emissions or reuse, as well as the use of shared water resources. Based on the monitoring data, technical measures will be continually modified or improved with the ecosystem approach.

3.7 Capacity building for the application of ecosystem approach to trans-boundary water environment management

As a relatively new concept for environmental management, the ecosystem approach has high requirements in terms of technology and management capabilities. Therefore, it is necessary to strengthen the capacity building for relevant institutions and personnel in order to effectively apply this approach to trans-boundary water environment management. The capacity building mainly includes improvement of the existing institutions and training of the personnel to be in line with the requirements of the ecosystem approach. It may also be needed to set up new institutions besides the improvements of the current ones. Personnel training should cover government officials, practitioners and managers, and the content of the training should include the concepts and procedures of the ecosystem approach, and the professional and technical knowledge related to trans-boundary water environment management.

References

Chen Xiaojing, Xiao Qiangang. Study on Basin Water Environmental Protection Act from the perspective of ecosystem management[C] // The Third Environment and Development Forum Proceedings ,2007:75-78.

Chen Zujun, Zhang Haiyan, Xu Guiquan,et al. 2011, Joint pollution control mechanisms for cyanobacteria control in cross- border lakes with Example of cyanobacteria control in Dianshan Lake[C]//China's first Lake Forum Proceedings,2011:161-166.

Cheng Ming. Comparative Study on Trans-boundary Water Ecological Sustainable Development Model, with Examples of Yanqing County and Huailai County in Guanting Reservoir basin [J]. Beijing Vocational College Journal,2009,23(3):18-24.

Yu Hong. Study on International Experiences on Trans-boundary Water Environment Management With Rhine River as An Example[J]. Information Development & Economy,2008,18(26): 74-76.

Hu Zhenpeng. Application of the Eco-system Approach to Management of Water Shortage in Panyang Lake[J]. The Yangtze River Resources and Environment,2010,19(2):133 - 138.

Huang Dechun, Chen Simeng, Zhang Haochi. Foreign Experiences on Trans-boundary Water Pollution Management and Enlightenment[J]. Water Resources Protection,2009(25).

Jing Chunyan, Huang Lei, Qu Changsheng, Cross-border Basin Environmental Management and Early Warning the European ExperienceS and Enlightenment[J]. Environmental Monitoring and Early Warning",2011,3 (1):8 -11.

Li YuanSheng, Hu Yi. From Hierarchy to Network: Watershed Management Mechanism Innovations[J]. Fuzhou Party School Journal,2010(2):36 - 40.

Liu Yong, Guo HuaiCheng, Huang Kai, et al. The Ecosystem Management in Lakes and Basins [J]. Acta Ecologica Sinica,2007,27 (12):5352-5360.

Tang GuoJian. Joint Effect: Field Study of Cross-border Water Pollution Management Mechanism - with "SJ Boundary Environmental Joint Meeting" as an Example[J]. Hohai University Journal,2010:45.

Tian Fujiao, Lu Jizhao. Constructon of Partition Index System Based on the Water Pressure on Ecosystem[J]. Environmental Science and Technology,2011,24 (10):191-195.

Wan Wei, Zhang Shiqiu, Zou Wenbo. Regional Environmental Management Mechanism: International Experiences and Enlightenments [J]. China Environmental Science Society Annual Conference Proceedings,2009:895-896.

Wang Yong. Inter-governmental Coordination Mechanism for Water Environment Protection In The Basin[J]. Social Science,2009(3):26-35.

Yi Zhibin, Ma Xiaoming. Inter-governmental Coordination Mechanism for Water Pollution Control In The Basin[J]. Social science,2009,(3):20-25.

Zhang Chao. Study on Management Mode of Cross-border Public Issues - With Cross-border Water Pollution Control As an Example[J]. Theoretical Discussion", 2007(6):140-142.

Zhang Lei, Research On Water Environmental Ecological System in North China[J]. Hebei Normal University Journal (Natural Science Edition),2002, 26(1):196-199.

Zhao Ming, Hu Xiquan. Analysis on Trans-boundary Water Pollution Control in China and Its Prospect[C]//the second chapter about environmental pollution control technology and development, Proceedings of Annual Academic Conference For Environmental Science Associate, 2009:285-288.

Team for Sustainable Development Strategy, Chinese Academy of Sciences. Report on China's Sustainable Development Strategy- Water: Management and Innovation[M]. Beijing:Science Press,2007:231.

Zhou Jinsong, Wu Shunze, Yu Xiangyong. Analysis on Conflicts With Trans-boundary Water Environmental Management and the Countermeasures, China's Environmental Science Society Academic Conference Proceedings, 2007:1230-1236.

Zhou Shengxian. Opportunities and Choices --An In- depth Reflection On the Songhua River Event [M]. Beijing: Xinhua Press,2007:51.

Zhu Demi. Establishment of the Iinter-departmental Cooperation Mechanism for Water Pollution Prevention and Control: with Taihu Lake Basin As An Example[R]. Chinese Administration Management, 2009(4):86-91.

Avramoski O. Working paper: Strategies for public participation in the management of trans-boundary waters in countries in transition,2001:7-8.

Hattah Lakes - restoring the Balance at One of Australia's Most Iconic and Environmentally Significant Lake Systems

Jennifer Collins, *Lauren Murphy*, *Nicholas Sheahan* and *Emma Healy*

Mallee Catchment Management Authority (CMA)

Abstract: Reviving the drought-ravaged Hattah Lakes has been a visionary project facilitated by the Mallee Catchment Management Authority (CMA), the Victorian Department of Sustainability and Environment and the Murray Darling Basin Authority along side the land manager Parks Victoria, as part of a sustained effort to restore better health to this system of semi-permanent freshwater lakes within Australia's largest and most important river system.

The Hattah Lakes system is an iconic environmental oasis in Australia's Murray Darling Basin. The lakes are part of the 48,000 hm^2 Hattah-Kulkyne National Park, 60 km south of Mildura, Victoria. The 18 km Chalka Creek connects to the Lakes to the River Murray. The lakes support populations of River Red Gums and many threatened and rare native plants and animals. Twelve of the 21 lakes are Wetlands of International Importance under the Ramsar convention and many of the bird species found at the lakes are recognised under the China-Australia Migratory Birds Agreement and/or the Japan-Australia Migratory Birds Agreement. This critical ecosystem has been threatened by the effects of river regulation, ongoing drought and low inflows. The natural flooding pattern has been reduced, leaving important flora and fauna struggling to survive.

The Mallee CMA has worked with a range of government agencies; indigenous stakeholders; not for profit organisations; corporate bodies; and community groups to make it possible to return environmental flows to Hattah Lakes, with the support of the wider community.

The Authority is now supporting a multi-million dollar works project at Hattah Lakes to provide a long-term solution to protecting and enhancing the system's ecological future. This is part of Australia's largest river restoration program and will involve the construction of a permanent pumping station, environmental regulators and stop banks. Construction began in early 2012 and will cost an estimated $30 million. At the maximum inundation level, 119,500 mL of water will be used to inundate 6,383 hm^2. Approximately half the water used will be returned to the river system.

This environmental works program will be an exemplary engineering achievement that will help a vital part of Australia's natural environment to survive; it is also a leading example of what can be achieved when all levels of government work together with Indigenous stakeholders and the wider community.
Key words: Australia, Hattah Lakes, environmental infrastructure

1 The Murray Darling Basin

The Murray - Darling Basin is one of the most socially, environmentally and economically significant areas of Australia. The basin is the catchment for the Murray and Darling rivers and their many tributaries. It covers an area of more than one million square kilometres, which is equivalent to 14% of Australia (MDBA 2008).

In total there are 23 river valleys in the basin, as well as important groundwater systems. The basin's average annual rainfall is 530,618 GL; a total of 94% of this rainfall evaporates (MDBA 2008).

The Murray Darling Basin generates 39% of Australia's income derived from agricultural production. In 2002, in response to the declining health of the river system, the Murray Darling Basin Authority (MDBA), along with member states, established The Living Murray (TLM) program to help restore the health of six environmentally significant sites along the River Murray. The project is a partnership of the Commonwealth, Queensland, NSW, Victorian, South Australian and Australian Capital Territory Governments (Mallee CMA 2010).

Each of the six sites chosen was selected to recognise high ecological, cultural, recreational, heritage and economic values. The Hattah Lakes was selected as one of the six Icon Sites. In 2004, under the 'First Step' decision, partner governments committed to an investment of $500 million to recover a long term average of 500 GL of water to improve environmental outcomes at the six Icon Sites. The TLM program is now focussing on designing and building environmental infrastructure to enhance the environmental outcomes that can be achieved through the use of the recovered environmental water.

2 Hattah Lakes

Hattah Lakes is an extensive wetland complex covering approximately 13,000 hm^2 within the 48,000 hm^2 Hattah-Kulkyne National Park. It is located in Victoria, Australia, on the River Murray (Fig. 1).

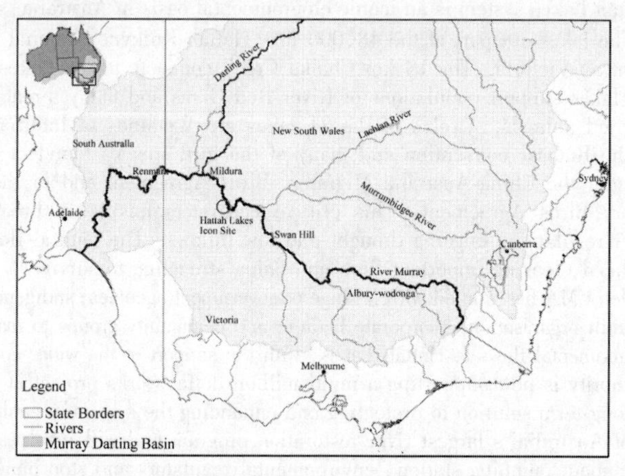

Fig. 1 Location of Hattah Lakes Icon Site, Victoria, Australia.

Hattah Lakes is recognised for its many environmental, social and economic values; in particular for its role as a refuge and breeding habitat for waterbirds and for its sites of Indigenous cultural significance. The system is comprised of more than 20 perennial and intermittent freshwater lakes, ranging in size from less than 10 hm^2 to around 200 hm^2. Twelve of the lakes are listed as wetlands of international importance under the Ramsar Convention (Mallee CMA 2009). Surrounding vegetation communities range from those that require frequent flooding, such as Red Gum (Eucalyptus camaldulensis) forest, to those that require only periodic inundation, such as Black Box (Eucalyptus largiflorens) and Lignum (Muehlenbeckia florulenta).

Hattah-Kulkyne National Park supports a high diversity of vertebrate fauna. These include: 225 species of native birds (including 47 waterbirds), 27 species of mammals, 38 species of reptile; and five species of frogs.

Fifty-seven of these fauna species are listed as threatened under Victorian (Flora and Fauna Guarantee Act 1988 (FFG Act) - and various advisory lists) and federal legislation (Environment Protection and Biodiversity Conservation Act 1999 (EPBC Act)). The lake system provides important breeding habitat, particularly for colonial nesting waterbirds. Sixteen waterbird species have been recorded breeding at the lakes and in excess of 20,000 waterbird individuals have been recorded present during flooding (DSE 2003). Twelve of these species are migratory and are protected under international migratory bird agreements for China (CAMBA) and Japan (JAMBA).

The lakes have been a focus for traditional Aboriginal society for thousands of years who used the area for hunting and gathering. More than 1000 Aboriginal archaeological sites have been registered with Aboriginal Affairs Victoria (SKM 2009b). These include burials, scarred trees, shell middens, artefact scatters, hearths and other topological sites.

2.1 Hydrology

The Hattah Lakes system is a complex of lakes and watercourses set within a wider floodplain vegetated by River Red Gum, Black Box and Lignum. Water first enters via Chalka Creek, a well-defined natural watercourse which diverges from the River Murray at Messenger's Crossing (Fig. 2). When river flow exceeds 36,700 ML/d, water flows through Chalka Creek and spills into Lake Lockie, 19 km from the River Murray. Lake Lockie fills first, then spills water into complexes of nearby lakes to the north (Lakes Mournpall, Yelwell and Yerang) and the south (Lakes Hattah, Bulla, Arawak, and Brockie) (Mallee CMA 2009).

These lakes are well defined basins that retain water for long periods after River Murray flow peaks recede. They provide persistent aquatic habitat that supports communities of native fish and waterbirds. At higher river flows, water also enters the Hattah floodplain from the northern continuation of Chalka Creek (Chalka Creek North), which joins the River Murray at 994 km river kilometres. Water spreads to additional lakes and eventually waters the fringing Red Gum woodland community. Lake Cantala is an isolated lake filled by an independent flow path, Cantala Creek. Lake Cantala fills when river levels exceed 45,000 ML/d at Euston (SKM 2004).

As river flows increase further, widespread floodplain inundation occurs. Flow in Chalka Creek North reverses, flowing north. The lakes with the highest commence-to-flow thresholds are filled and water spreads into the Black Box and Lignum communities at the outer limits of the floodplain. Very high and sustained flow peaks will inundate the Lake Kramen area in the south of the floodplain and the Lake Boolca area to the north.

2.2 Threats

The key threats to the values of the Hattah Lakes are the reductions in frequency and duration of flooding, as well as a shift in flooding seasonality, associated with diversion and regulation of flow upstream (Mallee CMA, 2010).

Tab. 1 Frequency of key flows under various scenarios (provided by MDBA)

River flow (ML/d)	Flood count (% of years with flow peaks above threshold)			Effective flood (% of flow exceeds the threshold for at least 3 months)		
Modelled pre-development	Modelled current	Median climate change scenario at year 2030	Modelled Pre-development	Modelled current	Median climate change scenario at year 2030	
40,000	82	47	37	48	20	11
60,000	59	31	22	21	7	3
75,00	47	23	12	16	7	3
100,000	36	12	8	19	8	4
150,000	18	6	2	4	2	1

Under pre-regulation conditions, inflows would have occurred in most years, with large flooding events occurring on average every five years. The majority of flow peaks would have occurred during July. However, upstream storages, locks and weirs installed along the River Murray during the 1920s and 1930s have led to significant changes in hydrology at Hattah Lakes. River regulation, along with climatic conditions (under current and 2030 medium climate change scenarios, floods are significantly less frequent and the longest spell between events is significantly greater than the pre development scenario), have led to a significant reduction in the frequency of medium and large flow peaks, while the occurrence of low flows (less than the inflow threshold at Chalka Creek) has increased (Tab. 1). Medium and large flow peaks are also delayed, typically arriving one month later than under pre-regulation conditions. As a result, significant areas of floodplain are not inundated as frequently as they would have been under pre-regulation conditions.

During the decade to 2010, the Murray Darling Basin also suffered severe and prolonged drought which, when combined with the effects of river regulation, resulted in environmental degradation at the Hattah Lakes. The lack of connectivity to the river and the complete drying of the system had detrimental effects on the ecosystem and the lake system's ability to act as a refuge. This stress was eased during the flooding of 2010/11; however this relief will be short lived in light of continuing river regulation and climate change conditions.

While the lake system would have periodically dried out under pre-regulation conditions, the higher incidence and potentially longer durations of drying poses an increased pressure and risk to flora and fauna populations. As a result, the lakes are less able to support local populations of aquatic fauna such as macro-invertebrates, fish, frogs and turtles, which then have to re-colonise when the lakes next receive inflows.

Prolonged dry spells have also threatened vegetation communities. Vegetation health has been monitored since 2003 and shown to be declining. Surveys in 2008/2009 (MDFRC 2009) examined the distribution, condition, age-structure and relative abundance across three River Red Gum communities. Foliage scores of less than 30% were common to all classes indicating the condition of River Red Gum across the site is generally poor. Black Box condition is also very poor, with more than 90% of trees classified as having <20% of the foliage carrying capacity (MDFRC, 2008). Understorey vegetation that depends on flooding was found to be absent.

2.3 Ecological objectives

The Hattah Lakes Environmental Management Plan (MDBC 2006) defined a set of specific ecological objectives for the Hattah Lakes Icon Site. A set of corresponding flow objectives were also developed, based on the water requirements of the floodplain vegetation communities and associated biota.

Site-specific ecological objectives include:

(1) restoring a mosaic of hydrological regimes, which represent pre-regulation conditions (to maximise biodiversity);

(2) maintaining, and where practical, restoring, the ecological character of the Ramsar site with respect to the Strategic Management Plan (DSE 2003);

(3) restoring the macrophyte zone around at least 50% of the lakes to increase fish and bird habitat;

(4) improving the quality and extent of deep freshwater meadow and permanent open freshwater wetlands so that species typical of these ecosystems are represented;

(5) maintaining habitat for the Freckled Duck, Grey Falcon and White-bellied Sea-eagle in accordance with Action Statements;

(6) increasing successful breeding events for colonial waterbirds to at least two years in 10 (including Spoonbills, Egrets, Night herons and Bitterns);

(7) providing suitable habitat for a range of migratory bird species (including Lathams snipe, Red-necked stint, and Sharptailed sandpiper);

(8) increasing distribution, number and recruitment of local wetland fish (including endangered Murray Hardyhead , Smelt and Gudgeon) by providing appropriately managed habitat; and maximising the use of floodplain habitat for recruitment of all indigenous freshwater fish.

The overarching flow objective of water management at Hattah Lakes is "to achieve the original

---

[1] Freshwater Meadow-These include shallow (up to 0.3 m) and temporary (less than four months duration) surface water, although soils are generally waterlogged throughout winter.

[2] Permanent Open Freshwater-Wetlands that are usually more than 1 m deep. They can be natural or artificial. Wetlands are described as permanent if they retain water for longer than 12 months; however they can have periods of drying.

[3] The FFG Act requires Action Statements to be prepared for all threatened species in Victoria. These are briefmanagement plans that provide some background information about the species and describe previous and planned actions to conserve the species.

species diversity, structure and function of the ecosystem by providing the hydrological environments required by indigenous plant and animal species and communities" (Ecological Associates (EA)2007).

The water requirements of the floodplain vegetation and associated wetlands have been defined by Ecological Associates (EA 2007 and EA 2009) (See Tab. 2).

Tab. 2 Water Requirements of Water Regime Classes

Water Regime Class	Required Water Regime	Outcome
Semi-permanent Wetlands	Water levels exceeds 50% of the retention level 80% of the time Wetland is dry less than 5% of the time Water level reaches the retention level in 30-50% of years for 12 weeks	Resident populations of large fish and small fish Drought refuge Frog, fish and waterbird breeding Emergent macrophyte recruitment
Persistent Temporary Wetlands	Water level falls to 50% of wetland depth in 50% of years Retention level should be exceeded for 12 weeks in 25% of years Retention level should be exceeded for 24 weeks in 25% of years	Support a broad zone of emergent macrophytes at the fringe of the wetland Mineralisation of organic matter Limit growth of Typha and Phragmites Emergent macrophytes growth and recruitment Frog, fish and waterbird breeding Breeding and feeding by small fish, frogs and reed-dependent waterbirds Major breeding events by flood-dependent fauna
Temporary Wetlands	Water reaches retention level in 40% of years for 24 weeks Full events normally commence in July and September Wetland is dry after 75% of filling events (i. e. 37.5% of years)	Waterbird breeding Emergent macrophyte growth Frog, fish and waterbird breeding Mineralisation of organic matter Limit growth opportunities for Typha and Phragmites Support emergent macrophytes dependent on seasonal inundation
Episodic Wetlands	Lakes reach retention level in 5 to 10% of years for 2 months Water depth exceeds 2 m for 1.5 to 2.5 years following full events Wetland remains dry for a median duration exceeding 5 years between full events	Initiate growth of emergent macrophytes at lake fringe Waterbird, fish and frog breeding Submerged aquatic plant growth Terrestrial plant community Mineralisation of organic matter Death of perennial aquatic plants
Fringing Red Gum Woodland Compliance threshold is 43.5 m AHD at Mournpall	Inundation of: 60 days duration in 15% of years 120 days duration in 20% of years 160 days duration in 15% of years (50% of years with an event)	Red Gum growth, productivity and recruitment Flood-dependent understorey growth Flood-dependent vegetation growth
Red Gum Woodland with Flood Tolerant Understorey Compliance threshold is 44.5 m AHD at Mournpall	Inundation of: 30 days duration in 20% of years 60 days duration in 20% of years 100 days duration in 10% of years (50% of years with an event)	Red Gum growth, productivity and recruitment
Black Box Woodland Compliance threshold is 45 m AHD at Mournpall	Inundation of 20 days duration in 10% of years 50 days duration in 10% of years (50% of years with an event)	Black Box growth, productivity and recruitment

3 Restoring the balance

To help restore the ecological balance to Hattah Lakes, short-term emergency and environmental watering events have been provided using temporary pumps since 2006; however temporary pumps alone don't have the capacity to deliver to all of the Ramsar-listed lakes, or to provide floodplain inundation. Under current conditions, 40,000 ML/d must be flowing in the River Murray before the lakes start to fill. Through The Living Murray Program, a package of works has been developed for the Hatah Lakes Icon Site to replace temporary pumping with a more suitable long term option for the lakes and floodplain. When the works are completed, the lakes may be filled under normal river operating conditions (e. g. approx 5,000 ML/d passing flows) during late winter-early spring, prior to the irrigation season (Mallee CMA 2009).

The $30 million works package aims to provide a water regime that addresses the set ecological objectives over as large a proportion of the lakes and floodplain as possible (Fig. 3). The works package includes:

(1) lowering sills in Chalka Creek to 41.75 m Australian Height Datum (AHD);

(2) building three regulators and three stop banks, and refurbishing an existing regulator; and

(3) building a permanent pumping station (capacity of up to 1,000 MG/d) on Chalka Creek at Messenger's Crossing.

The package of water management works has been designed to allow inflows to Hattah Lakes at the lowest practical flows in the River Murray; and to retain water in the lakes system to increase the duration of inundation.

The works have also been designed to enhance natural peaks in River Murray flow, but can also provide watering events independently of river flows if the period between natural inundation events is too long. The design will allow water managers to control the frequency (through sill lowering and pumping), duration (by regulators) and extent (by pumping and stop banks) of lake and floodplain inundation to meet the ecological requirements of the system. The works will be operated to provide an inundation regime tailored to the site's ecological requirements.

The package of works will be used to increase the frequency of inflows and enable inundation up to 45 m AHD across the central lakes and to 46.3 m AHD at Lake Kramen. Watering events will be delivered to fill any extended gaps between floods, protecting the lakes during drought. In the absence of natural floods, using pumps to deliver water to the lakes will use considerably less water than would be required to create a flood via river flows (Mallee CMA, 2010).

Investigations have found that the area of inundation is optimised in relation to the number and extent of stop banks at an elevation of 45 m AHD; at this level 5,358 hm^2 is inundated (not including Lake Kramen). To retain water at elevations higher than 45 m AHD, the size, number and cost of structures increases disproportionately to the additional inundation footprint.

Inundation areas were calculated by digitally mapping an elevation model (SKM, 2006), the Water Regime Class information (EA, 2007) (Tab. 2) and water volume requirements information (Fig. 2, Fig. 3). The volumes in Tab. 3 are those required to fill the system from a dry state and were an estimate of water use based on a simplified operating strategy including inundation to 43.5 m AHD once every three years, inundation to 45 m AHD once every eight years and inundation at Lake Kramen to 46.3 m AHD once every eight years.

Tab. 3 Estimated water use of the operating strategy

Elevation (m AHD)	Frequency	Volume per event (GL)	Volume per year (GL)	Net water use per event (GL)	Net water use per year (GL)
43.5	1:3 years	41	14	41	14
45	1:8 years	106	14	52	6.5
Lake Kramen (46.3)	1:8 years	13.5	1.7	13.5	1.7
Combined net annual water use					22.2

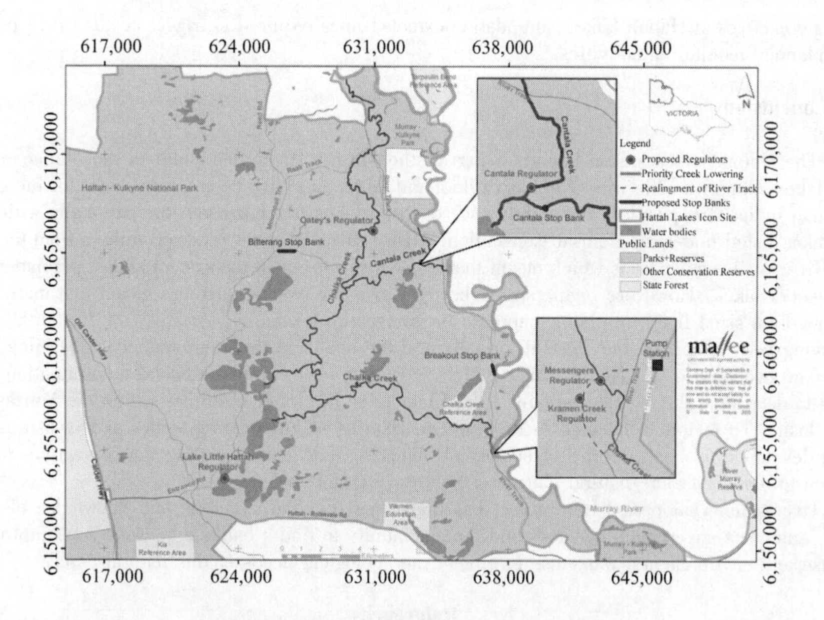

Fig. 2 Proposed works at Hattah Lakes

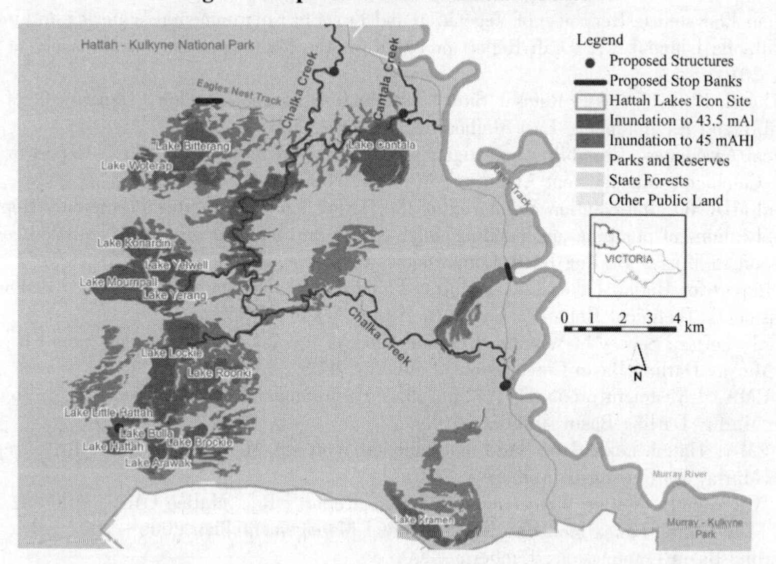

Fig. 3 Inundation extents achievable using the TLM works package.

4 Current status

Construction of the planned environmental works at Hattah Lakes began during March 2012. The construction of the permanent pump station, four regulators and three stop-banks is expected to take most of 2012. The operation of the works and measures will be informed by an Environmental Management Plan, which is currently under development. This plan will outline environmental watering objectives at Hattah Lakes, inundation extents/times required to trigger ecological responses, and planned monitoring activities.

5 Conclusion

The Hattah Lakes are an important part of the Murray River floodplain in the Mallee region. The lakes support a wide variety of native flora and fauna and have been a focal point for indigenous and non indigenous people. The significance of these lakes is recognised internationally with 12 of the lakes listed under the Ramsar Convention. Historically the lakes received inflows from the Murray River on a regular basis which meant that water was present in the system almost permanently in the deeper lakes. However, regulation of the river over the past century has meant that many small and medium sized floods are being captured in storages upstream. As a result, the Hattah lakes are receiving less water than they had historically and the health of the environment is declining.

The evolution of the Hattah Lakes project has taken place against a backdrop of a nation struggling to determine the most sustainable way of managing the water resources within the Murray Darling Basin. To restore a balance to this system a series of engineering works and measures have been developed to meet a series of ecological objectives and corresponding flow objectives for this system to deliver a more natural pattern of flows to the Hattah Lakes.

Over a ten year period this project has drawn together the expertise and knowledge of scientists, engineers government agencies and the community to find a balance between consumptive water use and environmental water use to protect the ecological needs of this natural asset.

References

Australian Ecoystems. Response of Vegetation and Frogs to Environmental Watering on Lindsay and Wallpolla Islands[R]. Draft Report prepared for the Mallee Catchment Management Authority, 2010.

DSE. Hattah-Kulkyne Lakes Ramsar Site - Strategic Management Plan, Department of Sustainability and Environment, East Melbourne, Victoria, 2003.

Ecological Associates. Feasibility Investigation of Options for the Hattah Lakes, Report to the Mallee Catchment Management Authority, Mildura, 2007.

EPA and MDFRC. Intervention Monitoring of the Hattah Lakes Icon Site: Preliminary Report[R]. Implications of pumping and ponding water on the development of diverse aquatic ecosystems. Report to Murray Darling Basin Commission, Canberra, 2007.

GHD. Report for Hattah Lakes Living Murray Floodplain Management Project [J]. Ecological Assessment. Technical Report to Goulburn Murray Water, 2009(a).

Kingsford, Porter. Survey of Waterbird Communities of the Living Murray Icon Sites [R]. Report to Murray Darling Basin Commission, Canberra,2008.

Mallee CMA. Investment proposal: Hattah Lakes Environmental Flows Project [R]. Prepared for the Murray Darling Basin Authority,2009.

Mallee CMA. Hattah Lakes Icon Site Environmental Watering Management Plan [R]. Prepared for the Murray Darling Basin Authority, 2010.

Mallee Waterwatch. Mallee Waterwatch 2008 Data Report [R]. Mallee CMA, Mildura, 2009.

MDBC. The Hattah Lakes Icon Site Environmental Management Plan 2006 ~ 2007 [R]. Murray-Darling Basin Commission, Canberra, 2006.

MDBC. The Living Murray Icon Site Condition Report October 2007[R]. Murray-Darling Basin Commission, Canberra, 2007.

MDBA. Explore the Basin, viewed February 3 2012, http://www.mdba.gov.au/explore-the-basin/about-the-basin, 2008.

MDFRC . The Living Murray Condition Monitoring at Hattah Lakes 2007/08 [R]. Report to Mallee Catchment Manatgment Authority, 2007.

MDFRC. The Living Murray Condition Monitoring at Hattah Lakes 2008/09 [R]. A MDFRC draft report prepared for the Department of Sustainability and Environment, 2009.

SKM. Hattah Lakes Water Management Plan – Background Report [R]. Report to the Mallee Catchment Management Authority, Mildura, 2004.

SKM. Hattah Lakes Living Murray Floodplain Management Project – Complex Cultural Heritage Management Plan Volume 1 [R]. Report to Goulburn Murray Water, 2009(b).

SKM. Hattah Lakes Water Management Works – Cultural Heritage Assessment[R]. Report to the Mallee Catchment Management Authority, Mildura, 2007.

Study of Integrated Methods for Assessing Urban River Region Ecosystem Services

Huang Liqun*, *Wang Chunyan*, *Yang Jue*, *Feng Qiao* and *Wu Xiaomeng

China Water International Engineering Consulting Co. , Ltd. , Beijing, 100044,China

Abstract: Urban river relative region to city should be considered with ecosystem services assessment, because its services are very important during the growth of city. Integrated assessment method for urban river was established based on value evaluation methods of ecosystem services. It includes qualitative analysis, quantitative analysis and localization analysis. The integrated assessment was used in Harbin region of the Songhua River. Based on current situation survey, the ecosystem services are classified as production supply, adjustability, culture and support function. Qualitative assessment gives description of the river ecosystem services. Quantitative assessment gives the value of the river ecosystem services by using different methods. Localization assessment makes district divide by three levels. The first level includes production supply district, adjustability district, culture district and support district. The second level is divided considering different ecosystem services. The third level is divided according to the factors such as terrain, gradient, etc. It shows that ecosystem services of Harbin region of the Songhua River are integrated. The value of adjust function is the largest one. Its ecosystem services distribute in 3 reaches, which are water body, exurban islands and urban islands. The value and localization of the ecosystem services of the river can help to develop moderately the nature resource of urban river and support to water ecological protection and restoration.

Key words: urban river, ecosystem services, integrated assessment, quantitative analysis, localization analysis

1 Introduction

The current evaluations of ecosystem service function are classified as qualitative description and quantitative evaluation. Starting from Daily's "Nature's Service: Societal Dependence on Natural Ecosystem"(1997), the ecosystem service evaluation developed into the functions of ecosystem service、brief history research, service value assessment, various bio system service, as well as the conception and development of regional ecosystem service at the beginning of the study development on ecosystem service.

During the same period, another way to evaluate ecosystem is to classify the types of ecosystem service before assessing the values . This idea was proposed by Costanza. Afterwards, among lots of specific results from different types of ecosystem in subsequent researches, the most comprehensive one is the wetland ecosystem service functional value assessment.

From the year 2001 to 2005, the MA (Millennium Ecosystem Assessment) project classified the ecosystem service into four groups: supply (such as water and food), adjustment (such as flood and disease control), culture (such as mental happiness, entertainment and culture profits), and support (such as the nutrients to maintain the earth living cycle). Based on functional evaluations of the four types of service, the structures and conditions of ecosystem service could be acquired. Meanwhile, human beings would realize the importance of ecosystem service through the quantification of values assessments.

As a result of the complexity of ecosystem as well as its function complicacy, the debates over ecosystem service assessment methods become more and more during the research development. For instance, the ecosystem service diversity, spatial heterogeneity, various utilizations of functions are all defined as complexity.

Commonweal project of Ministry of Water Resources "*Study on risk evaluation and monitor technology of sudden water pollution in the Yellow River*"(20 1001010)

To sum up, the focus of ecosystem service is mainly around the values assessments. To be honest, current research of ecosystem service on localization is incomplete, meanwhile, the future service value improvements that will be brought by urban development in urban river region are also ignored.

As a result of the large river basin, there exist various natural influential factors such as terrain, topography, climate, etc. Thus, the urban river region plays different roles in different zones. Especially in formation and development of cities, the urban river region provide the most significant environment, also, its performance connects location and function together. To sum up, the value assessment of its characteristic and values should not only assess the characters and values, but also analyze and estimate its location.

Integrated assessment method for urban river was established based on value evaluation methods of ecosystem services. It includes qualitative analysis, quantitative analysis and localization analysis. The integrated assessment was used in Harbin region of the Songhua River. The propose of integrated assessment is to determine the type and value of urban river ecosystem service and point out the main functions in relatively consistent spatial location in urban river region. Thus, it can help to develop moderately the nature resource of urban river and support to water ecological protection and restoration.

2 Study on ecosystem services integrated assessment method for urban river

2.1 Integrated Assessment

In this study on ecosystem services integrated assessment method for urban river, with the combination of qualification analysis, quantitative analysis and localization analysis, it makes the global and systematic assessment, meanwhile, the features, values and spatial locations in urban river ecosystem service can be determined. Based on current situations, qualitative assessment in integrated method gives description of the river ecosystem services. Quantitative assessment gives the value of the river ecosystem services by using different methods. And localization assessment determines the emphasized spatial locations. In order to make specific and clear types, values and spatial distributions of urban river region service, the study on ecosystem services integrated methods determine the targets and development directions. The target is to achieve sustainable application and the main directions are to protect and restore the function of support and adjustment in urban river region service system. Appropriate development of products and cultural functions are emphasized directions at the same time (Shown in Fig. 1).

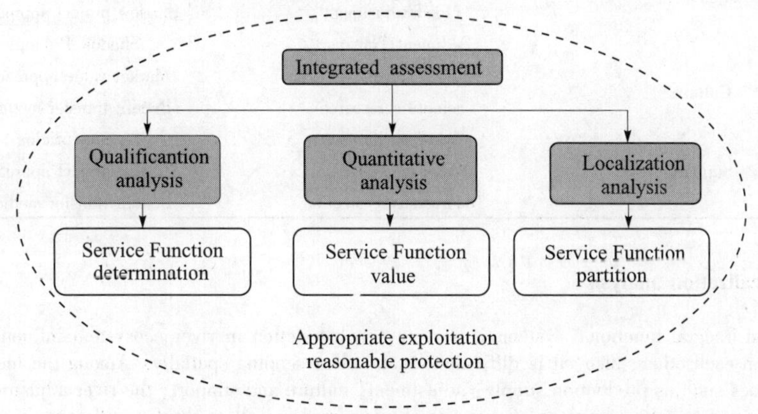

Fig. 1 Integrated assessment framework

2.2 Qualification analysis

During the ecology process of materials, energy and information exchanges in urban river service systems, the direct or indirect abilities to contribute to the society of ecosystems form the inner definition of urban river ecosystem service. Based on the described process mentioned above, the qualification analysis gives description of the river ecosystem services and makes basis for further quantification and localization analysis.

The qualification based on the recent survey in urban river region and the region spread whole catchment area, especially in river. Thus, the research survey area is mainly about the river area. Information collection, reference selection and surveys are all important research methods. Usually, the descriptions and analysis of providing, adjusting, culture and supporting are all based on surveys.

2.3 Quantification analysis

According to the conclusions from qualification analysis and information collection, quantification analysis quantifies the products supply, adjustment service, cultural service and support service in urban river region. In China, the business accounting methods of evaluating urban river region service are still in primary stage, mainly about introduction of foreign theory, imitation evaluating methods, accumulating cases, etc.

Due to the definition from RASMAR Convention, wetlands conclude rivers and lakes. As a result, the qualifications methods use the international assessment methods of wetlands as reference. Tab. 1 shows the urban river ecosystem service value assessment methods.

Tab. 1 Urban river ecosystem service value assessment methods

Service types	Content	Methods
Production supply	Water supply	Market value approach
	Aquatic products supply	
	Hydraulic electro generation	
	Shipping	
Adjustment	Microclimate adjustment	Shadow pricing
	Atmospheric constituent adjustment	
	Flood adjustment	Shadow project approach
	Sediment Transport	Shadow Pricing
Culture	Travel	Market value approach
	Scientific research	Benefit transfer method
Support	Water purification	Shadow pricing
	Water conservation	Shadow project approach
	Biodiversity Protection	Benefit transfer method

2.4 Localization analysis

As an integral functional systematic service, each function in river ecosystem influences and depends on each other; also, it is difficult to make clear scoping spatially. Among the mentioned four services such as production supply, adjustment, culture and support, the river adjustment and support service has the advantage of spreading uniformity, as well as the few difference in spatial. However, due to fact that products changes with areas, and the cultural service varies with exploitation content, the differences become larger.

Therefore, because of qualification and quantification analysis, the localization analysis of urban river ecosystem service is relative and conceptual in this study. What is more, as a huge and complex system, the localization in inner ecosystem service has to take several factors into considerations, for instance, spatial heterogeneity in ecosystem structure, different characters in every ecosystem, and current urban river resource development.

The amount of previous research results about localization analysis methods is very small, even the global unique localization is still lack. Based on MA (Millennium Ecosystem Assessment) and the results of qualification and quantification, localization assessment makes district divide by two levels. The first level includes production supply district, adjustability district, culture district and support district. The second level is divided considering significances in different ecosystem services, including water supply, aquatic products, biodiversity protection, flood adjustment and water purification, etc.

Finally, referring to the "National Ecological Functional Regionalization", the ecosystem is divided by three levels. Also, the division considers other factors such as different characters in ecosystem service, soil application, terrain, gradient, etc. Through the localization analysis, the main functional service spatial location is determined and afterwards, the appropriate measures in ecosystem partition can be proposed.

3 The integrated assessment used in Harbin region of the Songhua River

The study chose Harbin region as the urban river ecosystem research area. The river section is 122.7 km, ranging from Shuangcheng boundary to Dadingzishan navigation – power junction engineering as Fig. 2. Not only the river channel, river band, but also the underground water, wetland, river mouth, and other near shore area in floodplain are all included in the study area.

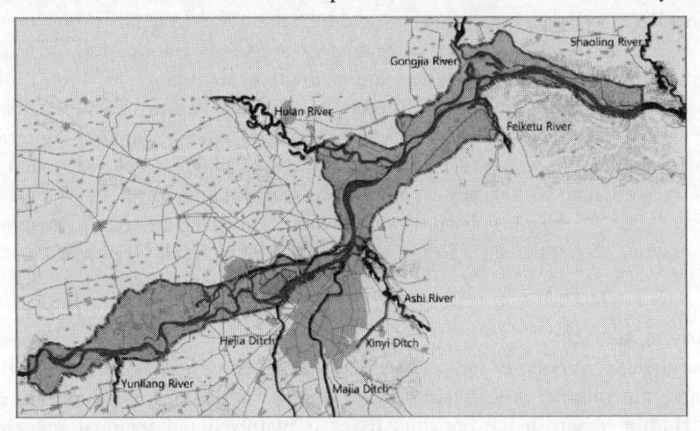

Fig. 2 The schematic picture of Harbin region in Songhua River

As a significant intersection area among Songhua River, Heilong River and Northeast Ecosystem Area, this study river section has lots of advantages: plain terrain and broad river, original landscape and many islands. Thus, it is a suitable habitat for animals and plants, as well as a connection corridor between left and right side. Also, the river runs through the city and plays an important role both in the development and ecosystem service of cities.

3.1 Qualification analysis

The improvements in ecosystem in Harbin region of the Songhua River contain two aspects. In one hand, the river can adjust water capacity and climate, improve water conservation, supply underground water, etc. All of these can make more contributions to maintain the stability in the ecosystem. On the other hand, for the changes in micro environment, the river can not only keep the biodiversity, create new lives, but also clean the pollutants. Meanwhile, as the Harbin region of the Songhua River, the river section has cultural services as well.

3.1.1 Production service supply

The production service supply is the so called direct service, which can exchange products directly in normal markets. All of water supply, aquatic products supply, hydropower and genetic resources belong to production service supply in Harbin region of the Songhua River.

The Songhua River is the main drinking water resource in Harbin, and provides abundant fresh water for agricultural irrigation, industry, and ecology environment. However, the water quality in the Songhua River is poorly Ⅳ Level in recent years due to the excessive contaminative water discharge upstream. Moreover, the water pollution event in November 2005 affected citizens' life and economy badly. To sum up, because of the pollution upstream and the weaker water quality, the fresh water supply ability of the Songhua River decreased.

Bio – productivity is a comprehensive reflection of material recycle and energy flow in the Songhua River ecosystem. For the existing fish in the Songhua River, there are seven families and thirty – seven species, which contribute to 35% of all fish in the Heilongjiang Province. To be specific, the cyprinidae species are 25, 3 loach species and one for each species such as Ecocidae, Osmeroidea, and catfish. It is easy to draw the conclusion that the aquatic products in this river region are abundant and cyprinidae is the most, loach comes in the second place.

It is well known that water is the cleanest energy. In the Songhua River, the fall in landform creates lots of potential energy. The constructions of hydroelectric power station base on the effective conversion between power and energy. Dadingzishan navigation – power junction engineering was finished in 2008. The design flood level is 117.33 m; checking flood level is 117.91 m, with a total storage of 17.3×10^{8} m^{3}. The six bulb turbine units has 11 MW for unit capacity, and 66 MW in total to make sure the 18 MW output ($P = 85\%$). The hydraulic power set is 3.32×10^{8} kW · h annual.

The Songhua River ecosystem creates a suitable habitat for livings. From microorganisms to higher propagation, these livings contain a huge amount of genetic information. In Harbin Region of the Songhua River, the propagation resources are incredible abundant: for birds, white stork and black stork belong to national first – class protection species; white spoonbills, white – fronted goose, whooper swans, mandarin ducks, black grouse, hazel grouse, single – top cranes, Falconiformes, ospreysshaped head and other birds of prey are second – class protection species. The plants formation in wetlands are mainly about Carex meyeriana (tower head pier), gray clock Carex, the rafting Carex, Carex lasiocarpa, water onion, bulrush, reed, Calamagrostis arundinacea, Typha, S. natans, Nymphoides , Ling, water lilies, lotus, spike – like Myriophyllum, Ibaraki algae, etc.

3.1.2 Adjustment service

The river adjustment service is indirect service that its value can not exchange directly in normal market. Thus, the value of adjustment service is called indirect application value. The adjustment service in Harbin region in the Songhua River is mainly about regional water amount, flood management, sediment transport and climate adjustment, etc.

Harbin region of the Songhua River belongs to the type of sluicing plain riverbed, such as some wetlands, beaches and shallow lakes. The whole river channel is natural huge flood water storage. In this section, the river not only adjusts the area water distribution, but also decreases the flood peak through flood storage and flood discharge and result in lower loss caused by flood. Meanwhile, river flows can carry sediment, scour channel sedimentation and clear cannels.

In Harbin region of the Songhua River, the phytoplankton, aquatic plants and plants on band can exchange oxygen and carbon dioxide with atmosphere and the process so important in maintain the dynamic balance of oxygen and carbon dioxide.

3.1.3 Culture service

Harbin region of the Songhua River has special landscape so that the travelling service can be well developed. Nowadays, the river section in Harbin has become river storage since the construction of Dadingzishan navigation – power junction engineering. The build Jinhewan Wetland Botanic Garden on Qianjin Beach, "Three Islands and A Lake", Hejia Beach, Zhengyang River Beach, as well as the famous Sun Island are all leisure resources.

The Songhua River is one of the top seven rivers in China, and is the biggest branch of Heilong River at the same time. As the river running through Haerbin city, it provides abundant aquatic resources, ecological resources and landscape resources. Meanwhile, the river section has high research value in studying wetlands protection and application, birds immigration, water pollution protection, etc.

3.1.4 Supporting service

Supporting Service is an essential process for other ecosystem service and the impact on human beings will reveal after a long period. The supporting services include water purification, water conservation, and biodiversity protection.

With the character of complex and biodiversity, Haerbin region of the Songhua River has unique and special wetland ecological features. The complicated ecosystem in wetlands create an organic whole and play many roles in primary production of atmosphere, the formation of oxygen and soil, as well as their nutrition recycle, water recycle, and ecological service. Moreover, it is equipped with the strong ability in pollutants degradation.

The length of the main stream in Haerbin region of the Songhua River is 122.7 km, and the runoff is 41×10^9 m^3/a. In most cases, the flowing period is between mid - April and mid - November and the average flow rate of main channel cross - section is 1 m/s. The main stream of the Songhua River has the ability to supply water for both underground water and surface water. Moreover, the various kinds of plants growing in wetlands can do water conservation through impoundment precipitation.

As the Harbin region of the Songhua River belongs to the type of sluicing plain riverbed, there exists lots of banco and swales formed by flows in sine curves and the short cutoffs. Thus, the upstream has large area of wetlands and creates a diverse aquatic environment and animal habitat. Meanwhile, the downstream has a high water quality with many lakes, water networks and abundant plants communities.

3.2 Quantification analysis

After quantification analysis, Tab. 2 shows the total value of Harbin region of the Songhua River is 15.315×10^9 yuan. The values of each service come in a sequence as: flood adjustment, fresh water supply, traveling, water conservation, shipping, biodiversity protection, climate adjustment, water purification, scientific research, aquatic products supply, sedimentation transport, hydroelectric power. Also, if it is classified by types, the rank is adjustment, products supply, support and culture.

Tab. 2 The values in ecosystem in Harbin Region of the Songhua River

Service types	Content	Value(unit: $\times 10^9$ Yuan)	Total(unit: $\times 10^9$ Yuan)
Production supply	Water supply value	1.562	1.876
	Aquatic products supply value	00.39	
	Hydraulic electro generation value	00.07	
	Shipping	0.268	
Adjustment	Atmospheric constituent adjustment value	0.163	10.312
	Flood adjustment value	10.171	
	Sediment transport value	0.032	
Culture	Scientific research value	1.493	1.562
	Travelling value	0.069	
Support	Water purification	0.111	1.565
	Water conservation	1.286	
	Biodiversity protection	0.168	
Total	15.315		

As shown in the Fig. 3, it is obvious that adjustment has a better service than the other three among four ecosystem services. Also, the flood adjustment contributes most with 10. 171 × 10^9 Yuan; the values of production supply, support and culture are nearly the same. From Fig. 4 we can see that the values of water purification and biodiversity protection are both small.

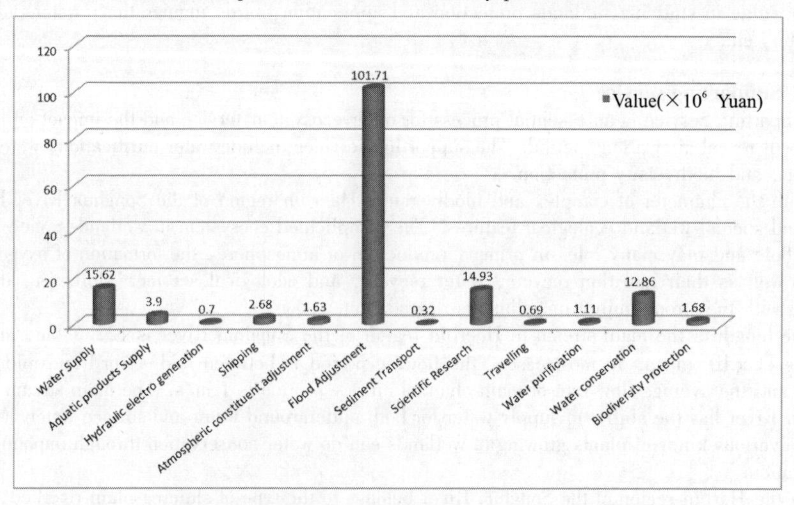

Fig. 3 Ecosystem service values in Harbin region of Songhua River

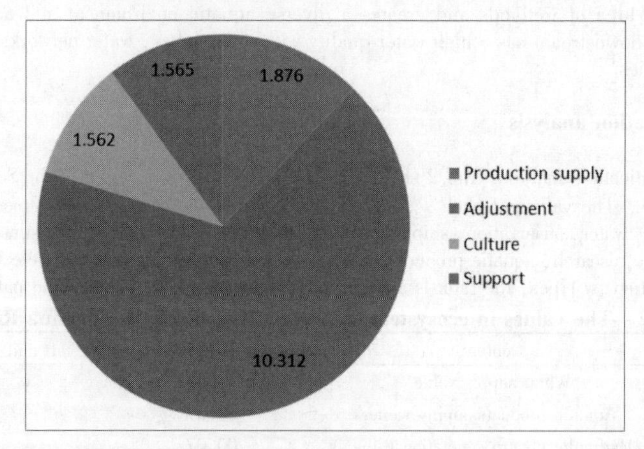

Fig. 4 Total values of ecosystem service in Harbin Region of the Songhua River

3.3 Localization analysis

With a comprehensive combination of factors such as soil application situations, densities of human activities, and locations, the localization analysis in Harbin Region in the Songhua River ensure the main function of each region after partition. In this study, Fig. 5 shows the soil application situations and Tab. 3 illustrates the partition with the considerations of current resource and environment situations in Harbin Region of the Songhua River.

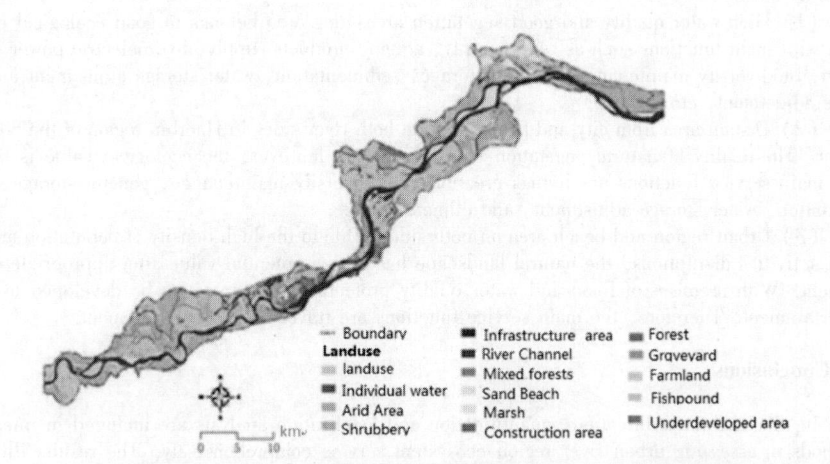

Fig. 5 Localization partition in Harbin Region of the Songhua River

Tab. 3 Ecosystem service partition in Harbin Region of the Songhua River

First level	Second level	Third level
Production Supply	Water supply	River channel area
		Individual water body
	Aquatic products supply	River channel area
		Fish pool area
		Individual water body
	Hydroelectric power	River channel area
	Shipping	River channel area
Adjustment	Climate adjustment	River channel area
		Wetlands area
		Forests area
	Flood adjustment	River channel area
		Wetlands area
	Sedimentation transport	Individual water body
		River channel area
Culture	Traveling	Construction area
		Sandy area
	Scientific research	Infrastructure construction area
Support	Water purification	Wetlands area
	Water conservation	Grass area
		Forests area
		Bush area
		Mixed forests area
	Biodiversity protection	Wetlands area
		Mixed forests area

In the three - level partition in Haerbin ecosystem services, there are mainly three types regions as listed:

(1) High water quality and good vegetation area: this area belongs to good ecological quality area with main functions such as water supply, aquatic products supply, hydroelectric power generation, biodiversity maintenance, genetic storage, sedimentation, water storage adjustment and climate adjustment, etc.

(2) Distant area from city and beach area on both river sides in Haerbin region of the Songhua River. The quality of natural vegetation area is general; however, the ecological value is better. The main service functions are livings creations, biodiversity maintenance, genetic storage, sedimentation, water storage adjustment, and climate storage.

(3) Urban region and beach area on both sides. Due to the high density of population and human activities disruptions, the natural landscape has a high potential value after appropriate development. With premises of flood and water quality protection, the area can be developed to make entertainment. Therefore, the main service functions are traveling and entertainment.

4 Conclusions

In this study, qualification, quantification and localization analysis are included in integrated methods in assessing urban river region ecosystem service comprehensively. The results illustrate the service of study area is overall and even can reach a value of 15.315×10^9 yuan in total. Generally, the service distribution is mainly in three regions: river body, rural beach and urban region beach. With the acknowledgements of river ecosystem service and localization, it is meaningful to the rule – making and adjustments of principles in protection and restoration of river ecosystem. In order to know more about dynamic changes in urban river region to make exact assessment, the suggestion comes about more combination of multi – disciplinary research and comprehensive consideration in every aspect in river ecosystem. Besides, due to the experiment result in this study, the periodical assessment of urban river ecosystem should be proposed to make rules, protective adjustments and restoration projects, as well as the comparative analysis and prediction of changes in time and spatial.

References

Daily, G. C. Eds. Nature's Service: Societal Dependence on Natural Ecosystems [M]. Island Press, Washington, 1997.

Costanza R, et al. The Value of the World's Ecosystem Services and Natural Capital[J]. Nature, 1997, 387(15): 253 – 260.

Bolund P. , Hunhammar S. Ecosystem Services in Urban Areas[J]. Ecological Economics, 1999 (29): 293 – 301.

Bjorklund J, Limburg K, Rydberg T. Impact of Production Intensity on the Ability of the Agricultural Landscape to Generate Ecosystem Services: an Example From Sweden [J]. Ecological Economics , 1999(29): 269 – 291.

Holmund C, Hammer M. Ecosystem Services Generated by Fish Populations [J]. Ecological Economics, 1999,(29): 253 – 268.

Wang V, Guo H. Cedar Ecosystem Service Value Assessment in China [J]. Forest Sciences, 2009, 45(4): 124 – 130.

Feng H X, Hou Y Z, Feng Z K. Temperature Adjustment of Ecosystem Service in Shandong Province. [J]. Forest Sciences, 2010, 46(5): 20 – 25.

Xie Y X, Zhang Y. Aquatic Environment and Ecosystem Service Value Assessment in Taizi River [J]. Environment Protection and Recycle Economics, 2011(2): 50 – 52, 61.

Richard T W, Wui Y S. The Economic Value of Wetland Services: a Meta – analysis[J]. Ecological Economics, Elsevier, 2001(37): 257 – 270.

Turner R K, Jeroen C J M, Brouwer R. Managing Wetlands: a Ecological Economics Approach [M]. Northhamton MA: Edward Elgar Pub. , 2003.

Bergh J C J M van den, Barendregt A, Gilbert A J. Spatial Ecological Economic Analysis for Wetland Management: Modeling and Scenario Evaluation of Land Use [C]. Cambridge University Press, 2004: 239.

Kirsten D S. Economic Consequences of Wetland Degradation for Local Populations in Africa[J]. Ecological Economics, 2005(53).

L. L. Wu, J. J. Lu, F. C. Tong. Wetlands Ecosystem Service Value Assessment in the Yangtze River[J]. 2003,12(5): 411 – 416.

K. Xin, D. N. Xiao. Wetlands Ecosystem Service Value Assessment in Panjin[J]. Ecology Journal,2002,22(8):1345 – 1349.

H. Zhang, L. Y. Zhang, J. Fu. Binhai Wetlands Classification and Ecosystem Service Study in Liaoning Province [J]. Wetlands Science, 2009, 7(4): 342 – 349.

MA. Ecosystem, Welfare. Biodiversity Comprehensive Study[M]. Beijing: China Environmental Science Press, 2005: 60 – 69.

Z. Y. Ouyang, T. Q. Zhao, X. K. Wang. Aquatic Ecology Service Analysis and Its Indirect Value Assessment. [J]. Ecology Journal, 2004, 24(10): 2091 – 2099.

Serafy S. Pricing the Invaluable: the Value of the World's Ecosystem Services and Natural Capital [J]. Ecological Economics, 1998(25): 25 – 27.

Aryes R. Special sections: Forum on Valuation of Ecosyst em Services: the Price – Value Paradox [J]. Ecological Economics, 1998(25): 17 – 19.

Opschoor J. B. The Value of Ecosystem Services: whose value [J]. Ecological Economics, 1998 (25): 41 – 43.

Z. W. Guo, Y. L. Gan. Several Scientific Questions About Ecosystem Service[J]. Biodiversity, 2003, 11(1): 63 – 69.

X. M. Tang, B. M. Chen, Q. B. Lu. The Ecological Zone Revision Methods in Ecosystem Service Value. [J]. Ecology Journal, 2010, 30(13): 3526 – 3535.

Monitoring, Evaluation and Mitigation: Eutrophication in Sutami Reservoir, Brantas River Basin, Indonesia

Harianto[1], ***Alfan Rianto***[2], ***Erwando Rachmadi***[3] and **Astria Nugrahany**[4]

1. Director of Technical Planning and Development Affairs
2. Chief of Business, Management and Technology Development Bureau
3. Chief of Management and Technology Development Unit
4. Staff of Business, Management and Technology Development Bureau

Jasa Tirta I Public Corporation, The Brantas and Bengawan Solo River Basins Management Agency, Jl. Surabaya 2A, Malang, 65115, Indonesia

Abstract: The outstanding issues for the present water quality management are deterioration of water quality and shortage of pollution control. Increased population, industries, and agriculture activities have caused water quality deterioration in many developing countries while less control due to insufficient budget and less awareness of some stakeholders have worsened the situation.

The Sutami Reservoir is located along the main course of the Brantas River Basin. It has a catchment area of 2,050 km^2 and having the largest storage capacity in the basin. The reservoir has functions as flood control, hydropower generation of 9.0×10^8 kWh/a, water supply for irrigation of 83,000 ha paddy field (or 2.6×10^6 m^3/a), domestic and industrial use of 3.0×10^8 m^3/a and recreation. Since its completion in 1972, Sutami Reservoir has supported regional economic development of East Java Province.

Present serious problem encountered in Sutami Reservoir is contamination of reservoir water by pollution flowing into reservoir from upstream cities, villages, agriculture lands, and factories paper mills and tapioca along the rivers. The lacks of effective measures in waste water treatment finally decrease the quality of water in the reservoir. Furthermore, this condition leads to potential pollution and eutrophication phenomenon in the reservoir. In Sutami Reservoir, the indication of pollution has revealed a few years ago but some significant indications happened in June 2001. At the end of February 2002, the similar case happened again and it became worst in March – April 2002. Water surface around dam site colored in blue and brown, clods contained pollution material revealed on every side of the reservoir. Then, blooms of blue – green algae have appeared, as a sign of more advanced eutrophication.

To cope with that problem, some measures have been done and proposed to be done in order to restore water quality in the reservoir. The short term program consist of inspection and law enforcement to upper reach of the Sutami Reservoir to dismantle and close some illegal waste disposal channels and conducting fish sow in the reservoir to increase the population of certain fish species. The medium term program consist of conducting research to establish recommendations for mitigating and controlling water pollution; construction of check dam to reduce erosion and sedimentation and conducting field visit to the students to improve community awareness. For long term program some structural and non – structural measures will be conducted to decrease effluent loads from industrial, domestic, agriculture and animal husbandry sources flowing into rivers.

Key words: water quality management, eutrophication, structural and non – structural measure

1 Introduction

The Brantas River basin, one of the largest river systems in Indonesia, is located in the eastern part of the Java Island, Indonesia, between 110°30′and 112°55′ East Longitude and 7°01′ and

8°15′ South Latitude. It covers catchment area of 11,800 km^2 in total and its main stream, the Brantas River, runs about 320 km long. The basin is located within the Inter – tropical Convergence Zone, in which the semiannual reversal of prevailing winds results in distinct wet (November – April) and dry (May – October) seasons. During the wet season there are around 25 rainy days per month, compared to seven or fewer during the dry season. Annual precipitation is around 2,000 mm on average, with roughly 80% occurring in the wet season. Mean annual temperature range from 24.2 ℃ at Malang City (445 m above sea level) to 26.6 °C at Porong in the Delta, and relative humidity varies seasonally between 55% ~95%.

The Brantas River itself originates in Mt. Anjasmoro, located northwest of Malang City and takes its way around the alluvial cone of extinct volcanoes such as Mt. Kawi, Mt. Butak et al. Gathering together many tributaries above the delta include the Lesti (Southeast), Ngrowo (Southwest), Konto (Central) and Widas (Northwest) Rivers along its traveling. At the confluence with the Ngrowo River in the Southwestern portion of the basin, the Brantas River turns north through the agriculturally productive plains region and finally east through the delta, also an important paddy growing area. Then, it finally bifurcates at Mojokerto City to the Porong River and the Surabaya River, both of which pour into the Madura strait, as shown in Fig. 1.

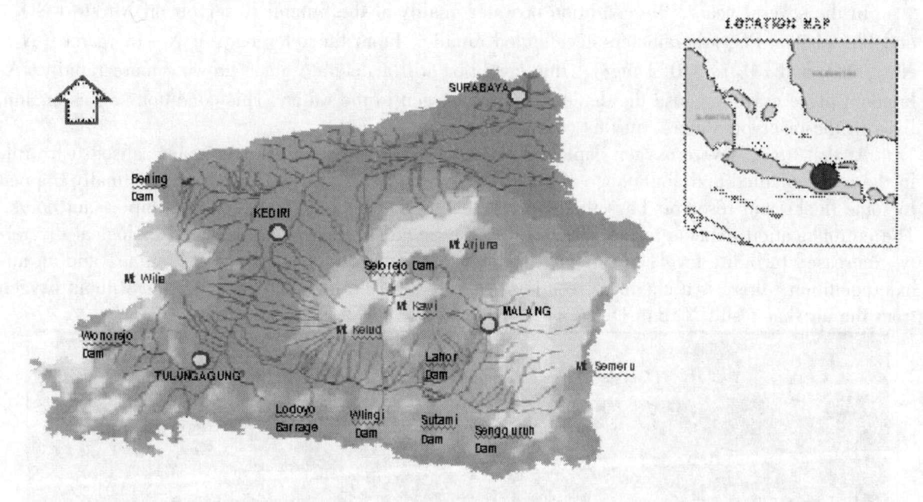

Fig. 1 Brantas River Basin

The Sutami Reservoir also known Karangkates Reservoir, the largest reservoir in the basin, is located about 40 km downstream of Malang City along the main course of the Brantas River. It is bounded on the north by Mt. Kawi and Mt. Butak on the south by a relatively low mountainous zone along the coast of the Indonesian ocean. This reservoir which completed in 1970 has catchment area of 2,050 km^2 and impounding approximately 7.9 km^2 area of Malang Regency. An 800 – m tunnels connects the Sutami Reservoir to Lahor Reservoir (completed in 1977). The Sutami Reservoir plays a vital role on the economic development in the region and for the country as a whole. The reservoir has functions as flood control, hydropower generation of 5.0×10^8 kWh/a, water supply for irrigation of 83,000 ha paddy field (or 2.6 MMC/a), domestic and industrial use of 300 MMC/a and recreation. In a normal condition, the Sutami Reservoir operated between elevation of 246.00 m (LWL) and 272.50 m (HWL) with effective storage remains 136 MMC in 2011 or about 54% compared to its initial effective storage. Nowadays, the operation and maintenance activities in the Sutami Reservoir is carried out by Jasa Tirta I Public Corporation, a State Owned Company acting as the Brantas and Bengawan Solo River Basin Management Agency.

2 Eutrophication Problem

Present serious problem encountered in the Sutami Reservoir is contamination of reservoir water by nutrients flowing into reservoir from agricultural, domestic and industrial sources from upstream cities and villages; agriculture lands; paper mills, sugar and tapioca factories along the rivers. The lacks of effective measures in wastewater treatment in the upstream finally decrease the quality of water in the reservoir. Futhermore, this condition leads to potential pollution and eutrophication phenomenon in the reservoir.

In Sutami Reservoir, the indication of pollution has revealed a few years ago butsome significant indications happened in June 2001. At the end of February 2002, the similarcase happened again and it became worst in March – April 2002. Water surface around the reservoir colored in blue and brown, clods contained pollution material revealed on every side of the reservoir. Then, blooms of blue – green algae have appeared, as a sign of more advancedeutrophication. Furthermore, in September 2002, water quality in the reservoir much deteriorated that noticed by dead fish and unpleasant odors disturbed and annoyed breathing system of people who live near the reservoir.

In the several years, the condition of water quality at the Sutami Reservoir on Nitrate (NO_3) and Phosphate (PO_4) parameters accelerated rapidly. From the reference, if N – inorganic (N – NO_3, NO_2 and PO_4) > 0.3 mg/L, this condition will accelerate algae growing more rapidly. Abundant algae growing cause the depletion of the oxygen in the water. This condition causes organic lives in the reservoir waters, in this case, fish and algae died.

At that time, severe oxygen depletion occurred in the Sutami Reservoir water caused fish within the reservoir died and abundance of algae, which float and suspend in the water, finally trapped in some flanks and reservoir bays then became flocks and decomposed resulting unpleasant odors. The eutrophication impacts to the ecosystem of the Sutami Reservoir are reduced biological diversity, increased turbidity level, accelerated sedimentation that reduced reservoir lifetime, and an anoxic condition (there is a chemical reaction between Nitrogen and BOD in the water without oxygen from the air that result N_2 dan O_2 gas).

Fig. 2 Eutrophication that causes fish in the Sutami Reservoir died in October 2006

Comprehensive researches have been done by several institutions to monitor the growth of algae, fitoplankton and zooplankton in the Sutami Reservoir. The institutions are Water Resources Research and Development Center (Puslitbang SDA), Limnology Center of Indonesian Institute of Science (LIPI) and Mathematics & Science Faculty (MIPA) of Brawijaya University, Malang and Jasa Tirta I Public Corporation in various time during 2002 ~ 2008. Results of the researches are summarized below.

2.1 Water quality in the Sutami Reservoir

From assessment of water quality in Sutami Reservoir done in March – May 2002 showed the existence of Mycrocystis, a kind of Blue Green Algae that its growth is influenced by excessive nutrient (nitrate and phosphate) in the reservoir. Nitrate is a major constituent of farm fertilizer and is necessary for crop production. During rainy season, varying nitrate amounts are washed from farmland into nearby waterways. Nitrates stimulate the growth of plankton and water plants that provide food for fish. This may increase the fish population. However if too much nitrate in the present, the process of eutrophication will be advanced. In such circumstances algae and water plants grow wildly, choke the waterway, and use up large amounts of oxygen. Many fish and aquatic organisms may die as a secondary effect.

2.2 Pollution sources of the Sutami Reservoir

As the main macro nutrients, phosphorus and nitrogen play an important role in the health and functioning of both terrestrial and aquatic ecosystems. An adequate supply of nutrients is essential for the production of food crops both for animal and human consumption. However, an excess supply of nutrients, particularly phosphorus (P) and nitrogen (N) compounds with subsequent losses to the aquatic environment can lead to adverse effects on both the ecology and uses of receiving waters. Nutrients sources can be broadly segregated into two categories; readily identifiable point sources (such as sewage treatment works) and diffuse sources (non point source, such as the run-off from agricultural land). The contributions, in any given catchment, of nutrient inputs from different sources, are vary and depend on the nature of the catchment, environmental conditions and the impact of human activities.

2.2.1 Domestic source

Direct discharge of sewage, not connected to treatment plants, from domestic sources in the upstream of the Brantas River basin, will eventually make its way into the Sutami Reservoir. The population in the upsream of Brantas River spread widely in Malang Regency and Municipality and Batu Regency is approximately 3.4 million. Domestic source from these areas can be divided into two kind of waste.

2.2.1.1 Liquid waste

Liquid waste consists of household detergent, often with high phosphate content add to the nutrients in sewage effluents. According to the population in the upstream of Brantas River, total potential pollution loads from domestic source is estimated at 30.80 t/d (N) and 5.59 /d (P).

2.2.1.2 Solid waste (Garbage)

Effluents leaching from solid waste (garbage) mainly come from Final Disposal Site in Supit Urang Village, Sukun Sub - Regency, Malang City, flow into Banger River, a tributary of Metro River and then flows to Brantas River. Until today, total potential pollution loads from leaching of solid waste in the basin has not been estimated yet. For total load of garbage that thrown directly into the river in Brantas River corridors from Pendem bridge to Gadang (from the research study conducted by Merdeka University, Malang 2002 & 2004) is estimated at 16.52 t/d.

2.2.2 Industrial source

In upstream of the Sutami Reservoir, there are many industries producing high nutrient levels that finally pour into the reservoir. Those are sugar - processing factories, tapioca factories, leather processing factories, paper factory, rubber - processing factory, cigarette factory, shrimp - processing factory, slaughter house. Total pollution loads from industry sector flow into the Sutami Reservoir estimated at 32.15 t/d (BOD) (the data was obtained from the monitoring of Jasa Tirta I Public Corporation Water Quality Laboratory and Water Resources Research and Development Center (Puslitbang SDA) during the Preliminary Research on the Sutami Reservoir Pollution Case, August 2002).

2.2.3 Agricultural source

Farming practices, including use of fertilizers and pesticides rich in N and P in the upstream of the Sutami Reservoir deposit increased amounts of these nutrients in the soil then run - off from these farms to reservoir believed as one cause of eutrophication there. Total pollution loads on N and P parameters from this source which flow into Sutami Reservoir estimated at 94.28 t/d and

1.05 t/d respectively.

2.2.4 Animal husbandry waste

Intensive animal husbandry in upper stream of the Sutami Reservoir managed by some large and small - scale corporations consists of 27,000 cattle, 15,700 pigs and 2 million poultry. Liquid wastes from these animals husbandry come from cage cleaning activities. Total pollution loads on N and P parameters from this sector which flow into Sutami Reservoir estimated at 10.24 t/d and 3.15 t/d respectively.

2.2.5 Sediment

Reservoir sedimentation in the Sutami Reservoir believed contributes significant nutrient. Deforestation in some area in the basin has worsened this condition. Total Nitrogen and total Phosphate from eroded soils flowing to Sutami estimated at 1,025 t/a (2,81 t/d) (N) and 102.5 t/a (0.28 t/d) (P).

2.2.6 Atmosphere

Combustion of fossil fuels in industrial and energy production and in transportation in the basin is an important source of gaseous oxides of nitrogen. Total pollution loads on N and P parameters from atmosphere, which consumed by algae in the Sutami Reservoir, estimated at 0.02 t/d and 0.004,1 t/d respectively.

3 Evaluation

3.1 Observation data

3.1.1 Water quality of the Sutami Reservoir

Water quality assessment was done on physical, chemical and algae parameters. The physical parameters that examined are temperature, brightness, pH, DO, conductivity and turbidity. For the chemical parameters are Nitrate and Phosphate, and for algae parameter are phytoplanton and zooplankton. The result of monitoring was compared to water effluent standards based on Indonesian Government Regulation Number 82 of 2001.

The other comparison is Citeria of Reservoir and Lake Eutrophication Level (Based on Organization for Economic Co - Operation and Development / OECD, 1982). Based on the monitoring data in Nitrate and Phosphate value, the Sutami Reservoir is categorized in Eutrophic waters. Tab. 1 summarizes measurement of Total Nitrogen (N), Phosphate (P), Chlorophyll and Transparency in the Sutami Reservoir compared to the criteria of Reservoir and Lake Eutrophication Level based on OECD criteria.

3.1.2 Algae Abundance

Algae parameters such as phytoplankton and zooplankton are observed in the Sutami Reservoir. In the latest observation in 2008, it was found 11 species of phytoplankton and 12 species of zooplankton. Phytoplankton was dominated by *Ceratium Sp.*, *Synedra Sp.*, and *Mycrocystis Sp.*, and zooplankton was dominated by *Anuraeopsis Sp.*, *Asplanchna Sp.* and *Brachionus Sp.*

In normal water environment, the diversity of algae was spread evenly between species. However if the diversity of algae was only 2 until 3 kinds of algae dominated, this indicates a poor water environment in where the algae domination can be an excessive blooming or algae bloom. If algae bloom happened, there is a possibility that oxygen depletion occurs and will cause fish and algae died. This is one of the eutrophication effects. From each time of observation, it can be noticed that *Ceratium Sp.* can be used as indicator that the water is polluted by organic material. *Synedra Sp.* and *Mycrocystis Spp.* can be used as indicator that the water contained high Nitrate level.

Tab. 1 Measurement of Total Nitrogen (N), Phosphate (P), Chlorophyll and Transparency in Sutami Reservoir compared to the criteria of Reservoir and Lake Eutrophication Level (Based on Organization for Economic Co – Operation and Development/OECD, 1982)

Category	Variable (Annual Mean Value)			
	Total Nitrogen (mg/L)	Total Phosphate (mg/L)	Chlorophyll a (mg/L)	Transparency (m)
Oligotrophic	0,661	0,008	0,0017	9,9
Mesotrophic	0,753	0,026	0,0047	4,2
Eutrophic	1,875	0,084	0,0143	2,4
Sutami	0,145 ~ 2,500[1]	0,005 ~ 0,360[1]	—	0,29 ~ 0,56[1]
Reservoir	0,068 ~ 6,697[2]	0,073 ~ 1,908[2]		0,23 ~ 1,48[2]
	0,000 ~ 5,191[3]	0,011 ~ 0,355[3]		0,20 ~ 0,85[3]

Note: 1. Measurement value for October 2004 – December 2004 by Jasa Tirta I Public Corporation and Mathematic & Science Faculty, Brawijaya University.
2. Measurement value for January 2005 – June 2005.
3. Measurement value for January 2006 – March 2006.

4 Mitigation

The mitigation efforts that have been done to solve or reduce the eutrophication problem in Sutami Reservoir by Jasa Tirta I Public Corporation in cooperation with local government and communities that have the same interest to handling this issue, consist of:

4.1 Research

Preliminary study on pollution problem in the Sutami Reservoir has been done. The objectives of the study were to understand conditions of pollution, to identify potential point sources of pollution, to establish recommendations for mitigating and controlling water pollution in short and long terms for the recovery of the Sutami Reservoir water quality.

4.2 Inspection and law enforcement

Agencies in Malang Regency and East Java Province i. e. Regional Government, Police Department, Forensic Laboratory and Laboratory of Environmental Health in May 2002, conducted inspection to upper reach of the Sutami Reservoir. They found some industries considered contribute significant effluent (tapioca factories) and some illegal disposal channels. By a coordination meeting among agencies, it was decided to dismantle and close some illegal disposal channels.

4.3 Algae harvesting and cleaning

On 16th October 2002, Jasa Tirta I Public Corporation, in co – operation with stakeholders have been done some efforts to cope with the eutrophication phenomenon involving participation of the community around the reservoir. On 16th October 2002, Jasa Tirta I Public Corporation in co – operation with Malang Regencial Office of Environmental Impact Management Agency and Provincial Fisheries Service disseminated the reservoir water pollution control program to fishermen from Sukowilangun Village, Sub District of Kalipare, Malang Regency. On 17th October 2002, the first pilot algae harvesting was implemented based on the recommendation given by the Water Resources Research Center in order to recover this reservoir pollution.

Fig. 3 Algae harvesting and cleaning

4.4 Fish sow to increase the population of certain fish species

To maintain ecosystem balance in Sutami Reservoir, Jasa Tirta I Public Corporation and Regional Government of Malang Regency conducted fish sow in 1998 ~ 2002 with 300,000 species of Oreochromis sp (*Nila Merah*) and Tillapia mossambica (*Mujaer*).

And then regularly conducting fish sow in 2003 (200,000), 2004 (200,000), 2005 (500,000), 2007 (190,000), 2008 (50,000), 2009 (150,000), 2010 (100,000) and 2011 (105,000) with species of Oreochromis sp (*Nila Merah*).

4.5 Construction of check dam and gully plug

To reduce the erosion and sedimentation in the reservoir and preserve the river environment, Jasa Tirta I Public Corporation along with the local government agency constructs many check dam especially in the upstream area. In 2010, PJT I constructed 13 check dams and 68 gully plugs and planted 861,576 trees. In 2011, PJT I plans to construct 16 check dams and 62 gully plugs and to plant 1.2 million trees.

4.6 Conducting field visit to the students

To improve community awareness especially school students to the river environment, a field visit was conducting in the river to gain lesson about the important of the river.

For the long term program, some structural and non – structural measures will be conducted to decrease effluent loads from industrial, domestic, agriculture and animal husbandry sources flowing into rivers. Some measures will be done in this stage consist of development of industrial waste water treatment, development of communal domestic wastewater treatment in Malang Regency and Municipality including Batu City, development of aeration system in the bottom layer of reservoir, modification or substitution of some industrial products (detergent) which containing high Nitrogen and Phosphate, development of artificial rain to increase inflow into the reservoir, development dredging activity to dredge sediment contains high organic matter and nutrients in specific location then embank using sanitary landfill system, relocation of industries that produce high effluent loads and develops wastewater recycling, implementation of waste water discharge license and implementation of clean production to industries.

5 Conclusions

(1) Present serious problem encountered in the Sutami Reservoir is contamination of reservoir water by nutrients flowing into reservoir from agricultural, domestic and industrial sources from upstream cities and villages; agriculture lands; paper mills, sugar and tapioca factories along the rivers. The lack of effective measures in wastewater treatment in upstream part of the Sutami reservoir

finally will deteriorate water quality in the reservoir.

(2) To cope with eutrophication problem in Sutami Reservoir, it requires active participation from all stakeholders particularly in Malang Regency and Municipality. Comprehensive prevention and mitigation programs to eutrophication should be determined, planned and implemented integrated with water quality management program in the whole basin.

References

Djaenuddin, Sudaryati Cahyaningsih, Putut Irwan Pudjiono, et al. Penanggulangan Blooming Algae di Waduk Sutami, Pusat Penelitian Kimia LIPI,2004.

Setijono Samino, Catur Retnaningdyah, Dwi Setyowati, et al. Monitoring Dinamika Komunitas Fitoplankton dan Zooplankton Di Waduk Sutami Malang, Fakultas MIPA, Universitas Brawijaya 2004 ~ 2006.

Soekistijono. Eutrofikasi di Waduk Sutami, Monitoring, Evaluasi dan Upaya Penanganannya, Perum Jasa Tirta I, 2005.

Soekistijono, Swasti Hendrati, Astria Nugrahany. Monitoring Dinamika Komunitas Fitoplankton dan Zooplankton Di Waduk Sutami Malang 2004 – 2006, Perum Jasa Tirta I, 2006.

Simon S. Brahmana, Rosihan F, et al. Penelitian Danau/Waduk Yang Terganggu Pemanfaatannya Oleh Eutrofikasi, Pusat Penelitian dan Pengembangan (Puslitbang) Sumber Daya Air, 2007.

Analysis and Study on the Status and Countermeasures of Ecological Environment of the Yellow River —Xiaolangdi Dam Environmental Impact Post – Assessment

Chen Kaiqi[1] , ***Yu Lian***[2] and ***Ge Huaifeng***[3]

1. Appraisal Center for Environment & Engineering, Ministry of Environmental Protection, Beijing, 100072, China
2. Yellow River Basin Water Resources Protection Bureau, Zhengzhou, 450004, China
3. China Institute of Hydropower and Water Resources Research, Beijing, 100038, China

Abstract: The issues of ecological and environmental protection about rivers have become the focus over the world. In the Yellow River, shortage of water resources is facing serious ecological and environmental problems because of high intensity human activities. This paper analyzed the water ecological environment status quo of the Yellow River Basin, and put forward the countermeasures; meanwhile this study took Xiaolangdi Hydropower Project Environment Impact Post Assessment for example, analyzed the post assessment's positive role in the ecological and environmental protection of the lower reaches of the Yellow River, and it will have a great significance in the management.

Key words: Yellow River, ecological environment, countermeasures, Xiaolangdi hydropower project, environmental impact post assessment

1 Forword

The Yellow River Basin, located in arid, semi – arid regions, blongs to water resource shortage basin. Now the Yellow River is facing floods, water scarcity, the mismatch of water supply and demand, soil erosion and water environment pollution, all these problems of the Yellow River are very typical representative in the major rivers of China. The Yellow River Watershed is of great strategetic status in the important energy and chemical industry bases of China. So far, the development of economy and society has far beyond water resources carrying capacity, which is becoming the key bottleneck of the Yellow River Basin's sustainable development. According to the correlation analysis, the Yellow River accounts for 2% of Water resources quantity of all the country, but it is providing for 12% of the population, 15% of the farmland, and nearly 8% of the wastewater amount. Therefore, river development shall be in a reasonable way, and make the balance development of socio – economy and eco – environment.

2 Eco – environmental problems of the Yellow River Basin

2.1 Water environment pollution

The Yellow River Water Resources Survey results in 2010 showed that: the evaluation river length of the main stream and tributary of Yellow River was 14,295.4 km, of which class Ⅰ ~ Ⅲ, Ⅳ ~ Ⅴ, inferior Ⅴ were respectively 6,324.6 km, 3,120.4 km, 4,850.4 km , and accounted for 44.2%, 21.9% and 33.9% respectively. The evaluation number of water function zones in 2010 was 203, and the water function zones qualifications rate of 43.8% with the number of 89. The number of provincial boundary section evaluated in 2010 was 30, and the qualifications rate of 50.0% with the number of 15. The number of important urban water supply sources (for drinking water) evaluated was 15, and a total of 13 could meet the requirements which accounted for 86.7%.

The pollution loadings of the Yellow River have far beyond water environmental carrying capacity of the whole river with the serious structural pollution and spacial pollution problems. Compared with the total emissions of the whole basin industries, COD emissions of Petrochemical, coal, paper and other industries accounted for more than 80%, but the urban sewage treatment rate is less than

50% , in a word high – polluting industries of the Yellow River Basin account for a large proportion; water environmental carrying capacity and effective utilization rate are all low because of the Yellow River Basin terrain conditions and urban layout, the Yellow River is using a 20% of assimilative capacity to bear about 90% of the pollutants.

The reasons for water environment deterioration of the Yellow River Basin include irrational economic structure, large pollution loadings into the water body, water scarcity, low water environmental carrying capacity and so on. In recent years, water quality deterioration of the Yellow River have been easing (mainly with water pollution control and the increase of river flow), but it did not show stable trend of improvement radically. According to the severe water environment problems, the Yellow River Conservancy Commission of the Ministry of Water Resources has prepared and reviewed the water pollution prevention plan and water conservation plan, and is aiming to strengthen the Provincial Water Pollution Control tasks and responsibilities, and establish the accountability system of the provincial leadership of water pollution. And also in accordance with the requirements of the water environment protection, scientific pollution prevention need to be implemented, and the positioning and layout of industrial development also need to be optimized.

2.2 Ecological issues

The Yellow River Basin is of great eco – types diversity, the source region, the upper and middle reaches, and the downstream river basin and river deltas of the Yellow River are all important ecological sensitive area. According to the "Eleventh Five – Year" scientific and technological support research projects of the Yellow River, the Yellow River wetlands are shrinking significantly, and the basin is facing the serious problems of water ecological imbalance. Research results showed that, the status quo of the Yellow River basin wetland area decreased by 10.2% and 15.8% respectively compared with 1996 and 1986; The recent fish surveys of main stream conducted in 2002 and 2007, but only 47 kinds of fish were found with endangered fish species of 3 kinds; compared with the survey results of 1980s, 125 kinds of fish were found with protected and endangered fish species of 6 kinds, Consequently, there is a great change in recent years.

The area of the source wetlands account for 40.9% of the total area of the basin wetlands, but the runoff of Lanzhou from above accounts for about 48% of the total amount of water resources in the Yellow River Basin. As a major source area of the Yellow River water, the ecological protection of the source regions has a pivotal role in the ecological safety. According to the research results, the area of the source wetland of the Yellow River decreased by 20.8% from 1986 to 2006, and which was much higher than the average degradation rate of basin wetlands. Therefore, the environmental protection issues in the source regions are needed to be solved timely and reasonably, including to strengthen the restoration and protection of the natural meadow ecosystem, to strictly limit hydropower development and so on.

The features of the upper and middle reaches of the Yellow River contains scarce rainfall, adequate evaporation, serious soil erosion, so many measures should be carried out including strict Yellow River water management, efficient water – saving agriculture, the control of the disorderly construction of the artificial water landscape with high water consumption, the optimization of the important energy and chemical industry base on the basis of the carrying capacity of water resources and water environment, evasive action of environmental risks.

As for the lower reaches of the Yellow River and estuarine areas, the flood control is the first, next is to strengthen eco – environment protection, there are many researches and works for example to conduct the scientific restoration and protection of the river ecosystem, the Yellow River Delta and coastal area, and to protect the stable development and succession balance of estuarine ecosystems.

3 Eco – environmental protection countermeasures of the Yellow River

In addition to the above reasons causing the ecological environmental deterioration of the Yellow River Basin, the construction and operation of water conservancy and hydropower project, such as Longyangxia, Liujiaxia, Wanjiazhai, Sanmenxia, and Xiaolangdi et al. , which aggravate the trend of water environmental and water ecological deterioration. Water conservancy engineering has played an important role on the side of the Yellow River harnessing, watershed electricity demand, eliminating water supply and demand imbalances, and realizing watershed flood prevention and control. On the other hand, water conservancy engineering has also made seriously water and sediment conditions unbalance problems of watershed level, therefore, the ecological water is difficult to stable sustain. As a result, the problems of river ecosystem damage and reverse imbalance are appeared in large numbers.

Facing for the state of eco – environmental deterioration in the Yellow River basin, the paper supplies the measures to alleviate the worsening of water environmental pollution, as well as the degradation of water ecological function. The measures are based on the Yellow River Basin of ecological environment status and the goal of coordinated development among ecological environment, economic and social benefits.

(1) The hydropower development in the Yellow River Basin should make ecological priority, overall consideration, moderate development, and ensure that the bottom line as principle. Developing hydropower actively in the premise of ecological protection is good for natural eco – environment protection, good ecological functions maintenance, as well as rich biodiversity conservation. In addition to this, the deterioration of eco – environment caused by hydropower development can be avoided.

(2) To strictly implement the system of environmental impact assessment, to strengthen planning environmental impact assessment and construction projects environment impact assessment. To strengthen the basin unified management of water resources and make the instream ecological water as a important index. Based on the most stringent water resources management system, to ensure the ecological flow of the key important cross – sections on the premise of flood control as much as possible to improve the satisfaction ratio, and finally to keep Yellow River from drying up .

(3) To sound the system of environmental management, in particular, the research on the environment impact post – assessment of the key water conservancy and hydropower projects, such as the Xiaolangdi Dam and the cascade development of the upper and middle Yellow River, all those could have a great significance in the ecological protection of the Yellow River. “Environment impact post – assessment between Longyangxia and Liujiaxia hydropower plant in the upper reaches of the Yellow River” was completed in 2008, so far, the research about environment impact post – assessment of the Xiaolangdi Dam is being done.

(4) To enhance the protection of natural wetland with the priority of source and estuarine wetlands, to build a scientific and effective legal system to protect the wetlands, to clear the principle of wetland protection in the watershed development, to carry out some measures in order to restore and protect the damaged wetlands.

(5) To strengthen the protection of indigenous and endangered fishes and species habitats, to protect the key sections in order to ensure the migratory pathway, to prohibit the instream sand mining in the fish spawning grounds and aquatic marginal wetlands, to implement the system of no – fishing zones and closed fishing season, to prohibit the over – fishing, to carry out marine breeding and releasing.

(6) To ensure the achievement of basin water quality objective, to coordinate of water resources protection and water pollution control work comprehensively, to establish a pollution control mechanism combined the basin with the region, to strengthen the protection of the centralized drinking water source, and to improve the water pollution control level in the round. Meanwhile, the economic development pattern and industrial structure with the circular economy concept need to be optimized; the construction of high water consumption and heavy pollution projects in the Yellow River basin shall be in control reasonably; and it still contains some other measures including to implement the dynamic management of water intake, water consumption and water return, to strengthen the management of oil erosion and non – point source; and to study the feasibility of implementing more stringent pollutant emission standards in the key control unit of basin.

(7) To strengthen the basic research of water ecological environment and improve the water environment monitoring system. There are some monitoring factors, such as hydrology, water quali-

ty, sediment, plankton, benthic animal and microbial ecological environment. The accumulated ecological environment data are the foundation for the basic research.

4 Research of ecological environment protection in the Yellow River——the environmental impact post evaluation of the Xiaolangdi Key Water Control Project

4.1 The necessity of environmental impact post evaluation of the Xiaolangdi Key Water Control Project

4.1.1 The Xiaolangdi Key Water Control Project's future development requirement that carry out eco - environmental impact post evaluation

The Xiaolangdi Key Water Control Project (hereinafter referred to as Xiaolangdi Project) is a large - scale comprehensive water conservancy project with the functions of "mainly in flood control, ice prevention and sediment reduction; taking into account water supply, irrigation and power generation; clear water storage and muddy water removement; comprehensive utilization; and turning harmful into beneficil" on the Yellow River. In the late of last century, the Yellow River was drying up, and the construction and operation of the Xiaolangdi Project played an important role in improving downstream eco - environment. It is the key project in the management and development of the Yellow River. Since its run in 1999, Xiaolangdi Project has gone through three processes: sediment control and deposition reduction in the early operation period, clear water storage and muddy water removing in the later operation period, and sediment regulation process implemented in every year since 2002. A large amount of data was accumulated during the runtime. Therefore, carrying out the eco - environmental impact post evaluation of the Xiaolangdi Project systematically can inspect the effectiveness of measures that were taken in the project environmental impact assessment, understand the Xiaolangdi Project's ecological cumulative effects and ecological benefit comprehensively, and have implications to achieve the sustainable development of the Yellow River Basin. At the same time, it is the development needs of the Xiaolangdi Project.

4.1.2 To provide technical support for fully implementing environmental impact post evaluation of water control project

Since the 1990s, some of the construction project environmental impact of retrospective assessment, environmental impact post evaluation, confirmatory evaluation and so on have been carried out gradually in China. But there are lack of clear defines in terms of content, technology, and methods. Some tentative evaluation work that has been carrying out can not meet the overall demand of the country in investment project post evaluation. In view of the lack of supporting regulations and technical specifications, post - evaluation work moves slowly, theoretical exploration and practice accumulation are far from enough, and the effects of post evaluation are not fully played out. Thus it is necessary to carry out the case studies about the Xiaolangdi Project eco - environmental impact post evaluation, and summarize a set of technical method system for eco - environmental impact post evaluation of water control project at the present stage. It will promote the full implementation of water resources and hydropower engineering eco - environment impact post evaluation.

4.1.3 The needs of pefecting the whole process management of water control project

Environmental management of traditional construction project is consisted of environmental impact assessment, "three simultaneous", and management system of environmental protection acceptance for construction project. Viewed from the life cycle management of construction project, environmental impact assessment is a pre - evaluation, and is also the main environmental management measures taken for the eco - environmental impact of water control project. While the "three simultaneous" and environmental protection acceptance is the management tools and legal guarantee of implementing pollution control and ecological protection measures which are referred in the environmental impact assessment report. Promulgation and implementation of "Environmental Impact

Assessment Law" enriches the connotation of construction project environmental management. The environmental impact post assessment proposed in the law is a post evaluation, which becomes the remedying and extending of the existing environmental impact assessment. The case research about Xiaolangdi Project eco – environmental impact post evaluation is not only playing an important role in the coordination between project development target and environmental protection target in the running period, the effective functioning of environmental protection projects, the pertinence of environmental protection measures of future water control project and so on, but also is of great significance to perfect the whole process management system in the planning – construction – completion – run different stages.

4.2 Research contents of the Xiaolangdi Project environmental impact post evaluation

The content of Xiaolangdi Water Control Project environmental impact post evaluation focuses on ecological system impact of the reservoir area, downstream river and the Yellow River Delta wetland in the runtime. Then, the post evaluation is trying to evaluate the actual impact imposed by water conservancy and hydropower engineering, verifies the effectiveness of environmental protection measures, environmental monitoring and environmental management. In that case, we can condense the technical methods of water conservancy and hydropower project environmental impact post evaluation.

4.2.1 Eco – environment impact and benefit analysis in the reservoir area and downstream river corridor of the Xiaolangdi Project

To calculate the impact factor changes before and after the construction of reservoir, and to analyze the impact of eco – environment factor under the different operation modes based on the long series data of hydrological regime, water quality, water temperature, aquatic organisms, terrestrial flora and fauna.

Benefit assessment is the core content of the post evaluation of investment projects. Water control project can leads to changes of ecosystem elements and functions, and thus have an impact on regional ecological value. Interference on the river ecosystem imposed by the water control project is both positive, but also negative. Compared value increments of the river ecosystem to the loss, the impact on river ecosystem function by the Xiaolangdi Project and the ecological benefits are both analyzed quantitatively.

4.2.2 Research of the impact on downstream wetland protection area in the operation of the Xiaolangdi Reservoir Project

There is a drop in frequency of floodplain flood in the Yellow River, which is caused by the artificial regulation of the Xiaolangdi Project. But it is difficult to improve the ecological status of wetland protection area relying solely on the initiative replenishment of water from the Yellow River. Therefore, there needs an in – depth analysis of water area under the typical wetland system, water level variation, wetland sedimentation volume, quantity and quality of flora – fauna and aquatic in the protected areas, and habitat environment of rare waterfowl in the different operation phase. In addition, effect of the reservoir operation on typical wetlands downstream development and succession under the conditions of sediment regulation is needed to be researched.

4.2.3 The effectiveness evaluation of environmental protection measures of the Xiaolangdi Project and adaptive management

Project environmental protection process evaluation includes three stages: the early work, project construction and implementation, operation and management. While environmental protection measures can be descripted as water environment, water ecosystem and terrestrial ecosystems three perspectives, and then analyse the effect of these mitigation measures. Environmental monitoring and environmental management plan should include the monitoring scheme and environmental management program effectiveness evaluation required in EIA report. In addition, we can analyse the affect of ecological system protection scheme on realizing ecosystem function compensation and ecological value with the ecological profit – loss assessment techniques method.

4.2.4 Adaptive management research of the Xiaolangdi Water Control Project

Adaptive management is a necessary condition for sustainable development, and it is used widely in the certain change and uncertain impact of water resource development, management and use. Therefore, we can apply strategies and measures of adaptive management, concise the Xiaolangdi project's adaptive management objectives which focus on sediment regulation, flood control and water supply, then comb and work out reservoir adaptive management program. After that, we can evaluation the effect of adaptive management program of the Xiaolangdi Reservoir on the downstream ecological system and important wetland protection areas, last, give a comprehensive description of implementation effect of the management program. Based on the key issues of the Xiaolangdi Reservoir, adaptive management improvement program is proposed, which provides a new, suitable ways and methods for optimal scheduling and management of the Xiaolangdi Reservoir in the context of climate change.

Acknowledgements

Financial supports are from the Ministry of Environmental Protection's special funds for scientific research on public causes (2010467060).

Study on Ecological Restoration Measures of the Heihe River Basin

Yang Lifeng*, *Bi Liming*, *Chen Na* and *Li Fusheng

Yellow River Engineer Consulting Co. Ltd, Zhengzhou, 450003, China

Abstract: Heihe River is the second-largest inland river basins in Northwest China. In order to curb environmental degradation and gradually restore the ecosystem in the Heihe River basin and according to the objective requirements of the basin population, resources, ecological environment and the economic and social development , for different ecological characteristics of the middle and lower reaches, study puts forward the main measures of ecological restoration. on the basis of analysis of the eco-environment present situation and existing problem Upstream ecological repair measures mainly includ: water conservation engineering and ecological immigrants engineering, etc; ecological restoration in the middle reaches of the measures includ: farmland protection forest system update, returning farmland to forest and grassland, wetland protection, natural vegetation in oases edge enclosure protection and saline-alkali soil transformation, etc; downstream ecological repair measures main includ: ecological water conveyance project, and ecological restoration project, ecological resettlement project and ecological restoration of the East Juyanhai, etc.

Key words: Heihe River, Ecological restoration, measure, research

The Heihe River is the second-largest inland river basins in Northwest China, flows through Qinghai, Gansu and Inner Mongolia (autonomous regions) provinces, the basin is bounded by south Chilien, north Mongolia people's Republic , east the Shiyang He basin, west Shule River basin , whose strategic position is very important. The middle reaches of the Zhangye city, located in the ancient Silk Road and today's Eurasian continental bridge, agricultural development has a long history, enjoyed "Golden Zhangye" of reputation; downstream of Ejina Banner border line long 507km, which has China important of defense research base, Ejina oasis of Juyan Delta area, can be both the natural barrier that block sand invasion, and protect ecological, and is the local people live relying on growth, important for national defense scientific research and construction of border.

1 Environment status and main problems of the Heihe River basin

1.1 Present situation of ecological environment

Upper reaches of Heihe River basin area is 10,000 km^2, which high mountain area accounted for 72%, river mesa accounted for 27%, oasis hills accounted for 1%, the area is located in alpine mountain, vegetation growth slow; middle reaches of Heihe River basin area is 25,600 km^2, which Gobi desert accounted for 31%, soil beach accounted for 29%, oasis accounted for 24%, low mountain hills accounted for 16%, the area light hot resources rich, artificial oasis development; downstream is basin ecological environment most vulnerable of area, basin land area 80400 km^2, main for Gobi desert and erosion residual mountain, its area accounted for 94%, which near half is desertification land, oasis only accounted for 6%. Heihe River Basin water quality is in good condition, but part of the water quality in the midstream has been a downward trend. According to the 2008 water quality evaluation results of Hongshui River, Liyuan River, Heihe and Shandan River, water quality of class Ⅰ ~ Ⅲ accounted for 80.9% of the river length of Evaluation, Ⅳ and Ⅴ accounted for 19.1%.

1.2 The main problems

Extreme drought in middle and lower reaches of Heihe River basin and regional water resources can't meet the needs of local economic development and ecological balance, historical contradictions about water affairs have been quite prominent. Since the 1960s to the 1990s, under the influence of climate and human activity, ecosystem deterioration basin on the middle and lower reaches exist in varying degrees. Due to population growth and economic development, overexploitation of water and land resources, since the 1960s, the amount of water flows into the lower reaches is gradually reducing, rivers and lakes dried up, trees, dust storms raging, and grassland degradation, ecological and progressive deterioration of the environment.

For the growing deterioration of the ecological environment in the Heihe River basin, in 2000 and 2001, the country has started a project of integrated water regulation of the Heihe River and Heihe River watershed, initial containment of eco-environmental deterioration in trends. However, subject to the constraints of the various conditions, ecological environment in the Heihe River basin are still very fragile currently. The upper reaches firstly is water conservation capacity remains weak, grassland degradation and land desertification are serious, damage by rats and poisonous weeds are rampant ;secondly over-grazing remain serious. Through recent governance, local awareness of the cadres and masses to grazing and grassland has been enhanced, the trend of over-grazing has been curb, overgraze phenomena persists, grassland burden is still heavier. The middle reaches, firstly is farmland shelterbelts and plantations local within the oasis region to appear for the purpose of protection appears the decline and death phenomenon; the second was serious degradation of natural vegetation in oases edge areas is serious; the third was the main stream of Heihe Ganlingao wetland degradation is evident. Fourthly, there are irrigation Salinization problems in part of the agricultural. The downstream, first is the scope and cover of natural forest grass are small, there is still a certain gap to achieve ecosystem restoration of the mid 1980s, overloading of natural grassland grazing is still exist; thirdly, co-environmental of East Juyanhai and surrounding is fragility.

2 Study on the treatment measures

2.1 Basic idea

Water resources act as a core element of ecological problems in the Heihe River basin, limited by the total amount of water resources, ecological construction can only follow the principle called limited goals, key breakthrough and do something or not.

The Heihe upstream origin area, with the goal of water source conservation, there should strengthen natural forest protection and natural pasture construction, intensify prevention and supervision, prohibit land reclamation, destruction of forests and grass and overgrazing. by delimiting some enclosure which has reached a certain scale, and supporting a certain number of feeding or half feeding, to promote change in the way of traditional grazing. The main way is natural repair ecology, be aided with artificial measures such as fencing and governance of poisonous weeds and rodents, in order to improve water conservation ability; in the Middle reaches, there can take various measures to rescue ecological forest, part of which is at death's door, to ensure the stability of ecological environment. According to the water function zoning, there should strictly control waste effluent standards, implement the pollution discharge permission and system total amount control of pollutants into the River. Downstream Ejina Banner, over-grazing and reclaiming wasteland is strictly prohibited. Through the measures such as oasis irrigation project, ecological migrants project, improvement of forest-grass irrigation methods and soil and water conservation, to curb environmental degradation and gradually restore the ecosystems.

2.2 Treatment goals

Ecosystem degradation in the Heihe River basin is a long-term accumulation process of natural and unreasonable human activity, we should fully realize the ecological construction is urgency and long-term, there should take comprehensive measures and phased governance. Before 2003, integrated water resources management and the systems of ecological construction and environmental protection should be establishment and improvement, in order to containment the deterioration of ecological environment; after 2003, the River basin eco-systems can gradually recover to the level in 1980s, achieving eco-system in a virtuous circle.

2.3 Control measures

The varied topography of Heihe river basin is high in the south and low in the north. In accordance with height above sea level and natural geographical features, it can be divided into three landform type area called the upstream of Chilien , the middle reaches of the corridor plains the lower reaches of Alxa high plains it is dominated by animal husbandry production in the upper stream area, irrigated agricultural area in middle reaches, the lower reaches is fade zone of runoff which belongs to hungriness rainless region and extreme arid sub-region. There need to take different ecological control measures for the different characteristics of the River basin.

2.3.1 Origin in the upper area

To aim at the main problems of ecological situation in the upper reaches of the Heihe River, according to ecological restoration goals, the control measures include water conservation projects, ecological migrants and soil and water conservation measures. Water conservation project include ecological construction of pastoral area and non-pastoral area; the ecological migration projects include immigration allocation and supporting forage construction projects. Soil and water conservation measures include forest for water and soil conservation, channel slope protection project, gully head protection works, wire gabion and gully control dam, etc.

(1) Water conservation project.

①Ban pastoral area.

To the upper reaches area where the elevation is 3400m above, with poor natural conditions and serious deterioration of grasslands, there should take measures for and breeding grass. There can prohibit people and livestock damage, to create conditions for the natural repair of the pasture, water conservation, by enclosures the boundaries of the grazing district. At the same time, combining with poisonous weeds and rodent pest control and artificial reseeding, the natural grassland of Non-pastoral area can restore.

Non-pastoral area grass control measures include artificial reseeding and grassland improvement for "the Black Beach" type deteriorated grassland and sandy grassland. There should govern the grassland that is harmed by rats and poisonous weeds, should increase grass cover and water conservation capacity.

In order to meet the Non-pastoral area ecological treatment measure ,to achieve bearing, no rebound, there need complementary development feeding facilities, including sheep shed, cowshed, etc. Through barn feeding, there can change the traditional grazing methods, protect the ecological environment; ensure the living standard of herdsman without lowering. At the same time, there need to construct fodder land, which used to supply the feeding requirement of banning grazing.

②Non-pastoral area .

For areas whose grassland has certain coverage, with relatively good natural conditions, should take fencing, rotational grazing, and other measures to prevent overgrazing, and combine measures such as poisonous weeds, rodent pest control and artificial reseeding, In order to promote the natural restoration of pastures.

③Ecological migration.

Calculated based on the analysis of grass livestock balance, status appropriate stocking rates in the upper reaches of Heihe River is 1.216×10^6 sheep, actually ,the stocking rate is 1.474×10^6 units of sheep, there are 258,000 sheep units overload ,overloading rate is up to 21.2% , overgrazing is serious. In order to achieve the Balance of grass and livestock, in addition to a series of engineering measures of ecological management, there can take measures of ecological migration, to achieve retire graze and livestock by means of ecological migrants. The retired livestock can be fed on fodder in barn to ensure that local animal husbandry economy development, by building Forage bases and Livestock barn and Fattening facilities.

(2) Soil and water conservation.

To prevent the harm such as water erosion-caused soil erosion, playing a slope, erosion protection, slow flow and deposition effects, reducing soil erosion, relief for landslide, collapse, etc, there need to arrange the construction of soil and water conservation forest in mountain areas; there are many tributaries in the upper reaches of the Heihe River, most of the tributaries have steep bank slopes. Whenever every rainstorm, the bank collapse seriously, grassland and forest on the shore collapsed, at the same time, due to tributaries has high gradient slope and large flow velocity, water cutting brush, there may result channel-bank slumping, grassland and forest damage. In order to protect the forest meadow and water conservation, there need to construct channel slope protection project. The Heihe upstream of Qilian County and Sunan County is located in Mountains surrounded by, most of the farmer live in groove road within or groove mouth, where is very easy to be harmed by flash floods and landslides. There may construct the gully head protection works, to prevent ditch head to go forward and Slope eroded cause by runoff scouring, reduce the runoff abration to Plateau surface. It can arrange to construct wire gabion to control rivers and storm flood, prevent the flow scouring riparian which cause new water and soil loss. There can arrange to construct check dam engineering, to reduce the river sediment from tributaries.

2.3.2 The middle reaches

In the middle reaches, ecological restoration project includes: regeneration of farmland shelterbelt, wetland restoration and protection, protection of oasis peripheral transition belt and soil and water conservation measures, etc.

(1) Measures of system update of farmland shelterbelts.

Farmland protection forest system is used to prevent natural disasters, improving the conditions of climate, soils, hydrology, and create an environment conducive to the growth of crops and livestock breeding, to ensure that the agriculture and animal husbandry have high and stable yield, and provides people living with a variety and effective artificial ecosystem, it is stability is most important to the oasis's security and stability. Because the production water diverted water from ecological water use, parts of the shelterbelt in midstream are short of water seriously. Existing shelterbelt system has greatly diminished the protective effect of the oasis. Therefore, we need to renovate the shelterbelt died of lack of water in the middle reaches, and rescue the manmade Shelterbelt that is at death's door by water replenishing.

(2) Safeguard of wetland restoration in Ganlingao.

Wetlands are important components of oasis ecosystem and are very sensitive to the outside human disturbance. Human activity such as digging drainage gutter, blocking the water channel, reclaim land from the marsh, etc, is the main reason causing the degradation of oasis and wetland. April 16, 2011, the state council approved the establishment of the Heihe wetland national nature reserve of Zhangye, wetland reserve area of which is 615,000 acres. In order to effectively consolidate the results of returning farmland to wetland, better protect existing wetland, there need fencing protection of the wetland nature reserve to reduce human interference and influence, that it can be natural restoration, wetland vegetation biomass may restore to the level of natural communities in 3 ~5 years.

(3) Peripheral oasis transition belt protection measures.

In order to effectively protect the existing oasis in middle reaches, there need to construct resistance sarin belt and Sand fixing piece forest surrounding the existing oasis, to stop the invasion of oasis peripheral sand. That is to build Enclosure belt surrounding the Oasis edge within a certain range. Construct the protection system which is oasis-center and from the edge to the periphery combine with "stop, strong, seal", at the desertification serious strip especially the wind gap at the edge of Badain Jaran Desert. Thus there can provide a good protective barrier for the oasis which is multi-level and omnibearing, also can weaken the oasis internal wind speed, reduced evaporation, reduced dust, achieve the purpose of protecting the existing oasis .

(4) Soil and water conservation measures.

For prevent storm and the flow scour caused soil erosion, there can arrange the construction of soil and water conservation forest; it can arrange the river revetment works, to protect riverside form down cutting by floods and whitewater which may cause landslide and collapse. In order to barrier sand erosion, and protect existing oasis, there can arrange construction forests for windbreak and sand-fixation. Combine with local sand control experience, and for improve the enthusiasm of the masses to control sand, the Sand break forest species should select Haloxylon ammodendron, which has a high value of sand-fixing effect and economic.

2.3.3 Downstream the Ejina Oasis

For ecosystem restoration in the lower reaches of the Heihe River retune to the level in the 1980 of the 20th century, measures of ecological construction projects include: ecological water transport engineer, ecological restoration and ecological migration project. Ecological water transport engineer includes the construction of canal system, river obedience, etc; ecological restoration including natural forest fencing, etc; and ecological migration project including immigration allocation and forage base construction, etc.

(1) Ecological water transport engineer.

Combined with distribution of oasis, topography, irrigation, there needs to plan to arrange some water supply works for east and west River oasis. The East River irrigated area: dredge up the reaches which is waterproof serious and low of conveyance efficiency at the east main canal end to Angci river sluice about 3km long. The west river irrigated area: channel improvement is mainly by blocking tributaries, strong main stream; straighten out the main stem, etc. And set the gate in a large braided river. Set the strobe in the larger braided river. And use of natural river setting diversion gate or irrigation ridge dam to transmission of water to the oasis along the river.

(2) Ecological restoration project.

Combined with artificial irrigation, fencing populus euphratica forest protection, are effective measures to promote the ecological restoration of natural vegetation. On the basis of the Heihe recent governance, planning to implement of fencing which is focused on the east River irrigated area including Suponaoer, Jargalangt, Gurinai and Guaizihu (Inner Mongolia Autonomous Region, China), etc, and parking area and Saihantaolai area of the west river irrigated area. At the same time, combined with oasis irrigation project construction, there are aided with artificial irrigation measures which is necessary in the fence in fences with Irrigation conditions. Dangerous sluices which is difficult to complete the task to restore oasis ecological implement maintenance and alterations.

(3) Ecological resettlement project.

In order to reduce the overload of natural grazing, curb the trend of ecological degradation of oases, and protect euphratica forest, there may implement ecological migration, and construct the fodder base which links with ecological resettlement project.

(4) Ecological restoration measures of east Juyanhai.

The east Juyanhai is shallow saucer-shaped, whose depth of water is shallow, water surface area and evaporation loss are large. Maintain a healthy life of the Heihe River must be linked to the eco-system, and the eco-system is inseparable from the Ejina oasis in Heihe River and the Terminal Lake, diverting water enters the east Juyanhai is an important indicator of water diversion work of Heihe River. According to the current shortage of water resources in the Heihe River basin, to repair Juyanhai and its surrounding ecosystem can not even necessary to single pursue the expand of the surface area of East Juyanhai, should be pursuing limited objectives, under the support of the limited water resources to construct the East Juyanhai oasis functional areas, makes the links between Juyan culture and surrounding ecological, developed the ecological scenic spots, played the oasis benefit. The East Juyanhai ecological restoration measures including:

Firstly, increase in water depth, reduce evaporation losses. In East Juyanhai region, the average annual precipitation is about 36 mm, the Surface evaporation is up to 2,155 mm. Based on the measured data, when the water-holding of East Juyanhai is 50×10^6 m^3, water surface area is about 32 km^2, depth is about 1 ~ 2 m. In order to guarantee the East Juyanhai does not run dry in flat or low years, reducing evaporation losses is the key, more feasible approach is to increase the water depth of East Juyanhai; the second is to implement artificial replant forests and grass. In order to play the ecological benefits as soon as possible, while the oasis natural repairs, complemented by the necessary artificial measures should replant vegetation. According to the scale of ecological restoration, taking into account local climatic conditions, soil type and other natural conditions, may arrange artificial reseeding of shrubs such as rose willow Chenopodiaceae Russian olive and herbaceous plants such as Sophora alopecuroide; thirdly, three can start enclosure and Fencing. Combined with nature condition around the East Juyanhai, to prevent human and livestock destroyed; fencing is effective measures to restore the ecological environment around. In order To create conditions to improve the self - repair capacity of the natural vegetation, there are Plans to implement fencing with grassland and forest around the East Juyanhai and its surrounding.

3 Conclusions

The ecological construction and environmental protection in Heihe River basin is a major part of the large-scale development of the Western region, not only related to the residents ' living environment and economic development, and also related to the environmental quality of Northwest and North China region, it is great event related to national unity, social stability, defense solid for national, and to start the eco-environment construction and protection in Heihe River basin is very necessary and urgent. For the upper and middle reaches of Heihe River basin of different ecological characteristics, through ecological restoration measures taken on the upper reaches of Heihe river water conservation district, artificial oasis in the middle of taking protection measures, measures of ecological restoration of Ejina oasis in the lower reaches of the Heihe River, ecosystem degradation in the Heihe River basin has been curbed, with the further increase of ecological control, watershed ecosystems will continue to improve. But we must also recognize clearly that problems of ecosystem degradation in the Heihe River basin are long-term accumulation process of natural and unreasonable human activity. We should fully understand the urgency and long term of ecological construction.

References

Heihe River Basin Management and Planning[R]; The Ministry of Water Resources of the People's Republic of China, August 2001.

Heihe Water Rresource Problems and its Countermeasures. Yellow River Conservancy Commission of the Ministry of Water Resources; November 2000.

Heihe Water Resources Development and Protection[R]. China Yellow River Engineering Consulting Co., Ltd. December 2007.

Liu Min, Gan Zhimao. Heihe River Basin Water Resources Development Ejina Oasis and countermeasures [J]. Journal of Desert Research, 2004 02(2).

Qi Shangzhong, Wang Tao, Luo Fang, et al. Heihe River Basin Desertification and Sustainable Development [J]. Research of Soil and Water Conservation, 2004 (2).

Sun Xuetao. Heihe River Basin Thinking of a Number of Issues of Sustainable Development of Water Resources [J]. Science &Technology Review, 2002 (6)

Quality of River Waters of Aral Sea Basin and Ways of Their Management

E. I. Chembarisov

Scientific Research Institute of Irrigation and Water Problems of Ministry of Agriculture and Water Management, Uzbekistan

Abstract: The problems of the Aral Sea desiccation and the considerable deterioration of the ecology in the Aral Sea basin and in other regions of Central Asia have now acquired global importance. An estimation of ecological condition of the Aral Sea basin has been made by using the methods of system analysis, budget, statistics, and cartography. Based on the volume of surface water and its quality in different regions of Aral Sea basin, the optimum variants of their use within irrigation systems have been proposed. Methods have been developed for decreasing the extent of river water pollution, reducing the number of collector-drainage water discharges into rivers and principles of hydroecological zoning of territories.

Key words: Quality of river waters, Aral Sea basin, Index of Water Contamination (IWC), hydroecological zoning of territories

1 Introduction

The independence of Central Asia Republics and the transition to a market economy, which has been accompanied by a breakage of the links between the CIS countries, and the general background of a regress in the economy of the area, have made the economical use and protection of water resources and introduction of new technologies even more important. A new approach is required using new technologies. In this, given the limited capacities in the region it is necessary to select priorities that provide a short - term effect and to use the experience, finance and technologies of reputable international organizations and companies to develop an integrated utilization of inter - republican water resources, giving an improvement of river water quality and the provision of good quality drinking water.

2 Existing knowledge on use and protection of water resources

The main problems of the region are low quality of drinking water and the ineffectiveness of purification facilities in cities, settlements and rural areas. Up to 20 ~ 30 years ago there was no serious drinking water supply problem as surface and ground water was not contaminated with toxic substances and ponds, wells, surface water could be utilized using primitive purification devices. A growth in river water mineralization levels and contamination has led to the degradation of near-river and near canal fresh water lenses.

Wells used by the rural population are contaminated by agrochemical-pesticides, nitrates, oil products with those in industrial zones being contaminated by toxic metals and organic components in addition.

Water resource monitoring is carried out by sector: Uzhydromet (Uzbek Meteorology Agency) monitors regularly water quality in water courses and reservoirs; Minzdrav (Ministry of Health, Sanitary & Epidemiology Service) monitors quality of drinking water; Goscomgeologia (State Committee on Geology) monitors quality of underground mineral and fresh drinking water, and Minzelvodhoz (Ministry of Agriculture and Water Management) is responsible for water allocation in river basins and water intake in canals for irrigation purposes. It also monitors quantities of collector-drainage water. Water quality and river discharges monitoring are carried out in the most effective way by Uzgydromet which has a modern analytical and methodological basis.

The total Aral Sea basin water resources are estimated by experts to be about 120 ~ 125 km^3, whilst the total current annual runoff collected according to calculations is about 33 ~35 km^3, i.e. about 30% of water resources. Between 21 ~22 km^3 of collector runoff is included within the Amu Darya basin, including the Karakum canal with its Murgab and Tejen irrigation areas. A further 13 ~ 14 km^3 is collected in the Syr Darya basin which has mean mineralization levels varying from 1.7 to 6.0 g/L.

3 Objective and scope

To achieve the necessary condition of optimal and harmonious development of the technical and economical level of Central Asian Countries necessary permanent data on the quality of water resources, which can be used for water are required so that future information about natural water quality have main practical significance for economic developing. The acceptable chemical content levels for the different uses of drinking, municipal use, agriculture, technical etc; Results of the research of the hydrochemical laboratory of Institute of Water Problems at Academy of Science (from 2005 to 2011 yy.) focused on solving the hydrochemical problems of this region. This included the design of systems for the prevention, limitation and removing of contamination of river waters upon which I want to concentrate in this paper and not the research that is being carried out into the examination of river waters. It is these investigations into solving hydrochemical problems.

4 Results

On the basis of an analysis of the "Bank of hydrochemical dates" (which included details of river water quality since 1990) interactive estimating of the modern contamination levels of rive waters were conducted using five classes of quality: good quality, satisfactory quality, bad quality, dangerous quality and highly dangerous quality.

The contamination level of river water quality was calculated using the method of Index of Water Contamination (IWC). At the present time, the level of contamination of surface water is estimated using the IC method (Index of Contamination) from Institute of Hydrochemistry. This index included six ingredients which exceeded a Maximum Admitted Concentration (MAC), considering oxygen content. But, in this form that index note chaff for practical demands.

From this reason was suggesting considering of all ingredient, which high than MAC, introduce coefficients K_1 and K_2. When Index of Contamination exchange from 0 to 1—water in good quality, 1 ~3 – satisfactory, 3 ~5 – bad, 5 ~1 – dangerous and more than 10 – very dangerous.

Results of investigation permitted to make map of "Distraction of river waters of Uzbekistan on quality of drinking water". Conducted analyze of dates about area and population with different water quality. 8% of area of country have quality of "good water", where living 10% of population, "satisfactory" quality area is 15%, where living 16% of population. Area with "bad" quality 41%, and population when living 50% and end cantor "dangerous" quality of water 35% of area, where living more then 24% of population of Uzbekistan. As this, points solving of drinking water quality very actual for territory of Uzbekistan. Investigation results showed "dangerous" water had not only in territory of low Amudarya, also in low Zeravshan, where situation with water contamination one of hard.

Calculation showed IC in Syrhendarya river – 2,8; Kashkadarya – 4,0; Zeravshan – 5.3; and Amydarya – 5,4 this number proved our conclusions.

For example, we can give detal investigation result's Zeravshan. Average year mineralization of river before Samapkand 0,3 g/L, after flow of into the high mineralization drainage water's increased to 0,5 g/L. In the part of river from Khatirchi to Navoi mineralization increased reason is waste waters from industry plants Kattakurgan and Navoi city, mineralization of river after flow of into the river water's from "Navoiazot" – 1,6 g/L.

Quality of water changing essentially line of river is water composition in exit in the mountain is sulphate – hydrocarbonate – magnezuim – calcium, at lower part is sulphate – magnesium – calcium – natreum composition.

From contamination chemicals pesticides have higher level than MAG. Maximum concentration of alfa hyxochoran (GHCG) had Khatirchi network station (6,2 MAC).

From higher metals chromium and zinc (ZN) had higher concentration. Higher concentration of this metals had collectors Siab and Chaganak. Also, in water of Zeravshan contented curium, which very dangerous for mans health.

Contamination level of river waters with organic elements estimated by BPK5 (biochemical use of oxygen during five days). This index higher than MAC 1,1 - 1,2 only in three networks. Water of Zeravshan contaminated with phenol also, concentration which higher than 3 - 7 MAC. Maximum concentration was river Amankutansay IC of river water had -5.3.

Chemical composition of Amudarya water in the flat area is composed under the influence of collector waters coming into the from irrigated areas of Surhandarya, Sherabad, Kashkadarya (through Southern collector), Zeravshan (through the Main Bukhara collector) river basin areas as well as from irrigated area of Turkmenistan left bank irrigation area (Chardjou oasis) Due to this the Amudarya water mineralization levels in midstream and especially in downstream areas are elevated up to 1.2 ~ 1.3 g/L. Water contamination is caused by nitrate nitrogen, oil products, phenol, copper, pesticide, etc. , the content exceeding MAC.

5 Conclusions

For the conservation and improvement of surface water quality in the Aral Sea basin there is a need to introduce integrate water safe measures:

(1) The organization of monitoring network responsible for water quality , with the study of charges for negative processes;

(2) The installation of a water safe zone and shelf line in water objects and the measurement of pollution and exhaustion to support sanitary conditions;

(3) To decrease the flow of collector - drainage water from rives irrigation fields by the application of progress possible of irrigation with using less volume of irrigation water, and more, full using these waters in place their forming (with estimate method of kvazidesalition and methods of kvazidesalition and methods of hydroelectric);

(4) To extend in practice water rotation systems of product water - supply, and wider counts ruction for treatment disaffection and render harmless of waste water;

(5) To introduce precept of payment off using of water resources, this will be stimulation for increase natural capacity of products uses will be try intriduce water save technology, for economy of water;

(6) Necessary introduce methods of economy stimulation for organizations-users, liberation from taxes for treatment installation now shocked live test;

(7) It is necessary to initiate serious research ad practical work on utilization and purification of these waters as the present time they impact on the environment causing contamination of river (and drinking) water, saltinization of pastures, creation of salty sewage lakes, etc. The problem is related with solution of problem of preservation of drying of the Aral Sea as well.

For best execute all problems for hydrochemical of basin of Aral Sea and ecology his basin necessary indicate efforts of specialists, hero in republic of Central Asia , and foregone specialists.

References

Chembarisov E. I. Environmental and Rational Use of Water Resources of Aral Sea Basin: Geographic basin of Nature and Use in Uzbekistan, Tashkent.

Chembarisov E. I. Hydrology Problem of Ecology Conditions of Aral Sea: International Congress by Aral Sea. Indiana State University, Bloomington, Indiana, USA.

Chembarisov E. I. Hydrochemistry of River, Collector and Drainage Waters in the Aral Sea Bbasin: NATO ASI Series 2. Environment,1994(12):115 - 120.

Current Situation and Problem Discussion of Fishpass Facilities for Chinese Hydraulic and Hydropower Engineering

Zhuo Junling[1] and ***Huang Daoming***[2]

1. Environmental Engineering Assessment Center of Environmental Protection Department, Beijing, 100012, China
2. Ministry of Water Resources and Water Engineering Ecological Research Institute of Chinese Academy of Sciences, Wuhan, 430079, China

Abstract: through thorough review of design and construction, operation management, effect monitoring and assessment, technical research, establishment of legal system and some other aspects for the fishpass facilities in China, systematically analyze the existing outstanding problems for the hydraulic and hydropower projects of China, such as the recognition of fishpass facilities is not sufficient, the scientific and technological support is not enough and the operation and supervision is not proper, to provide reference for the sustainable development of hydraulic and hydropower engineering.
Key words: fishpass facility, development status and problem

As an important public beneficial and fundamental project, the hydraulic and hydropower project plays a significant role in the good and fast development of economy and society for China. With the promotion of social and environmental protection awareness, people's attention are increasingly paid to the ecological protection of rivers, especially the adverse environmental effect of hydraulic and hydropower construction on the connectivity of rivers. Studying and building fishpass facilities for hydraulic and hydropower engineering and promoting protection for biodiversity of rivers have become the outstanding environmental problems need to be solved for the development of hydraulic and hydropower engineering.

Fishpass facilities refer to facilities or technical measures which can artificially assist fish to pass barriers. Simple fishways are constructed in France since 17th century, as an important measure to relieve the obstruction of fish migration, the fishpass facilities have gained integrated development. Not only the traditional fishway construction craft and technology are increasingly mature, but also the technology and craft of fish lift, fish gate, fish collection and transportation boat and downstream fishpass facilities are gradually rebuilt and perfected. Especially the artificially natural bypass which has gained more attention in recent years, in addition to achieve fishpass of upstream and downstream multiple targets, it also has relatively good ecological gallery functions, which attracts the attention of river managers and hydroelectric development industry. At present, the fishpass facilities have become an important means to recover the connectivity of rivers, and develop from the recovery of single engineering construction or local connectivity to recovery of connectivity for river, river basin even multinational cooperated river basin. Such as France built or rebuilt 500 fish ways from 1984 to 2001 (Food and Agricultural Organization), England and Welsh have 380 fish ways, there are 240 fish ways in the east coast of Canada (Clay, 1995), Japan built about 10,000 fish ways (Nakamura & Yotsukura, 1987), America started national fish way plan in 1999, and the countries in the Rhine river basin stipulated and are implementing trout action plan.

The construction of fishpass facilities in China starts late, for a long time, there are disputes on fishpass effects, selection of fishpass facilities type, even on whether to construct fishpass facilities. In the past two decades, with the rapid development of hydraulic and hydropower industry in the country, the ecological and environmental problems caused by hydraulic and hydropower river way construction projects are gradually severe. In order to solve the problem of harm to biodiversity and aquatic animals' genes caused by the obstruction of hydraulic and hydropower engineering to rivers, exploring hydraulic and hydropower engineering fishpass approaches based on ecological protection, recovering the connectivity of rivers to reduce the adverse effect of hydraulic and hydropower engineering on migrating fish to the utmost, have become unavoidable environmental problems for hydraulic and hydropower engineering construction.

1 Development status of fishpass facilities

Since 1958, China have planned to develop Fuchunjiang Qililong Hydropower Station and put forward fish way construction for the first time, the development of fishpass facilities construction of the country is mainly divided into three stages. The first stage is the initial development stage for fishpass facilities, which is from 1958 to the early stage of 1980s, it mainly used technology and experience of foreign fish way construction for reference, and mainly focused on the construction of fish way in East China coastal plain gate dam with low water head. The second stage is lag phase of development, in the 20 years since the early stage of 1980s, the "Gezhou Dam Fish Rescue Discussion" adopted released reproduction instead of fishpass facilities, fishpass facilities were barely built, relevant research stopped, and already built fishpass facilities were out of service or abandoned, except for adding fish way at the Suifenhe downstream divergence head works barrage. The third stage is a new stage for construction and development for fishpass facilities, after entering 21 century, the importance of fishpass facilities gained unprecedented attention. In 2006, the Environmental Engineering Assessment Center of the former State Environmental Protection Administration printed and issued letter of "*Environmental Impact Assessment Technical Manual for River Ecological Water, Low - temperature Water and Fishpass Facilities of Hydraulic and Hydropower Engineering Projects* (*trial*)" (environment assessment letter (2006) No. 4), promoted the development of plan, design, construction and relative technical research for fishpass facilities.

1.1 Plan, design and construction of fishpass facilities

After the initial design and construction of fish way in the construction of hydropower station in 1958, the Heilongjiang newly opened fish way was constructed and operated in 1960. With the rapid construction and development of fishpass facilities after that, the number of constructed fish ways until the early stage of 1980s was 40 to 60. The construction of fishpass facilities in the first stage is relatively simple, they are almost fish ways for low water head, and most of them are at the low water head gate dam in the east coast and downstream flat areas of the Yangtze River, the base slops of which are relatively gentle, the lift height is not big, generally less than 10m. The fish way construction mainly adopts clapboard, mainly includes three kinds: submerging portholes (Jiangsu Tuanjie river dam fish way, Yangkou north gate fish way), perpends (Jiangsu Doulong Harbor fish way, Guazhou fish way) and combined type, no separate downflow weir fish way is constructed, but the constructed combined type fish ways have many combination methods, such as downflow weir and perpend (Jiangsu Taiping gate fish way), porthole and perpend (Jiangsu Liuhe fish way) and combination of porthole and downflow weir (Hunan Yangtang fish way). The Hunan Yangtang fish way constructed in 1980 adopted the combined type of porthole and downflow weir, arranged fish way inlet which can adapt to downstream water level, constructed fish luring and collecting system with water replenishing function. It is the fish way with the most advanced fishpass facilities construction technology, most complete functions, complete testing and monitoring materials and relatively good fishpass effects in the first stage in the country, it is even the best in the constructed fishpass facilities so far.

In the recent 10 years, with the strengthening of people's environmental protection awareness, it is a common sense that the hydraulic and hydropower engineering construction shall take necessary measures to relieve the influence of dam obstruction. Since the rebuild of Chaohu gate fish way in 2000, the Beijing Shangzhaung gate fish way was built in 2006, the first national large scale fish way—Changzhou fish way (with length of 1,500 m, width of 5 m and depth of 3 m) was built in 2007, the Tibet Shiquanhe hydropower station fish way with the highest elevation (sea level elevation of 4,300 m) was planned and constructed. So far the number of fishpass facilities of various kinds which are built, rebuilt or planned and designed is more than 30, although most of the types also belong to fish way, fishpass facilities of various kinds are studied and designed, such as artificially natural bypass (Sichuan Angu hydropower station), fish lift (Xinjiang) and fish collection and transportation system (Chongqing Pengshui hydropower station).

1.2 Supervision management and operating effects

Most of fishpass facilities initially constructed by China are not large, most fishpass facilities do not have observing rooms, and the investigation of fishpass effects are mainly fishery investigations or statistical analysis for fishery harvests. Fishpass facilities with relatively large scale are usually equipped with observing rooms, such as the Doulong Harbor engineering fish way and the Yangtang engineering fish way, the Yuxi gate fish way for Anhui Chaohu has two observing rooms, one upstream and the other downstream. For this kind of fish ways, observation was performed at regular basis during the initial period of operation, and fishpass effects assessment was performed combined with the statistical materials for fishery harvests. Seen from the results of initial operation supervision, the operation effects of these fishpass facilities are relatively good. For example, the number of observed sea eel fry was 210,000 in two hours on April 27th, 1967 in the Doulong Harbor fish way, the number of crab fry on July 1st, 1969 reached 900,000, correspondingly, the output of river crab increased by several times even twenty times (chart), and the quantity of coilia increased by ten times. After the construction of the Yangtang fish way, through 2 to 3 years of operation experience, the mean annual quantity of fishpass is about 580,000, the mean fishpass quantity during April to June in 1980 was 2,623/h, the total fishpass quantity in 145 d and 2,814.3 h from 1981 to 1982 was 1,280,000. The category of fish mainly includes silver chub, bream, grass carp, carp, black carp, chub, siniperca chuatsi, xenocypris with yellow tail, plagiognathops microlepis, bare eye trout, yellow check carp and other categories, and the maximum size of passed fish is 90 cm. Observation was performed for 384 h from April to May in 1974 in the Anhui Chaohu Yuxi gate fish way, the quantity of fish passed for two fish ways was 189,411, and the speed was 544.3/h, the category included coilia ectenes, eel, river crab and other kinds, and they were mainly coilia ectenes (182,682) and eel fry (5,149). According to the different observing results in big and small fish ways and upstream and downstream observing rooms, it was found that eels mainly sail upstream along small fish way, while coilia ectenes like to pass through big fish way. During the operation process, it was found that the coilia ectenes in the big fish way can not sail upstream to be above 38 clapboards during some periods, based on which the flow rate can be overcome by coilia ectenes of different specifications and flow rate change of all the clapboards in the fish way are analyzed, to perform improvement for fish way.

For most already built fish ways, the supervision and assessment for fishpass effect have not been performed, and the knowledge of fishpass effect is relatively less. Especially after 1980s, the design, construction and relative research of fishpass facilities were in dead state, most scientific research personnel switched to another field or were lost, except for few fishpass facilities which were operated freely, most of them were long been neglected and in disrepair, abandoned and destroyed. At present, China lacks thorough assessment for the actual state of built fishpass facilities.

Tab.1 The river crab quantity before and after construction of fish way for Doulong Harbor engineering

	Before fish way construction	After fish way construction		
Year	1966	1967	1968	1969
Output (kg)	722.5	3,569	16,227	12,183

Most of the planned and designed fishpass facilities in recent years have not been constructed and operated yet, and the effect assessment result of few built fishpass facilities is less. At present, only the Zhejiang Caoejiang gate fish way was performed fishery investigation, some were found to be river crossing fish, but for detailed fishpass effect, further investigation and assessment shall be performed.

1.3 Technical research and model experiment

The fishpass facility construction in China is developed on the basis of referring to and assimilating foreign technology and experience, the artificial degree is very high, especially for the deign and construction of fishpass facilities during the initial development period. However, in the process of introduction and digestion, basic theory and applied technology research are also carried out, such as fish migration action habit, flow tendency habit, stream overcome capacity, hydraulics feature in the fish way and design and construction technology for clapboard fish way, and model experiment of hydraulics in the fish way and swimming capability of fish. Especially during the process of the Gezhou Dam Fish Rescue Discussion, relevant units and research personnel systematically investigated technology and experience of foreign fishpass facilities´construction and operation, and carried out a lot of relative research and model experiments about Chinese sturgeon and the four major Chinese carps. For example, the experiment studied the suitable flow rate and the maximum flow rate of the four major Chinese carps and other fishes. Through hydraulics study and model experiment, according to the action characteristics of fish in the country, took improvement measures of setting up rest room to perpend type fish way, researched and designed "trapezoid – rectangular comprehensive section" combined clapboard. Performed prototype observation and research for the Yangang fish way, the Doulong Harbor and the Yuxi gate, assessed fishpass effect, and put forward improvement measures.

At present, the relative researches and experiments on fishpass facilities are relatively active. In recent years, on the basis of researching main river fish resources and distribution characteristics in the country, mainly implement research and experiment for Chinese sturgeon, round mouth corieus heterodun, procrypris rabaudi, schizothorax and other migrating fish and action habit and swimming capability of rare fish. Adopt methods of physical model and numerical simulation to study hydraulics characteristics and flow distribution of different fishpass facilities, to improve the design of fishpass facilities. For the fishpass problem of hydraulic and hydropower engineering high dam for the country, implement design and construction technology research for fish collection and transportation system, fish lift and combined fishpass facilities, and more than ten patents for invention have been obtained. In order to study the effect of water turbine in hydropower station on downstream fish, select the Gezhou Dam No. 2 water turbine to carry out standard fishpass test, which indicates, for fish with length of 20 cm to 50 cm downstream to swim over the dam, the survival rate within 24 h can reach above 89%.

1.4 Policies, regulations and specifications

The 27th provision of Water Law of the People's Republic of China specifies that "To build permanent gate dam over river in aquatic animals' migration path, the construction units shall also build fishpass facilities at the same time, or take other compensation measures after approval by department authorized by the State Council". The 32nd provision of Fishery Law of the People's Republic of China specifies that "To build dam or gate in the fish, shrimp and crab migration path, if they have severe influence on the fishery resources, the construction units shall build fishpass facilities or take other compensation measures". The Chinese Aquatic Biological Resource Maintenance Executive Summary issued by the State Council in 2006 also specifies that "By adopting water gate transformation, building fishpass facilities and implementing irrigation and seeding, recover the ecological relation of river fish, and maintain completeness of rivers and lakes water area ecology". In 2006, the Environmental Engineering Assessment Center of the former State Environmental Protection Administration printed and issued letter of "*Environmental Impact Assessment Technical Manual for River Ecological Water, Low – temperature Water and Fishpass Facilities of Hydraulic and Hydropower Engineering Projects* (*trial*)" (environment assessment letter (2006) No. 4), put forward detailed requirements for the planning, design and construction of fishpass facilities during the work of environmental impact assessment for hydraulic and hydropower engineering construction.

In addition, the Nanjing Water Conservancy Science Institute edited a book named "*Fish*

Way", and the Eastern China Water Conservancy Institute edited a book named "*Hydraulic Manual*", the contents of which all play an important part in promoting the healthy development of fishpass facilities in China. At present, on the basis of systematic summary of research practice of fishpass facilities for China, and referencing foreign technology and experience, the water conservancy department is organizing and editing "*Design Guidelines for Hydraulic and Hydropower Engineering Fishpass*", and the hydroelectric industry is editing "*Design Specifications for Hydropower Engineering Fishpass Facilities*".

2 Main problems existing in the construction management of fishpass facilities

2.1 Insufficient knowledge of fishpass facilities construction

2.1.1 Insufficient knowledge of fishpass facilities construction necessity

Review the development progress of fishpass facilities in China, from the introduction in the 1950s and subsequent development, stagnation after the early stage of 1980s for more than 20 years and the rapid development in recent years, the continuous recognition and repeated discussion of the necessity of fishpass facility construction run through the whole period. So to speak, the recognition of necessity of fishpass facility construction is the key factor influencing its development. The most typical case is the Gezhou Dam Fishpass Discussion in the early 1980s, which resulted in the stagnation of more than 20 years for the development of fishpass facilities in the country. Actually, the connectivity of river is the most fundamental features for the river ecosystem, and is an important precondition for river to play its ecological function. Damming across river, rivers´and lakes' obstruction and other human activities severely destroy the connectivity of river, which is an important reason for ecological problems in the rivers and lakes of the country. Therefore, recovering the connectivity of rivers has become the basis for reducing the effect of hydraulic and hydropower engineering on ecological environment, maintaining health of rivers and protecting ecological environment. As an important measure to recover the connectivity of rivers, the importance of fishpass facilities has been acknowledged widely. Especially the maturity and application of artificially natural bypass technology, which strengthens the ecological corridor function of fishpass facilities. The insufficient awareness of the necessity of fishpass facilities construction is not denial to the function of fishpass facilities in the recovery of river connectivity, but mainly is determination for selection of subjects must pass dam, water area and dam must pass fish and doubt for fishpass effect. For example, which kind of fish must pass dam and whether all the dams need to build fishpass facilities. Especially for cascade developed water area, as the habitat change is relatively severe after cascade development, the original perch and reproduction habitats for fish are lost, and some experts doubt the fish passing purposes. The fish pass in high dam is a world wide problem, the constructed fishpass facilities in China are mostly low dams, the referenced experience of passing fish at high dam is rare, and the main subject for fishpass is cyprinid fish which have limited flow overcome capabilities, some experts doubt the effect of building fishpass facilities in high dam. The discrepancies of these recognitions or academic discussions are very normal, in a sense, these shall be beneficial to the research and practice for fishpass facilities. The international recognition of necessity for fishpass facilities construction is continuously expanded and deepened with strengthen of environmental protection awareness and deepen of scientific research. However, during the development process of national fishpass facilities, the divergence of these opinions caused stagnation of development, even retrogress. For example, the "Gezhou Dam Fish Rescue Discussion" was originally discussion about fishpass object and special project to rescue fish, but it resulted in the overall stagnation of fishpass facilities construction and research for more than 20 years. Therefore, promoting the recognition of necessity of fishpass facilities construction is the key to the healthy development for fishpass facilities.

2.1.2 Insufficient knowledge of complexity of fishpass facilities construction

Fishpass facilities are engineering technological measures built for fish to pass through dam, which need to meet the basic requirements for the action habits and physiological functions of fish.

Therefore, the planning, design, construction and operation of fishpass facilities involve fish ecology, fish ethology, hydraulics, hydrotechnics and other subjects, which requires the cooperation of fish biology and engineering technical personnel, to design and construct effective fishpass facilities. The categories and quantity of fish in China are large, the ecological habits are various, the span of longitudinal and latitudinal space is big, and the difference of river upstream and downstream ecological environment is obvious, the early stage work load is big and requirement is high, such as fundamental investigation and research and model experiment for the construction of fishpass facilities. In addition, it shall be continuously improved and perfected by testing, monitoring and assessment for fishpass effect during the process of operation and management, to achieve ideal effect. By simply simulating the construction technology for fishpass facilities of other countries, even by planning, designing and constructing fishpass facilities from experience, the fishpass effect can not be guaranteed. The early constructed fish ways in the country were started from simulating fish way technology from occident, most of them had a bad fishpass effect. The main reason for that is the insufficient fundamental investigation work, and the fishpass facilities were not designed and constructed based on the ecological habits and action features of fish in China. For example, the rebuilding and improvement of the Chaohu Yuxi gate fish way and the Qinghaihu fish way achieved good effect. This kind of example also exists abroad, for example, the early stage fish ways in Australia were mainly constructed by simulating western technology, they are under large scale rebuild in recent years.

2.2 Insufficient scientific and technical support

The construction of fishpass facilities in China is relatively late, and lag phase of more than 20 years occurred in the development process, the relevant fundamental research is very weak, and it is difficult to support the healthy development of fishpass facilities in the country.

There are 3,800 kinds of fish recorded in the country, the fish are in a great variety, the ecological types are various. Cyprinid fish takes an absolute advantage in the inland waters, which mainly are rivers and lakes migration types, the migration habits and countercurrent capacity of which are not stronger than that of trout, and the national reference of fishpass technology for trout are limited. Except for the ethological study for conventional fish (for example the four major Chinese carps), the knowledge of action learn ability and swimming capability of most rare fish is few.

Construction for fishpass facilities is a technology with very strong experience, the domestic fishpass facilities constructed and operated are just few. The constructed fishpass facilities mainly are low water head fish ways, the type and structure of which are relatively simple. The artificial natural bypass, fish lift and fish collection and transportation system are under research and design or construction, and most of the constructed fishpass facilities are under abandoned state, the results of monitoring test and effect assessment are few, and the experimental experience which can be referred to are few. In recent years, relative technical research and model experiment attracted certain attention, but the time is short, the research effort is weak, and the attention is insufficient, it is hard to meet the requirement for the rapid development of fishpass facilities construction.

2.3 Improper later stage supervision

2.3.1 Ambiguity of responsibility main body for operation management

As an environmental protection project which can relieve the effect of hydraulic and hydropower engineering on aquatic ecology, the later stage monitoring of fishpass facilities involves water conservancy, hydroelectric, environmental protection, fishery and other departments. According to the environmental protection principle of who destroys, who recovers, the hydraulic and hydropower engineering construction units surly are the responsibility main body for operation and management. However, fishpass does not belong to the developing task of most hydraulic and hydropower engineerings, hydraulic and hydropower departments do not have definite management responsibilities, they do not have professional administration organization and professional technical personnel, not only effective management can not be realized, but also the fish ways can become sacrifice when the monitoring is not proper and the usage is in conflict with the engineering task. While the fishery and environmental protection functional departments can not participate in the key management and monitoring, and can not execute industrial management responsibility, which result in the misplacement and loss of management.

2.3.2 Imperfect compensation mechanism

Currently, there is no perfect mechanism for ecological compensation constructed in China, as ecological environmental protection project, for the corresponding operation and management fee for fishpass facilities, there are relatively large differences among the process methods of various industries. For hydropower station, the operation and management fee for fishpass facilities is included in the operating cost of power station, and power loss shall be born by enterprises. As the operation, management and maintenance of fishpass facilities are not related with the responsibility of enterprise, and there is electric quantity loss during the operation process, enterprises do not have enthusiasm for the operation and management of fishpass facilities. The construction fee of fishpass facilities for key water control project belongs to state investment, but fishpass does not belong to the developing task of engineering, and the operation and management fee can only be born by the local government or operation and management units. While most of the key water control projects are public, the economic benefit is limited, and the operation and management funds can not be implemented, even become the burden of water management units, most of the constructed fishpass facilities are abandoned, and even destroyed.

2.3.3 Deficiency of professional and technical personnel

At present, the fishpass facilities are mainly operated and managed by hydraulic and hydropower management units, which actually do not belong to the main management scope, there are no management mechanism and formation, and no corresponding professional technical management personnel. And the relative technical research for fishpass facilities in China is weak, the effective operated and managed fishpass facilities are few, and the cultivation capacities for professional technical personnel are insufficient. While the technical requirements for monitoring and assessment of operation and fishpass effect, the maintenance and improvement of fishpass facilities are very high, and the professional technical personnel are insufficient, which severely limit the operation and management of fishpass facilities.

Management Information System for Pollution Control in Taihu Watershed—A Case Study of Changzhou Municipality and Wuxi City

Pang Yong[1,2] and *Xie Rongrong*[2]

1. Key Laboratory of Integrated Regulation and Resource Development on Shallow Lakes of Ministry of Education, Hohai University, Nanjing, 210098, China
2. College of Environment, Hohai University, Nanjing, 210098, China

Abstract: Taihu Watershed Management Information System (TWMIS) for pollution control, which is based on Global Information System (GIS) and water models, is an intelligent information system which can calculate water environmental capacity and allocate pollution loads spatially and temporally in Taihu watershed. Water environmental capacity here is total maximum pollution loads in certain rivers, which concerns with the aquatic system features, water function, hydrological data, correlation coefficient and the reciprocating of the river and is computed on the base of 1 - D unsteady mathematical model in Taihu watershed. Afterwards, Total Maximum Daily Loads (TMDL) technique is introduced to allocate water environmental capacity. TWMIS integrates the water environmental capacity model to GIS platform, together with administrative boundaries, hydrographic net and pollution loads information, then an acceptable discharge level of ambient quality standards is limited. Thus, TWMIS is a powerful watershed pollution management tool for pollution discharge, which provides technique support for pollution control and scientific management.

Key words: Taihu watershed, water environmental capacity, GIS technology, basic geographical information, management information system

1 Introduction

For ages, people have used rivers as refuse sites to dispose of the waste which they have generated. As the environmental pollution problems have been gaining importance worldwide, environmental capacity, a more popular saying in Europe and USA as assimilation capacity or maximum allowed discharge, is raised. Nonetheless, environmental capacity is a hot issue currently.

The concept of environmental capacity was described firstly in Japan in 1970s for the pollution control in atmosphere, since then, many versions of environmental capacity definition have been proposed by scholars at home and aboard (Li et al., 2007; Xia, 1996), however in this paper, environmental capacity is defined as pollutant - carrying capacity which is acceptable maximum level of ambient quality standard. Water environmental capacity, a branch of environmental capacity (GESAMP, 1986), is total maximum pollution loads in certain rivers, which concerns with the aquatic system features, water function, hydrological data, correlation coefficient and the reciprocating of the river. In the area of water environmental capacity, water quality mathematical model is a core element. Up till now, a considerable number of water quality models have appeared, such as Streeter - Phelps, Thomas, Camp - Dobblins, QUAL - II. etc (Bowles and Greeney, 1979, Zielinski, 1988; Barnwell et al., 2004). Over the past few years, plenty of researches have been carried out, with regards to water environment capacity, are to control total amount to achieve in advance deciding environment objective (Liang and Wang, 2005). Meanwhile, calculation method of the water environment capacity has been furthered (Pang and Lu, 2010). Nowadays, allocation of pollution loads spatially and temporal is a research focus, however, Total Maximum Daily Loads (TMDL) technology, is the earliest example to allocate the daily loads on the perspective of whole watershed, enacted in the federal clean water act of United States in 1972. The TMDL establishes the allowable loadings for specific pollutants that a water body can receive without exceeding water quality standards thereby providing the basis to establish water quality based controls (USEPA, 2001; NRC, 2001). However, in China, it is still a start which is limited by the monitoring system and the management dispute in watershed level, meanwhile, an affective watershed management information system for pollution control is to be in highly demanded by environmental managers.

In this paper, we aims to resolve the problems for pollution control in Taihu watershed, Changzhou Municipality and Yixing City, located at the upstream of Taihu watershed, is defined as a typical demonstration area here. With two methods, river environmental capacity is calculated; the smaller value of the two is the environmental capacity in the demonstration area. According the spatial allocation principle, the environmental capacity is allocated to each control unit. Then based on the temporal allocation principle, Total Maximum Monthly Loads (TMML) of each calculating unit are obtained. Finally Taihu Watershed Management Information System (TWMIS) for pollution control is established, which is a powerful watershed pollution management tool for pollution discharge, which provides technique support for pollution control and scientific management.

2 Research area

Taihu watershed, is located in the lower part of the Yangtze River Delta, has a complicated river network system with a river density of 3.24 km/km^2 (Qin et al., 2007). In Taihu Basin, the greatest amount of rainfall occurs in summer and the smallest in spring. Land use in the basin consisted of farmland (60%), forest (14%), urban area (13%), and water bodies (13%) in 2000. However, in recent years, with farmland use decreasing from about 60 to 51% between 2000 and 2008, and urban land use increasing from about 13% to 22% (Xu et al., 2009).

Demonstration area, includes Changzhou Municipality and Yixing City, is located at the northwest of Taihu Lake and occupies a total area of 6,423 km2 (Fig. 1). It is rapidly urbanizing area. For the purpose to determine pollution responsibility spatially, control unit is defined in this study. Control unit is the spatial range affected by the pollutions, which is the basic unit for pollution control. According to the basis of water ecology function, integrity of hydrographic net, feasibility of management and implementation and controlling point of the rivers, 13 control units in Changzhou Municipality and 6 control units in Yixing City were settled in this study(Fig. 1).

3 Methodology

3.1 Water environmental capacity model

In the calculation of water environmental capacity, considering the pollutant source location is the key problems, so two models for water environmental capacity were derived in this study. The pollutant source location was not included in 0 - D model, while which was consider in 1 - D model. Finally, the smaller value of the two is the water environment capacity for demonstration area.

3.1.1 0 - D water environmental capacity model setup

(1) Basic water environmental capacity model.

0 - D water environmental capacity model is a calculating method based on 0 - D water quality mathematical model, without considering the location of the water pollution, which is as follows:

$$W = \sum_{j=1}^{n} \sum_{i=1}^{m} \alpha_{ij} \times W_{ij} \quad (1)$$

$$W_{ij} = Q_{0ij}(C_{sij} - C_{0ij}) + KV_{ij}C_{sij} \quad (2)$$

Where, W is the water environmental capacity of trace compounds; W_{ij} is water environmental capacity of trace compounds in minimum time and space interval, the minimum time interval is a day and the minimum space interval is the distance between two calculation points; n is the number of time interval; m is the number of space interval; Q_{0ij} and V_{ij} are designed hydrological conditions which are derived from water quality mathematical model; C_{sij} is target water quality for water use purpose of trace compounds; C_{0ij} is the water quality of trace compounds in upstream; K is degradation coefficient of trace compounds; α_{ij} is non - uniform coefficient, $\alpha_{ij} \in (0, 1]$.

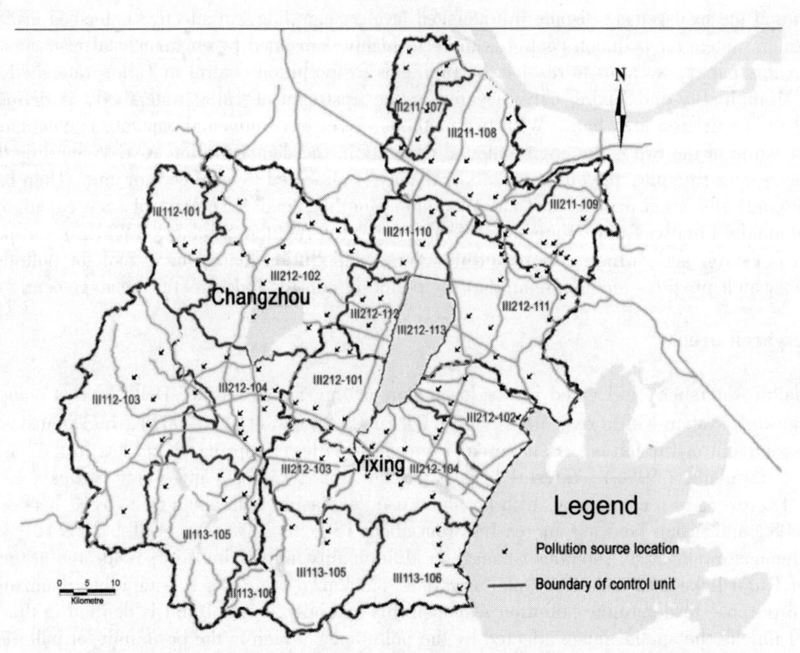

Fig. 1 Demonstration area (Changzhou Municipality and Yixing City) in Taihu watershed and control units in the demonstration area

In Taihu watershed, most rivers are reciprocating flow rivers, hence, the effect should be considered as:

$$W = \frac{A}{A + B} W_P + \frac{B}{A + B} W_N \tag{3}$$

$$W_P = Q_{01}(C_s - C_{01}) + KVC_s \tag{4}$$

$$W_N = Q_{02}(C_s - C_{02}) + KVC_s \tag{5}$$

Where, A is the days of positive flow; B is the days of negative flow; W_P is the water environmental capacity of positive flow; W_N is the water environmental capacity of negative flow.

(2) Parameters preparation.

0 - D model is considered that the pollution for the rivers mixed completely, however, it is ideal condition and in fact the mixed degree is affected by the volume of water body. Hence, non - uniform coefficient is raised to redress the result. Generally, there is a strong negative correlation between non - uniform coefficient and water surface area. Non - uniform coefficient for rivers and lakes are presented in Tab. 1 and Tab. 2.

Tab. 1 Non - uniform coefficient for rivers

River width (m)	Non - uniform coefficient	River width (m)	Non - uniform coefficient
<30	0.7 ~ 1.0	200 ~ 500	0.3 ~ 0.4
30 ~ 100	0.5 ~ 0.7	500 ~ 800	0.3
100 ~ 200	0.4 ~ 0.6	>800	0.1 ~ 0.3

Tab. 2 Non – uniform coefficient for lakes

Water surface area (km^2)	Non – uniform coefficient
≤5.0	0.6 ~ 1.0
5 ~ 50	0.4 ~ 0.6
50 ~ 500	0.11 ~ 0.4
500 ~ 1,000	0.09 ~ 0.11
1,000 ~ 3,000	0.05 ~ 0.09

3.1.2 1 – D water environmental capacity model setup

(1) Model setup.

1 – D water environmental capacity model is a calculating method based on 1 – D model, with considering the location of the water pollution. Typical control section is polluted water section which is located at close to the entrance of Taihu Lake. Afterwards, the relationship between source of pollutants and water quality conditions of the control section is established. Finally, total amount of pollution is the water environmental capacity of the calculating units.

1 – D steady model of water quality for controlling section is as follows:

$$C = \frac{W + Q_0 C_0}{Q_0} \exp\left(- K \frac{x}{86,400u} \right) \tag{6}$$

Where, C is the concentration of trace compounds in control section; C_0 is the concentration of trace compounds in upstream; Q_0 is the water volume of upstream; x is the distance between the location of the control section and pollutant source; u is velocity of the flow.

(2) Parameters preparation.

The degradation coefficient of trace compounds is fixed according to the mathematic model in Taihu watershed, namely degradation coefficient of chemical oxygen demand (COD) is 0.09 ~ 0.15 d^{-1} and degradation coefficient of ammonia nitrogen ($NH_4 - N$), total nitrogen (TN) and total phosphorus (TP) are 0.05 ~ 0.08 d^{-1}, 0.05 ~ 0.12 d^{-1} and 0.04 ~ 0.05 d^{-1}, respectively.

3.2 Spatial allocation for pollution loads

Spatial allocation for pollution loads is to allocate the water environmental capacity to each control unit according to the realistic river length, water area and the target water quality for the ecological function.

$$W_j = \alpha_j \times W \tag{7}$$

$$\alpha_j = \frac{\sum_{i=1}^{5} \left(S_j \times \frac{l_{ij}}{\sum_{i=1}^{5} l_{ij}} \times C_{si} \right)}{\sum_{i=1}^{5} \left(S \times \frac{l_i}{\sum_{i=1}^{5} l_i} \times C_{si} \right)} \tag{8}$$

Where, W_j is the water environmental capacity in each control unit; W is total water environmental capacity in the demonstration area; α_j is weight coefficients for district distribution; C_{si} is the target water quality of trace compounds; S_j is the water area in each controlling unit; l_{ij} is river length in the unit; S is the sum of water areas in all control units; l_i is the sum of river length in all control units; j is the number of controlling units.

3.3 Temporal allocation for pollution loads

Temporal allocation for pollution loads is to allocate the water environmental capacity to each month in the light of research on TMDL. Hence, Total Maximum Monthly Loads (TMML) is obtained in this study, which is to allocate the environmental capacity of each calculating unit monthly.

$$W_i = \lambda_i \times W \tag{9}$$

$$\lambda_i = \frac{A_i}{A} \tag{10}$$

Where, W_i is TMML in each control unit; W is water environmental capacity in each control unit; λ_i is weight coefficients for monthly distribution, which is decided by the rainfall and discharge; A_i is the monthly hydrological index, generally it is monthly rainfall, if there is no rainfall information for control unit, monthly water volume is adopted; A is annual hydrological index correspondingly.

4 Results

According to Eq. (1) ~ Eq. (6), the environmental capacity is calculated, and then the pollution control is allocated spatially and temporally. As presented in Tab. 3 to Tab. 6, The total pollution control of COD is 58,894.2 t/a, with NH_4^+-N, TN and TP are 3,808 t/a, 6,054.6 t/a and 386.6 t/a, respectively.

Tab. 3 Pollution control of COD spatially and temporally in demonstration area

(Unit: t)

Code	Jan.	Feb.	Mar.	Apr.	May	Jun.	Jul.	Aug.	Sep.	Otc.	Nov.	Dec.	Sum.
Ⅲ112 - 103	432.8	110.2	104.8	140.9	292.7	838.4	253.7	250.6	273.2	592.6	68.2	83.5	3,441.6
Ⅲ212 - 104	192.2	80.2	94.2	94.5	187.3	470.5	270.3	211	200.4	369.1	91.7	63.3	2,324.7
Ⅲ113 - 105	492.4	191.9	178.9	210.6	282.5	530.9	217.1	220.4	284.5	450.7	157.5	174.7	3,392.1
Ⅲ113 - 106	68.7	62	63.7	67.6	70.1	73.2	77.7	74.7	68.6	70.9	72.7	75.3	845.2
Ⅲ112 - 101	200.9	111.8	105.4	116.4	193.4	354.5	157.5	121.7	155.2	211.6	90.4	99.5	1,918.3
Ⅱ212 - 102	369.6	247.2	292.1	293.9	775.9	1,165.6	836.9	725.7	855.5	902.5	217.9	198.2	6,881
Ⅲ211 - 110	135.4	104.6	124.1	127.8	364	560.9	412.2	334	414.5	436.5	99.6	83	3,196.6
Ⅲ212 - 113	315.8	217.5	239.3	224	535	1152.3	667.5	438.1	585.7	920.5	300.3	172	5,768
Ⅲ212 - 112	42.7	22.8	24	23.9	55.9	134	72.7	47.7	62.4	110.2	31.2	17.4	644.9
Ⅲ211 - 107	41.6	51.3	71.1	98.5	71.5	182.2	209.1	136.2	156.8	109.4	13.1	35	1,175.8
Ⅲ211 - 108	87.5	81.4	100.5	119.1	266.7	335.6	333.9	302.4	317.4	281.4	72.5	63.2	2,361.6
Ⅲ211 - 109	89.1	77.9	76.6	89.2	313.9	289.5	298.7	407.9	370.9	244.6	103.1	59.4	2,420.8
Ⅲ212 - 111	259.2	224.4	299.2	283.7	846.7	1,102	837.6	853.2	946	838.3	280.9	196	6,967.2
Ⅲ212 - 102	66.6	53.6	59.4	51.2	95.4	197.9	139	78.9	102.3	170	70.9	48.4	1,133.6
Ⅲ212 - 104	330.7	200.6	220.3	181.4	386.6	981.8	599.5	252.1	425.5	852.6	289.2	168.7	4,889
Ⅲ212 - 101	274.1	113	107	128.5	234.1	763.6	366.1	307.4	285.1	632.2	152.4	88.7	3,452.2
Ⅲ212 - 103	449.9	185.3	186.3	188.9	371.4	1,035.8	458.7	294.6	413.9	914.4	238.9	145.4	4,883.5
Ⅲ113 - 105	247.7	40.2	28.4	51.1	115.2	295.8	91.2	46.9	143.5	292.9	24.3	24.2	1,401.4
Ⅲ113 - 106	160.1	73.8	66	74.8	135.3	420	132.8	76.5	196.1	351.9	57	52.4	1,796.7
Sum.	4,257	2,249.7	2,441.3	2,566	5,593.6	10,884.5	6,432.2	5,180	6,257.5	8,752.3	2,431.8	1,848.3	58,894.2

Tab. 4 Pollution control of NH_4^+ – N spatially and temporally in demonstration area (Unit: t)

Code	Jan.	Feb.	Mar.	Apr.	May	Jun.	Jul.	Aug.	Sep.	Otc.	Nov.	Dec.	Sum.
Ⅲ112 – 103	31.4	8	7.6	10.2	21.3	60.9	18.4	18.2	19.9	43.1	5	6	250
Ⅲ212 – 104	15	6.3	7.4	7.4	14.6	36.8	21.1	16.5	15.7	28.9	7.2	4.9	181.8
Ⅲ113 – 105	24.5	9.6	8.9	10.5	14.1	26.4	10.8	11	14.2	22.4	7.8	8.6	168.8
Ⅲ113 – 106	3.3	3	3.1	3.3	3.4	3.5	3.8	3.6	3.3	3.4	3.5	3.7	40.9
Ⅲ112 – 101	12.5	6.9	6.5	7.2	12	22	9.8	7.6	9.6	13.1	5.6	6.1	118.9
Ⅱ212 – 102	26.9	18	21.2	21.4	56.4	84.7	60.8	52.7	62.2	65.6	15.8	14.2	499.9
Ⅲ211 – 110	9.8	7.6	9	9.3	26.4	40.7	29.9	24.3	30.1	31.7	7.2	6.2	232.2
Ⅲ212 – 113	22.9	15.8	17.4	16.3	38.9	83.7	48.5	31.8	42.5	66.9	21.8	12.5	419
Ⅲ212 – 112	3.1	1.7	1.7	1.7	4.1	9.7	5.3	3.5	4.5	8	2.3	1.2	46.8
Ⅲ211 – 107	3	3.7	5.2	7.2	5.2	13.2	15.2	9.9	11.4	8	1	2.4	85.4
Ⅲ211 – 108	6.2	5.7	7.1	8.4	18.7	23.6	23.5	21.2	22.3	19.8	5.1	4.2	165.8
Ⅲ211 – 109	6.5	5.7	5.6	6.5	22.8	21	21.7	29.6	26.9	17.8	7.5	4.3	175.9
Ⅲ212 – 111	18.8	16.3	21.7	20.6	61.5	80.1	60.8	62	68.7	60.9	20.4	14.3	506.1
Ⅲ212 – 102	4.1	3.3	3.6	3.2	5.8	12.1	8.5	4.8	6.3	10.5	4.4	3	69.6
Ⅲ212 – 104	17.2	10.5	11.5	9.5	20.1	51.2	31.2	13.1	22.2	44.4	15.1	8.7	254.7
Ⅲ212 – 101	14.3	5.9	5.6	6.7	12.2	39.8	19.1	16	14.9	32.9	7.9	4.6	179.9
Ⅲ212 – 103	23.4	9.7	9.7	9.8	19.4	54	23.9	15.4	21.6	47.6	12.5	7.4	254.4
Ⅲ113 – 105	11.4	1.8	1.3	2.3	5.3	13.6	4.2	2.2	6.6	13.4	1.1	1.1	64.3
Ⅲ113 – 106	8.3	3.8	3.4	3.9	7.1	21.9	6.9	4	10.2	18.3	3	2.8	93.6
Sum.	262.6	143.3	157.5	165.4	369.3	698.9	423.4	347.4	413.1	556.7	154.2	116.2	3,808

Tab. 5 Pollution control of TN spatially and temporally in demonstration area (Unit: t)

Code	Jan.	Feb.	Mar.	Apr.	May	Jun.	Jul.	Aug.	Sep.	Otc.	Nov.	Dec.	Sum.
Ⅲ112 – 103	51.8	13.2	12.6	16.9	35.1	100.4	30.4	30	32.7	71	8.2	9.8	412.1
Ⅲ212 – 104	25.4	10.6	12.4	12.5	24.8	62.2	35.7	27.9	26.5	48.8	12.1	8.6	307.5
Ⅲ113 – 105	82.6	32.2	30	35.3	47.4	89.1	36.4	37	47.7	75.6	26.4	29.4	569.1
Ⅲ113 – 106	11.7	10.5	10.8	11.5	11.9	12.4	13.2	12.7	11.7	12.1	12.4	12.9	143.8
Ⅲ112 – 101	26.6	14.8	14	15.4	25.6	47	20.9	16.1	20.6	28	12	13.2	254.2
Ⅱ212 – 102	36.7	24.5	29	29.2	77	115.7	83.1	72	84.9	89.6	21.6	19.7	683
Ⅲ211 – 110	11.7	9.1	10.7	11.1	31.5	48.5	35.7	28.9	35.9	37.8	8.6	7.1	276.6
Ⅲ212 – 113	35.3	24.3	26.8	25.1	59.8	128.9	74.6	49	65.5	102.9	33.6	19.2	645
Ⅲ212 – 112	4.3	2.3	2.4	2.4	5.7	13.6	7.4	4.8	6.3	11.2	3.2	1.8	65.4
Ⅲ211 – 107	4.5	5.5	7.6	10.6	7.7	19.5	22.4	14.6	16.8	11.7	1.4	3.6	125.9
Ⅲ211 – 108	8.3	7.7	9.5	11.2	25.2	31.7	31.5	28.6	30	26.6	6.9	5.8	223
Ⅲ211 – 109	7.8	6.9	6.7	7.8	27.6	25.5	26.3	35.9	32.6	21.5	9.1	5.1	212.8
Ⅲ212 – 111	23	19.9	26.6	25.2	75.2	97.9	74.4	75.8	84.1	74.5	25	17.5	619.1
Ⅲ212 – 102	6.8	5.5	6.1	5.3	9.8	20.3	14.3	8.1	10.4	17.4	7.2	5.6	116.8
Ⅲ212 – 104	26.2	15.9	17.5	14.4	30.6	77.8	47.5	20	33.7	67.5	22.9	13.3	387.3
Ⅲ212 – 101	22.1	9.1	8.6	10.3	18.8	61.4	29.5	24.7	22.9	50.9	12.3	7.1	277.7
Ⅲ212 – 103	37.9	15.6	15.7	15.9	31.3	87.2	38.6	24.8	34.9	77	20.1	12.2	411.2
Ⅲ113 – 105	26.3	4.3	3	5.4	12.2	31.4	9.7	5	15.2	31.1	2.6	2.5	148.7
Ⅲ113 – 106	15.6	7.2	6.4	7.3	13.2	41	13	7.5	19.1	34.4	5.6	5.1	175.4
Sum.	464.6	239.1	256.4	272.8	570.4	1 111.5	644.6	523.4	631.5	889.6	251.2	199.5	6,054.6

Tab. 6 Pollution control of TP spatially and temporally in demonstration area (Unit: t)

Code	Jan.	Feb.	Mar.	Apr.	May	Jun.	Jul.	Aug.	Sep.	Otc.	Nov.	Dec.	Sum.
Code	Jan	Feb	Mar	Apr	May	Jun	Jul	Aug	Sep	Otc	Nov	Dec	Sum
Ⅲ112-103	3.14	0.8	0.76	1.02	2.13	6.09	1.84	1.82	1.98	4.31	0.5	0.61	25
Ⅲ212-104	1.72	0.72	0.84	0.84	1.68	4.21	2.42	1.89	1.79	3.31	0.82	0.56	20.8
Ⅲ113-105	2.45	0.95	0.89	1.05	1.41	2.64	1.08	1.1	1.42	2.24	0.78	0.89	16.9
Ⅲ113-106	0.33	0.3	0.31	0.33	0.34	0.35	0.38	0.36	0.33	0.34	0.35	0.38	4.1
Ⅲ112-101	1.25	0.69	0.65	0.72	1.2	2.2	0.98	0.75	0.96	1.31	0.56	0.63	11.9
Ⅱ212-102	2.69	1.8	2.12	2.13	5.64	8.47	6.08	5.27	6.21	6.56	1.58	1.45	50
Ⅲ211-110	0.98	0.76	0.9	0.93	2.64	4.07	2.99	2.43	3.01	3.17	0.72	0.6	23.2
Ⅲ212-113	2.29	1.58	1.74	1.63	3.89	8.37	4.85	3.18	4.25	6.69	2.18	1.25	41.9
Ⅲ212-112	0.31	0.17	0.17	0.17	0.41	0.97	0.53	0.35	0.45	0.8	0.23	0.14	4.7
Ⅲ211-107	0.3	0.37	0.52	0.72	0.52	1.32	1.52	0.99	1.14	0.79	0.1	0.21	8.5
Ⅲ211-108	0.61	0.57	0.71	0.84	1.87	2.36	2.35	2.12	2.23	1.98	0.51	0.45	16.6
Ⅲ211-109	0.65	0.57	0.56	0.65	2.28	2.1	2.17	2.96	2.69	1.78	0.75	0.44	17.6
Ⅲ212-111	1.88	1.63	2.17	2.06	6.15	8.01	6.08	6.2	6.87	6.09	2.04	1.42	50.6
Ⅲ212-102	0.59	0.48	0.53	0.46	0.85	1.76	1.24	0.7	0.91	1.51	0.64	0.43	10.1
Ⅲ212-104	1.72	1.05	1.15	0.94	2.01	5.12	3.12	1.31	2.22	4.44	1.51	0.91	25.5
Ⅲ212-101	1.43	0.59	0.56	0.67	1.22	3.98	1.91	1.6	1.49	3.29	0.79	0.47	18
Ⅲ212-103	2.34	0.97	0.97	0.98	1.94	5.4	2.39	1.53	2.16	4.76	1.24	0.72	25.4
Ⅲ113-105	1.14	0.18	0.13	0.23	0.53	1.36	0.42	0.22	0.66	1.34	0.11	0.08	6.4
Ⅲ113-106	0.83	0.38	0.34	0.39	0.71	2.19	0.69	0.4	1.02	1.83	0.3	0.32	9.4
Sum.	26.65	14.56	16.02	16.76	37.42	70.97	43.04	35.18	41.79	56.54	15.71	11.96	386.6

5 Management information system for pollution control

On the rationale for water environmental capacity, Taihu Watershed Management Information System (TWMIS) for pollution control is established. The Framework of TWMIS (Lv, 2007), includes four modules namely query module, calculating module of water environmental capacity, allocation module of water environmental capacity and statistical analysis module, is presented in Fig. 2.

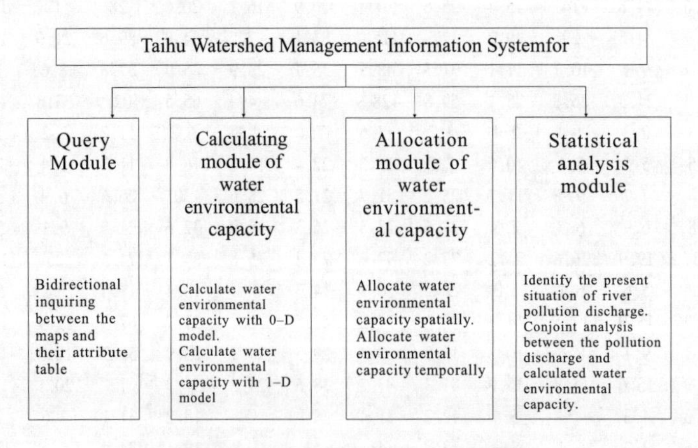

Fig. 2 The framework of TWMIS

(1) Query module: bidirectional inquiring between the maps and their attribute table;

(2) Calculating module of water environmental capacity: calculate water environmental capacity with 0 – D model and 1 – D model;

(3) Allocation module of water environmental capacity: allocate water environmental capacity spatially and temporally;

(4) Statistical analysis module: identify the present situation of river pollution discharge and conjoint analysis between the pollution discharge and calculated water environmental capacity.

6 Concluding remarks

This paper develops an allocation procedure of environmental capacity to highlight the spatial and temporal pollution control. The main contribution of the paper is to establish TWMIS based on the rationale of water environmental capacity. TWMIS integrates the water environmental capacity model to GIS platform, together with administrative boundaries, hydrographic net and pollution loads information, then an acceptable discharge level of ambient quality standards is limited. Thus, TWMIS is a powerful watershed pollution management tool for pollution discharge, which provides technique support for pollution control and scientific management.

Acknowledgements

Part of this work was supported by National Water Special Project of China (2012ZX07506 – 006 & 2012ZX07506 – 007) and National Natural Science Foundation of China (51179053).

References

Barnwell Jr., Thomas O, Brown, et al. Importance of Field Data in Stream Water Quality Modeling Using QUAL2E – UNCAS[J]. Journal of Environmental Engineering, 2004, 103(6): 643 – 647.

Bowles D S, Greeney W J. Steady State River Quality Modeling by Sequential Extended Kaman Filters[J]. Water Resources Research, 1978, 14(1):84 – 86.

GESAMP. Environmental Capacity. An Approach to Marine Pollution Prevention[R]. Rome: Food and Agriculture Organization of the United Nations, 1986.

Liang B, Wang X Y. The Current Situation and Expectation of Our Country Water Environment Contamination Gross Control[J]. Journal of Capital Normal University (Natural Science Edition), 2005, 26(1): 93 – 97.

Li S Q, Li X L, Wu Y C, et al. Actuality and Prospect on Water Environment Capacity Research [J]. Journal of Architectural Education in Institutionsof Higher Learning, 2007, 16(3): 58 – 60.

Lv J. Calculate the Water Environmental Capacity of Surface Water in Hangzhou Base on GIS, Jiangsu Province[D]. Nanjing: Hohai University, 2007.

National Research Council. Assessing the TMDL Approach to Water Quality Management[R]. Committee to Assess the Scientific Basis of the Total Maximum Daily Load Approach to Water Pollution Reduction, Water Science and Technology Board, 2001.

Pang Y, Lu G H. Theory and Research of Water Environmental Capacity[M]. Beijing: Science Press, 2010.

Qin B, Liu Z, Havens K, et al. Eutrophication of Shallow Lakes With Special Reference to Lake Taihu, China[R]. Springer, Dordrecht, The Netherlands, 2007.

United States Environmental Protection Agency. Protocol for Developing Pathogen TMDL's. EPA 841 – R – 00 – 002[R]. Office of Water (4503F), Washington, DC, 2001.

Xia Q. Controlling of Contamination Gross[M]. Beijing: Environment Science Press, 1996.

Xu J, Wang J, Liang T, et al. Analysis of Land Use Changein Taihu Basin in the Past 18 Years [J]. Geospatial information, 2009, 7(4):48 – 52.

Zielinski P. A. Stochastic Dissolved Oxygen Model[J]. Journal of Environmental Engineering, 1988, 114(1): 74 – 90.

Effects of Hg^{2+} on Microbial Community Structures of Aerobic Activated Sludge

Dai Ning[1,2], *Wang Jinsheng*[3,4], *Teng Yanguo*[3,4] and *Su Jie*[3,4]

1. Bureau of Hydrology, MWR, Beijing, 100053, China
2. Groundwater Monitoring Center, MWR, Beijing, 100053, China
3. College of the Water Sciences, Beijing Normal University, Beijing, 100875, China
4. Engineering Research Center of Groundwater Pollution Control and Remediation, the Ministry of Education, Beijing, 100875, China

Abstract: The effects of Hg^{2+} on microbial community structures of aerobic activated sludge were investigated by nested PCR-DGGE techniques. The results showed that the microbial community of each group might have obvious change compared with the group without Hg^{2+} in the inflow, especially when the Hg^{2+} concentration was 46.87 mg/L in the inflow, and the microbial community changed obviously with the extend of operation time. The variance of microbial species become obvious in each activated sludge group with the extend of operation time. Compared with the group without Hg^{2+} in the inflow, the dominant microbial species of each activated sludge group changed, and the dominant microbial species of the group whose concentration of Hg^{2+} increased gradually in the inflow had the biggest changes after 60 d. Hg^{2+} in the inflow may stimulate the growth of actinomycetes after an operation period of 30 d.

Key words: Hg^{2+}, aerobic activated sludge, microbial community structures, PCR-DGGE

1 Introduction

The activated sludge process, which uses activated sludge for the transformation of organic and inorganic pollutants, has contributed greatly to the improvement of the aquatic environment worldwide, and is still the most widely used process for the treatment of municipal wastewater because of its low operation cost and high performance. Different kinds of active sludge treatment processes for municipal sewage were originally designed for removing organic matter and other nutrient elements (e. g., nitrogen and phosphorus). Heavy metals enter wastewater from a variety of domestic and industrial sources, and can also be effectively removed, which is regarded as an additional benefit. However, ionized heavy metals can dramatically affect the performance of biological systems.

Mercury is one of most studied metals in environmental and human health research. It has a very complex biogeochemical cycle, can biomagnify along the food web and is toxic, impacting environment and man(US EPA, 2000). The impact of mercury on environmental bacterial communities has been studied in various ecosystems, including soil, sediment and aquatic environments, but less been reported in activated sludge. Because of these concerns, the effects of Hg^{2+} on microbial community structures of aerobic activated sludge were investigated by nested PCR-DGGE techniques with group-specific 16srDNA primers, and the microbial community of the activated sludge were analyzed. Further more, the dominant microbial species of the activated sludge with different Hg^{2+} concentration in the inflow were identification. These results may provide a reference to the normal operation of wastewater treatment plant under the Hg^{2+} in inflow conditions.

2 Materials and methods

2.1 Reactor and operation

The reactor was inoculated using 1.5 L fresh activated sludge taken from the aeration tank of operating wastewater treatment plant in Beijing. The composition of synthetic wastewater was shown in Tab. 1. The reactor was aerating for 23 h per day, deposited for 0.5 h, and outflow for 0.5 h. Chemical oxygen demand (COD), NH_3-N, TP, mixed liquor suspended solid (MLSS), and sludge volumetric index (SVI) were measured using standard methods (SEPAC, 2002).

Tab. 1 The composition of synthetic wastewater

Nutrition component	Concentration (mg/L)	Trace element	Concentration (mg/L)
Glucose	1500	H_3BO_4	150
Peptone	100	$ZnSO_4 \cdot 7H_2O$	120
NH_4Cl	400	$MnCl_2 \cdot 7H_2O$	120
$CaCl_2$	200	$CuSO_4 \cdot 5H_2O$	30
$FeSO_4 \cdot 7H_2O$	40	$NaMoO_4$	65
$MgSO_4 \cdot 7H_2O$	30	$NiCl_2$	50
K_2HPO_4	70	$CoCl_2 \cdot 6H_2O$	210
KH_2PO_4	30	KI	30
Trace element	1.0mL/L		

2.2 The different Hg^{2+} inflow concentration contrast groups

5 groups of experiments were carried out with different Hg^{2+} concentration in the inflow. The first one was without any Hg^{2+} in the inflow with an operation period of 60 d (activated sludge sample numbered group 1); the Hg^{2+} concentration of group 2 was shown in Tab. 2 (activated sludge sample numbered group 2); the Hg^{2+} inflow concentration of group 3 was 2.34 mg/L (activated sludge sample numbered group 3); group 4 was with a Hg^{2+} concentration of 4.69 mg/L (activated sludge sample numbered group 4); and group 5 was with a Hg^{2+} concentration of 46.87 mg/L (activated sludge sample numbered group 5). The microbial communities of activated sludge samples were analyzed using denaturing gradient gel electrophoresis method (DGGE) 30 d and 60 d later, respectively.

Tab. 2 The Hg^{2+} inflow concentration of the group 2 (mg/L)

Days	$C_{Hg^{2+}}$	Days	$C_{Hg^{2+}}$	Days	$C_{Hg^{2+}}$	Days	$C_{Hg^{2+}}$	Days	$C_{Hg^{2+}}$	Days	$C_{Hg^{2+}}$	Days	$C_{Hg^{2+}}$	Days	$C_{Hg^{2+}}$
1	2.34	9	7.03	17	11.70	25	16.38	33	21.06	41	25.74	49	30.42	57	35.10
2	2.34	10	7.03	18	11.70	26	16.38	34	21.06	42	25.74	50	30.42	58	35.10
3	2.34	11	7.03	19	11.70	27	16.38	35	21.06	43	25.74	51	30.42	59	35.10
4	2.34	12	7.03	20	11.70	28	16.38	36	21.06	44	25.74	52	30.42	60	35.10
5	4.69	13	9.36	21	14.04	29	18.72	37	23.44	45	28.08	53	32.76		
6	4.69	14	9.36	22	14.04	30	18.72	38	23.44	46	28.08	54	32.76		
7	4.69	15	9.36	23	14.04	31	18.72	39	23.44	47	28.08	55	32.76		
8	4.69	16	9.36	24	14.04	32	18.72	40	23.44	48	28.08	56	32.76		

2.3 Microbial community of the activated sludge analysis

Samples were placed in sterile polypropylene tubes, stored at −20 ℃ before whole-cell PCR and DNA extraction (< 48 h).

The community characteristics of the sludge were analyzed using denaturing gradient gel electrophoresis method (DGGE).

(1) Preparation of sludge sample and genomic DNA extraction. Firstly, the frozen sludge samples are melted by warm bath and 100 mg (wet weight) sample are taken into a clean centrifuge tube (1.5 mL), and centrifugate at 12,000 r/min for 5 min. Secondly, wash the samples for 2 times with PBS buffer and after centrifugations, extract the genomic DNA of sludge samples with the Small Amount of Bacterial Genome DNA Rapid Extraction and Purification Kit (Shanghai Sangon Biotech Co., Ltd., China). Finally, detect the extraction effect by 0.8% agarose gel electrophoresis.

(2) PCR amplification of sludge sample DNA. The universal primer in 16S rDNA V3 District for most bacteria are used. The length of amplified fragment is about 250 bp. The primers were shown in Tab. 3. PCR reaction system is as follows: the materials include of 10 × Ex Taq buffer (Mg^{2+}) 5 μL, 2.5 mmol/L dNTP 1 μL, 20 pmol/L *pros*and *cons* primer each 0.5 μL, template DNA 100 ng, Ex Taq 0.2 μL and then add water to make up 20 μL. The study utilizes TD-PCR: pre-deformation at 94 ℃ for 4 min, then at the same temperature deformation for 40 s, further to annealing at 56 ℃ for 60 s, follow by extension at 72 ℃ for 30 s. Every cycle of the 35 cycles decrease annealing temperature by 0.1 ℃ until to 51.5 ℃ and then the final extension is at 72 ℃ for 7 min. Finally, detect the extraction effect by 0.8% agarose gel electrophoresis.

Tab. 3 The PCR Primers

Primers		
F357 - GC	5' - CGC CCG CCG CGC GCG GCG GGC GGG GCG GGG GCA CGG GGG GCC TAC GGG AGG CAG CAG - 3'	250 bp
R518	5' - ATT ACC GCG GCT GCT GG - 3'	

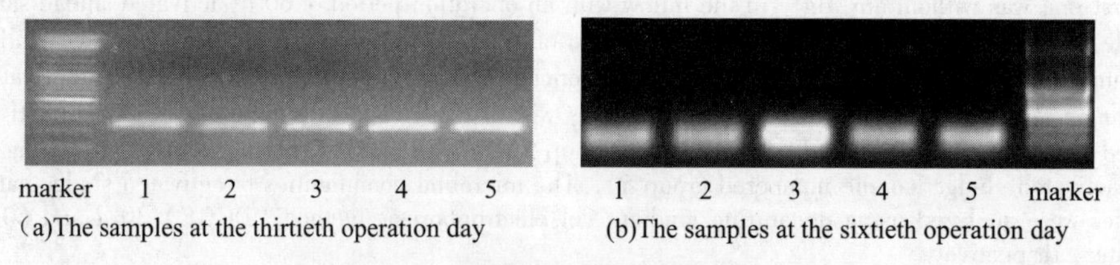

(a)The samples at the thirtieth operation day (b)The samples at the sixtieth operation day

Fig. 1 The PCR amplification

(3) Optimize the condition of DGGE gel electrophoresis and cloning sequencing for the purpose strip was done by Shanghai Sangon Biotech Co., Ltd. DGGE gel electrophoresis of PCR is carried out by the gene mutation detection system. The parameters are as follows: the concentration of polyacrylamide gel is 8% (bisacrylamide : acrylamide = 37.5 : 1), denaturing gradient is from 30% to 60% (100% denaturant with 7 mol/L urea and 40% deionized formamide mixture). The electrophoresis was run at 60 ℃, 4.0 h at 180 V. After electrophoresis, the gels were stained for 30 min with ethidium bromide and photographed on a UVI transillumination table with a Gel Documentation Systems

2.4 Recovery of bands from DGGE gels and sequence analysis

The processing of the DGGE gel was done with the BIORAD software Quantity One 4.3.0. DGGE fingerprints were automatically scored by the presence or absence of co-migrating bands, independent of intensity.

DNA was recovered by cutting the objective band of the gelatin map, then re-amplified and cloned by the SK1135 vector, finally sent to the Shanghai Sangon Biotech Co., Ltd China for sequencing. PCR amplification was the same, and the primers were shown in Tab. 4. For each band selected, only the middle portion of each band was excised. Then the slices were placed in 2-ml sterilized screw-cap polypropylene tubes and 1 μL TE buffer was added. The DNA was allowed to passively diffuse into water at 16 ℃ overnight. 0.2 pmol of each PCR product was subjected to agarose gel electrophoresis to check product recovery and to estimate product concentration. 10 L of each reaction mixture was also subjected to DGGE analysis to confirm the melting behavior of the band recovered. The remaining PCR products (0.2 pmol) were sent to Shanghai Sangon Biotech Co., Ltd. China Company for sequence analysis.

Tab. 4 The PCR Primers

Primers		
F357	5' - CC TAC GGG AGG CAG CAG - 3'	170 bp
R518	5' - ATT ACC GCG GCT GCT GG - 3'	

3 Results and discussions

3.1 Results of microbial community structure sequencing

Bacterial community structures in each contrast group activated sludge were investigated by PCR-DGGE method, and the results are shown in Fig. 2. A cluster analysis based on the values of Dice coefficients was performed using the Pearson correlation (1926) and visualized in a dendrogram (Fig. 3). It can be seen that the Dice coefficient for the microbial communities of each contrast group compared with the no Hg^{2+} inflow concentration group were 66.3% (group 3), 63.6% (group 4), 63.1% (group 2), and 49.7% (group 5) at the thirtieth operation day. It was suggested that the microbial community of each groups might have obvious change, especially when the Hg^{2+} inflow concentration was 46.87 mg/L (group 5). Furthermore, the Dice coefficient of each group declined to 41.2% (group 3), 37.1% (group 2), 35.7% (group 5), and 31.9% (group 4) at the sixtieth operation day. It meant that the microbial community of each groups changed even more obviously with the operation time extending. At the sixtieth operation day, the Dice coefficient of group 2 was higher than group 5 and group 4 meant the method that increasing inflow concentration of Hg^{2+} gradually has less impact of the microbial community of aerobic activated sludge.

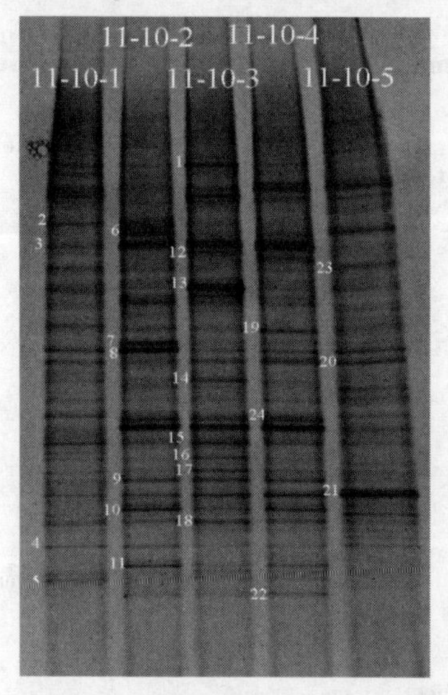

(a) The samples at the thirtieth operation day

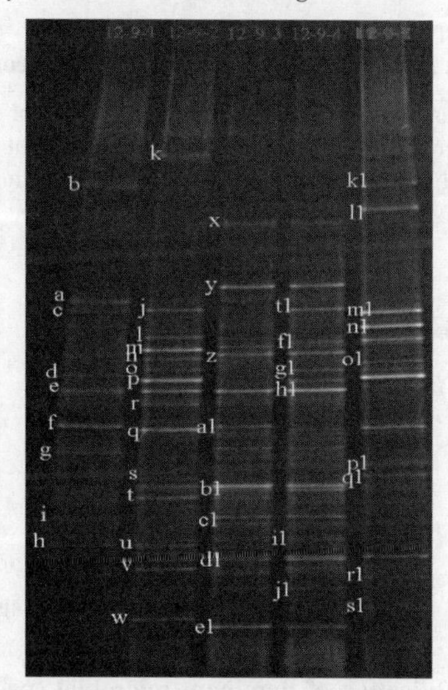

(b) The samples at the sixtieth operation day

Fig. 2 DGGE fingerprint of sludge samples under different Hg^{2+} inflow concentration contrast groups

The cluster analysis was used to analysis microbial community similarity (Fig. 4). It showed

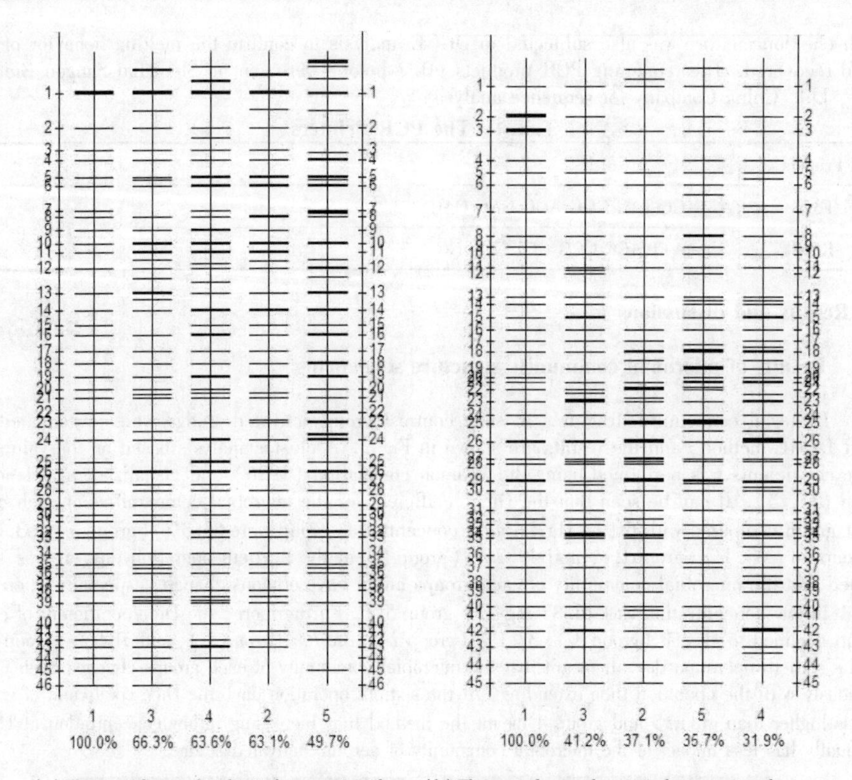

(a) The samples at the thirtieth operation day (b) The samples at the sixtieth operation day

Fig. 3 Comparison of actinomycetic communities between the contrast groups

that the clustering tree of 30 d operation were not similar with the one of 60 d operation. The changes of microbial community structures in different groups were not same with the operation time extending. This results were similar with cluster analysis.

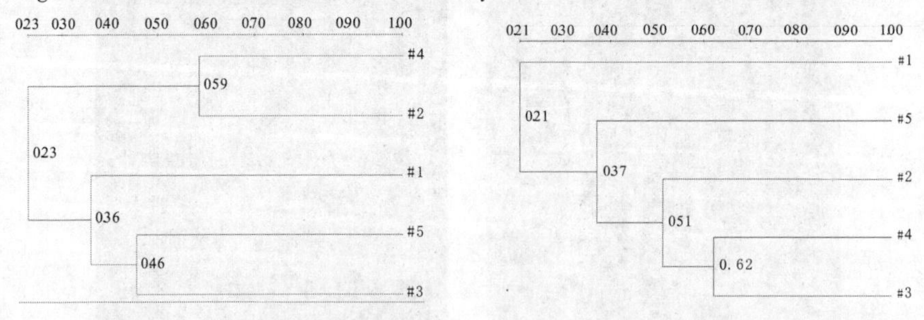

(a) The samples at the thirtieth operation day (b) The samples at the sixtieth operation day

Fig. 4 Cluster analysis of the DGGE profile

3.2 Identification of dominant microbial species

Sequencing results of bands were aligned in the GenBank database to obtain the closest relative strains (Tab. 5). The similarities between most bands and corresponding sequences were larger than 94%, which indicated that they may belong to the same genus. The alignment results of 31

OTU indicated that there were 4 OTU that can not be affiliated to known species after 30 d operation, but 7 OTU that can not be affiliated to known species after 60 d operation. It meant the variance of microbial species was more obvious with the operation time extending.

Tab. 5 Partial 16S rDNA sequence similarity of bands 1 - 31 on DGGE profile

Band	Closest sequence (GenBank number)	Similarity (%)	Putative division
2	Pseudomonas plecoglossicida XJUHX-16 (EU194334)	96.8	Pseudomonas
3	uncultured bacterium SedUMA11 (FJ849528)	99.4	Bartonella
5	uncultured bacterium M0509_52 (EU104137)	90.1	Crocinitomix
6	Novosphingobium sp. HZ11 (AY690709)	100	Sphingomonadales
7	Sphingobium xenophagum AJ1 (AB099636)	95.7	Sphingobium xenophagum
8	Sphingomonas melonis Acj 102 (AB480750)	100	Sphingomonas melonis
9	Brucella melitensis biovar Suis 1997003632 (AY513506)	100	Brucella
10	Microbacterium aoyamense (T) KV-492 (AB234028)	100	Microbacterium
11	Pseudomonas psychrophila (T) E-3 (AB041885)	100	Pseudomonas?
13	Brucella melitensis biovar Suis 1997003632 (AY513506)	100	Proteobacteria
15	Rhodococcus erythropolis DCL14 (AJ131637)	100	Actinomycetales
21	Uncultured bacterium G19-71 (GQ487970)	82.1	Gammaproteobacteria
22	Arcicella sp. MG83 (AJ746140)	95.9	Arcicella
23	Thiocapsa marina (T) 5811 (AF112998)	83.3	Chromatiaceae
24	Ochrobactrum sp. M231 (EU604246)	93.9	Rhizobiales
a	Uncultured bacterium (AY302113)	100	Terrimonas
b	Mucilaginibacter sp. DR-f3 (GU139696)	66.7	Mucilaginibacter gossypii
e	Xanthomonas axonopodis S53 (AB101447)	100	Thermomonas
g	Uncultured bacterium pLW-42 (DQ067033)	78.1	Polyangiaceae
k	Acinetobacter johnsonii P152 (AF188300)	96.1	Acinetobacter johnsonii
m	Uncultured bacterium M0509_29 (EU104115)	100	Marinilabiaceae
p	Sphingobium xenophagum AJ1 (AB099636)	100	Sphingobium xenophagum
w	Uncultured bacterium M0509_29 (EU104115)	100	Marinilabiaceae
x	Uncultured bacterium bacteriap48 (AF402983)	75.5	Thiovirga
y	Flavobacterium degerlachei LMG 21474 (AJ441005)	92.2	Flavobacterium degerlachei
b1	Uncultured Polyangiaceae bacterium Amb_16S_1127 (EF018495)	87.1	Nannocystineae
c1	Uncultured soil bacterium HSB NT53_E03 (DQ128816)	82.9	TM7_genera_incertae_sedis
e1	Uncultured Arsenicicoccus sp. AV_4R-S-I05 (EU341177)	98.2	Actinomycetales
i1	Uncultured Burkholderiales bacterium SHBZ945 (EU639298)	84.0	Burkholderiales
m1	Pseudomonas sp. LAB-36 (AB051702)	100	Pseudomonas
n1	Flavobacterium sp. WX3 (EF601819)	96.2	Flavobacterium pectinovorum

After 30 days operation, the dominant microbial species of group 1 were Proteobacteria and Bacteroidetes. However, microbial community changed in different groups. The dominant microbial species of group 2 and group 3 were Proteobacteria and Actinobacteria, and was Proteobacteria only of group 5. It was predictedthat Hg^{2+} in the inflow may stimulate the growth of Actinomycetes during 30 days operation. After 60 d operation, the dominant microbial species of group1 were still Proteobacteria and Bacteroidetes. Different from the results of 30 d operation, the dominant microbial species of group 3 were Proteobacteria, Bacteroidetes, Actinomycetes, and TM7, which changed a lot. It showed that the method that increasing the inflow concentration of Hg^{2+} gradually may cause the microbial community changed a lot.

4 Conclusions

(1) The microbial community of each group might have obvious change compared with the group without Hg^{2+} in the inflow, especially when the Hg^{2+} concentration was 46.87 mg/L in the inflow, and the microbial community changed obviously with the extend of operation time;

(2) The variance of microbial species become obvious in each activated sludge group with the extend of operation time;

(3) Compared with the group without Hg^{2+} in the inflow, the dominant microbial species of each activated sludge group changed, and the dominant microbial species of the group whose concentration of Hg^{2+} increased gradually in the inflow had the biggest changes after 60 d;

(4) Hg^{2+} in the inflow may stimulate the growth of Actinomycetes after an operation period of 30 d.

References

Principi, P., Villa, F., Bernasconi, M., Water Res. 40(2006): 99 – 106.

Ewa Lipczynska – Kochany, Jan Kochany, Chemosphere, 77 (2009): 279 – 284.

Jean – Baptiste Ramond, Thierry Berthe, Robert Duran, Research in Microbiology, 160 (2009): 10 – 18.

Tatiana A. Vishnivetskaya, Jennifer J. Mosherl, Anthony V. Palumbo, Appl. Environ. Microbiol, published online (2010).

E. Shoham-Frider, G. Shelef, N. Kress, Marine Environmental Research, 64 (2007): 601 – 615.

Hsu, H., Sedlak, D. L.. Environ. Sci. Technol,2003(37): 2743 – 2749.

SEPAC (State Environmental Protection Administration of China). Standard Methods for the Examination of Water and Wastewater (4th ed.). Beijing: Environmental Science Press, (2002).

Song Zhiwei, Pan Yuejun, Zhang Kun. Journal of Beijing Environmental Sciences,2010 (22): 1312 – 1318.

Liu Xinchun, Zhang Yu, Yang Min. Journal of Environmental Sciences, 2007(19): 60 – 66.

Heuer H, Smalla K. Application of denaturing gradient gel electrophoresis (DGGE) and temperature gradient gel electrophoresis (TGGE) for studying soil microbial communities[M]. In: Modem soil microbiology. New York: Marcel Dekker(1997).

Beaches Ecological Protection and Safety Utilization Based on the Xiaolangdi Reservoir Flood Control of the Yellow River Lower Reach: Kaifeng Area as a Case Study

Qin Mingzhou[1,2], *Zhang Pengyan*[1,2] and *Yan Jianghong*[1,2]

1. Institute of Natural Resources and Environmental Science, Henan University; Kaifeng, 475004, China
2. College of Environment and Planning, Henan University, Kaifeng, 475001, China

Abstract: In this article, the Yellow River beach land in the Kaifeng area is used as a case study to explore beach utilization issues after Xiaolangdi Reservoir flood control. Based on the latest SPOT2.5 images for 2007 and ArcGIS and ERDAS remote sensing image processing software, along with field surveys, the Yellow River channel boundary and beach land use within Kaifeng City are mapped in detail. Then, employing the TM images of annual highest flood peaks between 1992 and 2007, and data from the hydrological observation stations along the Yellow River, the river channel flood fringe line for this typical 15-year period is outlined for flow levels up to 8,000 m^3/s. In conjunction, the submerged the Yellow River beach areas within Kaifeng City area under different regularly occurring flood flows are drawn up, overlaid by background data. Using a mathematical model of the Yellow River downstream runoff and sediment transport, with cross section data for the river channel after 2004, various high level floods with flows over 4,000 m^3/s are simulated and the potential submerged areas within Kaifeng Yellow River beach region are identified. To shed light on land resource management issues, as well as ecological safety requirements for floods transiting within the river channel, the current land utilization situation and existing problems of the lower reach of the Yellow River are discussed. Finally, based on the outcome of the research, together with the incorporation of new socialism village construction standards, four zones of ecological protection and safety use to avoid floods are suggested for the Kaifeng Yellow River beach region. At the same time, some important measures are proposed for using beach land resources and ecological safe respectively while avoiding flood risk.

Key words: the Yellow River beach region, land ecological safety, new socialism village construction, flood risk, Xiaolangdi Reservoir flood control

1 Introduction

The lower reach of the Yellow River from Taohuayu in Zhengzhou to its estuary is about 786 km in length. Because the lower reach is located on the North China Plain, the vertical gradient of the river bottom is very small, with a total head drop of only 95 m (YRCC, 2009a). The upper river section in Henan Province is a typical wandering river course (Qian et al., 1961) with the distance between banks varying from 5 km to 20 km, but usually around 7 km (Yang et al., 2007). The following section in Shandong Province is narrower, down to 400 m at its narrowest point, and because of a slowing velocity and the resulting deposition, the riverbed is raised every year (Xu, 1992; Xu, 2004). Different from other rivers, the Yellow River beach area is formed of sedimentation from the wandering river course within the two river embankments (Liu, 1999). In conjunction with the slow subsidence (0.9 mm/annum) of the North China Plain (Xu et al., 2003), the lower Yellow River reach gradually formed the world-famous "Suspended River" or "River above the Ground", which is generally about 5 m higher than the ground outside the river embankments, with sections up to 10m higher (Liu et al., 2008). The beach area of the lower reach is about 3,500 km^2, accounting for 84% of the total area of the lower river course (Huangbu et al., 2003; Yang et al., 2006).

The latest statistics show that there were 1.8 million permanent residents and 0.24×10^6 hm^2

of cropland in the lower beaches, belonging to 1,819 villages, 43 counties, 15 cities, and the two provinces of Henan and Shandong. As an important part of the river course, the Yellow River beach has multi-functions: reducing flood peaks, depositing sediments, dividing and storing flood waters and protecting the dams, as well as flood control. Thus, the beach land resources simultaneously have three major missions for society based on economics and ecology both inside and outside the river course: the socio-economic development of its own beach area, the social and ecological safety of the beach and bank areas, and the Yellow River flood control. These conflicting tasks seriously restrict the stable and efficient use of beach land. Lacking compensatory support through funding, policies or projects, floodplain land use efficiency is quite low and socio-economic development faces many difficulties. The beach area has become a poverty zone across the two provinces of Yu (standing for Henan Province) and Lu (standing for Shangdong Province) (Wang et al., 2007; Wang et al., 2006).

Based on incomplete statistics, from 1949 to 2003, the lower Yellow River beaches were flooded more than 30 times. After the 1950s, in order to protect riverside land from flood disasters, four periods of massive embankment rebuilding, repairing, heightening and thickening were carried out in the lower reach. For the beaches, protecting the river course by dams, and transiting the river flow more effectively by straightening bends was used to stabilize the downstream river course, as well as to protect beach and embankments (Qin et al., 2007; Zhang et al., 2003). After the 1980s, the state began addressing the problem of residents' safety in beach areas by increased investment in floodplain safety projects such as establishing some villages on platforms to avoid flooding, relocating villages outside the embankments, and providing a means of escape to avoid floods in particularly dangerous regions.

Recently, some important projects to harness the Yellow River were carried out, such as strengthening dikes by depositing sediments behind them, and regulating river flow and scouring sediments. Especially after the Xiaolangdi Reservoir began operation in 2002 and the continuous regulation of river flow and scouring sediments over nine periods, the downstream river channel flood capacity was greatly improved. Flows of less than 4,000 m^3/s did not overflow the normal river channel (YRCC, 2009b). The flood-protection standard for downstream embankments was also enhanced to "*once in a millennium*", that is, the flood flow safely passing Huayuankou, the first hydrological observation station, was raised from 2.2×10^4 m^3/s once in one hundred years to 3.4×10^4 m^3/s once in one thousand years.

The security of the beach areas has increased to an unprecedented level, basically with no flooding in the average year. This has supplied an enhanced level of safety for land use but also raises new challenges in determining the limits of safe beach use and the safety of residents' life and property. Many scholars, experts, and government agencies have proposed constructing flood control works, heightening the elevation of village platforms (Hu, 2007), completing escape roads and relocating low-lying beach villages (Wang et al., 2006; Zhang et al., 2003; YRCC, 2009b; Hu, 2007; Wang et al., 2007; Mark, 2005), as well as developing grassland and husbandry industry (Duanmu et al., 2003), establishing an organic agricultural production base and generally improving the policies supporting beach area development including raising compensation standards (Wang et al., 2007; Wang et al., 2006). However, given our national condition in which significant conflict exists over human land use practices outside the beach area, that the land area per capita within the beach region is larger than that outside, and since all the people who live in the beach were never relocated outside, it is impossible to use the river beach land just for flood control neglecting other purposes (Jeffrey et al., 2009; Jeffrey et al., 2010). Based on the reality of this historical situation, the safe use of floodplain land and reasonable land development is an unavoidable imperative.

2 Case study area and data sources

2.1 Introduction of study case

The Kaifeng Yellow River Floodplain case study area is located in the northern section of

Kaifeng. It is bounded on the south by the Yellow River embankment, on the north by the Yellow River water's edge, on the west by the junction of HuihuiZhai of Shuidao township and Zhongmu County, and on the east by the junction of Yuezhai of Guying township in Lankao County, on the border with Shandong province. The straight-line length of the first section of the Kaifeng Floodplain, comprising the Ming and Qing' old river course, is about 52 km; the river channel is that of a typical wide and shallow, wandering river with a distance between banks of 5.5 ~ 12.7 km and a river channel of 1.5 ~ 7.5 km in width. Below the first section, from the Lankao east groin head, the following length is about 7 km, and is the river section formed by the Tongwaxiang changing its channel and occupying the Daqing channel in 1855; the distance between the two banks is 1.4 ~ 20 km, and the river channel is 1.0 ~ 6.5 km wide (Wang et al., 2006). The floodplain region belongs to Kaifeng County, Lankao County, 11 townships in 3 districts of Kaifeng City, and 88 administrative villages. The population of residents who live in the beach areas is 156.8 thousand, the total land area is 31,139.53 hm^2, the arable land is 12,727.78 hm^2, resulting in an average of 1.22 mu (1 mu = 1/15 hm^2) of arable land per person. Being located in the middle of the North China Plain, the historical birthplace of Chinese civilization, and adjacent to modern cities, the problems arising from the coexistence of man and water, and the threat of floods, still exist.

2.2 Beach region boundaries

The lower Yellow River beaches are a special type, different from those in ordinary rivers, and lacking the typical characteristics of floodplain sediment and non-dualistic structure. Before the Xiaolangdi Reservoir was completed, they were often subjected to river erosion and flooding. After the Xiaolangdi Reservoir was completed and the regulation of downstream flows was initiated, the beach region was able to stabilize because of the absence of flooding. Therefore, the current medial boundary is established by the currently approved minimum flood flow, and the lateral boundary is clearly established by the Yellow River dyke. Definition of the medial boundary is different from that of a general river because of the high sediment and changeable channel characteristics of the lower Yellow River. In view of this, confirmation of the minimum flood flow is important for the safe use of flood plain land because the water level associated with this flow establishes the boundaries for the maximum available land resources in the flood plain. This confirmed minimum flow is 4,000 m^3/s, the basis for which is as follows.

Based on the analysis of data for multiple years of flooding from 1992 ~ 2007, the frequent flood flow risk design is listed at 4,000 m^3/s. This is determined by Yellow River Conservancy Commission (YRCC) at Huayuankou, the first hydrological station in the lower Yellow River floodplain (Huo et al., 2009). According to the 20-year series of hydrological statistics from 1986 to 2005, 11 years of flood flows over 4,000 m^3/s occurred at Huayuankou. However, since the Xiaolangdi Reservoir became operational, there have been no flows over 4,000 m^3/s in the lower Yellow River (Huang et al., 2006).

Based on all of the above, 4,000 m^3/s appears to be a minimum safe downstream flow, and thus is a reliable basis for the analysis of floodplain land use.

2.3 Data source and analysis

Geographic Information System (GIS) mapping technologies were applied to remote sensing (RS) images obtained by the Thematic Mapper (TM) satellite to extract the river flood transit fringe line for October 16, 1992, September 29, 1994, August 17 and October 4, 1996, September 7, 2002, October 24, 2003, September 22, 2004, April 7, 2006 and June 29, 2007. Using the June 26, 2006 (3700) TM image as background data, Fig. 1 shows these flood regimes in four typical years during this nearly 20 year period.

With Huayuankou as the control section entrance for flow processing, and Lijin as the exit, floods of different orders of magnitudes where simulated using a two-dimensional water-sediment mathematical model of the lower Yellow River. Topographic data for the beach regions was obtained from 1∶10,000 topographic maps for 1998 produced by the YRCC. The location of training works, anti-flood-danger projects, levee boundaries and additional information was measured using GPS. These data, along with large scale data from after the 2004 flood, were used to produce the maps in Fig. 2 which show submerged areas for floods of different orders of magnitude. These data were then superimposed on the maps of the flood transit region over the past 20 years, and 1∶50,000 scale maps were produced showing the pattern of submerged areas for floods of different orders of magnitude (Fig. 1).

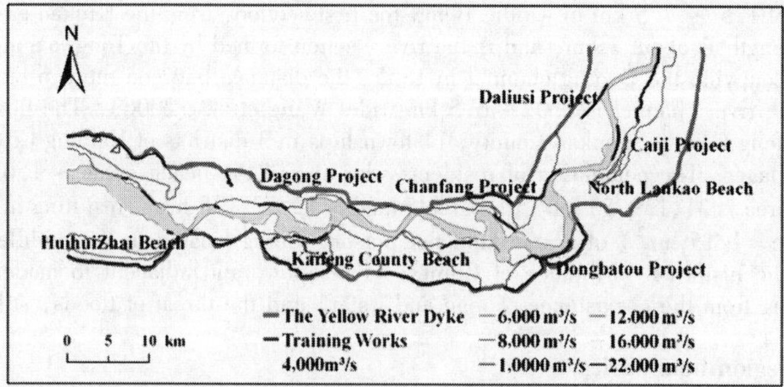

Fig. 1 Superpimposed map of Kaifeng Yellow River flood plain for different orders of magnitude of flow

2.4 Current status of land use

Using the superimposed map produced above, the water level line for the 4,000 m^3/s flow was drawn on a 2007 SPOT2.5 remote sensing image, which was then overlain on a 1∶10,000 land use map for 2005 to produce a current (2007) 1∶10,000 land – use map. Six types of land use were identified (cultivated land, forest land, garden plots, water area, residential quarters and industrial-mining land, and unused land) and their areas classified and calculated by pixels feature using field survey and remote sensing mapping. Based on these statistics, the total land area of the Kaifeng Yellow River Floodplain is 31,139.53 hm^2, the main components of which are cultivated land and beach, accounting for 40.87% and 35.77% respectively. The remaining areas are rural residential land, brick-tile kilns and so on. The areas for all land use types are shown in Tab. 1.

Tab. 1 Current land use in Kaifeng beach

Classifications		Area(hm^2)	Of beach(%)
1. cropland		12,727.78	40.87
2. garden		295.33	0.95
3. forest		828.87	2.66
4. residential & industrial land	Subtotal	4,652.62	14.94
Inside:	Village	2,322.75	7.46
	Brick kiln	1,970.30	6.33
	Other	359.57	1.15
5. water body	Subtotal	12,535.36	40.26
Inside:	Ponds	1,335.16	4.29
	Wildland	11,139.57	35.77
	Ditch&channel	60.63	0.19
	Water conservation buildings	1.97	0.01
6. unused land		99.57	0.32
Total		31,139.53	100

Note: 1. Other land is mainly composed of rural idle land of 273.35 hm^2, industrial-mining land of 69.55

hm^2, special used land of 16.68 hm^2.

2. Unused land is composed of waste grass land of 31.30 hm^2, sandy land of 68.27 hm^2.

Based on the analysis above, Tab. 2 shows estimates of the areas of different land use types which would be submerged under different flood flows.

Tab. 2 Analysis of land use types, other than arable land, which would be submerged at different flow levels

(Unit: hm^2)

Submerging beach flow of m^3/s	Idle land	Brick kiln	Pond	Wildland	Grassland	Sandy land	Total
4,000 ~ 6,000	11.34	847.90	763.28	8,180.75	3.23		9,806.50
6,000 ~ 8,000	58.46	745.90	324.42	2,450.97	21.05	17.90	3,618.70
8,000 ~ 10,000	127.13	322.59	67.16	395.37	0.5	10.66	923.41
Total	196.93	1,916.39	1,154.86	11,027.09	24.78	28.56	14,348.61

3 Problems of ecological security of land use in beach regions and planning

3.1 Problems of ecological security

The above analysis of the current status of land use and flood inundation risk, combined with the requirements of "Planning for New Countryside Construction," the national program which aims to improve rural development, demonstrates that problems with land use in beach regions still remain.

(1) Flood risk for low-lying villages.

(2) Imperfect infrastructure and poor farming conditions.

(3) Harmful brick-tile kilns and borrow pits.

3.2 The spatial planning of ecological protection and land-use

In accordance with the characteristics of the Yellow River beach land, combined with land use control principles and land management practices, the following beach zoning land classification and program management system is proposed.

3.2.1 Highest-risk buffer band of ecological protection closest to the river

This band is closest to the main river channel, extending approximately 100 ~ 250 m from the 4,000 m^3/ s flow water line which is the core area of the Yellow River wetlands. Prohibit any form of exploitation and utilization.

The situation of the river channel has become relatively more stable after recent water and sediment diversions, and it is difficult for a 4,000 m^3/s current to overflow the river channel. However, in order to avoid flood risk, because of the unique Yellow River bank silt characteristics, these 200 m areas along the river channel are not suitable for farming. This buffer band ought to guard against the risk of river erosion, and protect the security of farming on higher terrain. Together with the river channel, the band closest to the river channel is the drainage channel for small or mid-sized floods, as well as the major flood drainage channel. So it is important to create this buffer zone in order to alleviate risk to farming on higher terrain from river bank erosion.

3.2.2 Zone of land suitable for cultivation near river

This zone is located outside the high-risk buffer band closest to the river, roughly between the 4,000 m^3/s and 6,000 m^3/ s flow water levels. The use of these beaches is basically safe if it can be controlled and used according to the "farmland" standard. Large areas of land within the region are appropriate for farming and can play multiple roles in exploiting and using beach land, such as reducing the flood peak, flood detention, and reducing the flood water level. However, there are a large number of brick kilns in this zone which should be resolutely stopped and the existing settlements relocated as soon as possible.

After implementation of land use category management in the flood plain, according to the plans, a production embankment is to be constructed at the boundary between the high-risk buffer band closest to the river and the zone of land suitable for cultivation. In principle, this production embankment should be constructed and guarded by the people of the floodplain themselves, according to the required standards. Key parts, such as water import and export gates, are directly managed by the Yellow River maintenance department in order to facilitate flood control operations. It is possible to better resolve the contradictory goals between flood control and flood plain development and utilization through the policy guidance and effective management of these production embankments.

3.2.3 The zone of relatively stable land use

This zone is located outside the zone of arable land near the river, approximately between the 6,000 m^3/s and 8,000 m^3/s flow water levels. The area is stable all year, with thick soil, few villages, and sporadic distribution of agriculture production facilities. It is recommended that the zone should be governed in accordance with the "basic farmland" standard in order to improve land use efficiency and increase the income and improve the living standards of the Yellow River floodplain population. This is under the premise that they comply with relevant laws and regulations and abide by the requirements for flood prevention.

A further advantage of less land pollution of the Yellow River beach area can be obtained if the forms of agricultural production established in this area are based on organic agriculture, green food production, ecological agriculture and high-quality raw material supply.

4 Conclusions

As a result of the favorable geographic conditions of the Yellow River beaches, and the good ecological environment coupled with the unique advantages of excellent tourism resources in the Yellow River itself, Kaifeng could also establish agricultural tourism in the beach region. This could be formatted as a natural, fresh, simple agricultural park with integrated tourism and the participation of the population as a whole. With a local flavor and a cultural connotation, an agricultural park could effectively demonstrate the potential of the Yellow River beach land, and improve the efficiency of its use.

Acknowledgment

This publication has been funded under the Chinese national "11th Five Year Plan" for science-technology support of programs for village-town land use assessment and sustainable use, key technology integration and pilot projects (2006BJA04 -3). Its content does not represent the official position of the Chinese government and is entirely the responsibility of the authors. We are grateful to Ronald Briggs, our American adjunct professor, for his editing of the draft of the text.

References

Ding C R. Policy and Praxis of Land Acquisition in China[J]. Land Use Policy, 2007(24): 1 - 13.

Duanmu L M, Cheng G: Discuss on Comprehensive Management and Development Measures of Yellow River Floodplain in Henan [J]. China Water Resource, 2003(11): 66 - 67. (in Chinese)

Gao J, Liu Y S. Determination of Land Degradation Causes in Tongyu County, Northeast China Via Land Cover Change Detection [J]. International Journal of Applied Earth Observation and Geoinformation, 2010(12): 9 - 16.

Hu Y S. Study on Floodplain Security Construction and Compensation Policy of the Yellow River [J]. Yellow River, 2007, 29(5): 1 -2. (in Chinese)

Huang S S, Yang Z Q, Wang Y. Probability studies of flood plain areas in the lower reaches of the Yellow River [J]. China Water Resources, 2006(18): 6 -8. (in Chinese)

Huangbu X F , Zhang C X , Liu X H, et al. Study of Hanging River Stability Assessment the Yellow River in Henan Section [J]. Yellow River ,2003, 25(4): 12 – 14. (in Chinese)
Huo F L, Lan H L. Study of Mitigation Measures and Analysis of Flood Risk at Floodplain in Yellow River Downstream [J]. Water Conservancy Science and Technology and Economy , 2009, 15(2):135 – 137. (in Chinese)
Jeffrey J, Opperman, Gerald E, et al. Sustainable Floodplains Through Large – Scale Reconnection to Rivers [J]. Science, 2009(326): 1487 – 1488.
Jeffrey J, Opperman, Ryan L, et al. Ecologically Functional Floodplains: Connectivity, Flow Regime and Scale [J]. Journal of the American Water Resources Association, 2010, 46(2): 211 – 226.
Li G Y. Discussion About Long Term Harness and Safety of the Yellow River [J]. Yellow River, 2001, 23(7): 1 – 4. (in Chinese)
Li W, Feng T T, Hao J M. The Evolving Concepts of Land Administration in China: Cultivated Land Protection Perspective [J]. Land Use Policy, 2009(26): 262 - 272.
Liu S K. The Yellow River Floodplain and Sub – detention Basin Risk Analysis and Mitigation Measures — the Yellow River Harnessing and Exploitation of Water Resources [M]. Zheng zhou: Yellow River Conserrancy Press,1999.
Liu Z C, Qin Y C, Jin S. Study in Treatment of Lower Yellow River Channel and Its Beach Area's Problem [J]. Progress In Geography, 2008, 27(2): 32 – 38. (in Chinese)
Mark H. Nature Makes them Lazy: Contested Perceptions of Place and Knowledge in the Lower Amazon Floodplain of Brazil [J]. Conservation and Society, 2005, 3(2): 461 – 478.
McCarthy J. Rural Geography: Globalizing the Countryside [J]. Progress in Human Geography , 2008(32): 129 - 137.
Qian N, Zhou W H, Hong R J. The Characteristics and Genesis Analysis of the Braided Stream of the Lower Yellow River [J]. Acta Geographica Sinica, 1961, 27(12): 1 – 27. (in Chinese)
Qin M Z, Richard H J, Yuan Z J, et al. The effects of Sediment – Laden Waters on Irrigated Lands along the Lower Yellow River in China [J]. Journal of Environmental Management , 2007(4): 858 – 865.
Wang X P , Wang L. Yellow River Floodplain: a Special Land [J]. China Water Resources, 2007(9): 2 – 10. (in Chinese)
Wang Y, Zhang X Y, Song G S. Issues Related to Flood Plain Areas in Lower Reaches of the Yellow River [J]. China Water Resources, 2006(18): 3 – 5.
Wang Z Y , Yang J S , Zhang G C. Studies on Compensation Policy to Flood Plain in Lower Reaches of the Yellow River [J]. China Water Resources, 2007(18): 1 – 2,5. (in Chinese)
Xu F H Study Oftraining Issues on Meandering River in Lower Yellow River [J]. Yellow River, 1992(12): 25 – 26. (in Chinese)
Xu J X. Tendency of Sedimentation in the Lower Yellow River Influenced by Human Activities [J]. Journal of Hydraulic Engineering, 2: 8 – 16. (in Chinese)
Xu J X , Sun J Sedimentation rate change in the lower Yellow River in the past 2,300 years [J]. Acta Geographica Sinica, 2003, 58(2), 247 – 254. (in Chinese)
Yang J S , Xu J X , Liao J H. The Process of Secondary Suspended Channel in the Lower Yellow River under Different Conditions of Runoff and Sediment Load [J]. Acta Geographica Sinica, 2006, 61(1); 66 – 76. (in Chinese)
Yang J S, Xu J X, Wang Z Y, et al. Influencing Factors of the Shrinkage of the Braided Reach of the lower Yellow River [J]. Geographical Research, 2007, 26(9):915 – 921. (in Chinese)
YRCC(Yellow River Conservancy Commission). Dictionary of the Yellow River (downstream). http://www.yellowriver.gov.cn/ziliao/hhcd/x/; 2009a.
YRCC(Yellow River Conservancy Commission). Sediment regulation achieved remarkable results in the Yellow River. http://www.yelowriver.gov.cn/; 2009b.
Zhang J M , Li J P. Analysis on Economic Development of Yellow River Floodplain in Henan Province [J]. China Water Resources, 2003, 8(A): 62 – 64. (in Chinese)

Healthy Evaluation of Water Eco-environment for Region

Zhang Huaxing[1], *Wang Linwei*[2], *Zhang Ran*[3] and *Zhang Shaofeng*[4]

1. Yellow River Conservancy Commission, Zhengzhou, 450003, China
2. Yellow River Engineering Consulting Co., Ltd., Zhengzhou, 450003, China
3. College of Hydrology and Water Resources of Hohai University, Nanjing, 210098, China
4. Yellow River Water Resources Protection Institute, Zhengzhou, 450004, China

Abstract: Based on studying the meaning of water eco-environment health, combination of qualitative and quantitative analysis is used to establish healthy evaluation index system of water eco-environment. Taking Shenyang City for an example, adopting multi-level fuzzy comprehensive evaluation to assess the water ecological environment health of Shenyang in 2010, and the results show that the comprehensive evaluation of water ecological environment health overall rating value is 0.468,7, in unsafe to basic security transition stage. On the basis of the membership of each index the water ecological environment health in Shenyang was diagnosed.
Key words: water ecological environment health, qualitative and quantitative, evaluation indicator, multi-level fuzzy comprehensive evaluation

1 The present situation of the water eco-environment in China

In 2010 the national water resources development and utilization rate is 20%. In which 48% of the north, the south 13%, the highest area of the Haihe River has been as high as 120%. Because of the shortage of water resources and the water pollution, water ecological environment problems have become increasingly prominent. At present the China facing the water ecological environment problems can be concluded to the following four aspects: discontinuous flow and lakes atrophy; excessive extraction of groundwater; water environment deterioration and aquatic biodiversity decrease; and the serious soil erosion.

2 The connotation of the water eco-environment health

At present the domestic and overseas scholars about the connotation of water ecological environment health did not form a unified understanding. The China encyclopedia did not give a clear explanation too. Most of the scholars from China think that the water ecology refers to water environment factors on the influence of biology and biological to adapt to all kinds of water conditions. The water environment refers to all kinds of water body on earth's surface, including the oceans, rivers, lakes, swamps, glacier and shallow groundwater water, etc. Compared, water ecology emphasis the dynamic relationship between aquatic organisms and inorganic environment (Wang Hao, et al., 2010).

The human is not reasonable development to water resources that directly affect the water ecological environment of the basic service functions and human sustainable use of water resources. Human activities have become the most important influence factors in water ecological environment health. So the real meaning of water ecological environment health is not a simple water ecological system and water environment safety, it should not only to a certain extent to meet the needs of human social and economic development, and to ensure the health of the water ecological environment. So the water ecological environment health content should include the following three aspects: the first is to guarantee the most basic human survival and development, and social economic development water not threatened; the second is to make water guarantee of a certain quantity and quality, to ensure the survival of aquatic organisms and diversity, to ensure the safety operation of the water ecological system and water environment; the third is to ensure that the water ecological environment system can provide long-term natural resources and a variety of functions of living environment for human.

3 The establishment of the evaluation index system

Constructing healthy evaluation index system of water eco-environment should be based on the connotation of the water eco-environment health, for the purpose to maintain water ecological environment health as the final purpose. In order to fully reflect the water ecological environment situation, the establishment of the evaluation index system needs the primary selection, screening and the establishment two stages. The principles and the method of selection are different in each stage. Water ecological environment health characteristics requirements the health evaluation results of water ecological environment that must embody the integrity, hierarchy and dynamic. So the primary indicators should be fully considered for the various influential factors of research object, to avoid missing (Geng Leihua and Bian Jinyu, 2008).

3.1 The primary selection of index system

The primary selection is to build a complete evaluation index system, in this stage, the number of the parameters and the intrinsic relationship of index, index of the operability is not to do too many requirements. Therefore, the primary selection of index system should follow the comprehensive scientific principles, strive to comprehensive. Give priority to the method of synthesis and analysis, frequency statistics method as auxiliary, construct a complete evaluation index. The primary selection including 139 indexes, involved the four aspects: social economic indicators, water resources development and utilization index, water ecological environment index and construction protection index.

3.2 The screening and the establishment of index system

The primary selection of index system just gives a general range to reflect water ecological environment health, but it is not necessarily completely science and necessary. Too many indicators may lead to repeat and mutual interference. Therefore, we must screening and optimization the primary indexes.

The principle of the index screening on the basis of not losing comprehensive, as far as possible to reduce the number of indicators. With the screening method of the qualitative and quantitative, through the expert system, theoretical analysis and correlation analysis to determine the basic form of the index system, and then the principal component analysis is used to quantitative analysis to determine the final form of the index system. Finally the following 28 indexes are selected. See Tab. 1.

Tab. 1 Evaluation index system of the water ecological environment health for region

Target	Criterion layer		Basic indicators	Index
Water eco-environment health	Social economic system	Social system	Water structure coefficient (0.062)	Water deficient ratio (0.031)
				Population density (0.023)
				Urban water supply pipeline leakage rate (0.010)
		Economic system	GDP unit water (0.040)	The utilization coefficient of irrigation water (0.013)
				Population density (0.023)
				The water-reusing rate of industry (0.013)

Continued Tab. 1

Target	Criterion layer		Basic indicators	Index
	Water resources system	Water resources condition	Drainage density (0.026)	
		Water resources development and utilization	Surface water utilization rate (0.048)	Groundwater index (0.048)
		Ecological index in the river	The satisfaction of minimum ecological water requirement in the river (0.079)	Lake eutrophication ratio (0.042) Hydrophilic landscape construction area, effect and accessibility
			The diversity index of large invertebrate benthic animal (0.079)	Wetland dynamic evolution analysis (0.024)
Water eco-environment health	Water eco-environment system	Ecological index outside of the river	The satisfaction of ecological water requirement outside of the river (0.113)	Flood control standard (0.037) Ecological water consumption per capita (0.037) Public green area per capita in built up area (0.037)
		Water environmental index	The water quality compliance rate of water function (0.067)	The water quality compliance rate of ground water (0.039) Urban per capita wastewater (0.016)
		The land use status indicators	The area ratio of water and soil erosion (0.053)	The land use dynamic evolution analysis (0.017)
	Construction protection system	Water environment management index	Sewage treatment (0.022)	Soil and water conservation area ratio
		Water ecological construction index	Forest coverage rate (0.012)	Protection investment accounts for GDP (0.012)

4 Healthy evaluation of water eco-environment for Shenyang

4.1 The method of multilayer fuzzy comprehensive evaluation

According to the data acquisition of Shenyang, the information of the indicators that soil and water conservation area ratio and hydrophilic landscape construction area, effect and accessibility can not get. Consider to delete the two indicators. This evaluation finally chooses 26 indicators.

Water ecological environment health was restricted and affected by numerous factors, including quantitative indicators and the qualitative indexes, quantitative indicators can get accurate numerical value through the statistical analysis and mathematics calculation, and qualitative indicators are difficult to use a numerical value to describe, have certain fuzziness, such as flood control standard, the satisfaction of minimum ecological water requirement in the river, etc. And the concept of water ecological environment health has certain fuzziness, and hierarchical level is difficult to have a definite range. For this kind of fuzziness strong problem, can use the method of multilayer fuzzy comprehensive evaluation to evaluate (Xie Jijian, et al., 2000).

The advantages of the fuzzy comprehensive evaluation can be a very good solution to solve the fuzzy comprehensive evaluation problem, more suitable for more influence factors and complicated system, can be more comprehensive response the quality of the evaluation system; but there are two of the more obvious disadvantages: first the fuzzy comprehensive evaluation can not eliminate the evaluation information between the problem of overlapping, and the second there is no more system and generally accepted methods to determine the membership degree. About the problem of overlapping information, in the front of index screening time, we have been through the combination of qualitative and quantitative methods, as far as possible to deal with the repeat information, the selection indexes relatively independent.

In this study indexes is divided into discrete index and continuous index at the time to determine the membership degree, for discrete index, according to index of the related data and present situation using expert evaluation determine its membership; and the continuous indicator can be divided into the bigger the better and the smaller the better, introduce the related formula to calculate.

Because the healthy evaluation of water ecological environment involves more indexes, and has obvious hierarchy, should to all levels of empowerment, analytic hierarchy process (AHP) can reasonable to all levels of empowerment, and AHP and fuzzy comprehensive evaluation can also be a good combination (Han Yuping and Ruan Benqing, 2003). Therefore, this research takes fuzzy comprehensive evaluation method as a basis, adopting the method of multilayer fuzzy comprehensive evaluation to establish the model of multi-level fuzzy comprehensive optimization, and evaluate the healthy of water ecological environment in Shenyang. See Fig. 1.

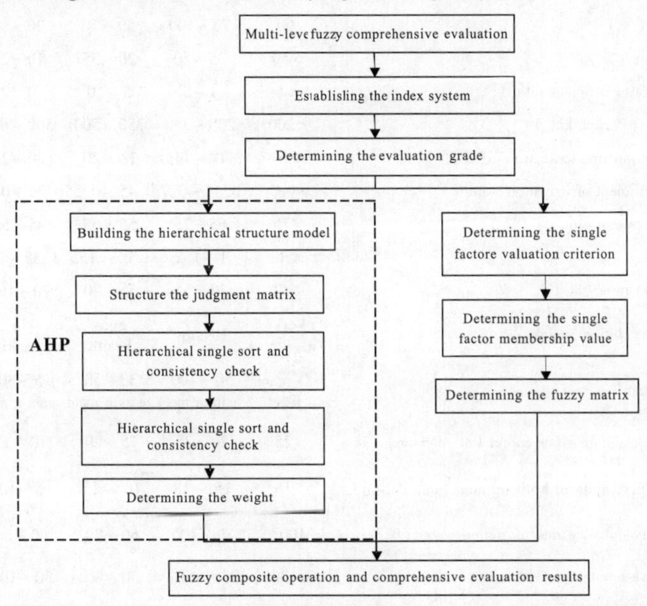

Fig. 1 The structure diagram of multi-level fuzzy comprehensive evaluation

4.2 The determination of evaluation criteria

The determination of evaluation criteria refer to the following four: ① for the indicators have national standards or trade standards specification can be used directly, such as the national water environmental protection standards; ② some indicators can be reference to international or domestic advanced area index, such as the water-reusing rate of industry, GDP unit water; ③ should be combined with the actual situation and regional development planning to determine, such as "The planning outline of water ecological environment system protection and restoration in Shenyang" and so on; and ④ combined with expert advice. The evaluation criteria will be divided into five levels of optimal (level Ⅰ), good (level Ⅱ), general (level Ⅲ), poor (level Ⅳ) and very bad (level Ⅴ) (See Tab. 2).

Tab. 2 Standard values of evaluation

Evaluation index	Level Ⅰ	Level Ⅱ	Level Ⅲ	Level Ⅳ	Level Ⅴ
Basic indicators					
Water structure coefficient	0 ~ 0.2	0.2 ~ 0.4	0.4 ~ 0.6	0.6 ~ 0.8	0.8 ~ 1
GDP unit water (m^3/yuan)	> 150	120 ~ 150	90 ~ 120	50 ~ 90	< 50
Drainage density	≥0.8	0.6 ~ 0.8	0.4 ~ 0.6	0.1 ~ 0.4	≤0.1
Surface water utilization rate (%)	< 10	10 ~ 20	20 ~ 30	30 ~ 40	> 40
The satisfaction of minimum ecological water requirement in the river (%)	Very satisfy	Satisfy	Basic meet	Dissatisfy	Very dissatisfy
The diversity index of large invertebrate benthic animal	> 3	2 ~ 3	1 ~ 2	0 ~ 1	0
The satisfaction of ecological water requirement outside of the river (%)	100	70 ~ 100	55 ~ 70	40 ~ 55	< 40
The water quality compliance rate of water function area (%)	100	70 ~ 100	55 ~ 70	40 ~ 55	< 40
The area ratio of water and soil erosion (%)	0	0 ~ 10	10 ~ 20	20 ~ 35	> 35
Sewage treatment (%)	100	70 ~ 100	50 ~ 70	30 ~ 50	< 30
Forest coverage rate (%)	> 50	35 ~ 50	20 ~ 35	10 ~ 20	< 10
Expansion indexWater deficient ratio (%)	< 1	1 ~ 5	5 ~ 10	10 ~ 20	> 20
Population density (Person/km^2)	< 200	200 ~ 350	350 ~ 500	500 ~ 600	> 600
Urban water supply pipeline leakage rate (%)	< 10	10 ~ 14	14 ~ 20	20 ~ 24	24 ~ 28
The utilization coefficient of irrigation water	> 0.7	0.55 ~ 0.7	0.45 ~ 0.55	0.35 ~ 0.45	< 0.35
The water-reusing rate of industry (%)	> 75	65 ~ 75	55 ~ 65	45 ~ 55	< 45
Groundwater index	≤1	1 ~ 1.2	1.2 ~ .135	1.35 ~ 1.5	> 1.5
Lake eutrophication ratio (%)	< 40	40 ~ 55	55 ~ 70	70 ~ 100	100
Wetland dynamic evolution analysis	Very healthy	Healthy	Sub-health	Ill health	Very unhealthy
Flood control standard	100 years a meet	50 ~ 100 years a meet	30 ~ 50 years a meet	5 ~ 30 years a meet	5 years a meet
Ecological water consumption per capita (m^3/Person)	> 25	20 ~ 25	15 ~ 20	10 ~ 15	< 10
Public green area per capita in built up area (m^2/Person)	> 18	15 ~ 18	10 ~ 15	5 ~ 10	< 5
The water quality compliance rate of ground water (%)	100	80 ~ 100	60 ~ 80	40 ~ 60	< 40
Urban per capita wastewater (m^3/Person)	0 ~ 30	30 ~ 50	50 ~ 80	80 ~ 100	> 100
The land use dynamic evolution analysis	Very healthy	Healthy	Sub-health	Ill health	Very unhealthy
Protection investment accounts for GDP	> 3.5	3 ~ 3.5	2 ~ 3	1 ~ 2	< 1

4.3 Healthy evaluation of water eco-environment for Shenyang

4.3.1 Multilayer fuzzy comprehensive evaluation

(1) Determining the evaluation factor set.

Assume the kth subsystem have m indexes, evaluation factor set can be said for$U_k = \{u_{k1}, u_{k2}, u_{k3}, \cdots, u_{km}\}$, this evaluation in a total of four subsystems, so($k=1,2,3,4$).

(2) Determining the evaluation set.

The evaluation level of all index are 5 levels, so the k th subsystem comments to collect$V_k = \{v_{k1}, v_{k2}, v_{k3}, v_{k4}, u_{k5}\}$.

(3) Establish a fuzzy relation matrix.

Determining the membership in comment set of the evaluation index, get the fuzzy relation matrix of the kth subsystem.

$$_kR = (r_{ij})_{m\times 5} = \begin{bmatrix} r_{11} & \cdots & r_{15} \\ \vdots & & \vdots \\ r_{m1} & \cdots & r_{m5} \end{bmatrix} \tag{1}$$

In which, r_{ij} is the i th index to the j th grade of membership in the kth subsystem.

According to the different characteristics of the indicators and the specific requirements of different subsystems to determine the different membership functions. In the healthy evaluation of water ecological environment, according to the basic principle of membership, adopting the expert evaluation method and formula the membership can be determined. For the discrete variable we adopted the experts marking method to determine in Tab. 1, such as the satisfaction of minimum ecological water requirement in the river, the diversity index of large invertebrate benthic animal etc.

For continuous variables using formula calculation, for the indicators, the greater the value, the water ecological environment is poor, adopting the drop half trapezoidal distribution function to determine their membership, then the five grades membership functions of each index are as follows:

$$U_{\mathrm{I}} = \begin{cases} 1 & 0 \leqslant x \leqslant A_1 \\ \dfrac{A_2 - x}{A_2 - A_1} & A_1 < x < A_2 \\ 0 & x \geqslant A_2 \end{cases} \tag{2}$$

$$U_{\mathrm{II}} = \begin{cases} 0 & x \leqslant A_1 \text{ 或 } x \geqslant A_3 \\ \dfrac{x - A_1}{A_2 - A_1} & A_1 < x < A_2 \\ 1 & x = A_2 \\ \dfrac{A_3 - x}{A_3 - A_2} & A_2 < x < A_3 \end{cases} \tag{3}$$

$$U_{\mathrm{III}} = \begin{cases} 0 & x \leqslant A_2 \text{ 或 } x \geqslant A_4 \\ \dfrac{x - A_2}{A_3 - A_2} & A_2 < x < A_3 \\ 1 & x = A_3 \\ \dfrac{A_4 - x}{A_4 - A_3} & A_3 < x < A_4 \end{cases} \tag{4}$$

$$U_{\mathrm{IV}} = \begin{cases} 0 & x \leqslant A_3 \text{ 或 } x \geqslant A_4 \\ \dfrac{x - A_3}{A_4 - A_3} & A_3 < x < A_4 \\ 1 & x = A_4 \\ \dfrac{A_5 - x}{A_5 - A_4} & A_4 < x < A_5 \end{cases} \tag{5}$$

$$U_{\mathrm{V}} = \begin{cases} 0 & 0 \leqslant x \leqslant A_4 \\ \dfrac{x - A_4}{A_5 - A_4} & A_4 < x < A_5 \\ 1 & x \geqslant A_5 \end{cases} \tag{6}$$

The formula $U_{\mathrm{I}}, U_{\mathrm{II}}, U_{\mathrm{III}}, U_{\mathrm{IV}}, U_{\mathrm{V}}$ are the membership function of the indicators, x is a specific value, A_1, A_2, A_3, A_4, A_5 are evaluation level standard.

For the indicators, the greater the index, water ecological environment is not safe, adopting up half trapezoidal distribution function, then the five grades membership functions of each index are as follows:

$$U_{\mathrm{I}} = \begin{cases} 1 & x \geqslant A_1 \\ \dfrac{x - A_2}{A_1 - A_2} & A_2 < x < A_1 \\ 0 & 0 \leqslant x \leqslant A_2 \end{cases} \tag{7}$$

$$U_{\mathrm{II}} = \begin{cases} 0 & x \geqslant A_1 \text{ 或 } x \leqslant A_3 \\ \dfrac{A_1 - x}{A_1 - A_2} & A_2 < x < A_1 \\ 1 & x = A_2 \\ \dfrac{x - A_3}{A_2 - A_3} & A_3 < x < A_2 \end{cases} \tag{8}$$

$$U_{\mathrm{III}} = \begin{cases} 0 & x \geqslant A_2 \text{ 或 } x \leqslant A_4 \\ \dfrac{A_2 - x}{A_2 - A_3} & A_3 < x < A_2 \\ 1 & x = A_3 \\ \dfrac{x - A_4}{A_3 - A_4} & A_4 < x < A_3 \end{cases} \tag{9}$$

$$U_{\mathrm{IV}} = \begin{cases} 0 & x \geqslant A_3 \text{ 或 } x \leqslant A_5 \\ \dfrac{A_3 - x}{A_3 - A_4} & A_4 < x < A_3 \\ 1 & x = A_4 \\ \dfrac{x - A_5}{A_4 - A_5} & A_5 < x < A_4 \end{cases} \tag{10}$$

$$U_{\mathrm{V}} = \begin{cases} 0 & 0 \geqslant A_4 \\ \dfrac{A_4 - x}{A_4 - A_5} & A_4 < x < A_5 \\ 1 & 0 \leqslant x \leqslant A_5 \end{cases} \tag{11}$$

The formula $U_{\mathrm{I}}, U_{\mathrm{II}}, U_{\mathrm{III}}, U_{\mathrm{IV}}, U_{\mathrm{V}}$ are the membership function of the indicators, x is a specific value, A_1, A_2, A_3, A_4, A_5 are evaluation level standard.

(4) Using analytic hierarchy process (AHP) to determine weight vector${}_k w_i$.

The calculation results see Tab. 1.

(5) Fuzzy composite operation, get the evaluation results.

$$ {}_kB = {}_kw \otimes {}_kR \tag{12} $$

In which, ${}_kw = (w_1, w_2, w_3, \cdots, w_n)$, $\otimes$ is fuzzy composite operator, ${}_kB$ is the membership which belonging to an evaluation grades after the kth system comprehensive evaluation.

(6) Comprehensive evaluation and calculation the overall rating value.

$$ B = w \otimes R \tag{13} $$

Where, $w = (w_1, w_2, w_3, w_4)$, $\otimes$ is fuzzy composite operator, B is the membership which belonging to an evaluation grades after the healthy comprehensive evaluation of water ecological environment. Give each level a score, and calculation the overall rating value.

4.3.2 The results of multilayer fuzzy comprehensive evaluation

For discrete index, the experts rating determined, the results as shown in the Tab. 3, for continuous index can be divided into positive index (the bigger the better) and negative indicators (the smaller the better), positive index whose membership using Eq. (7) to Eq. (11) to determine, negative index whose membership using Eq. (2) to Eq. (6) to determine, the results as shown in the Tab. 4 and Tab. 5.

Tab. 3 The membership values of dispersion indexes

Project	The satisfaction of minimum ecological water requirement in the river	The diversity index of large invertebrate benthic animal	Flood control standard	Wetland dynamic evolution analysis	The land use dynamic evolution analysis
U_{I}	0	0	0	0	0
U_{II}	0	0	0.25	0	0
U_{III}	0.20	0.70	0.50	0.50	0.60
U_{IV}	0.60	0.30	0.25	0.50	0.30
U_{V}	0.20	0	0	0	0.10

Tab. 4 The membership values of negative indexes

Index layer	U_{I}	U_{II}	U_{III}	U_{IV}	U_{V}
Water structure coefficient	0	0.3	0.7	0	0
Surface water utilization rate (%)	0	0.368	0.632	0	0
The area ratio of water and soil erosion (%)	0	0	0.994	0.006	0
Water deficient ratio (%)	0	0	0.696	0.304	0
Population density (Person/km^2)	0	0	0.129	0.871	0
Urban water supply pipeline leakage rate (%)	0	0.340	0.660	0	0
Groundwater index	0	0.257	0.743	0	0
Lake eutrophication ratio (%)	0	0	0	0.467	0.533
Urban per capita wastewater (m^3/Person)	0	0	0.388	0.612	0

Tab. 5 The membership values of positive indexes

Index layer	U_{I}	U_{II}	U_{III}	U_{IV}	U_{V}
GDP unit water(m^3/yuan)	0.238	0.762	0	0	0
Drainage density	0	0	0.28	0.72	0
The satisfaction of ecological water requirement outside of the river (%)	0.080	0.920	0	0	0
The water quality compliance rate of water function (%)	0	0	0	0	1
Sewage treatment (%)	0	0.605	0.395	0	0
Forest coverage rate (%)	0	0.3	0.7	0	0
The utilization coefficient of irrigation water	0	0.320	0.680	0	0
The water-reusing rate of industry (%)	1	0	0	0	0
Ecological water consumption per capita (m^3/Person)	0.4	0.6	0	0	0
Public green area Per capita in built up area (m^2/Person)	0	0.13	0.87	0	0
The water quality compliance rate of ground water (%)	0	0	0	0.802	0.198
Environmental protection investment accounts for GDP	0	0	1	0	0

Using the Eq. (12) to calculate the membership of each system, finally using the Eq. (13) to calculate the comprehensive membership of water ecological environment health in Shenyang, and according to the Tab. 6 to calculate the comprehensive evaluation value, the calculation results is 0.468,7. Results show that Shenyang water ecological environment health in the transition phase that unsafe to basic security.

Tab. 6 The rating value of water eco-environment

Grade	Very safe	Basic safety	Transition phase	Unsafe	Extremely unsafe
Score	0.8 ~ 1	0.6 ~ 0.8	0.4 ~ 0.6	0.2 ~ 0.4	0 ~ 0.2

4.3.3 The problem diagnosis of water ecological environment in Shenyang

According to the final score value, combined with the four subsystems and the target specific to analyze the reason, resulting in a stage of transition reason mainly consists of the following five aspects: ① the shortage of water resources, the development and utilization of large degree; ② sewage discharge more, water environmental problems; ③ pumping of groundwater, water quality compliance rate is poor; ④ soil erosion condition did not improve, management could not catch up the speed rate of destruction; and ⑤ water ecosystem worsening, rivers biodiversity decrease.

5 Conclusions

(1) Water ecological environment involves many aspects, multiple disciplines and affected by many factors, this study more consideration to water conservancy science, ecology, environment and resources shortage research. So in the future the theoretical research should be taken into account in all aspects, various disciplines, and further to deepen the basic theory of water ecological environment subject.

(2) Due to obtain material conditions is limited, some water ecological environment influential indicators cannot be considered, such as non-point pollution intensity, controlling soil and water loss rate, hydrophilic landscape construction effect accessibility, habitat condition, etc.. In addition, in the index selection, should as far as possible with the aid of remote sensing image and

some advanced monitoring equipment, makes the index data more accurately, the research, while in the index wetland dynamic evolution analysis and land dynamic evolution two indicators to use some of the remote sensing image data, but just stay in the area of the simple change analysis, and no further refinement to remote image, analysis of the regional landscape spatial pattern.

(3) Although the evaluation standards of considering the specific situations and development planning material in Shenyang City, but in practical application is still not enough accurate, pending further study.

References

Wang Hao, et al, Water Ecosystem Protection and Restoration Theory and Practice [M]. Beijing: China Water Conservancy and Hydropower Press,2010.

Geng Leihua, Bian Jinyu. The Evaluation Index System Study of Water Resources Rational Allocation [M]. Beijing: China Environmental Science Press,2008.

Xie Jijian, et al. The Fuzzy Mathematics Method and Lts Application [M]. Wuhan: Central China University of Science and Technology Press,2000.

Huang Changshuo, et al. China's Water Resources and Water Ecological Security Evaluation [J]. Yellow River, 2010,32(3):14 – 16.

Gao Yongsheng, Wang Hao, Wang Fang. Establish the Evaluation Index System of the River Health Life [J]. Water Scientific Progress, 2007,18(2):252 – 257.

Han Yuping, Ruan Benqing. The Multi-level Multi-objective Fuzzy Optimization Model in the Application of Water Safety Evaluation [J]. Journal of Resources Science, 2003,25(4):68 – 71.

Tu Min. The Method of River Health Assessment Based on the Water Quality Percentage of Water Function[J]. Yangtze River,2008, 39(23):130 – 133.

Reflections on the Ecological Water Regulation of Heihe River

Zhang Jie and ***Wang Daoliang***

The Administrative Bureau of Heihe River Basin, Lanzhou, 730030, China

Abstract: In 2007, according to the requirements of construction of Heihe's ecological civilization, the concept of ecological water regulation was proposed the first time, and made remarkable achievements through five years of exploration and practice, the tail of East Juyuan River never ran dry for seven years. Through the summary of ecological water regulation system, this paper analyses the response relationship between water regulation mode and downstream ecological system recovery, and finds out the key factor of restraining ecological water regulation, which are lagged basic research, roughness of regulation indicator, and lack of engineering measurement, etc.

For these key constraint factors, engineering and non – engineering measurements are proposed through several aspects, such as enhancing the research on rule of water need of ecological water system, perfecting ecological regulation indicator system, and setting up ecological monitoring and evaluation system as well.

Key words: Heihe River, regulation of ecological water, main stream, reflection

1 Background of regulation of ecological water of Heihe River

1.1 The emerging of the concept of regulation of ecological water of Heihe River

Since 1990s, regarding the serious deterioration of ecological system of Heihe River Basin, and obvious contradiction of water problem, government implemented integrated water resources management and regulation of Heihe River, and arranged large scale governance of basin. Through 12 years of practice and exploration of regulation, water regulation started from emergency water regulation, and developed from regular regulation, which begun to curb the trend of deterioration of Heihe River's downstream ecological environment, and eased water contradiction of water between regions. In 2007, according to the requirement of building harmonious society, carrying out concepts of ecological civilization and new idea of the Ministry of Water Resources and also the new object of "four important turnings" from YRCC, the idea of regulation of ecological water for Heihe River is firstly proposed, the key point is to regulate water according to demand rule of down stream's water resources, persisting in building and protection of basin's ecosystem, and using scientific management of water resources, and rational allocation, efficient usage.

1.2 The goal of regulating eco water

With the demand of water resources of Heihe River downstream ecosystem, in practicing one can continuously explore, initiate the idea and pattern of regulation, as well as realizing the transformation and leap. Recent objective goal of eco water regulation: first of all, to realize "two guarantees", Firstly one, to guarantee the water flow of 9.5×10^8 m^3 for Zhengyi gorge when Yingluo gorge's water inflow attains 1.58×10^9 m^3 en average; the second is to guarantee conveying water to East Juyuan River, then, optimizing the allocation of water resources to ensure all the water used to eco construction coming from Ejina oasis, realizing maximum of eco return, as well as maintaining and improving the eco system of downstream and Weilu. Long term goal will establish long term mechanism of eco water regulation of Heihe River, realizing long term objective of water division of the State Council, scientifically, rationally arranging the water of basin for life, production and eco water, attaining 9.9×10^8 m^3 of eco water for whole basin, gradually increasing the volume of water entered in downstream area and Juyan River, recovering and reconstruct maximum downstream eco system, boosting the development of basin's economy harmoniously, maintaining healthy lives of Heihe River, as well as establishing Heihe River harmonious basin.

2 Main approach of eco water regulation of Heihe River

2.1 Proposing the idea of eco water regulation

To obtain the maintenance of basin's related departments, through conference of water regulation of different classes, one can introduce this idea of Heihe River water regulation, and emphasize the idea of eco water regulation, we received positive responses from different parts. At the meantime, when we work out the annual proposal of water regulation, we take into account that maintaining, improving downstream and WeiLu eco system, as well as scientifically, rationally allocating downstream water volume as regulation objective. And through medias' promotion and report of eco water regulation, good atmosphere for water regulation will be established.

2.2 Formulating proposal of Heihe River eco water regulation, establishing criterion system for Heihe River eco water regulation

With the principle of maximization of eco return, and incorporating the situation of Heihe River Basin, based on the analysis of practice of water regulation and arrangement of obtained result, we can formulate the proposal of Heihe River eco water regulation, and point out precise requirement and criterion of eco water regulation for upper, middle and down streams, which can be used as basis of carrying out Heihe River eco water regulation.

2.2.1 Eco requirement for upper and middle streams

For upstream area, we can stop the trend of devolution of natural grass and reduction of forest, increase the ability of self - recovering of natural grass through ban of grazing, increase of plants and forest for the beginning area of Heihe River.

Eco regulation requirement for irrigation area of up and down streams: firstly, to guarantee the water flow of 9.5×10^8 m^3 for Zhengyi gorge when Yingluo gorge's water inflow attains 1.58×10^9 m^3 en average; secondly through adjustment of industrial structure and planting structure, one can optimize allocation; of water resources, reduce water volume of agricultural irrigation, gradually increase ratio of usage of eco water, prevent from desertification, finally, through the regulation of water on the ground and underground, one can control rationally underground water level, and prevent from alkalization.

2.2.2 Eco regulation requirement of downstream

The importance of eco water regulation consists in the downstream, the main criterion system of eco water regulation of downstream Ejina oasis contains 5 aspects:

(1) Water volume of Langxin Mountain.

The section of Langxin Mountain is the section to enter Ejina oasis, en average, the criterion of eco water regulation, increased to 0.5×10^9 m^3 from the current 0.487×10^9 m^3.

(2) Langxin Mountain section flowing process and water allocation criterion.

Annual water allocation: Langxin Mountain section water ratio between winter, spring and summer, autumn adjusted to 5:5 from 4:6.

The allocation ratio between east and west water volume is 7:3, for those years with large volume of water, this ratio is adjusted to 6:4.

(3) Criterion for underground water level.

From current data, on 2004, the surface of underground water less than one meter of Ejina oasis is around 1,000 km^2, the surface of underground water less than two meters is around 1,200 km^2, for less than three meters equals to 1,400 km^2. Different plants' water demand differs. The underground water level for populous is, 1 m to 3 m for infant period, 3 m to 5 m for mature period, if the level is between 5 m and 8 m, populous become deteriorated, and if it is bigger than 10 m, it is difficult for populous to live; for willow, acacia thorn, this kind of shrub, they can grow when the level is less than 5 m, which become deteriorated between 5 m and 7 m, from 7 m to 8 m, gravely deteriorated, and hard to live if it is bigger than 8 m; for herbaceous plant, they can grow at 2 m to 4 m, and hard to live if it is bigger than 4 m. Carrying out the eco water regulation, the surface of Ejina oasis's underground water level less than 3 m attains 1,500 km^2, to guarantee the plants' water demand of main oasis area is not less than their lest demand.

(4) Water criterion of East Juyuan River.

With analysis, to maintain a certain level of surface of water and water volume of East Juyuan River, one has to add about 50×10^6 to 60×10^6 m^3 water to East Juyuan River, and to maintain water surface around 25 km^2 to 35 km^2, as the annual compensating criterion of East Juyuan River.

2.3 Enhancement of administration of water usage and coordination

2.3.1 Enhancement of intensive regulation work, increase of downstream water volume in spring

Regarding the growth of downstream plants, spring is the key period for plant's growth, increasing the volume of water entering downstream plays an important role. As a result, in conjunction with industrial adjustment of middle stream, reducing plants harvested in summer, increasing plants harvested in autumn, through coordination, one can enhance the water management and centralize regulation, continuously increase regulating period in spring, days of closure augments from 15 d on 2007 to 35 d on 2011, which increases water flow of Zhengyi gorge efficiently, enlarges irrigation area of eastern and western region of downstream. This is a breakthrough for regulation in spring in terms of quality and quantity.

2.3.2 Emphasizing of regional optimal allocation of basin's water resource

As a controlling factor for maintaining and improving eco environment of Heihe River Basin, water resource has to be satisfied for balance and development. For optimal allocation in Ejina oasis region of downstream, with limited water resource, according to water inflow of Langxin Mountain, one has elaborated eco water's allocation plan of Ejina oasis, arranging reasonably water allocation for eastern river, western river and east main channel of Langxin Mountain, taking use of east river and east main channel, the priority is to convey water to east Juyan river, and then guarantee the water demand for Weilu region, using established construction, allocate water to oasis' edge area, eco fragile area, and grass that is not irrigated last year, irrigate from region to region, regulate precisely, enlarge grass irrigation area maximally, in order that the improvement and recovery of Ejina oasis can develop in terms of space.

2.3.3 Enhancing the control of regulation process

In practicing regulation, according to water's variation from upstream and irrigation water situation of middle stream, implementing the measurement "all line closed, centralized discharging", "limiting coming water, equally discharging", as well as regulation of blood period, one can regulate smoothly the water's coming process for a whole year, and enhance the control of regulation process, increase water flow of Zhengyi gorge efficiently, and relieve the situation of lack of water for downstream, compensate efficiently water volume for east, west river, Nalin River as well as east Juyan River.

2.3.4 Enhancing survey and supervision of eco water regulation

In regulation, one can survey , supervise, and indicate Ejina Qi to enhance the management of eco water, allocate rationally coming water, all the water regulated to downstream must be used to improve and recover the eco environment, control and reduce agricultural irrigation area of downstream, and realize maximization of eco return.

3 The effect of Heihe River eco water regulation

3.1 Water flow of Langxin mountain increases, days of absence of water flow decreases

Refereed as water flow of different time periods and flowing days of LangXin Mountain for recent five years, in April and May of each year, the average water flow is 34 $\times 10^6$ m^3 cubic meter, it increases 79% every 7 $\times 10^6$ m^3 compared with 2000 ~ 2006. The percentage of water flow for spring and summer of Langxin Mountain is adjusted from 1.9:2.8 to 2.6:3.6. For the last five years, water flowed all east and west rivers in spring, the edge of oasis, ecologically fragile area have been irrigated, the plants died a few years ago came to recover. In spring, the volume water flowed to downstream increases, which satisfies the water demand for plants in growth period, it plays an positive role in helping the plants grow. At the meantime, through eco water regulation, the volume of water entered to Ejina oasis increases, water flow of Langxin Mountain is 6.18 $\times 10^8$ m^3 on average for the last five years, which increases by 25% compared to 2000 to 2006, and by 75% without regulation. Flowing days is 221 d on average, 25 days more than 2000 till 2006, 100 d less than without regulation, which has prominent eco effect, see Tab. 1.

Tab. 1 Comparison of water flow at different time of Langxin Mountain

(unit: $\times 10^8$ m^3)

year	Closed time	days	Langxin Mountain				Annual flowing days
			April and May	Spring and Winter	Autumn and summer	Whole year	
2007	4.6 ~ 4.20	15	0.51	2.37	4.11	6.49	211
2008	4.2 ~ 4.21	20	0.42	3.09	3.90	6.99	229
2009	4.1 ~ 4.24	23	0.13	2.58	4.21	6.79	212
2010	4.8 ~ 5.4	26	0.27	2.51	2.32	4.83	219
2011	4.3 ~ 5.8	35	0.37	2.52	3.28	5.81	234
2007 ~ 2011 annual average		24	0.34	2.61	3.57	6.18	221
2000 ~ 2006 annual average		5	0.07	1.86	2.77	4.63	196

3.2 The water for life, production, and eco usage for whole basin begins to be rationally allocated, the trend of eco environment deterioration has been limited, and environment begins to be improved

Firstly one has to satisfy the water demand to live, at the meantime, one also guaranteed the usage of water for research for national defense, which guaranteed the water demand for launch and experiments of spacecraft, rationally arranged the water of production and eco usage for downstream. The level of underground water for Ejina oasis and related area increases, east river area augmented by 0.48 m, west river area augmented by 0.36 m, and 0.48 m for east Juyan River area. Populous and willow, which are on the edge of death, have been protected, annual growth of the diameter of Populous increased by 2.72 mm, growth period prolonged by 15 d to 20 d. The surface of oasis based on Populous, grass, and shrub increased by 40.16 km^2, the species of animals and plants also increased, the trend of eco environment deterioration has been limited, and environment begins to be improved.

3.3 East Juyan River did not dry out for 7 years, and the eco environment around has been improved prominently

From 20th august 2004, East Juyan River did not dry out for 7 years. The surface of east Juyan River attained 42.8 km^2, the accumulated added water volume achieved 3.18×10^8 m^3 under regulation. Now, around the lake, one can frequently see some rare animals, such as, swans, wild gooses, and yellow ducks, etc. And once disappeared bullhead which is a specific fish in Heihe River emerged again, and the biggest bullhead fished is of length of already 30 cm, the biggest hairy crab weighted 50 g. The water in the lake is clean, bulrush waves, around the lake, the surface of grass based on bulrush and Yan Zhuazhua attains 50 km^2, the variety of species and plants' covered area increased prominently, which tends to positively alter, at the meantime, East Juyan River maintains a certain level of surface and volume of water, which increased the ratio of water in the earth around the lake, and added underground water, efficiently expanded the eco tract of Ejina oasis, prevented the desert from moving, reduced the frequency and magnitude of sand storm, the eco system of Wei Lu recovered gradually.

3.4 Stimulating the development of tourism industry, pulling local economy

The underground water level increased, the eco environment based on Populous has been better off, "Guai Forest" encountered another growth period, the tourism is becoming more attractive. As reported, from 2007, "Ejinaqi Autumn Populous eco tourist festival" accommodated 107,000 tourists around the world, the tourist revenue attained 5.94×10^8 yuan, as the tourist revenue increased gradually, tourism became the main industry of Ejina, Autumn Populous eco tourist festival also became one of the three festival activities of Inner Mongolia Autonomous Region. Water regulation not only prevented the Ejina oasis from deterioration, also improved the life quality of soldiers of downstream Dongfeng area, stabilized the troop, stimulated national defense construction, stimulated solidarity of nation and stationarity of borderland, prospered populous tourism and economy, pulled the development of local economy, augmented the revenue of herdsmen and farmers, improved life, living and production environment, social employment was better off, people lived in good situation, social security is guaranteed, stimulated solidarity of nation and social stabilization.

4 Existing problems

With exploration and carrying out eco water regulation, one has got some good results, but along with the proceeding of work, some problems arose.

4.1 Measure of regulation is sole, it is lack of maintenance and guarantee from law, economy and projects for eco regulation

Eco water regulation is highly relied on the maintenance and guarantee of law, economy, as well as projects. From the administrative point of view, on one hand, the responsibility does not taken everywhere, on the other hand, institution of basin does not have direct and efficient administrative limit measure to local government. From the point of view of law, for those problems, such as, responsibility is not clear, confused right, administrative system is not perfect, etc, there is not operational legal basis, so if some illegal actions arose, there is not corresponding punishment basis and measure, which weakened the effect of administration and supervision; from economic point of view, for the usage of water out of criterion, because of lack of adjustment and restrict of economic leverage, it caused the low efficiency and waste of usage of water resource; from the point of view of project, the planned main project still did not commence, especially lack of construction of transfer and store, one cannot adjust efficiently the natural water resource. Eco regulation consists in timing of regulation according to plant's growth law, so it emphasizes the control of flowing process, then the problem of sole measure of regulation is highlighted, which restricted the proceeding of eco water regulation of Heihe River.

4.2 Basic research did not satisfy the demand of eco water regulation

The realization of regulation requires a lot of basic research, which is the basis of realizing creativity and breakthrough. After implementing the recent management planning, although we have done some basic research, as it lacks a lot of document before, then these research are used to fill the document needed before, or the research was carried out because it is urgent, for example, the distribution and growth law of natural plants as populous, the influence of water process to plant's growth, the demand of eco water of Weilu lake and oasis artificial eco system is not well known and mastered, it lacks of a scientific and complete criterion pattern of eco water regulation, then it is hard to satisfy the demand of eco regulation.

4.3 Lack of system of supervision for eco environment

Developing the eco environment supervision is the basis for implementing eco protection, environment governance and evaluation, as well as the basis for carrying out eco water regulation and ecological evaluation. Recently, the established administrative system consists in regulation, realization of remote control, construction of planning, eco environment supervision function is not included in construction of this system, one cannot master in time the distribution of downstream plants and situation of earth, underground water level, as well as dynamic information of ecological variation, it cannot well supervise and control eco environment, and it is not convenient to actual demand of regulation.

5 Forward outlook

To guarantee the implement of eco water regulation, realize the goal of regulation and plan of each phase, for the next step, one can use the following measure.

5.1 Pushing "synthetical planning of Heihe River Basin", especially the implementation of main project for example, HuangCang temple construction, etc. establishing a long term mechanism of eco water regulation

In order to continuously utilize water resource of Heihe River, improve and fix the basin's eco system, one has to rely on investing in the second period governance of Heihe River. Through implementing synthetical planning of Heihe River Basin, one can construct project system and non-project system for unified water governance and regulation, complete measure like project, law, economy, administration as well as scientific technology, construct long run mechanism of water regulation, based on reasonable allocation, efficient utilization as well as ecological repairing of water resource, one can realize scientific governance, rational allocation, efficient utilization and efficient protection of water resource, maintain healthy lives, the coordination and sustainable development of population, resource, environment as well as economy and society.

Now, eco regulation consists in timing of regulation according to plant's growth law, so it emphasizes the control of flowing process. But the regulation of Heihe River only depends on administrative measure in "all line closed, centralized discharging", the fulfillment of Zhengyi gorge discharging goal and allocation of water resource mainly consist in natural water process of upstream, in order to accomplish the goal of the State Council, supply sufficient water for downstream eco recovery, one needs to construct some main projects, for example, Huangcang Temple and Zhengyi gorge main stream construction, especially the most important construction—— Huangzang temple water control project must be constructed as soon as possible.

5.2 Carrying out basic research like water demand law for populous growth, to maintain eco water regulation

To allocate rationally limited water resource, maintain and improve eco system of Heihe River downstream and Weilu, realize the maximization of ecological return, one needs to do the research about criterion system of eco water regulation and ecological water demand law, which provides theoretical foundation and technical maintenance of eco water regulation.

5.2.1 Research on water demand law for populous growth

In order to rationally allocate water resource of Heihe River, maximally protect populous, one has to do the research on its water demand for growth, study its growth characteristics and water demand law, underground water level, bloom period of populous origin, and then one can have water demand criterion of different periods.

5.2.2 Research on water demand law of eco system of Heihe Weilu

In order to maintain and improve eco system of Weiju, one has to do the research on its water demand, investigate and state the actual situation and related characteristics of Weiju, study the water demand law of eco system of Weilu.

5.2.3 Research on criterion system of eco water regulation

To construct rational system of water regulation criterion, it is the basis of implementing eco water regulation, one has to research the system of water regulation criterion, provide the requirement of water demand for Langxin Mountaion water flow and flowing process, identify the requirement of underground water level of different plants, give the water demand criterion of plant's recovery around east Juyan lake, identify the surface of downstream oasis and grass for sustainable development.

5.3 Establishing supervision and evaluation system, which provides theoretical foundation

Enhancing the ecological supervision, especially for downstream of Heihe River and Weiju, using the technology such as remote control, remote supervision, mastering the supervision information like Ejina oasis irrigation, earth information, underground water level, variation of plants. One can supervise specific ecological environment spatially and temporally, which can provides information for eco water allocation. At the meantime, one has to establish eco effect supervision and evaluation system, then adjust and optimize regulation plan, it can provide information for eco water regulation and rationally evaluation of regulation effect.

5.4 Accelerating legislation process, to provide legal guarantee for water regulation

Eco water regulation involves all kinds of connections, including dam, national resources, agriculture, forestry, electricity, etc. In order that economic society and ecological construction can realize sustainable development, one needs different departments cooperate. The actual law of Heihe River is only available for ministry of water resources, it does not have any restrictive effect on other departments and industries, then we need to formulate regulations on Heihe River Basin, clarify and specify the responsibility, right of related departments, and restrict other's action, which provide legal guarantee for water regulation.

Significance of European Ecological and Biological Monitoring Methods to the Establishment of China's Aquatic Ecological Monitoring System

Wen Huina[1], ***Martin Griffiths***[2], ***Mu Yizhou***[1] and ***Yang Wenbo***[1]

1. Yellow River Water Resource Protection Bureau, Zhengzhou, 450004, China
2. EU – China River Basin Management Programme

Abstract: The EU Water Framework Directive is one of the most far reaching environmental initiatives undertaken by the EU to date. It sets new objectives for all rivers and lakes based on ecological and biological indicators of river health. Biological indicators have been developing in Europe over the past 30 years or so, but the EU Water Framewoik Directive is the first time that these have been adopted into mainstream water management. The wealth of knowledge on ecological and biological monitoring that has been built up in Europe over the past 10 years since the adoption of the EU Water Framework Directive. China's new round of national water resources protection planning guideline which was issued by Ministry of Water Resources in September 2012 stipulates that the principle of this new round planning is the integration of quantity, quality and ecology. However, the research on river and lake ecological protection was just started in China, the pilot results have not been integrated and applied in the routine water resource management and especially the aquatic ecological monitoring system as the basis and key element for ecological protection planning has not been established in China's main river basins. This paper drew together knowledge of the EU in the aspects of surface water classification and assessment method, statistical elements in monitoring programme design, methods of field sampling site selection and sampling technology in order to assist China to build a technical standard system specifically to China's national realities and natural status.
Key words: EU Water Framework Directive, ecological, biological, monitoring

1 Comparison of European ecological and biological monitoring development and China's current status

The history of ecological and biological research in Europe is over 30 years. The EU Water Framework Directive (WFD) which published in 2000 has set up the target of achieving the good status. Ecological status is the main indicator for the health of water ecological system. In order to achieving the requirement of WFD, the member states have to develop ecological and biological monitoring. In the recently 10 years of implementing WFD, EU has issued the supporting ecological and biological monitoring standards including the classification of surface water, biological indicator selection, statistical assessment method and so on.

Since 1990s China has started to pay attention to ecological protection in river management and has conducted various researches on river health assessment indicator and methods. In 2010, the Ministry of Water Resources issued the Health Assessment (Pilot) Outline of Important National Rivers and Assessment Indicators, Standards and Methods of River Health (for Pilot Work), officially launching the national – level river health assessment pilots led by the water conservancy administrative departments. This work has an important epoch – making significance in China's river management and protection. However, due to the late start of water ecology assessment, extreme lack of historical monitoring data and almost no system monitoring capability results in unknown ecological background and formidable challenges in the further development. Therefore, it's of practical significance to learn from developed countries of the advanced technology, and build a technical standard system specifically to China's national realities and natural status.

2 Surface water classification schemes

According to the EU WFD, surface water quality is expressed as Chemical Status and Ecological Status. The Chemical status is based on the concentration of priority substances. There are European standards for all priority substances. Based on the principle of one out - all out the Chemical Status can be good or bad. There are no European standards for the Ecological status; each Member State has to develop its own ecological assessment method and ecological standards. Fig. 1 provides a schematic view of this.

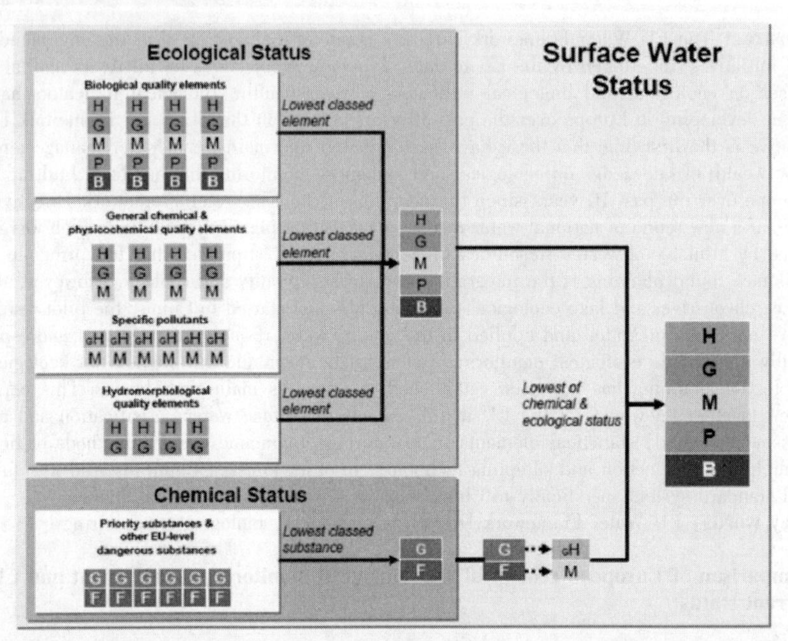

Fig. 1 Schematic representation of how results for different quality elements are combined to classify ecological status, chemical status and overall surface water status

China's current surface water assessment approach is limited by one single indicator - water quality which needs to be improved. The nationwide pilot assessment of the health of rivers and lakes, currently being done by the Ministry of Water Resources, aims to break away from the traditional methodology and adopt a holistic approach for a better picture of the ecological aquatic situation so as to provide more reliable data for better water resources management and protection plan. However, the results of the pilot study have not been applied in the routine monitoring system, and in most of the background investigation programmes conducted by the monitoring agencies there is no ecological and biological element added. One of the possible causes is technical obstacle, people have no knowledge of ecological and biological monitoring and no idea of what to do in ecological investigation, the other cause may be the traditional constrained thought which only concerns the quantity and quality. Therefore, it is very necessary to disseminate the river&lake health assessment pilot results to the monitoring agencies and enhance the training to the monitoring staff, make people realize the significance of ecological monitoring and be aware of the future monitoring direction. New round of national water resources protection planning requires to formulate the ecological monitoring system based on the routine system. The ecological monitoring relates to professional knowledge such as hydrology, ecology and statistics, as well as the accumulation of history data, it lays dramatic challenges to the current very low level of aquatic monitoring ability in China.

3 Biological quality element and indicator selection

Biological indicators have been developing in Europe over the past 30 years or so, but the EU Water Framework Directive is the first time that these have been adopted into mainstream water management. Different Biological quality elements are appropriate to the different surface types, for rivers and lakes, the normal biological quality elements are benthic invertebrates, fish, phytoplankton, macrophytes and phytobenthos.

Operational monitoring only needs to assess parameters indicative of the quality elements most sensitive to the pressures to which the water bodies are subject. Tab. 1 provides an expert judgement view (from UK Technical Advisory Group) as to which elements are most appropriate for specific pressures. The table provides a means of focusing monitoring effort to aid efficient use of resources. Where more than one element is sensitive to a pressure e. g. all are sensitive to eutrophication, expert opinion should be employed to choose the most sensitive elements for the category of water concerned.

Tab. 1 Quality elements sensitive to the pressures affecting rivers (From UK TAG, 2005, 12 A paper)

Source pressure	Category of effect	Exposure pressure	Macrophyte	Phytcbenthos	Macro – invertebrates	Fish	Morphology	Hydrology	General physico. chenical	Specific pollutants	Priority substances	Priority hazardous substances
Nutrient enrichment	Pnimary effect on biology	Change in nutrient concentration in defined water body. Enhanced biomass, changes to other primary producers	×	×				×	Nutrient suite			
Organic enrichment	Primary effect on biology	increased organic ennchment, change in biological community structure			×			×	Organic suite			
Annex 8 and annes 10 pollutants	Primary effects on sediment and water quallty	increased concentrations of contaminants (water column and sediments)			×			×	General suite	×	×	×
Hydrological	Primary effect on biology	Changed water levels from asbtraction; attered flow regime impac ting biology	×	×	×	×	×	×	General suite			
Morphological	Primary effect on biology	Riparian and channel modification, Aitered sediment characteristics (e. g. size), smothering and damage to river bed	×		×	×	×	×				
Acidification	Primary effedft on biology	Change in ANC & pH; change in biological community & toxicity synergies		×	×	×			Acidification suite			

The use of multiple metrics can improve confidence in the final classification. For example, when assessing phytoplankton, biomass is an important metric because it determines the overall amount of phytoplankton which in turn influences light penetration and oxygen concentration in a water body. Taxonomic composition is also an important metric of phytoplankton because it shows when highly undesirable species, such as cyanobacteria and other opportunistic taxa, are starting to dominant the phytoplankton community.

4 Reference conditions

The main goal of stream assessment according to the Water Framework Directive is to classify a stream stretch into an Ecological Quality Class ("high", "good", "moderate", "poor" or "bad"), which is defined by its deviation from a stream – type specific reference condition (AQEM consortium, 2002). So, the basic principle of ecological water quality assessment in the WFD is to compare the actual situation of a water body with the reference conditions of the relevant water type. When the reference conditions are met, the quality is defined as "good status".

The general definition of Good Status is given in the WFD: "There are no, or only very minor, anthropogenic alterations to the values of the physico – chemical and hydromorphological quality elements for the surface water body type from those normally associated with that type under undisturbed conditions. The values of the biological quality elements for the surface water body reflect those normally associated with that type under undisturbed conditions, and show no, or only very minor, evidence of distortion. These are the type – specific conditions and communities."

The reference values are derived from equivalent sites that have no or minimal alterations as a result of human interference or pressure. UK TAG recommends that 'reference conditions should reflect a state in the present or in the past corresponding to very low pressure, without the effects of major industrialisation, urbanization and intensification of agriculture, and with only very minor modification of physico – chemistry, hydromorphology and biology.'

Reference values may be determined using:

(1) networks of reference sites;;

(2) modelling approaches;

(3) or, where 1 and 2 are not possible (even in combination), expert judgement;

(4) network of reference sites.

In the AQEM project criteria for the selection of reference condition sites (AQEM consortium, 2002):

A reference stream should fulfil all requirements necessary to allow a completely undisturbed fauna to develop and establish itself. Therefore, "reference sites" should not only be characterized by clean water but also by undisturbed stream morphology and near – natural catchment characteristics. Though it is impossible for many stream types to find sites in such a pristine condition, AQEM has defined the following criteria, which should be met by "realistic" reference sites:

4.1 Basic statements

(1) The reference condition must be politically palatable and reasonable.

(2) A reference site, or process for determining it, must hold or consider important aspects of "natural" conditions.

(3) The reference conditions must reflect only minimal anthropogenic disturbance.

4.2 Land use practices in the catchment area

In most countries there is anthropogenic influence within the catchment area. Therefore, the degree of urbanisation, agriculture and silviculture (forestry) should be as low as possible for a site to serve as a reference site. No absolute minimum or maximum values have been set for the defining reference conditions (e. g. % arable land use, % native forest); instead the least – influenced sites with the most natural vegetation are to be chosen.

4.3 River channel and habitats

(1) The reference site floodplain should not be cultivated. If possible, it should be covered with natural climax vegetation and/or unmanaged forest.

(2) Coarse woody debris must not be removed (minimum demand: presence of coarse woody debris).

(3) Stream bottoms and stream margins must not be fixed.

(4) Preferably, there should be no migration barriers (affecting the sediment transport and/or the biota of the sampling site).

(5) Only moderate influence due to flood protection measures can be accepted.

4.4 Riparian vegetation and floodplain

Natural riparian vegetation and floodplain conditions must be retained, making lateral connectivity between the stream and its floodplain possible; depending on the stream type, the riparian buffer zone should be greater or equal to 3 × channel width.

4.5 Hydrologic conditions and regulation

(1) No alterations of the natural hydrograph and discharge regime should occur.

(2) There should be no, or only minor upstream impoundments, reservoirs, weirs and reservoirs retaining sediment; no effect on the biota of the sampling site should be recognisable.

(3) There should be no effective hydrological alterations such as water diversion, abstraction or pulse releases.

4.6 Physical and chemical conditions

There should be:

(1) no point sources of pollution or nutrient input affecting the site;

(2) no point sources of eutrophication affecting the site;

(3) no sign of diffuse inputs or factors which suggest that diffuse inputs are to be expected;

(4) "normal" background levels of nutrient and chemical base load, which reflect a specific catchment area;

(5) no sign of acidification;

(6) no liming activities;

(7) no impairments due to physical conditions; especially thermal conditions must be close to natural

(8) no local impairments due to chemical conditions; especially no known point - sources of significant pollution, all the while considering near - natural pollution capacity of the water body;

(9) no sign of salinity.

4.7 Biological conditions

There must not be any:

(1) significant impairment of the indigenous biota by introduction of fish, crustaceans, mussels or any other kind of plants and animals;

(2) significant impairment of the indigenous biota by fish farming.

In many cases, e. g. some lowland stream types or larger streams, no reference sites meeting the criteria above are available. For these stream types the "best available" existing sites, which meet most of the criteria should only be a starting point; the description of reference communities should be supplemented by evaluation of historical data and possibly the biotic composition of comparable stream types, e. g. streams of a similar size but located in different ecoregion (AQEM consortium, 2002).

As said, the reference values should be based on information obtained from sites at which the quality element concerned is in reference condition (i. e. at high status). UK TAG states that this does not mean that at these sites the quality element will be entirely unaffected by human activities. However, it does mean that alterations to it are expected to be minor. There are relatively few sites at which all quality elements are in reference conditions and from which data suitable for establishing reference values are available. Consequently, reference values can be derived from sites at which the quality element concerned is estimated to be in its reference condition but other elements at the sites may not be so. These sites can also be in water bodies within which there are other sites at which the quality element may not be in its reference condition.

4.8 Temporally based reference condition

Instead of actual reference conditions a temporally based reference condition may be used. Temporally based reference conditions may be based on either historical data or paleo - reconstruction, or a combination of both approaches. Both of these approaches are commonly used in areas where human - induced stress is widespread and unperturbed references are few or lacking entirely.

For example, paleo – reconstruction of past conditions may be determined either (i) directly, based on species presence/absence from fossil remains or (ii) indirectly, using relationships between fossil remains and inference to determine other values such as the reference pH situation. One of the strengths of a paleo – approach is that it can often be used to validate the efficacy of other approaches if the conditions are stable (CIS guidance 10, 2003).

Another advantage is that recent step – changes in ecological status are more easily determined. A further strength of paleo – reconstruction is that if strong relationships exist between land use and ecosystem composition and function, a predictive approach (hindcasting or extrapolating dose – response relationships) may be used to predict quality elements prior to major alterations in land use (e. g. pre – intensive agriculture).

Both of these approaches share some of the same weakness. They are usually site and organism – specific, and hence may be of limited value for establishing type – specific values. Regarding paleo – reconstruction, caution should be exercised in unequivocal reliance on this method as providing the definitive value, as choice of the calibration dataset used to infer ecological status may result in different values. Regarding the widespread use of historical data, it may be limited by its availability and unknown quality (CIS guidance 10, 2003).

5 Statistical elements in monitoring programme design

Risk, Precision and Confidence are key concepts in the design of monitoring programmes and are principles that underpin the EU WFD. These factors determine the shape and intensity of the monitoring programmes and determine aspects such as:

(1) number of waterbodies included in monitoring programmes;

(2) number of stations required to assess the status of each waterbody;

(3) the frequency at which parameters will have to be monitored.

Choosing levels of precision and confidence will set limits on how much uncertainty (arising from natural and anthropogenic variability) can be tolerated in the results of monitoring programmes. In terms of monitoring it will be necessary to estimate the status of water bodies and in particular to identify those which are not of good status or good ecological potential or are deteriorating in status. Thus status will have to be estimated from the sampled data. This estimate will almost always differ from the true value (i. e. the status which would be calculated if all water bodies were monitored and sampled continuously for all components that define quality).

The level of acceptable risk will affect the amount of monitoring required to estimate a water body's status. In general terms, the lower the risk of misclassification desired, the more monitoring (and hence costs) required to assess the status of a water body. It is likely that there will have to be a balance between the costs of monitoring against the risk of a water body being misclassified.

Misclassification implies that measures to improve status could be inefficiently and inappropriately targeted. It should also be borne in mind that in general the cost of measures for improvement in water status would be orders of magnitude greater than the costs of monitoring. The extra costs of monitoring to reduce the risk of misclassification might therefore be justified in terms of ensuring that decisions to spend larger sums of money required for improvements are based on reliable information on status. Further, from an economics point of view, stronger criteria should be applied to avoid a situation where water bodies fulfilling the objective are misjudged and new measures applied. Also it should be noted that for surface water surveillance monitoring, and all groundwater monitoring, sufficient monitoring should be done to validate risk assessments and test assumptions made.

The Directive does not specify the levels of precision and confidence required from monitoring programmes and status assessments. This perhaps recognises that achievement of too rigorous precision and confidence requirements would entail a much – increased level of monitoring for some, if not all, Member States

The key principle applied to monitoring is that actual precision and confidence levels achieved should allow meaningful assessments of status in time and space to be made.

6 Field monitoring methods

Field sampling methods, quality assurance and guidance are well established for chemical monitoring and flow monitoring. However ecological and biological monitoring and field monitoring are relatively new. Methods have been developed to ensure additional quality and consistency to meet the needs of the EU WFD. These are given in outline below and are based on the UK Environment Agency Operational Instructions. These are supplemented with other European guidance from the projects STAR, AQEM and FAME.

6.1 Sample site selection

Sampling sites cover a 'sampling area' and a larger 'survey area'. The whole site should be broadly similar in its physical characteristics.

Physical characteristics of the sampling site should be as natural as possible, similar throughout the section and representative of the river stretch. Predictions from RIVPACS or other tools are based on the fauna expected under natural conditions.

Aim to avoid sites that are:

(1) close to artificial influences, such as dams, bridges, fords, weirs or livestock watering areas. If this is not possible, the site must represent the reach as a whole. Record any artificial influences on the field data form and take them into account in data analysis;

(2) immediately downstream of confluences or discharges where waters are not fully mixed;

(3) close to the influence of in-stream lakes and reservoirs;

(4) on stretches subject to dredging or regular weed removal;

(5) in isolated habitats, such as in riffles when they are uncommon in the reach, isolation can cause invertebrate assemblages to be less diverse, so RIVPACS and other tools are likely to predict a more diverse fauna than the site could support;

(6) on braided or divided channels, if the site cannot be located elsewhere, such as on a fully braided river, sample within the largest natural channel;

(7) predominantly on bedrock, as it is difficult to sample the invertebrate fauna.

The survey area must be the length of seven channel widths, either side of the sampling area up to a maximum of 50 m. Fig. 2 illustrates a sampling site and the locations of the survey area and the sampling area. This reduces differences between samples taken on different occasions that may not be in the exact location sampled previously. Surveys that require more intensive sampling, such as those for conservation purposes or those requiring replicate samples, can also use this larger area.

6.2 Choosing the sampling methods

The preferred method for macro-invertebrate sampling is the three minute pond net sample, plus a one minute manual search. This is referred to as a kick sample, though it will often include sweep sampling.

If the preferred method is not possible use deep water methods, either:

(1) Dredge sampling - naturalists dredge.

(2) Or an airlift sampler.

Fig. 3 shows the decision logic in how to choose the most appropriate invertebrate sampling method.

7 Conclusions

Drawing from the European experience, China needs to carry out the biological background investigation to collect the basic scientific data and set up ecological database. Since the biological monitoring is quite complicated covering the biology and statistics it is very important to organize the

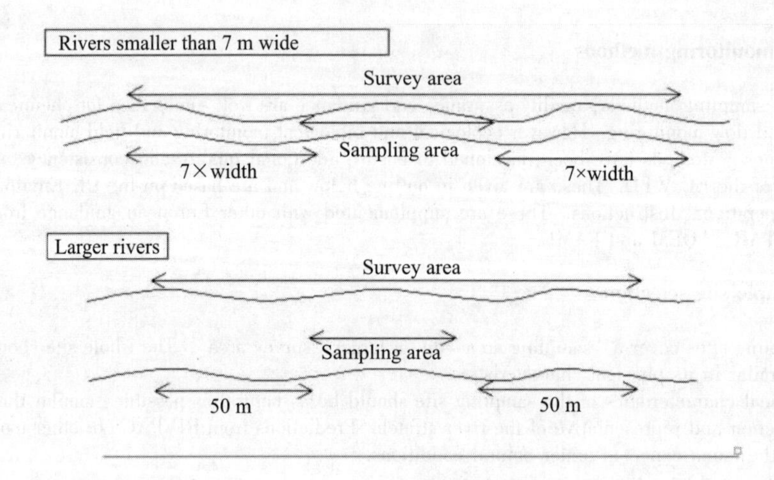

Fig. 2 Sampling site and locations of survey areas

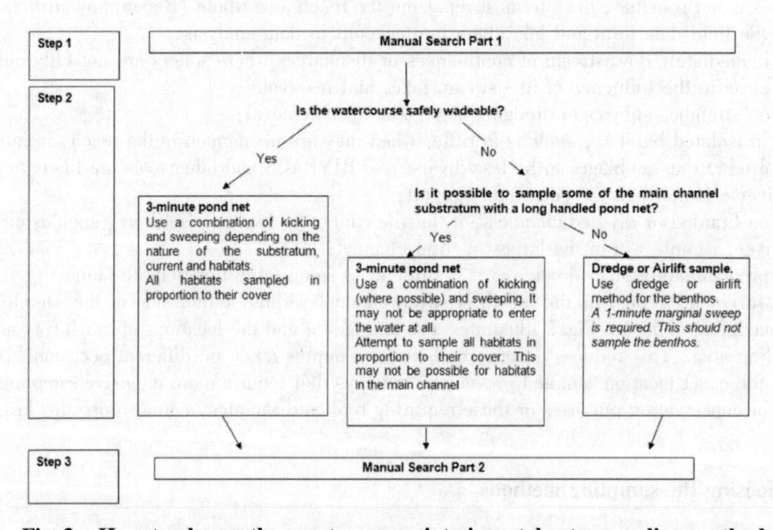

Fig. 3 How to choose the most appropriate invertebrate sampling method

experts to formulate the ecological and biological monitoring standards to ensure the reasonable confidence of the data. The selection of species for indicators in different eco – regions of China requires major further research. Start from international practice and adapt to fit regional ecology and also to the agreed system in China for defining class boundaries, reference conditions and maximum potential objectives. Also adaptation of sampling techniques to Chinese rivers, especially techniques suitable for very large and deep rivers or those with very heavy sediment loads.

Acknowledgements

We gratefully acknowledge the significant contributions from the EU – China River Basin Management Programme team members, from Mr. Reinder Torenbeek, Mr. Simon Spooner and Mr. Dong Zheren.

Research of the Pollutants Tracer Experiment in the Middle Reaches of the Yellow River

***Zhang Jianjun*[1], *Xu Ziyao*[2], *Huang Jinhui*[1], *Yan Li*[1], *Chen Wen*[3], *Cheng Wei*[1] and *Yu Zhenzhen*[1]**

1. Yellow River Water Resources Protection Institute, Zhengzhou, 450004, China
2. School of Civil and Environmental Engineering, Cornell University, Ithaca, New York, 14853, USA
3. DHI Water and Environment, Shanghai, 200032, China

Abstract: Tongguan is located in the middle reaches of the Yellow River, has become one of the most heavily polluted reaches, due to receiving the polluted water of Weihe River. During the non - flood seasons, the pollutants with high concentration from the Weihe River could be regarded as the "tracer" to study the partial - mixing and dispersion features of Tongguan Reach in the Yellow River. In May 2011, 14 sections were set in Tongguan Reach from Weihe River confluence to downstream 8 km, and 3 to 6 hydrographic measuring / water quality sampling points were set at every section. Some parameters have been measured, including hydrology and water quality elements. The mixing characteristics of the pollutants and the water quality degradation have been obtained by a tracer experiment. The result is shown that the diffusion and degradation rate were quick in the Tongguan Reach due to the effects of some factors such as channel conditions, degradation characteristic of pollutants. Ammonia and nitrogen were mixing evenly after the downstream of 8 km from the Weihe confluence. Based on the measured data, a nonlinear least squares procedure in Matlab was applied to estimate the diffusion coefficients. The analysis results show that the longitudinal diffusion coefficient E_x was 31 m^2/s and the transversal diffusion coefficient E_y was 0.26 m^2/s in the middle reaches of the Yellow River.

Key words: the middle reaches of the Yellow River, tracer experiment, mixing length, diffusion coefficients

1 Introduction

Weihe River is the largest tributary of the Yellow River and also contributes high pollution load to the Yellow River. Pollutant concentrations in Yellow River and Weihe River during the non - flood seasons are quite different. When water from Weihe River with higher pollutant concentration flow into Yellow River, the distribution of pollution belt on right bank of Yellow River in Tongguan section becomes evident, while the flow direction of mainstream in this reach of Yellow River changes significantly, leading to the vagrant plane positions and complex boundary conditions, which causes great difficulties in the mathematical simulation, because it is difficult for mathematical model to simulate the water pollution accurately only by some empirical coefficients. In this study a tracer experiment was carried out, high concentration waster of Weihe River in the non - flood season was used as the tracer to observe the concentration distribution of pollutant along and across the Yellow River after the Weihe River's afflux, which could analyze the mixing characteristics of pollutant in this reach and the degradation characteristics of water quality, and calculate the pollutant diffusion coefficient as well. This study play an important role in forecasting and evaluating the water pollution in Longsan Reach of the Yellow River, laying a foundation for the follow - up establishment of mathematical model for the water environment in the middle reaches of Yellow River.

2 General situation of reach

Longmen - Sanmenxia Reach of about 240 km long is in the middle reaches of the Yellow River. It is generally divided into Longmen - Tongguan Reach and Tongguan - Sanmenxia Reach. Longmen - Tongguan Reach of 121 km long and averagely 8.5 km wide is braided with moving beds and moving channels, the river trunk often swings to an extent over 5 km. Tongguan – Sanmenxia Dam is 119 km long. Its Tongguan – Dayudu section belongs to the naturally braided channel with wide and shallow main channel; the Dayudu – Fengzuo in the middle segment as the transitional channel has characteristics of both reservoir and channel, the channel is less changed compared to the upper reaches, but still changes largely compared to the lower reaches; and the following sections to Sanmenxia Dam have a form of a sharp channel with a high river bank, the main channel is relatively curved, being a stable bend river. Weihe River flows into the Yellow River, not far from the upper reach of Tongguan. The main stream of Yellow River after Weihe River flowing into turns 90° to east, a sharp change in the flow direction. Longmen – Sanmenxia main stream is located at the junction of Shanxi – Shaanxi – Henan provinces, so the sensitive water quality reach is the key to control the water quality of the Yellow River transitionally. Since Weihe River receives the production and living waste water from the economically and socially developed urban such as Baoji, Xianyang, Xi' an and Weinan, its water quality is relatively poor basically and worse than Grade V, which impacts the water quality of mainstream in the Yellow River significantly after flowing into with high concentration of waste water, or even the water security of the Sanmenxia Reservoir and Xiaolangdi Reservoir should be threatened greatly. Therefore, it is always the key issue of the Yellow River Basin water resources protection and management in recently years for the water quality forecast, emergency response, early warning and forecasting of water pollution incidents management in the Longmen – Sanmenxia Reach.

3 Theoretical background

At the location of the wider channel, especially in Sanmenxia Reservoir, because pollutant may not be able to completely mix at the cross – section, pollutant concentration obtained by a one – dimensional model could not meet the requirements of emergency response, early warning and forecasting. Therefore, a two – dimensional steady – state model will be applied to stimulate the lateral distribution of pollutant concentration in these reaches.

In the two – dimensional steady – state water quality stimulation, it is assumed that the pollutant concentration at any specific locations will not change over times, and the governing equation is given as follows:

$$u \frac{\partial C}{\partial x} + v \frac{\partial C}{\partial y} = E_x \frac{\partial^2 C}{\partial x^2} + E_y \frac{\partial^2 C}{\partial y^2} - k_1 C \tag{1}$$

where, C is average concentration in the vertical direction; u, v are average flow velocity in x and y direction; E_x, E_y are comprehensive diffusion coefficients in x and y direction; k_1 is decay rate.

The assumption of the steady – state model means a continuous and stable discharge from the pollution source, which is rare in the real world. But it is still feasible to stimulate the complete mixing process on river cross section in short term.

The transversal velocity in the straight and smooth channels is very small, therefore, the Eq. (1) can be simplified as follows:

$$u \frac{\partial C}{\partial x} = E_x \frac{\partial^2 C}{\partial x^2} + E_y \frac{\partial^2 C}{\partial y^2} - k_1 C \tag{2}$$

The river might be divided in to grids as shown in Fig. 1. The interval on x axis is L_x and that on y axis is L_y.

When the water depth and the flow velocity do not change much, the difference equation can be changed as follows:

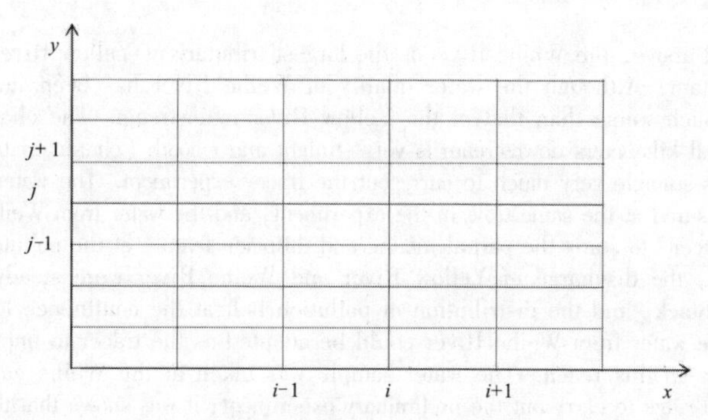

Fig. 1 Grid generation

$$\frac{\partial C}{\partial x} = \frac{1}{L_x}(C_{i+1,j} - C_{i-1,j})$$

$$\frac{\partial^2 C}{\partial x^2} = \frac{1}{L_x^2}(C_{i+1,j} - 2C_{i,j} + C_{i-1,j}) \tag{3}$$

$$\frac{\partial^2 C}{\partial y^2} = \frac{1}{L_y^2}(C_{i,j+1} - 2C_{i,j} + C_{i,j-1})$$

Substituting the Eq. (3) into Eq. (2):

$$a_1 = \left(\frac{E_x}{L_x^2} + \frac{u}{2L_x}\right) \qquad a_2 = \left(\frac{E_x}{L_x^2} - \frac{u}{2L_x}\right)$$

$$a_3 = \frac{E_y}{L_y^2} \qquad a_4 = \left(\frac{2E_x}{L_x^2} + \frac{2E_y}{L_y^2} + k_1\right)$$

Then the Eq. (3) can be transformed to:

$$a_1 C_{i-1,j} + a_2 C_{i+1,j} + a_3(C_{i,j+1} + C_{i,j-1}) - a_4 C_{i,j} = 0 \tag{4}$$

$C_{i,j}$ and u can all be measured in a tracer experiment. As is shown in Eq. (5), the value of diffusion coefficient can be derived by the least square technique.

$$f(E_x, E_y) = \sum_i^n \sum_j^m (C_{i,j}^{ob} - C_{i,j}^{cal})^2 \tag{5}$$

where, $C_{i,j}^{ob}$ is the measured values; $C_{i,j}^{cal}$ is the calculated values.

4 Testing program design

4.1 Program ideas

The transversal mixing and longitudinal dispersion in natural rivers are two important mixed transport processes of pollutant. When the pollutant flows into river in form of continuous point source, the migration and diffusion process are consist of the vertical mixing, the transversal mixing and the longitudinal transport process. For the wide and shallow rivers, the transversal mixing process lasts longer while the longitudinal mixing is shorter and the vertical mixing may be ignored compared with these two, it means that distribution of pollutant concentration in the vertical direction is even.

The method of the field tracer experiment is that the tracer substance is put in the river, then the pollution into the river could be simulated, the change of the tracer substance could be tracked to determine the tracer concentration, and obtain the actual concentration of tracer, finally the diffusion capacity of the river could be analyzed and identified. At present, the commonly used tracers are rhodamine, potassium permanganate and so on. Considering the economy, effect and applicability of the radioisotope, fluorescent tracers, and this kind of tracers should not be adopted in this reach.

As mentioned above, the Weihe River is the largest tributary of Yellow River and contributes high load of pollutant. Although the water quality of Weihe River has been improved in recent years, it is still much worse than that of the Yellow River mainstream. The channel from Weihe confluence to several kilometers downstream is very straight and smooth (characteristic shallow – broad channel), and it is suitable very much to carry out the tracer experiment. The water quality and flow field would be measured at the same time in the experiment, and the water from Weihe River would be adopted as the "tracer" to study the partial mixing and diffusion feature of the pollutant.

In May 2011, the discharge of Yellow River and Weihe River were steady, water body of Weihe River was black, and the distribution of pollution belt at the confluence location was obvious, therefore, the water from Weihe River could be adopted as the tracer to implement the pollutant diffusion study in this reach. The water sample was taken at the Weihe confluence and the Yellow River Iron Bridge to carry out the preliminary experiment, it was shown that the ammonia nitrogen in Weihe River was dozens of times higher than Yellow River while COD was only about two times. So the ammonia nitrogen data were the most suitable to calculate the diffusion coefficient.

4.2 Section layout and monitoring factors

The tracer experiment was divided into two parts: the hydrographic survey and the water quality survey, which were carried out simultaneously. The monitoring sections are shown in Fig. 2, Tab. 1 and Tab. 2.

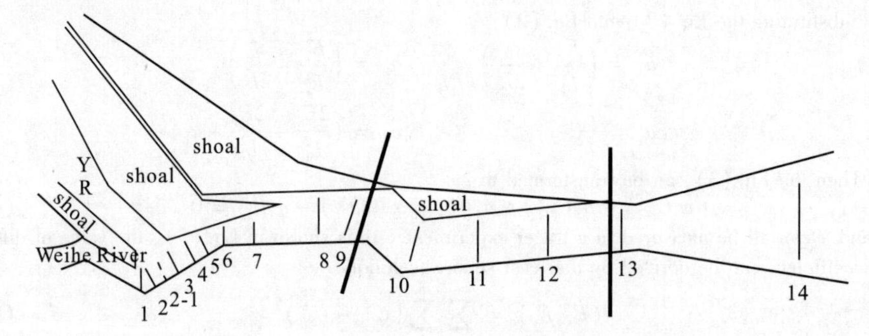

Fig. 2 Layout diagram of monitoring sections

Tab. 1 Hydrographic survey

Serial No. of the monitoring sections	1	2	3	4	5	6	7	8	9	10	11	12	13	14
Distance from Weihe confluence (m)	0	100	400	600	800	1,000	1,300	1,800	2,000	2,500	3,000	4,000	5,000	8,000
Vertical distribution	Determined after measurement of the river width by the rangefinder													
Measurement points on vertical line	Velocity was measured at 20% and 80% of the water depths by the wading rod. Previous statistic data show that the mean value of the longitudinal velocity at these two points can represent the mean velocity													
Measurement devices	GPS, rangefinder, wading rod, current meter, stopwatch													
Measured content	Water level, water depth, longitudinal velocity, river width, distance from right bank to measurement points													
Calculated content	Area of cross section, discharge													

Tab. 2 Water quality survey

Serial No. of the monitoring sections		1	2	2-1	3	4	5	6	7	8	9	10	11	12	13	14
Distance from Weihe confluence (m)	-100 (YR)	0	100	250	400	600	800	1,000	1,300	1,800	2,000	2,500	3,000	4,000	5,000	8,000
Number of measurement points	1	3	3	3	4	5	4	4	5	5	5	5	6	5	8	4
Distance from right bank (m)	M	R M L	5 30 50	5 30 65	5 20 30 60	5 50 100 150 200	10 20 50 100	5 50 100 130	5 50 100 150 200	5 50 100 150 200	5 50 100 150 200	30 50 100 150 200	5 50 100 150 200 300	5 50 100 150 190	5 50 100 150 200 240 280 310	60 120 180 240
Measurement devices	Multi - parameter water quality instrument, water collector, sampling bottle															
Measurement content	Water temperature, Electrical conductivity, Dissolved oxygen, Oxidation reduction, Potential, pH, Turbidity, Ammonia nitrogen, COD_{Cr}, COD_{Mn}															
Sampling method	10 ~ 15 cm underwater															
Method of water quality analysis	The Chinese National Standards of Water Quality Analysis															

Note: L, M, R mean left, middle and right.

5 Results and Analyses

5.1 Measurement results

5.1.1 Results of the hydrographic survey

According to the measured hydrographic data, the mean velocity of the cross - section and velocity distributions are shown in Fig. 3. The Yellow River and the Weihe River intersect at right angles. The reach 0 ~ 600 m down from Weihe confluence had turbulence, the transversal velocity was high. It can be seen from Fig. 3 that the discharge at 800 ~ 5,000 m down from Weihe confluence (section 5 ~ 13) was relatively steady. On the other hand, the changes in the water depth and flow velocity were smaller and suitable for measurement and calculation of parameters.

5.1.2 Results of the water quality measurement

The measured water depth and the concentrations of COD, ammonia nitrogen and other pollutants distributions at each section downstream Weihe confluence are shown in Fig. 4. It can be known from Fig. 4 that the concentration gradients of ammonia nitrogen at each section were clear, this is because the ammonia nitrogen concentration in Weihe River was several dozen times higher than the Yellow River, while the concentration of COD was only about two times.

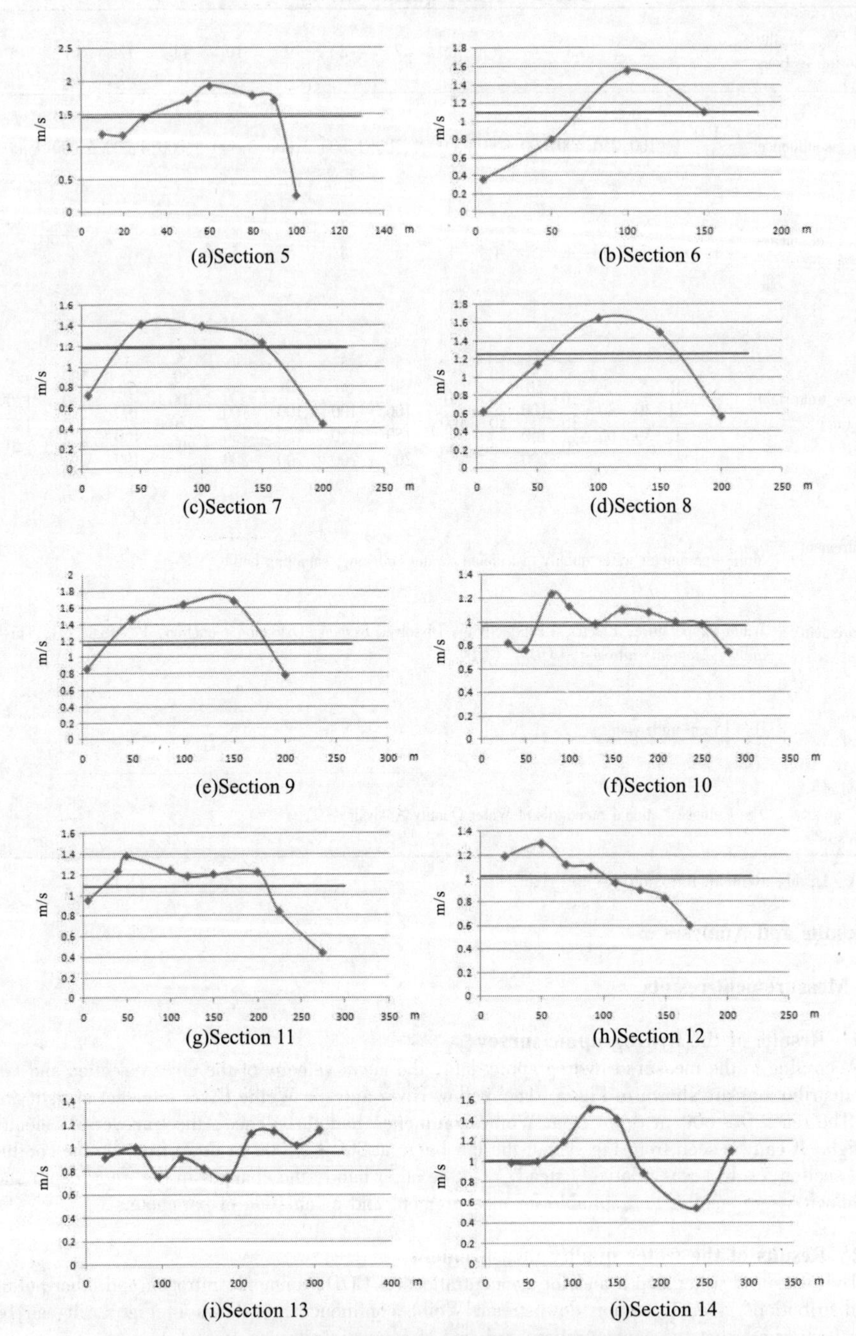

Note: the blue line means velocity distribution, the red line means average velocity.

Fig. 3 Velocity distributions at each section

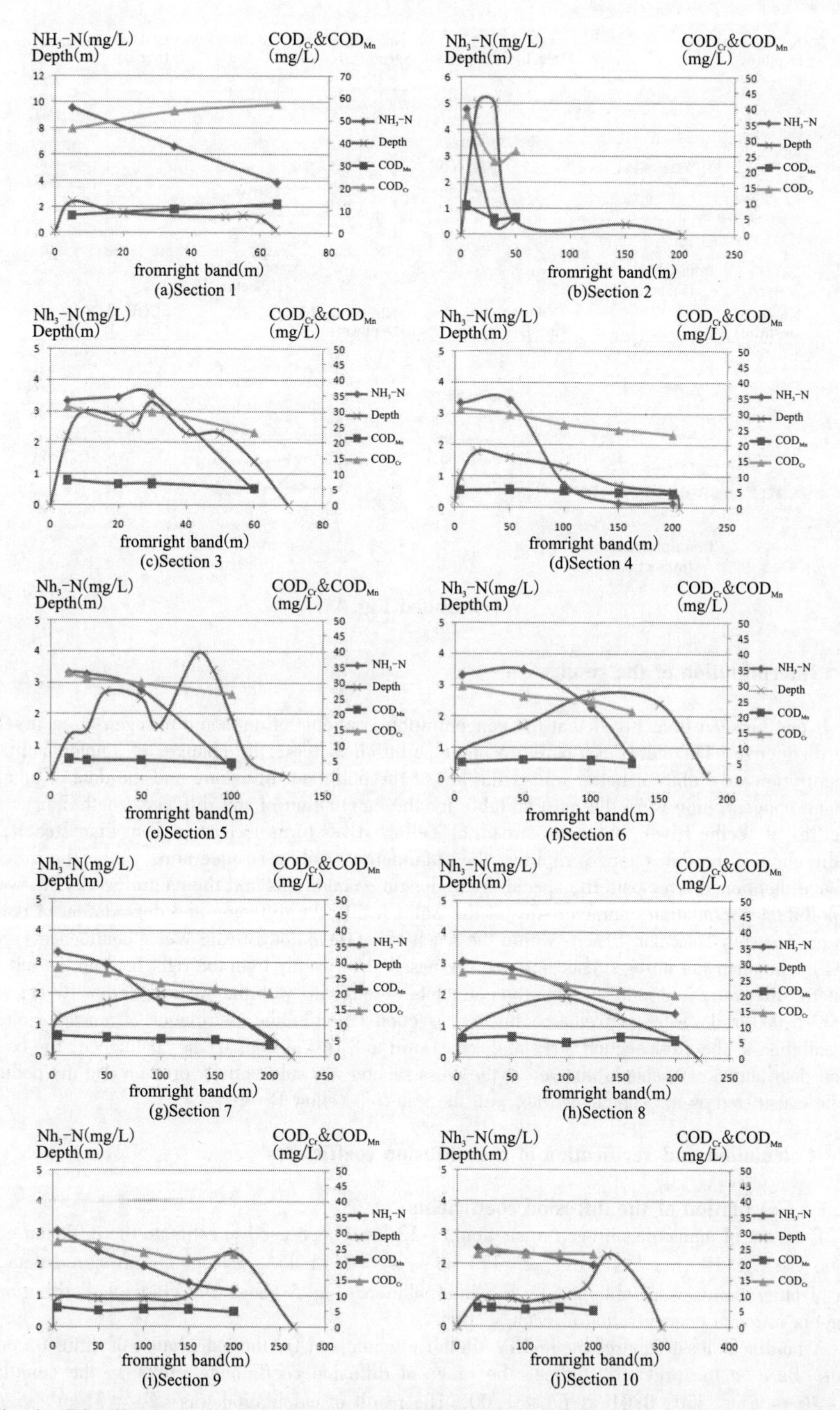

Fig. 4 Distribution of water depth and pollutant concentration at each cross - section

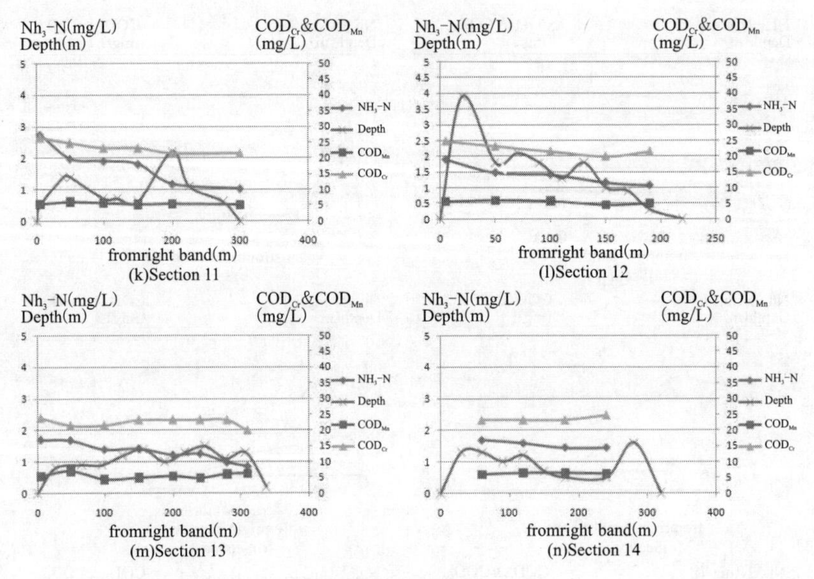

Continued Fig. 4

5.2 Interpretation of the results

It can be seen from Fig. 4 that the concentration gradients of ammonia nitrogen at each section were distinctive. Therefore, compared to other pollution factors, the changes of ammonia nitrogen concentration were able to better reflect the law of the pollutant diffusion, and the data of ammonia nitrogen concentration were the most suitable for the calculation of the diffusion coefficients. After the afflux of Weihe River, the main stream of Yellow River turns to east in Tongguan Reach, and the direction of the flow changes rapidly, the planimetric position is uncertain. Under the influence of above mentioned river pattern, special hydrodynamic conditions and the control works, as well as the pollutant degradation characteristics of the water itself, the diffusion and degradation of pollutants were faster in Tongguan Reach: within the reach of 3,000 m downstream Weihe confluence (section 1 ~ 11), the ammonia nitrogen concentration decreased significantly from the right bank to the left bank in lateral direction, it is indicated that the pollutants were mixing with the water of Yellow River; within 4,000 ~ 5,000 m downstream Weihe confluence (section 12 ~ 13) the distribution of ammonia nitrogen concentration at the cross section was stabilized; Down to 8,000 m from Weihe confluence, the concentration distribution of ammonia nitrogen at the cross section was substantially uniform and the pollutants can be considered as the completely mix with the water of Yellow River.

5.3 Calculation and verification of the diffusion coefficients

5.3.1 Calculation of the diffusion coefficients

The data of ammonia nitrogen in section 5 ~ 13 have been used to estimate the diffusion coefficients. Set $L_x = 100$ m; $L_y = 10$ m; $u = 1.1$ m/s; $k_1 = 0.15$ d^{-1}. Except the measured data, the data of other points were obtained by the interpolation. Fig. 5 shows the gradient distributions of ammonia nitrogen concentration at each section.

A nonlinear least squares method in Matlab was adopted for the calculation of diffusion coefficients. Base on the previous studies, the range of diffusion coefficients values in the calculation was: $20 \leq E_x \leq 300$, $0.01 \leq E_y \leq 1.00$. The result of calculation was: $E_x = 31$ m^2/s, $E_y = 0.26$ m^2/s. The result of calculation is shown in Fig. 6.

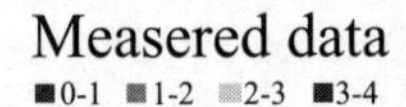

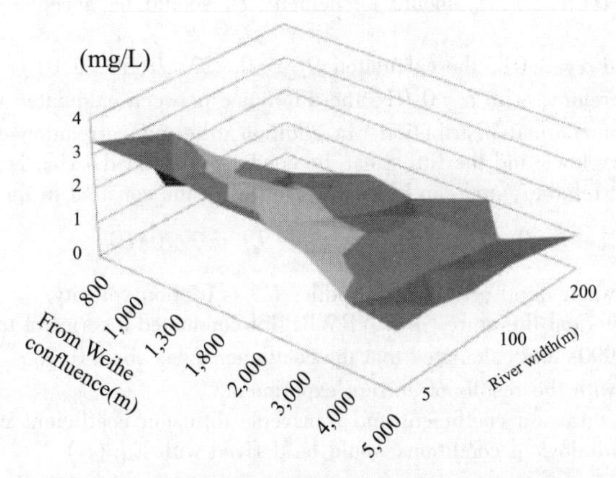

Fig. 5 Ammonia nitrogen concentration changes by measurement

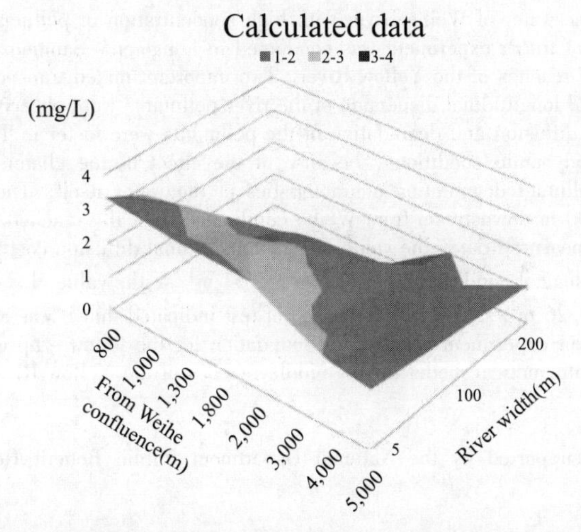

Fig. 6 Ammonia nitrogen concentration changes by calculation

5.3.2 Goodness of fit test

From Fig. 5 and Fig. 6, a good fitting degree can be seen. In addition, this study carried out the goodness of fit test by using $K-S$ method. The data of water quality mainly obey the logarithmic normal distribution. Thus as long as the difference of calculated value and measured value obeys the logarithmic normal distribution, the fitting degree can be identified as good.

Set:

H_0: the difference obeys the logarithmic normal distribution;

H_1: the difference does not obey the logarithmic normal distribution;

X: difference sequence;

$F_0(X)$: theoretical logarithmic normal distribution function;

$F_n(X)$: empirical distribution function;

$D_{\max} = \max |F_n(X) - F_0(X)|$;

n: number of data;

α: significance level.

When , $D_{max} \geqslant D(n,\alpha)$, H_0 should be denied, H_1 should be accepted with α. Otherwise should be contrary.

When $n=42$ and $\alpha=0.01$, the calculated $D_{max}=0.222$. $D(50, 0.01)=0.225$, the value larger than D_{max}. Therefore, with $\alpha=0.01$, the difference between calculated value and measured value obeys logarithmic normal distribution. In addition, the autocorrelation coefficient of differences sequence is very low, and the fitting can be considered as good. That is, the calculated diffusion coefficients are reliable, and can be adopted to derive the constant in the empirical formula.

$$E_x = 0.005^2 \frac{u^2 W^2}{HU^*}, \qquad E_y = 3.70hU^* \tag{6}$$

where, H is average water depth; W is river width; U^* is friction velocity.

According to historical literature [1], YRWRPBSI conducted a series of tracer experiments in the Yellow River in 1990s and calculated that the coefficient E_x is about 0.005^2 ($=0.000,025$). It is roughly consistent with the results of current experiment.

The longitudinal diffusion coefficient and transverse diffusion coefficient in the other reaches under the different hydrological conditions would be derived with Eq. (6).

6 Conclusions

In this study, the water of Weihe River with high concentration of pollutants was used as the tracer, and a pollutant tracer experiment was conducted in Longmen – Sanmenxia Reach, where is located in the middle reaches of the Yellow River. Two important mixed transportation processes – transversal mixing and longitudinal dispersion of the river pollutants were observed. The experiment results show that the diffusion and degradation of the pollutants were faster in Tongguan Reach under the special hydrodynamic conditions, because of the effect of the channel and river control works, as well as pollutant degradation characteristics of the water itself. The ammonia nitrogen concentration at 8,000 m downstream from Weihe confluence basically reached a completely mixed state. Based on the measured data, the value of E_x – longitudinal diffusion coefficient in the experiment under the hydrological conditions at that time was 31 m^2/s, the value of E_y – transversal diffusion coefficient was 0.26 m^2/s. The $K-S$ method of test indicated that it was a good fitting effect. The result of this tracer experiment would lay a foundation for the follow – up establishment of the water environment mathematical model in the middle reaches of the Yellow River.

Acknowledgements

This work was supported by the National Department Public Benefit Research Foundation (201001011).

References

Hao Fuqin, Zhang Jianjun, Huang Jinhui, et al. Study on Key Technology of Yellow River Water Quality Forecasting[R]. 2002.

Guo Zhenren, Xu Zhencheng. Simulation and Analysis for Pollutants Tracing in Great Rivers[J]. Research of Environmental Sciences, 1990 (4): 43 – 47 (in Chinese).

Li Yawei, Zhang Xuliang. Simple and Accurate Field Tracing Test Method of River Dispersion Coefficient[J]. Research of Environmental Sciences, 1990 (2): 63 – 64 (in Chinese).

Zhou Kezhao, Yu Changzhao, Zhang Yongliang. Study on Calculation Method for Longitudinal Dispersion Coefficient Tracer Experiment of Natural Stream[J]. Research of Environmental Sciences, 1986 (3): 314 – 325 (in Chinese).

Evaluation of Improvement in Lake Water Environment by Ecological Water Transfer

Zhang Mo[1], **_Mao Jingqiao_**[1] and **_Zhang Xian'e_**[2]

1. State Key Laboratory of Hydrology – Water Resources and Hydraulic Engineering, Hohai University, Nanjing, 210098, China
2. North China University of Water Resources and Electric Power, Zhengzhou, 450011, China

Abstract: Water transfer, from relatively clean water resources to an objective lake, is commonly considered to be an effective way to improve water environment by redistributing the water resources. Scientific estimation of water transfer quantity and quantitative evaluation of potential improvement of water environment are core issues of the lake environmental management. This study reports a simple evaluation method on the performance of water transfer to shallow lakes, based on a zero – dimensional water quality model. The method has been judged by means of a three – dimensional model and measured data. Finally, the simple model is applied to a typical water transfer project of an urban lake; the performances of this project are quantitatively predicted for different typical years.

Key words: water transfer, water environment, water quality model, water resource allocation

1 Introduction

There are more than 20,000 lakes in China, occupying 1/10 of the world natural lakes, and the total area is about 80,000 km^2; about 2,300 of them are larger than 1 km^2, with a total area of more than 70,000 km^2, accounting for about 0.8% of the land area of China; the total water storage of lakes is more than 7.0×10^{11} m^3 and freshwater storage is 2.25×10^{11} m^3. Lakes provide basic water resources for human life, society development and economic development. However, with the rapid growth of economy and population, the demand of water resources was terribly increased. Most lakes, especially urban lakes, are often badly influenced and controlled by human activities, irrational utilization have lead to a series problem of water environment such as water pollution, eutrophication and withering. Urban lakes refer to shallow lakes in the urban districts or suburbs of large or medium city, e. g., West Lake in Hangzhou, Kunming Lake in Beijing, Xuanwu Lake in Nanjing, and East Lake in Wuhan. As an important part of urban ecosystem, the urban lakes act as function of tourism, entertainment, drainage, climate modulation, and environment improvement. Due to its limited self – purification ability, large quantity of pollutant input and irrational utilization may lead to serious water environment deterioration. For these reasons, engineering measures are needed to restore the ecosystem services. Water transfer is one of the most effective and practical way besides pollution source control, sediment dredging and economical control.

Water transfer, from relatively clean water resources to an objective lake, is an economical and fast way to improve the water environment. This method makes full use of the self – purification capability of water by redistributing the water resources, and the recovery of fluvial – lacustrine system is beneficial to the environmental improvement of aquatic habitats. There have been some successful engineering examples reported, e. g, flushing of Veluwemeer Lake in Netherlands, water diversion from the Yangtze River to Taihu Lake, water diversion in Xuanwu Lake, and water diversion in West Lake.

As a kind of predictable issues, the predictable evaluation of water transfer project is of acknowledged difficulty. To meet the engineering effect and economy requirement, scientific estimation of water transfer quantities and quantitative evaluation of improvement of water environment are the core issues of the water transfer planning. The study reports a simple evaluation method on the performance of water transfer to shallow lakes, based on a zero – dimensional water quality model.

2 Water quality model

Mathematical model is an important tool in the water environmental management. According to the spatial dimension, it is divided into zero – dimensional, one – dimensional, two – dimensional, and three – dimensional models, and the select of model type should take into account geographic features, purposes, as well as computational cost. Multi – dimensional hydrodynamic or water quality models can simulate the flow and pollutant concentration accurately and provide finer temporal and spatial distributions, which are demonstrated in several researches. However, this kind of model is complex, requires a number of basic data, and often relies on advanced computer hardware, hindering its wide application. Urban lake can often be described as well – mixed water body considering its small volumes and simple geological boundary conditions, and the pollutants are easily mixed under the action of flow and waves. Therefore, the simple box model would be a good choice to analyze the general tendency of water quality instead of complex multi – dimensional model.

The well – mixed model assumes that waters are completely mixed reactor, so the pollutants keep a constant concentration in space. The model is simple in structure, yet proves successful in practice. Based on the well – mixed model, this study establishes a zero – dimensional water quality model for lakes, using chemical oxygen demand (COD) and total phosphorus (TP) concentrations as controlling indicators.

2.1 COD concentration

According to the mass balance principle, the variation of pollutant in a certain period is equal to the total pollutant inflow minus pollutant outflow and the loss amount (degradation, settlement, etc.). Then we have the fundamental equation of well – mixed lake

$$V\frac{\mathrm{d}C}{\mathrm{d}t} = Q_iC_i - Q_oC_o - KVC \tag{1}$$

where, V is lake water volume, m^3; C is the common pollutant concentration, mg/L; Q_i and Q_o are the total inflow and outflow charge, m^3/d; K is the attenuation coefficient.

If $t=0$, then $C_0 = C$, average pollution load $w_L = Q_iC_i$. The attenuation of pollutants includes degradation and settlement, so the average COD concentration model is presented as follow.

$$\frac{\mathrm{D}L}{\mathrm{d}t} = -\left(k_1 + \frac{f_pv_L}{H} + k_f\right)L + \frac{w_L}{V} \tag{2}$$

where, L is COD concentration, mg/L; v_L is the settling velocity of particulate organic matter, m/d; f_p is the organic decomposition rate; H is water depth, m; w_L is the average COD load, g/(m^3 · d); k_1 is the average degradation rate, 1/d; k_f is hydraulic purification rate.

Thus in the steady state, COD concentration is expressed as

$$L = \frac{w_L/V}{k_1 + f_pv_L/H + k_f} \tag{3}$$

2.2 TP concentration

Plankton bloom is one of the characteristics of eutrophication, and algal growth is closely related to nutrient concentrations in the environment. Algal growth rate is usually described as a first order process, whose rate coefficient is usually constrained by a function of temperature, light and nutrient. The TP concentration model is derived from the evolution of chlorophyll concentration.

$$\frac{\mathrm{d}Chla}{\mathrm{d}t} = \left(\mu_{\max}g(I)g(T)g(P) - rg(T) - \frac{v_s}{H} - k_f\right)Chla \tag{4}$$

where, $Chla$ is the chlorophyll concentration, μ max is the maximum algae growth rate, 1/d; $g(I)$ is the light intensity impact on algal growth, $g(T)$ is the temperature impact on algal growth, $g(P)$

is the phosphorous concentration impact on algal growth; r is the algae metabolism rate, 1/d; v_s is algae settling velocity, m/d.

The phosphorous concentration impact on algal growth is associated with TP concentration (P) and half saturation constant for phosphorous (k_P), as is described in the Monod equation

$$\mu = \mu_{\max}\left(\frac{R}{k_s + R}\right)$$

where, k_s Ris the half saturation constant; μ is the actual growth rate of algae; R is the nutrient concentration.

So the phosphorous concentration impact on algal growth can be presented as $g(P) = P/(k_P + P)$.

If we apply this function to Eq. (4) and identifying $r' = rg(T) + \frac{v_s}{H} + k_f, \mu' = \mu_{\max} g(I) g(T)$, we obtain the following zero – dimensional result.

$$P = \frac{r' k_P}{\mu' - r'} \tag{5}$$

In this model, TP concentration depends only on the ecosystem mass balance principle and hydraulic retention time, thus avoiding the errors caused by estimating the total phosphorus load.

3 Results and discussions

3.1 Parameters and simulation

Using a case study of a water transfer project for Yundonghai Lake, and by applying the zero – dimensional water quality model established above, this study makes a simple effect evaluation of water transfer based on different typical years.

Yundonghai Lake is a typical urban lake with a total area of 9.157 km^2. It is located in the Pearl River Delta, close to the confluence of West River, North River and Suijiang River, and is only 2 km away from the Sanshui City. The region terrain is tilted from northwest to southeast, belonging to the half – hilly areas; the eastern part is low – lying, distributing with farmland and fish ponds. Ancient Yundonghai Lake was the largest lake at the Pearl River Delta in Qin Dynasty, but the water area decreased in Ming Dynasty due to reclamation. Moreover, in the 1950's, Yundonghai was once drained and transformed to farmland, and the lake didn't appear until 1990's. In order to develop the modern ecological tourism industry and strengthen ecological protection, the local government planned a project that drew clean water from North River to Yundonghai Lake, using overflow weirs and other controlling constructions to form circulating flow, and avoiding stagnant water regions. After the impoundment, the total area of Yundonghai Lake will reach 1×10^6 km^2, and water residence time will be about one month.

Fig. 1 is location of Yundonghai Lake and the monitoring stations (A – D).

According to the climatic condition of the Pearl River Delta, we calculate in conditions of two typical years: normal year and dry year (once in a decade). Also, a typical year includes the flood season (May to September) and non – flood season (November to March). Moreover, according to the land types and designed annual runoff process of different return period, the runoff of each typical year is estimated through hydraulic models. Finally, the average inflow for each entrance can be estimated.

Considering the actual situation of Yundonghai Lake, the required parameters are valued refers to literature value and a three – dimensional water quality model. The design area of Yundonghai Lake is 8,698 mu, with a total volume of 11.6×10^6 m^3, and average water level is 2 m ($H = 2$). With diverting water flow of 2.7 m^3/s, the water changes every 30.11 days in flood season for the normal year ($k_f = 0.033$), and about 34.24 days for the dry year ($k_f = 0.029$). Estimate the COD load according to inflow rate and water quality condition, that is 25,460 kg/d in flood season for the normal year ($w_L = 2.19$ g/($m^3 \cdot$ d)), and 19,065 kg / d for the dry year ($w_L = 1.64$ g/($m^3 \cdot$ d)). The relevant parameter values and literature values see Tab. 1.

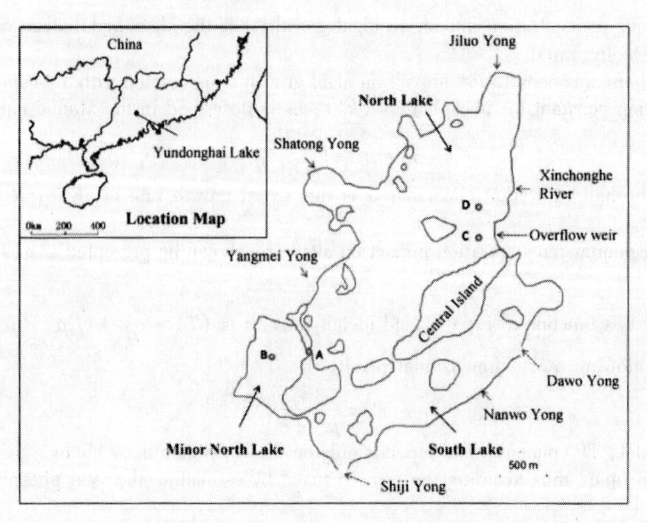

Fig. 1 Location of Yundonghai Lake and the monitoring stations (A – D)

Tab. 1 Values of parameters used in the model

Parameter	Description	Literature Values	Yundonghai Lake
v_L	Settling velocity of particulate organic matter(m/d)	0 ~ 0.2	0.005
f_p	Organic decomposition rate	0.5	0.5
K_1	Average degradation rate(1/d)	calibrated	0.23
$g(I)$	Light intensity impact on algal growth	0.1 ~ 0.3	0.15
r	Algae metabolism rate(1/d)	0.03 ~ 0.59	0.1
k_p	Half saturation constant for phosphorous(mg/L)	0.004 ~ 0.163	0.006
v_s	Algae settling velocity(m/d)	0.05 ~ 0.89	0.15
$g(T)$	Temperature impact on algal growth	$1.066^{(T-20)}$	1.06610
$g(P)$	Phosphorous concentration impact on algal growth	$1.08^{(T-20)}$	1.0810
$\mu_{\max}$	Maximum algae growth rate(1/d)	0.55 ~ 5.0	1.2

After model calculations, we have the predicted average COD concentration after water transfer in flood season from Eq. (3).

For the normal year,

$$L = \frac{2.19}{0.023 + 0.05 \times 0.5/2 + 0.033} = 38.25 \text{ mg/L}$$

For the dry year,

$$L = \frac{1.64}{0.023 + 0.05 \times 0.5/2 + 0.029} = 30.79 \text{ mg/L}$$

We calculate the TP concentration from Eq. (5)

For the normal year,

$$P = \frac{0.006 \times (0.1 \times 1.08^{10} + 0.15/2 + 0.033)}{1.2 \times 0.15 \times 1.066^{10} - (0.1 \times 1.08^{10} + 0.15/2 + 0.033)} = 0.113 \text{ mg/L}$$

For the dry year,

$$P = \frac{0.006 \times (0.1 \times 1.08^{10} + 0.15/2 + 0.029}{1.2 \times 0.15 \times 1.066^{10} - (0.1 \times 1.08^{10} + 0.15/2 + 0.29)} = 0.090,6 \text{ mg/L}$$

Compare the water quality monitoring data before transfer (see Tab. 2) with model predictions (after transfer), it is shown that the COD concentrations of point A, B, C, D (location see Fig. 1) are 51.50, 69.80, 56.50, 383.8 mg/L respectively, while the average concentration predicted is 38.25 mg/L; the TP concentration data are 0.30, 0.08, 0.04, 4.31 mg/L respectively, while the average concentration predicted is 0.113 mg/L. Clearly, the water quality of Yundonghai Lake after the water transfer is expected to be significantly improved.

3.2 Verification and discussion

For the further validation of model accuracy and reliability, we compare its results with the numerical results of 3D model that has been well validated by measured data. The measured data are obtained from four monitoring stations located in different part of the North Lake and the Minor North Lake, as is shown in Fig. 1.

The 3D model contains the hydrodynamic module and the water quality module. The hydrodynamic equations belong to shallow water equations, and the water quality module selects COD and TP concentrations as controlling indicators. The boundary conditions include the free surface, bottoms, shore boundary, and open boundaries (e.g. controlling facilities). By comparing the 3D model calculations with the measured data (before transfer, see Tab. 2), we find that the 3D model simulates the water quality of the lake with an acceptable accuracy. So, we set it as baseline for water quality predict of Yundonghai Lake.

Tab. 2 Comparison of three - dimensional model calculations and the measured data

(Units: mg/L)

Location	Time	Measured data		Three - dimensional model	
		COD	TP	COD	TP
A①	2010.7	51.50	0.30	49.07	0.21
B②	2010.7	69.80	0.08	54.52	0.25
C④	2010.7	56.50	0.04	34.38	0.21
D③	2010.7	383.8	4.31	40.19	0.18

Note: ① ~ ④ are the monitoring stations A - D separately in Fig. 1.

Furthermore, according to Tab. 3, we conclude that the zero - dimensional model is reliable; the average error between the zero - dimensional model and the 3D model is only around 5%.

Tab. 3 Comparison between zero - dimensional model and 3D model on the average value of controlling indicators.

Indicator	Condition	Three - dimensional model	Zero - dimensional model	Relative error (%)
COD (mg/L)	Normal year	36.87	38.25	3.74
	Dry year	29.42	30.79	4.66
TP (mg/L)	Normal year	0.102	0.113	10.78
	Dry year	0.086	0.091	5.35

4 Conclusions

The design and assessment of ecological water transfer project belongs to the prediction study.

By establishing a zero – dimensional water quality model for a typical urban lake, and through estimating the future conditions in typical years, this study provided a simple method of effect evaluation for water transfer. Moreover, the predictive performance of the zero – dimensional model was judged by the numerical results of a 3D model that has been well validated by measured data and proved reliable. The evaluation method can be used for the fast estimation of engineering effects as well as for the preliminary validation of multi – dimensional models. The study provides a scientific tool for the environmental impact assessment of water transfer projects for lakes and may help to minimize the casualness and blindness in water environmental planning.

Acknowledgements

This research is supported by the Specialized Research Fund for the Doctoral Program of Higher Education (20110094120012) and the National Natural Science Foundation of China (Grant No. 41001348).

References

Liu Z H, Chen Y Q, Liang Z B. Current Situation of the Water Environmental Management of Major Lakes in China[J]. Environmental Protection, 1998(12): 9 – 10.

Peng J J, Li C H, Huang X H. Causes and Characteristics of Eutrophication in Urban Lakes[J]. Ecologic Science, 2004(4): 370 – 373.

He Y, Li Y T. Exploring Possibility to Regain Polluted – lake water Quality by Means of Bioremediation and Water Diversion[J]. Journal of Safety and Environment, 2005(1): 56 – 60.

Zhang D N. Analysis of Environmental Benefit for Water Diversion Project of Xuanwu Lake[J]. The Administration and Technique of Environmental Monitoring, 1995. (3): 17 – 18.

Ma J L. Annual Cyclical Changes of Nitrogen, Phosphorus and Chlorophyll a Concentrations in West Lake Before/After the Water Diversion[J]. Journal of Lake Science, 1996(2): 144 – 150.

Yu C, Ren X Y, Ban X, et al. Application of Two – dimensional Water Quality Model in the Project of the Water Diversion In East Lake, Wuhan[J]. Journal of Lake Sciences, 2012(1): 43 – 50.

Missaghi S, Hondzo M. Evaluation and Application of a Three – dimensional Water Quality Model in a Shallow Lake With Complex Morphometry[J]. Ecological Modelling, 2010, 221(11): 1512 – 1525.

Mao J Q, Chen Q W, Chen Y C. Three – dimensional Eutrophication Model and Application to Taihu Lake, China[J]. Journal of Environmental Sciences, 2008(3): 278 – 284.

Chen Y C, Zhang B X, Lu Y L. Analysis and Prediction of Eutrophication for Miyun Reservoir [J]. Journal of Hydraulic Engineering, 1998(7): 13 – 16.

Ruan J R, Cai Q H, Liu D K. A Phosphorus – phytoplankton Dynamics Model for Lake Donghu in Wuhan[J]. Acta Hydrobiologica Sinica, 1988(4): 289 – 307.

Caperon J, Meyer J. Nitrogen – limited Growth of Marine Phytoplankton – Ⅱ. Uptake Kinetice and Their Role in Nutrient Limited Growth of Phytoplankton[J]. Deeep – sea Research, 1972, 19 (9): 619 – 632.

Li F, Li S, Luo P, et al. The Changes of the Yundonghai Wetland in Sanshui and its Protection and Utilization[J]. Tropical Geography, 2005(2): 146 – 150.

Jorgensen S E, Endoricchio G B. Fundamentals of Ecological Modeling[M]. 2nd ed. Elsevier, Netherlands, 1994.

Mao J Q, Lee J H W, Choi K W. The Extended Kalman Filter for Forecast of Algal Bloom Dynamics[J]. Water Research, 2009, 43(17): 4214 – 4224.

Mao J Q, Zhang X E. Numerical Assessment of Water Quality Improvement by Water Transfer for Yundonghai Lake Restoration From Impoldering[J]. Journal of Hydroelectric Engineering, 2012.

Study on Ecology-oriented Water Resources Rational Allocation in Qinghai Lake Basin

Zhao Maihuan[1], *Wu Jian*[1], *Xiao Sujun*[1] and *Zhang Yue*[2]

1. Yellow River Engineer Consulting Co. Ltd., Zhengzhou, 450003, China
2. North China University of Water Resources and Electric Power, Zhengzhou, 450003, China

Abstract: During recent 50 years, the water level of Qinghai Lake showed gradually downtrend, the water storage was continuously decreasing, the lake area was shrinking, and water salinity was increasing. The changes have been a threat to the ecological environment and the people's livelihood around the Qinghai Lake basin. Based on water use status and the economic and social development in the Qinghai Lake basin, as the goal of zero growth in water demand, in accordance with the principle of the priority to ensure the urban and rural domestic water and the water quantity of major cross- sections of the river while to strictly control the water use in industrial, agricultural and other sectors of water and properly use the groundwater and exploit unconventional water resources, the scheme of water resources rational allocation in the Qinghai Lake basin was proposed. By the plan, the whole amount of water supply will maintain 1×10^8 on m^3 at the planning years in Qinghai Lake basin, including the surface water decreasing from $1.009,8 \times 10^8$ m^3 in the base year of to $0.958,1 \times 10^8$ m^3 in 2030 while appropriately increasing groundwater and other water sources use. Applying the ecology-oriented water resources allocation will slow down the downtrend water level of Qinghai Lake.

Key words: Qinghai Lake basin, ecology-oriented, water resources allocation

1 Overview of Qinghai Lake basin

Qinghai Lake basin is located in the northeastern Tibetan Plateau, functioning as a tie between the hub area of the eastern and western Qinghai Province and Qingnan region and the main channel to Hexi Corridor in Gansu Province, Tibet and Xinjiang. The Qinghai Lake basin reaches the Sun and the Moon Ridge to the east connecting with the Xining City-owned Huangyuan county and neighbors to the west The Amu Paganini library Hill and the Qaidam Basin, Sahara Lake basin. It reaches to the north the Chase Hill ridge, next to Datong basin, and it researches Qinghai Nanshan ridge to the south next to Chaka-republican Basin. The drainage area of Qinghai Lake is 29,700 km^2, covering three states and four counties which are administratively attached to Gangcha County and Haiyan County in Haibei Tibetan Autonomous Prefecture of Qinghai Province (hereinafter referred to as the Haibei Prefecture), Tianjun County in Haixi Mongolian Tibetan Autonomous Prefecture (hereinafter referred to as Haixi Prefecture) and the Republican County in Hainan Tibetan Autonomous Prefecture (hereinafter referred to Hainan Prefecture).

Qinghai Lake basin is a home for a total population of 1.074×10^7 with the average population density being 3.6 people / km^2, the urban population being 3.04×10^4 and the urbanization rate being 28.4%. Qinghai Lake basin is a place whose priority is given to livestock production with a small amount of farming production industries and whose agricultural population is in a high proportion, reaching 7.7×10^4. According to the survey, the number of the livestock on hand in the Qinghai Lake Basin is $0.028,07 \times 10^8$, of which large animals are 449,000 and small ones are $0.023,58 \times 10^8$. If one large animal is converted as five sheep units of livestock and a small one as a sheep unit, the number of the livestock on hand in the Qinghai Lake basin is 0.046×10^6 sheep units and each herdsmen own 59 sheep units. With the western development and ecological environmental governance, the basin has been implementing the reforestation project since the year of 2000 and by the year of 2007, about 24-acre arable land has been retained in the basin. Qinghai Lake basin has a relatively weak industrial base and its main industries is of small scales, low yield and less water consumption, including coal mining, lead and zinc mining and dressing, meat processing, building materials and chain link fence manufacturing. The GDP in Qinghai Lake basin in

2007 totaled 9.84 × 10^8 Yuan with 9,163 Yuan per capita. Three industrial structures in this basin are 40.3: 13.1: 46.6.

The river network of Qinghai Lake basin was in significantly asymmetric distribution. Northwest river networks are developed and have big runoff; southeast river networks are sparse which are mostly seasonal rivers and have small runoff. There are more than 50 rivers flowing into Qinghai Lake of which major rivers are Buha River, Quanji River, Shaliu River, Hargai River and Heima River , etc. (see Tab. 1) .

Tab. 1 Hydrological characteristics table of the major rivers in the Qinghai Lake basin

The name of the river	The name of the control station	The location of the control station		Catchment area (km^2)	River length (km)	Natural rivers Runoff ($\times 10^8$ m^3)
		east longitude	northern latitude			
Buha River	Estuary of Buha River	99°44′12″	37°02′13″	14,337	272	7.821
Quanji River	Shatuo Temple	99°52′35″	37°13′37″	567	63	0.221
Shaliu River	Gangcha	100°07′49″	37°19′20″	1,442	85	2.507
Hargai River	Hargai	100°30′16″	37°14′25″	1,425	86	1.308
Heima River	Heima River	99°47′00″	36°43′23″	107	17	0.109

Note: the Material is quoted from 1956 ~ 2000.

2 Current situations and existing problems of developing and utilizing water resources in Qinghai Lake basin

2.1 Current situations of developing and utilizing water resources in Qinghai Lake basin

From the year of 1956 to the year of 2000, the average annual rainfall is 354.5 mm and annual water surface evaporation is 957.8 mm in Qinghai Lake basin. Annual surface water resources is 1.781 × 10^6 m^3, non-doubly counted amount is 3.82 × 10^8 m^3 between groundwater and surface water and total amount number of water resources is 2.163 × 10^9 m^3.

The total water supply of the Qinghai Lake basin in 2007 was 0.922,9 × 10^8 m^3, of which 0.907,8 × 10^8 m^3 is from surface water, 0.015 × 10^8 m^3 is from groundwater and 4,200 m^3 is from other water supply sources. The water is used for agriculture, forestry, animal husbandry and fisheries livestock is 0.887,6 × 10^8 m^3, accounting for 96.18% of the total water consumption. The water used for industries, building industry and the tertiary sector is 0.017,3 × 10^8 m^3, accounting for 1.88% of the total water consumption, domestic water (including urban life, rural life) is 1.78 × 10^8 m^3, accounting for 1.93% of the total water consumption, and ecological water use is 20,000 m^3, accounting for 0.02% of the total water consumption .

The total water resources of Qinghai Lake basin are 2.163 × 10^9 m^3, of which 0.922,9 × 10^9 m^3 have been developed with its development and utilization degree being 4.27% that is relatively low. The river districts above the Shanghuancang of Buha River have the lowest degree of development and utilization, while Hargai River District has the highest degree.

2.2 Main problems existing in water resource system in Qinghai Lake basin

2.2.1 Declining lake level, shrinking lake surface and increasing lake salinity

According to monitoring data of water level of the Qinghai Lake from 1956 to 2007, the water level have been generally tending to decline, falling by 3.62 m over 52 years at the annual decreasing speed of nearly 7 cm with lake areas narrowed by 314 km^2 and storage capacity reduced by about 1.5 × 10^{10} m^3. Qinghai Lake is being split into a large numbers of small lakes from a single large plateau lake with "a big one and several small ones". Meanwhile, the water-level decline of the Qinghai Lake has led to increased lake salinity. It has been estimated that the Qinghai Lake salinity rose to 15.6 g / L in 2008 from 12.5 g / L in 1962. The increased Qinghai Lake salinity to some extent has affected the growth and development of aquatic food organisms and Qinghai Lake naked carp, threatened the habitat of the fish resources, thereby affecting the circulation of Ichthyornis symbiotic ecosystem.

2.2.2 Water use in low efficiency

In 2007, per-capita water consumption in the Qinghai Lake basin was 859 m^3, which was 1.53 times as much as the average level (566 m^3) in Qinghai Province and 1.92 times as much as the national average (451 m^3). The ten thousand GDP consumes 938 m^3 water, which is 2.37 times as much as the average level (395 m^3) in Qinghai Province and 4.10 times as much as the national average (229 m^3). At the same time, irrigation projects in Qinghai Lake basin were largely built in the 1950s to the 1970s under low designing standards with insufficient engineering support, serious aging and low efficient irritation. Current annual utilization coefficient of irrigation water is 0.31, lower than 0.60 to 0.70 of the small and medium-sized irrigation area section. And compared to Qinghai Province and the national average, water use in the Qinghai Lake basin is still extensive, efficiency is still low, waste is still very serious, water-saving management and water-saving technology are still relatively backward, which is in stark contrast with water shortage situation with the double pressure from development and the ecology.

2.2.3 Overgrazing grassland and gradually worsening ecological environment

Qinghai Lake basin takes the grassland ecosystem as the principle thing with grassland area being approximately 15,300 km^2 in 2007, of which low-coverage grassland area was 0.42 km^2, accounting for 27.5% of the present grassland area. According to the statistical analysis, the total livestock in Qinghai Lake basin is up to 0.046×10^8 sheep units, while bearing capacity of the basin in theory is 0.024×10^8 sheep units, overloading $0.023,6 \times 10^8$ sheep units with overloading rate being113%. Excessive grazing and trampling by livestock have made the natural grass not recuperate, the grass layer height reduced, the proportion of good forage significantly reduced, poisonous weeds increased and grassland degradation trend intensified. With the wake of grassland degradation, wetland environmental quality has been declining within the basin. With the original wetland hills arising, drying, and peat exposed, wetland plants have been gradually replaced by mesophyte and water conservation function has been declining. Since the 1990s, the wetlands in Qinghai Lake has showed a decreasing trend, reducing by 107.20 km^2 from 1989 to 2000 and by 34.61 km^2 from 2000 to 2005. Meanwhile, ecological environment in Qinghai Lake region is very fragile and the tourism development has imposed potential pressure to the ecological environment of the Qinghai Lake.

3 To confirm inside and outside water requirement of Qinghai Lake river ways based on eco-services of the Qinghai Lake

3.1 Loss situations of Qinghai Lake water and the causes

During the past 50 years, water level of Qinghai Lake has been presented a more obvious fluctuating downward. Taking gagging station of Shatuo Temple in Qinghai Lake as an example, the average annual water level is 3,196.79 m in 1956, 3,193.17 m in 2007. From 1956 to 2007, it ascended in a small number of years, descended in most years and it descended in general. The water level of the lake altogether descended 3.62 m with the average annual decline being 0.07 m, lake area narrowed by 314 km^2 with an average annual narrowing of 6.4 km^2 and storage capacity reduced by about 1.5×10^{10} m^3. The descended water level of Qinghai Lake has made Lake Surface shirked and new sub- lakes are continuously born. At present, four relatively big sub- lakes have been isolated. From north to south, they are Gahai, the new Gahai (Milton Lake), Haiyan Bay (not completely isolated from Qinghai Lake) and the Ear Sea. According to the equilibrium analysis of the Qinghai Lake, from 1959 to 2000, average lake recharge into the Qinghai Lake was 3.759×10^9 m^3 for many years, annual average total water consumption was 4.12×10^9 m^3, i. e. an annual average deficit amount of Qinghai Lake is 3.61×10^8 m^3.

By virtue of analyzing impacts on water-level changes by hydrological and meteorological changes in Qinghai Lake basin, inter-annual fluctuations of the Qinghai Lake are positively synchronized with the fluctuations of precipitation and inflowing lake surface runoff and are negatively synchronized with the lake evaporation. Although lakes evaporation slightly tends to decline, the amount still exceeds the recharge. In the past 50 years, the average annual water consumption by the national economy in Qinghai Lake basin has been ranging from 0.6×10^8 m^3 to 0.9×10^6 m^2, accounting for 3% to 5% of the surface water resources, therefore, the impact on the water level of the Qinghai Lake by the national economy is not very significant. Therefore, the dropped water level of Qinghai Lake is a result of both climate change and human activities with the former being dominate, but the indirect effects of human activities on the lake level can not be ignored.

Since climatic environment of the basin can not be effectively changed at present, it will undoubtedly be an effective measure to rationally develop and utilize water resources, to strengthen water resources conservation and protection, and to modestly reduce the destruction of human activities to the underlying surface, to improve the surface water-conserving ability.

3.2 The analysis of water requirement of river outside in Qinghai Lake area

Ecological environment of Qinghai Lake basin is relatively fragile, the more water used in national economy, the less water used in ecology. The water requirement by the national economy is predicted, satisfying the requirements of constructing a water-saving society, taking the sustainable use of water resources as the goal, highlighting the ecological protection of the Qinghai Lake basin, under the premise of full water conservation and livestock balance, according to local water resources development conditions and engineering layout as well as with reference to water use by relevant areas and areas with high efficiency of water use.

The total annual average water requirement in the Qinghai Lake basin has declined to 1.02×10^8 m^3 in 2030 from 1.11×10^8 m^3 of the base year, water requirement is stable with a slight decline during the planning period. Water consumption by ten thousand Yuan GDP in Qinghai Lake basin is down to 123 m^3 in 2030 from 1,130 m^3 of the base year and that of ten thousand Yuan GDP in 2030 is still higher than the national average (69 m^3/ million). Per-capita water consumption in Qinghai Lake basin is down to 771 m^3 in 2030 from 1,035 m^3 of the base year. See Tab. 2.

Tab. 2 Forecasted water requirement by life, production and ecology outside Qinghai Lake Basin River (Unit: ten thousand m^3)

Target year	Water requirement by residents' living (10^4m^3)	Water requirement by Livestock (10^4m^3)	water requirement for production				Water requirement by ecology (10^4m^3)	Total water requirement (10^4m^3)	Water requirement per capita (m^3)	Water consumption by ten thousand Yuan GDP (m^3)
			Industries (10^4m^3)	buliding industry (10^4m^3)	Agriculture, forest (10^4m^3)	Total (10^4m^3)				
Base year	181	1,093	22	151	9,664	9,837	6	11,116	1,035	1,130
2020	240	779	55	240	9,207	9,502	10	10,531	867	300
2030	307	970	96	498	8,314	8,908	16	10,200	771	123

3.3 The analysis of water requirement of the river inside of Qinghai Lake area

Rivers flowing into Qinghai Lake are mainly the Buha River, Shaliu River, Hargai River, Quanji River and Heima River. In line with water resources and its development conditions in the Qinghai Lake basin, considering the needs of water resources allocation in the basin, three sections are selected as calculation sections for inside required water with the hydrological Buha estuary station of the Buha River, the hydrological Gangcha station of Shaliu River, and hydrological Heima station of Heima River. Since there is no long-term hydrologic monitoring data about Hargai River and Quanjie River, inside water requirement of Hargai River and Quanji River is determined by the ratio of annual multi-year water-coming processes of Hargai River and Quanji as well as that of Shaliu River, considering that it, together with the Shaliu River, in the northern Qinghai Lake and has relatively similar underlying surface conditions and runoff processes. Because of the limited material, inside water requirement is calculated with hydrology methods. It means that ecological water requirement for the main sections of inflow rivers in critical period is determined by the maximum value by comparing Tennant method and lasting curve flow method. See Tab. 3.

Tab. 3 Ecological water requirement for the main sections of Inflow Rivers in critical period

River name	Section name	Jun (m^3/s)	July (m^3/s)	Aug (m^3/s)	Sept (m^3/s)	Water quantity from June to September ($\times 10^4$ m^3)	Water quantity in the whole year ($\times 10^4$ m^3)
Buha River	hydrometric station at estuary of the Buha River	10.2	24.7	23.1	16.5	19,697	20,900
Shaliu River	Gangcha hydrometric station	3.6	6.7	6.5	5.1	5,777	5,843
Hargai River	In-lake section of Hargai River	2.0	3.6	3.5	2.8	3,166	3,202
Quanji River	In-lake section of Quanji River	0.7	1.3	1.3	1.0	1,160	1,174
Heima River	hydrometric station of Heima River	0.3	0.3	0.2	0.1	239	242

4 Rational water allocation scheme in Qinghai Lake basin

According to water requirement of each target year, with partitioned water resources zone including counties as units, the supply and demand analysis is conducted for the typical year. In the course of calculations, top priority should be given to the urban and rural domestic water and basic ecological and environmental water, to ensure that the discharged volume of the main section of inflow rivers in critical period, to co-ordinate arrangements for water use in industry, agriculture and other sectors. And meanwhile, the surface water supply project capacity of Qinghai Lake basin is taken into consideration and the groundwater extraction should be appropriately increased .

According to the principle of water supply and demand analysis and allocation of water resources, water resource allocation result of the Qinghai Lake basin is shown in the Tab 4. Starting from the supply and demand calculation and allocation result, the total water supply of the Qinghai Lake basin has been basically maintained 1.00×10^8 m^3, of which surface water supply reduced to $0.958,1 \times 10^8$ m^3 in 2030 from $1.009,8 \times 10^8$ m^3 of the base year with the proportion of the total water supply decreased from 98.5 % to 93.9%; underground water supply has increased to $0.059,4 \times 10^8$ m^3 in 2030 from 0.015×10^8 m^3 of the base year with the proportion of the total water supply increased from 1.5% to 5.8% ; other water supply sources have increased to 250,000 m^3 from 4,000 m^3 of the base year. Water shortage will be scheduled to be basically solved in the target year.

In the base year, the depletion volume of surface water allocated outside river ways in the Qinghai Lake Basin is 0.706×10^8 m^3, reducing by $0.029,8 \times 10^8$ m^3 in 2030 in comparison with $0.676,2 \times 10^8$ m^3 decreased in the target year of 2030 (See Tab. 4). The depletion volume of surface water in the target year of 2030 accounted for 3.8% of the total surface water resources; groundwater extraction volume is $0.059,4 \times 10^8$ m^3, not exceeding groundwater available to be extracted. Seen from regional distribution, sound urban and rural drinking water security system will be established in areas lacking in water at present in Hargai River, Shaliu River and eastern bank of the lake to speed up the repair and renovation of the irrigation canal system, to vigorously develop water-saving irrigation, and to meet the reasonable requirements of the economic and social development for water resources in order to improve the ability to deal with special drought with the

Tab. 4 Result table of allocated water resources in different target years in Qinghai Lake basin Unit: $\times 10^4$ m^3

water resources zone	target year					2020					2030				
	allocated output of supplying water				depletion amount of Surface water	allocated output of supplying water				depletion amount of Surface water	allocated output of supplying water				depletion amount of Surface water
	Surface water	Underground water	others	subtotal		Surface water	Underground water	others	subtotal		Surface water	Underground water	others	subtotal	
The zone above Shanghuancang in Buha River	91.2	0.9	0	92.1	85.6	121.3	39.4	0	160.7	108.2	163	48.3	0	211.4	138.6
he zone below Shanghuancang in Buha River	486.6	88.4	0.2	575.2	405.5	1072.9	188.3	4.2	1265.3	749.6	1155.1	261.6	12.2	1428.8	803.8
The south zone of lake	161.4	3.3	0	164.6	153.7	95.2	24.9	0	120.1	76	140.6	29.9	0	170.5	101.8
DaoTang River zone	146.2	1.2	0	147.4	116.4	101.1	12.9	0	113.9	68.1	138.1	15.1	0	153.2	86.6
The east zone of the lake	475.8	0	0	475.8	338.6	352.7	16.6	0	369.3	248.9	337.3	19.7	0	357	240.3
Hargai zone	4047.8	2.1	0	4049.9	2761.2	4449.8	35.6	0	4485.4	3108	4313.3	47.3	0	4360.6	3047.4
Shaliu River zone	3958.2	50.1	0.2	4008.5	2685.6	3490.1	144.2	4.1	3638.5	2428.8	2983.9	154	12.6	3150.5	2096.2
Quanjie River zone	730.5	4.3	0	734.8	513.3	362.7	15.5	0	378.2	256.7	350.1	18.3	0	368.4	247.4
Qinghai Lake basin	10097.6	150.2	0.4	10248.2	7060	10045.7	477.4	8.3	10531.4	7044.3	9581.4	594.1	24.8	10200.3	6762.1

basic ecological and environmental water use is ensure. In the areas with basically balanced supply and demand of water resources, like areas over and below Shanghuancang of Buha River and Quanji River, on the basis of guaranteeing safe urban and rural drinking water and achieving balanced "people-water-grass-animal" relation, man-made water consumption should be strictly controlled, and meanwhile attention should be paid to protect water resources to ensure that the inflowing water quantity and quality.

5 Conclusions

Based on analysis and evaluation of current development and utilization of water resources in the Qinghai Lake basin, main problems are discussed in the water resources system of the Qinghai Lake basin. In line with balance analysis of the lake water quantity, the annual average water deficit of the Qinghai Lake is 3.61×10^8 m^3, which is the combined result of climate change and human activities with the former being dominate but the indirect effects of the latter on the lake level can not be ignored. Considering current water use situations and future economic and social development in Qinghai Lake basin, based on eco-services by Qinghai Lake, to maintain coordinated development of the basin water resources-economy and society-eco-environmental system is taken as the main line to vigorously promote water-saving society and strive to improve water use efficiency and the allocation of water resources capacity. And the future ecological environment construction and sustainable and stable needs of water resources for economic and social development will be guaranteed through rationally curbing demand. To implement the configuration program will play an critical role in constructing water-saving society, easing conflicts between needs and supply of basin water resources and basically realizing the balance of "water, grass, animal", and supporting the sustainable development of the basin's economic and social aspects.

References

An Integrated Planning for Water Resources in Qinghai Lake Basin[R]. The Yellow River Engineering Consulting Co., Ltd., 2011.

"Eco-environmental Protection and Restoration for the Qinghai Lake Basin" Editorial Board. Eco-environmental Protection and Restoration for the Qinghai Lake Basin [M]. Xining: Qinghai People's Publishing House, 2008.

Su Xiaoling, Kang Shaozhong, Shi Peize. Rational Allocation Model and Application for Ecology-riented Water Resources in Arid Areas[J]. Journal of Hydraulic Engineering, 2008, 39 (9): 1111-1117.

Yang Shumin, Shao Dongguo. The Theoretical Framework and Mathematical Model for Optimizing the Configuration of Regional Ecology-oriented Water Resources[J]. Water Resources Research, 2005, 26 (3): 1-4.

Wang Hao. Rational Allocation Model for the Northwest Water Resources Oriented Towards the Ecological and Economical Construction [R]. China Water Resources and Hydropower Research Institute, 2005.

The Study of Ecological Environment Evolution and Countermeasures in Source Area of the Three Rivers

Cui Changyong*, *Yang Guoxian*, *Jing Juan* and *Yang Libin

Yellow River Engineering Consulting Co. , Ltd. , Zhengzhou, 450003, China

Abstract: Source area of the Three Rivers is in the west of China which is located in the hinterland of Qinghai Tibet Plateau and the birthplace of the Yangtze River, the Yellow River and the Lancang River. It is also China's main source of fresh water supplies. The Qinghai Tibet Plateau's unique natural environment decides its vulnerability of the ecological system. By the dual effects of nature and human activities, the self – restraint ability of area water source drops, the pasture degrades, and the eco – environment is in a worsening situation. In order to improve the self – restraint ability of Source area of the Three Rivers and to alleviate its trend of ecological degradation, we analyze its eco – environmental evolution and the causes of eco – environmental changes and put forward some corresponding measures for protection by use of RS technology combining with the typical field – investigation.

Key words: Source area of the Three Rivers, the ecological environment evolution, the RS technology, counter measures

1 The situation of Source area of the Three Rivers

Qinghai province's Source area of the Three Rivers is located in China's western Qinghai Tibet plateau hinterland, in the south of Qinghai province. Source area of the Three Rivers has an altitude of 4,200 m in average is the birthplace of the Yangtze River, the Yellow River and the Lancang River, which is also known as the "Chinese water tower" and even "Asian water tower". The three rivers convey water to the downstream about 42.4×10^9 m^3 in average annual, not only is it China's main supply source of fresh water, but is China's lifeblood of social and economic sustainable development.

The Qinghai Tibet plateau's unique natural environment decides its vulnerability of the ecological system. In recent years, due to the comprehensive effects of the natural factors and human activities, Source area of the Three Rivers appears problems of grassland – degradation, land – desertification, frozen soil – degradation, eco – environment is in a worsening situation. At the same time, the lakes in Source area of the Three Rivers atrophy, the self – restraint ability of area water source dropps, water production has a significant descend. January In 2005, The CPC Central Committee and State Council approved The overall planning of Source area of the Three Rivers nature reserve's ecological protection and construction in Qinghai, August in 2005, it put into practice, through the planning's gradually implement, the local area of Source area of the Three Rivers in the soil and water loss have been better, lakes and wetland habitat have improved, the function of water source self – restraint has resumed, but the trend of the ecological degradation is still not get fundamental curbed.

2 The changes of eco – environment

In order to improve the self – restraint ability of Source area of the Three Rivers and to alleviate its trend of ecological degradation, by the aid of TM, ETM satellite image survey in source area of the utilization of land. Based on the typical field survey and remote sensing survey results of land use in the source area, we analyze the source area's ecological and environmental conditions.

2.1 The situation of land use

Source area of the Three Rivers in Qinghai has the most widely distributed grass, next is desert and bare land. Yellow River source area covers an area of 70,834 km^2 grass, accounting for

72.56% of the total area; Desert and bare land is 12,346 km^2, accounting for 12.65% of the total area; Other land (farmland, forest land, wetland, construction land, etc) accounted for 14.79% of the total area. The pripary cover in the Yangtze River source area is overall grass, accounted for 58.06%; next is the desert and bare land, accounted for 26.34%; other land accounted for 16.06%. Lancang River source area's grassland area accounts for 58.02%; next is the desert and bare land accounted for 24.96%; other land accounted for 17.02%. The land utilization condition of Source area of the Three Rivers sees Tab. 1.

Tab. 1 The situation of land use (Unit: km^2)

Water basin	Plowland	Forest land	The grassland in middle and high coverage level	The grassland in low coverage level	Wetland	Constru cted land	Desert and bare land	Total
Yellow River	441	6,638	46,066	24,768	7,314	45	12,346	97,618
	0.45%	6.80%	47.19%	25.37%	7.49%	0.05%	12.65%	100%
Yangtze River	16	4,413	50,660	42,097	15,009	6,212	41,345	159,753
	0.01%	2.76%	31.71%	26.35%	9.40%	3.89%	25.88%	100%
Lancang River	109	3,717	15,478	6,187	2,524	3	9,321	37,340
	0.29%	9.95%	41.45%	16.57%	6.76%	0.01%	24.96%	100%

2.2 The situation of the changes in eco – environment

Relying on the situation of land utilization, we can choose land use types of the land in desertification, lakes, the grassland in middle and high coverage level etc. which are closely related to water conservation self – restraint, point 1989 ~ 2000 and 2000 ~ 2005 in two time periods, analyze the source area's eco – environmental evolution. The source area of land use remote sensing interpreter results showed that, from 1989 to 2005, in this 16 years, the reverse evolution process of the Yellow River's eco – environment is obvious; The Yangtze River's source area in the 1980s to the 20th century, at the beginning, its eco – environment is in degradation situation, and then, in improving signs; Lancang River source area is basically stable. The change situations of the land use of The Source area of the Three Rivers in Qinghai which is closely related to water source self – restraint see Tab. 2.

Tab. 2 The change situationsof the land use of The Source area of the Three Rivers in Qinghai which is closely related to water source self – restraint (Unit: km^2)

Project		1989	2000	2005	Changes 1989 ~ 2000	2000 ~ 2005	1989 ~ 2005
The grassland in middle and high coverage level	Yellow River	46,832	46,730	46,066	– 102	– 664	– 766
	Yangtze River	50,321	50,292	50,660	– 29	368	340
	Lancang River	15,450	15,447	15,478	– 2	31	29
	Source area of the Three Rivers	112603	112,470	112,205	– 133	– 264	– 397

Continued Tab. 2

Project		1989	2000	2005	Changes		
					1989 ~ 2000	2000 ~ 2005	1989 ~ 2005
Lakes	Yellow River	1,608	1,548	1,569	-60	21	-40
	Yangtze River	1,388	1,302	1,375	-86	73	-13
	Lancang River	32	31	31	0	0	0
	Source area of the Three Rivers	3,028	2,881	2,975	-146	94	-53
Desert	Yellow River	6,380	6,507	6,520	127	13	140
	Yangtze River	14,872	14,904	14,878	32	-26	6
	Lancang River	6,128	6,135	6,136	7	1	8
	Source area of the Three Rivers	27,380	27,546	27,535	166	-12	154

2.2.1 The changes' situation of land in desertification

In 1989 ~ 2005, the area of land in desertification in the Yellow River source area is in an increasing trend, which is accumulated increase to 140 km^2, its eco – environmental degradation is obvious. The Yangtze River source area desertification of land has increased at first, and then is in decreasing trend, its total variation is not big. Lancang River source area desertification land's variation is not big.

2.2.2 The changes situation of the lakes

In 1989 ~ 2005, lake area of the Yellow River source area firstly decreased, secondly is in an increasing trend, but the increment is small (restored 34%) and, in general, lakes area reduces 40 km^2. The Yangtze River source area lake area also decreased firstly, secondly is in an increasing trend, but the increment is large (restored 85%) and, in general, lakes area reduces 13 km^2. The lake area of Lancang River source area has few changes.

2.2.3 The changes' situation of the grassland in middle and high coverage level

In 1989 ~ 2005, the grassland in middle and high coverage level's area of the Yellow River source area continues to be in a decreasing trend, it reduces accumulated 766 km^2. The grassland in middle and high coverage level's area of the Yangtze River source area firstly decreased, secondly is in an increasing trend, and the amount is bigger (more than the decrease), its area increased totally by 340 km^2. The grassland in middle and high coverage level's area of Lancang River source area has few changes.

3 The analysis of the causes of eco – environmental changes

The influence factors of eco – environmental changes have two kinds, one are natural factors and the other are human factors. Natural factors include precipitation and changes of the temperature, human factors include excessive feed, the abuse of cutting trees, the abuse of digging, etc. The climate factors like precipitation and the temperature are the main driving force to the eco – environmental changes of the Source area of the Three Rivers, and adverse human activities aggravate the trend.

3.1 The analysis of natural factors

3.1.1 Precipitation

(1) The change situation of precipitation in the Source area of the Three Rivers.

According to the statistics, the Yellow River source area has the rainfall of 487 mm in average year (1956 ~ 2009). In statistical period of time, the Yellow River source area's rainfall is basic stability, having a slight decrease after the 1990s. The 1990s' average rainfall is 478 mm, average rainfall from 2000 to 2009 is 480 mm.

Based on Zhimenda hydrological station's precipitation data in the Yangtze River mainstream, the Yangtze River source area in 1956 ~ 2009 years of average rainfall is 358 mm. The Yangtze River source area's rainfall in statistical time is basic stability, but it is on the rise in recent years. The 1960s' average rainfall is 352 mm, 1970s' average rainfall is 353 mm, 1980s' average rainfall is 379 mm, 1990s' average rainfall is 341 mm, and average rainfall from 2000 to 2009 is 375 mm.

According to Xiangda hydrological station about its above area's rainfall materials of the Lancang River, the Lancang River source area in 1956 ~ 2007 years of average rainfall is 493 mm. In statistical times, in addition to the 1980s, the Lancang River source area's rainfall is slightly abundant; its rainfall in other times changed little, been stable. The 1960s' average rainfall is 480 mm, 1970s' average rainfall is 483 mm, 1980s' average rainfalls is 516 mm, 1990s' average rainfall is 488 mm, the average rainfall from 2000 to 2007 is 518 mm.

River source area's changes in rainfall since the 1950 s see Fig. 1.

(2) The analysis of how precipitation affects the eco – environmental changes. Water is the source of life, the primary production of plants in the source area nearly has linearity with rainfall, if precipitation increases, the amount of plants increase, and vice versa. In statistical times, the Yangtze River source area's and the Lancang River source area's rainfall are basiclly stable, the amount of precipitation is in an increasing trend in recent yeas, the corresponding eco – environment get better , total changes are basically stable. And the Yellow River source area's rainfall decreased after 1990 s, the corresponding eco – environment appeared degraded trend. A positive correlation. is between changes in rainfall and eco – environmental changes.

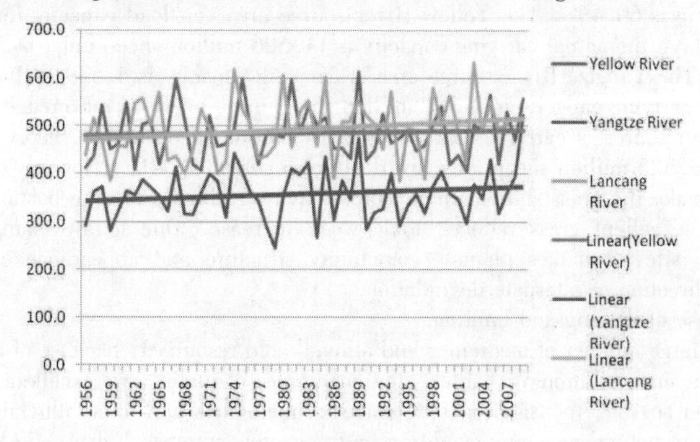

Fig. 1

3.1.2 The temperature

(1) The source area temperature changes.

According to the relevant research, The years average temperature's trend rate in in the source region is 0.37 ℃ / above 10 years from 1961 to 2009. From the spatial distribution of the years average temperature changes to see, the trend rate of the climates in Zeku、JiuZhi and Yushu are 0.40 ℃ / above 10 years, the temperature is significantly raising, the rising temperature in Maxin、Guinan、Tuotuo river and Wudaoliang is worst significant, but climate trend rate is still as high as 0.21 ~ 0.29 ℃ / 10 years. Nearly 49 years, average temperature's increasing trend in the source region is very significant, since the 1960 s, most of the region present sustainable rising condition; its increasing extent is the most obvious after getting into the 90s.

(2) The analysis of how the temperature affects the eco – environmental changes. The temperature change's drive function on the eco – environmental change mainly reflects in the following aspects: one is the generally increased temperature in the source area play an important role in the trend of the significantly increased amount of evaporation, evaporation capacity increases will enhances evaporation of the wetland water, causing lakes water reduced、lakes water flow inside soil and salinization, causing swamp plants die and the problems of changing the high – cool swamp alpine meadows into the high – cool alpine grassland and alpine meadow; Two is to make the frozen soil vestigial, and frozen soil degradation brings a series of eco – environmental deterioration problems, such as permafrost's degradation can cause degeneration of soil、plants and pasture and cause freeze – thaw desertification, regional groundwater's levels fall, springs mouth's position move down, marsh wetlands reduce, the hot – melted lake ponds disappear, water resources reduce and environment variation sustainable takes place, it will alse form the harm such as the secondary bare land which is the type of "Black soil " and high – cool grassland's desertification.

3.2 The analysis of human factors

None reasonable human activities also exacerbate the source area's eco – environmental damage. Since the 1950 s, with the increase in population of the source area, number of cattle grows, grassland overload to graze, which lead the grass productivity to reduce, and grassland to degenerate, the function of water self – restraint and water and soil conservation to lower. In addition, unplanned mining in the local area also exacerbate the process of grassland degradation and land desertification.

(1) Grassland overloads to graze.

Grassland overload to graze is a common phenomenon in Source area of the Three Rivers, the Yellow River source areas is the most common. According to 2007 statistics analysis, the pasture in Source area of the Three Rivers can load capacity for 12.602 million sheep unit in theory, but its actual situation is for the 20.212 million number of cattle sheep unit, overload 7.61 million sheep unit, overload ratio is 60.4%. The Yellow River source area can load capacity for 5.922 million sheep unit in theory, the actual carrying capacity is 11.506 million sheep unit, overload ratio is as high as 94.3%; The Yangtze River source area's carrying capacity is 4.539 million sheep unit in theory, its actual carrying capacity is 5.381 million sheep unit, overload ratio reach to 18.5%; the Lancang River source area's carrying capacity is 2.14 million sheep unit in theory, its actual carrying capacity is 3.325 million sheep unit, overload ratio gets to 55.4%. Not only can grassland overload to graze make the grassland's primary productivity declined, but the pasture's quality become to be bad, excellent grass reduce, toxic weed increase. Due to the change of dominant plants' resources site, it makes plants' community structure and appearance change, pasture plants go to the direction of alternate degradation.

(2) The abuse of digging and hunting.

There are a large number of medicines and alluvial gold resources, because of being located in highland, environment pollution is lighter, the medicines' quality are excellent, so, in every spring and summer season, thousands people excavate medicines and mine alluvial gold. In addition, by the limit of natural economy condition and economic interests' drive, in order to get the fuel and food, local nomads fell thickets vegetation, mine residential areas' turf, kill wild animals, these actions damage the vegetation, make species resources in reduction and eco – environment in deterioration.

4 The study of countermeasures

Combined with Source area of the Three River's remote sensing survey results of land use, refer to Qinghai province's soil and water conservation and eco – construction planning, do water conservation ability reduction area to comply with "water conservation, eco – protection, reduce disturbance, give full play to the nature's self – repairing ability", we can take the small watershed as a unit, mainly to the water resource conservation which is closely related to forest land, meadow of lakes and swamps and take protective measures, with the use of ditch (canal) protection and other small soil and water conservation project construction.

4.1 The protection measures to grassland

(1) Grassland sealing grow.

Fence sealing grow divided into two ways: full ban and half ban (grazing in turns and division). Full ban is aiming at the people area of relative concentration and poor vegetation of the grassland in low coverage level for the way of pasture vegetation recovery and improving its ability to take water conservation in the all year as soon as possible. Half ban (grazing in turns and division) is mainly aiming at the people area of relative low, relatively good vegetation of the grassland in low coverage level, taking the way of grazing in turns and division. Fence sealing grow grass: 38,197 km^2, of which, the Yellow River basin's is 14,366 km^2, the Yangtze River basin's is 22,091 km^2, the Lancang River basin's is 1,740 km^2. See Tab. 3.

(2) Grassland adding sow.

Combined with fence sealing grow to adding sow to the grassland in low coverage level (coverage generally under 20%), serious degeneration grassland and black soil beach, the main there they has are: needle spear, with alkali grass, precocious shocks, the stars grass on a. splendens, etc. Adding sow grass area is 845.4 km^2, of which, the Yellow River basin's is 413 km^2, the Yangtze River basin's is 348.9 km^2, the Lancang River basin's is 83.5 km^2, see Tab. 3.

4.2 Forestland protection measures

(1) Forestland sealing grow.

Degradation forestland like deforested, seedling to fire places, according to forest terrain and the distribution plot, partition and delimit banned scope with the way of full ban in the all year to facilitate afforestation. Forestland sealing grow area is 27 km^2, including 23 km^2 in the Yellow River basin, 4 km^2 in the Yangtze River basin. See Tab. 3.

(2) Forestland construction.

Water self – restraint forests (water and soil conservation forests) when it is under construction should be fully considered the site conditions, under the premise of non – cause "Govern mentality destruction", we can choose adaptable, root developed, rapidly growing fine native tree species to afforestate. Mainly in the hillsides to build shelter forest of soil and water conservation to provide water and soil conservation, prevent and control soil erosion and improve eco – environment, improve the ability of water conservation. At here, mountain shrub is primary to forests' construction, take the way of shrubs combined with arbors. Main shrubs species are seabuckthorn、strange willow, caragana intermedia etc.; The main arbors species have P. cathayana, spruce elm birch, etc. Area Forest construction area in Source area of the Three Rivers is 148.2km^2, of which, 77.6 km^2 in the Yellow River basin, 34.6 km^2 in the Yangtze River basin, 36.0 km^2 in the Lancang River basin, see Tab. 3.

(3) Lakes' and swamps' protection.

For the protection of the lakes and swamps, mainly through the sealing grow, vegetation restore, increasing vegetation coverage, improving the function of soil water storage and the function of alpine cool swamp meadow's water conservation. This time, we take protection for the lakes and swamps outside the natural protection zone in The Source area of the Three Rivers, its totaling area is 8,810 km^2, of which, 2,873 km^2 in the Yellow river basin, 5,619 km^2 in the Yangtze River basin, 318 km^2 in the Lancang River basin. See Tab. 3.

4.3 Small water source projection measures

In addition to water source self – restraint measures like grass protection, forestland protection, lakes and swamps protection . Small water source projection is also one of the effective measures in eco – protection. Small soil and water projection (Cereal workshop, protective wall and so on) layout in upstream channel in general, which has the role of making sand sluggish and keeping water clear, has an important role to prevent the channel's erosion slice, stable ditch bed, consolidate ditch slope, prevent pasture soil vegetation, stop the eco – environmental degradation, strengthen water conservation.

Combined with Qinghai province's soil and water conservation and eco – construction planning achievements, we have constructed 3,378 seats stopped engineering; Ditch protection engineering 343 km, see Tab. 3.

Tab. 3 Eco – environmental protection measures in Source area of the Three Rivers

Measures	The Yellow River	The Yangtze River	The Lancang River	The total
Grassland sealing grow(km^2)	14,366	22,091	1740	38,197
Grassland adding sow(km^2)	413	348.9	83.5	845.4
Forestland sealing grow(km^2)	23	4		27
Forestland construction(km^2)	77.6	34.6	36.0	148.2
Lakes sealing grow(km^2)	2,873	5,619	318	8,810
Cereal workshop, head of ditch protection(Seat)	1,850	1,040	488	3,378
Protective wall(km)	173	98	72	343

5 The peroration

(1) Climate factors like precipitation and the temperature are the main driving factors of the changes of the source area's eco – environment, but adverse human activities aggravate this trend. Precipitation increase, eco – environment get better, precipitation reduce, eco – environmental deteriorate. Rising temperatures make evaporation increase, cause lake water reduce, the frozen soil degenerate, vegetation deteriorate.

(2) Since the 1950 s, the Yellow River source area's rainfall declined, eco – environment present degradation trend. The Yangtze River source area's and the Lancang River source area's rainfall are basic stability, in recent years, the trend has risen, the corresponding eco – environment tends to get better, the change of all are basically stable.

(3) In view of natural factors source area is the main factor of eco – environmental degradation, eco – environmental control should adhere to the principles: "water conservation, eco – protection, reduce disturbance, give full play to the nature's self – repairing ability".

(4) Source area of the Three River's eco – environmental control should also consider local farming and animal husbandry which are their primarily economic conditions; we should make the best study of policies and mechanism to promote Source area of the Three River's eco – environmental protection.

References

Zhao Chuanchuan, Yang Xiaoyang, Zhang Fengcheng. Effect of Climate on Desert Land Vegetation

Biomass in Sanjiangyuan Region[J]. Research of Soil and Water Conservation. 2008,15(3),175 –177.

Dai Sheng, Lin Ling. Analysis on Climate Changing Features in Source Land of the Three Rivers from 1961 to 2009[J]. Qinghai Meteorology. 2011,01,20 –26.

Liu Chaihong, Su Wenjiang, Yang Yanhua. Impacts of Climate Change on the Runoff and Estimation on the Future Climatic Trends in the Headwater Regions of the Yellow River[J]. Journal of Arid Land Resources and Environment. 2012, 26 (4):97 –100.

Ji Lingling, Shen Shuanghe, Guo Anhong. Review on the Climate Changes in Source Land of the Three Rivers and their Impacts on Wetland[J]. Jinlin metorology. 2009,(1):15 –17.

Jin Huijun, Wang Zhaoling, Lu Lanzhi. Features and Degradation of Frozen Ground in the Sources Area of the Yellow River, China[J]. Journal of Glaciology and Geocryology. 2010,32(1):10 –17.

Zhang Senqi, Wang Yonggui, Zhao Yongzhen. Permafrost degradation and Its Environmental Sequentin the Source Regions of the Yellow River[J]. Journal of Glaciology and Geocryology. 2004,26(1):1 –6.

Wang Qiji, Lai Dezhen, Jing Zengchun. The Resources, Ecological Environment and Sustainable Development in the Source Regions of the Yangtze, Huanghe and Yalu Tsangpo Rivers[J]. Journal of Lanzhou University(natural sciences). 2005,41(4),50 –53.

Wu Wanzhen. Exploration of Eco – environmental Problems in Three – river – source Natural Reserve. Anhui Agriculture Science. 2010,38(35),20286 –20288.

Study on the Eco-environmental Water Demand for the Downstream Lakes and Wetlands of the Bayin River Basin

Dong Dianhong, Wang Jichang, Li Xuhui, Wang Jing and ***Jiang Libing***

Yellow River Engineering Consulting Co. ,Ltd. Institute
of Environmental & Resettlement Engineering, Zhengzhou, 450003, China

Abstract: Northwest arid and semi-arid areas in China are the regions with water resources shortage and more sensitive and fragile system of eco-environment. Study on the problems of lakes and wetlands, such as shrinking degeneracy problem at the river terminal lakes and wetlands, etc, has been extensively attended. It is of typicality and signification to carry out the regional ecological water demand studies for the Qaidam Basin in the northwest area where the Ecosystem is very fragile and its own recover ability is also relatively poor. Based on climatic characteristics and variation characteristics of runoff in the Bayin River watershed, and focused on Lake area change and Lake ecological water resources demand calculated by the hydrological characteristics and the lagoon of Lake evaporation rate law, ecological water demand of Lake wetland of the Bayin River watershed has been estimated. The purpose is to provide useful information and scientific basis on maintaining ecological balance and economic system for sustainable development, and achieving the rational allocation of water resources in the Bayin River basin.
Key words: lakes and wetlands, ecological water demand, system of eco-environment

1 Introduction

Over the past century, series problems of eco-environment are caused on lakes and wetlands by the impact of global climate changes and increased human activities, lake area reduction and water level decline, deterioration of water quality, overall number of lakes reducing, and recession of lake ecosystems due to the impact of global climate change anomalies and other natural factors and socio-economic status. This has aroused widespread attention , Therefor the study of ecological water requirement of lakes and wetlands has very important significance.

In China, many scholars and experts have carried out the specific analysis and studies on the ecological water demand for lakes, and also calculation. Among them, the more representative persons are Jia Baoquan who adopted Evapotranspiration minus rainfall estimation method of ecological water supply volume according to the size of the contribution rate of the lake on the oasis economy and the physical circumstance of sustainable development of the lakes and oasis; Xu Zhixia et al. that they researched and justified the ecological water demand for lakes at the lowest water level according to that the requirement of, taking the Nansi Lake for example, no severely degradation in maintaining the function of lake hydrology and terrain subsystem; Li Xinhu et al. those developed a calculation model and method based on water balance theory to aim at the characteristics of closed lakes and throughput lakes; Tan Xiaoming who starting from the point of view of project construction on the ecological impact of downstream lakes and according to the river hydrology characteristic of the Yangtze River and Dongting Lake has analyzed and estimated the minimum ecological water demand for the Dongting Lake during the dry season; and Tang Kewang who has estimated the ecological water demand for the Hongjiannao lake on the northern Shaanxi Province, considering the influence factors based on the dynamic variation of lakes.

In this paper, based on a brief overview on the development level of the research on ecological water demand, to take the terminal lakes and wetlands in the Bayin river lower reaches in Qaidam Basin, a cold and arid area in the northwest China as the object, in order to protect the basic function of river ecological environment as the ultimate goal, the research and analysis on ecological water demand of closed lake in inland plateau has been tried to carried out. And the estimation on ecological water demand of the lakes and wetlands is also undertaken preliminary. The paper is very important operation significance on the guidance of reasonable utilization and protection of water resources on Bayin river basin and protection of ecological environment, to provide a scientific basis.

2 General situation of the lakes and wetlands in the lower river basin

The terminal lakes of Bayin River on the Qaidam Basin include the Keluke Lake, Tuosu Lake and Gahai Lake.

The Keluke Lake in the northeast of Qaidam Basin, about 40 km to the southwest of Delingha city, is the largest freshwater lake in the Qaidam basin. The Bayin River empties its self into the northeast of Keluke Lake, and the Balegeng River into the southwest of the Lake. The Lake empties its self into the southwest of Tuosu Lake which is a salt water lake and the Lake surface is broad and vast open. The lake water is mainly replenished by the Keluke Lake water via a connected channel with each other, and the water in the Lake relies on water surface evaporation in vertical way. The Keluke Lake is surrounded reed, and lived numerous of plankton, fish and waterfowl. The Tuosu Lake is a typical inland salt lake, sporadic growth with white spines, chionese tamarisk, sand sagebrush, sasa and moss on its lakefrond, and on its island, there are many migratory birds inhabited such as gull, wild goose et al.

Bayin River downstream from Yikeshu to Keluke Lake is a flat and wide valley, forming a large area of natural wetland and swamp swarm, which is 35 km of length in the east and west direction, 10 km in the south and north direction, among them a perennial water surface area on wetlands reaches 40 km^2, and composite community of thickets and meadow in 150 km^2, Formed a unique lake ecological system in the reach from Keluke to Tuosu Lakes. Gahai Lake, a salt water lake, is located in southern Delingha City. (see Fig. 1)

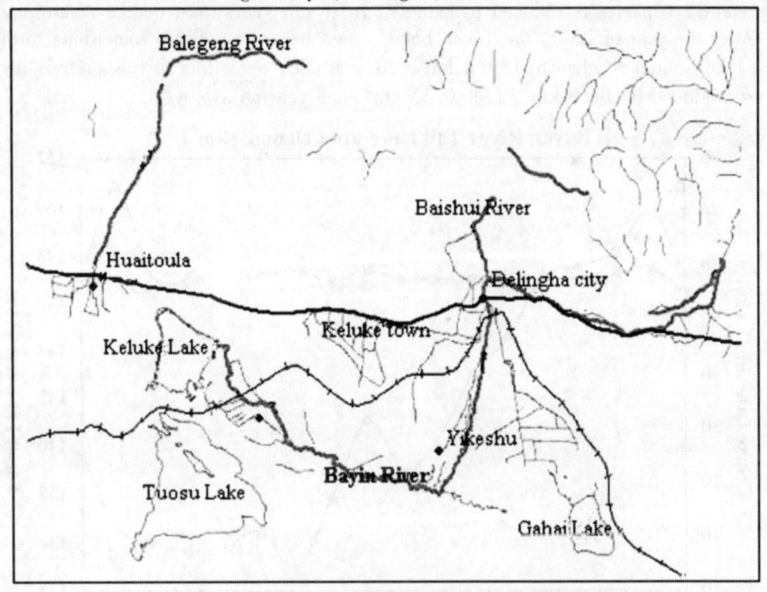

Fig. 1 Location of the Bayin River Basin

3 Analysis of downstream Tail lake area change and the reasons for

According to the information, water area of the Gahai Lake changes less, but largely in the Keluke and Tuosu lakes, therefore only the Keluke Lake and Tuosu Lake are analyzed

In order to analyze the area changes of downstream terminal lakes, the remote sensing data of water area from 1990 to 2006 of the Keluke and Tuosu lakes on the lower Bayin River have been analyzed and interpreted by the 2007 Remote Sensing and Monitoring and Ecological Assessment Center of the Qinghai Provincial Meteorological Sciences Research Institute.

In order to analyze the rule of lake area changes, the analysis on image graph of 18 time phase in total in November or December in 1989 ~ 2006 selected is carried out, and the years changes of water surface area in two lakes shown in Fig. 2. The changes of water surface area of Keluke Lake in 1989 ~ 2001 is small, fluctuating within 54.9 ~ 59.6 km^2. Since 2001, the area of Keluke Lake changes drastically, the areas in 2002 and 2005 reach 69.1 km^2 and 74.9 km^2 respectively, those are beyond the years average area of 60 km2, because of plentiful rainfall in that year. In 1989 ~ 1999, water surface area of the Keluke Lake declines in different extent, it is caused by that local economic development and city life water demand increase ceaselessly, and the specific impact remains to be further study. Because the Tuosu Lake water is supplied by the Keluke Lake water, the water surface area of the Tuosu Lake from 1990 to 2000 has also been shrinking, shrinking from149.8 km^2 to 135 km^2. After the year of 2001, the water surface area of the Lake picks up again gradually, and to the year of 2006, the area rise up to 139.8 km^2.

According to survey topographic map in 1972, the area of the Tuosu Lake is 151.7 km^2; and to the remote sensing image data, that is 149.8 km^2 in 1990, and from 1972 to 1990, the area is decreased 1.9 km^2, reducing by 0.106 km^2 each year on average, and the lake storage reduces the amount of 58,000,000 m^3. From 1990 to 2000, the Lake water area is decreased from149.8 km^2 to 135 km^2, reducing 14.8 km^2, 1.35 km^2 each year on average. Water storage capacity is reduced by 490,000,000 m^3, reducing by 45,000,000 m^3 each year on average, in which, the value of 135 km^2 is the minimum value since the data been measured. The lake area declines quickly, and water storage capacity is reduced expeditely. In recent years, the degree of atrophy has increased. After the year of 2000, the Tuosu Lake's area began to pick up somewhat, to the end of 2006, the lake's area reaches to 139.8 km^2. As a whole, according to the analysis above, the Lake's area reduced by 11.6 km^2, i.e. 0.35 km^2 each year on average.

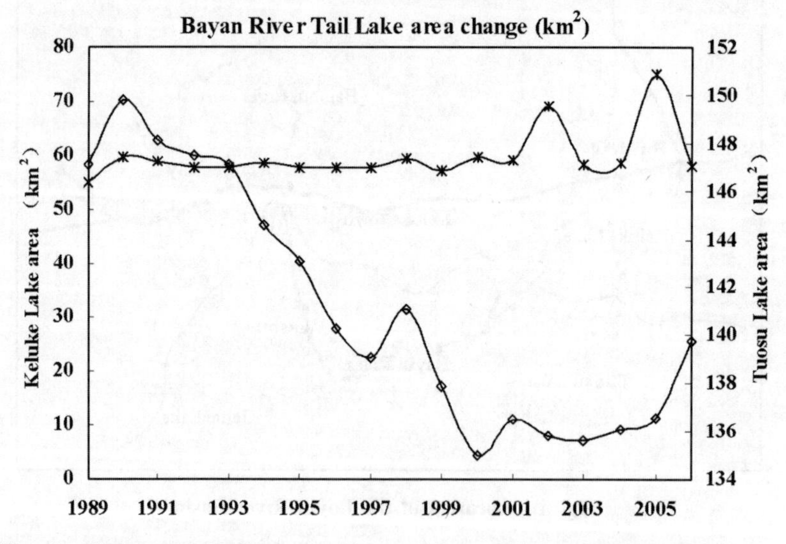

Fig. 2 Years water surface area changes of the Keluke and Tuosu lakes

The water surface area changes of the Keluke and Tuosu lakes present an apparent recession phenomenon, there are several reasons for this: Firstly, lakes in the atrophic degeneration are a common phenomenon in Qinghai-Tibet Plateau, under the background of global climate warming in recent decades, the Qinghai – Tibet Plateau is located in the earth's climate sensitive areas, appeared apparent drying and warming; The Qaidam Basin located in the hinterland of Qinghai-Tibet Plateau basin where climate warming is significant, and where the temperature rises higher than the average level of the Qinghai plateau, especially after the 1990's of the 20th century (i. e. in 1991 ~ 2000), the temperature increases are the most obvious, and reached a maximum value of average temperature of that decade. It is also shown that the years (in 1990 ~ 2006) rainfall (the length limit is no longer displayed) at Delinha compares with average annual rainfall of 180 mm, exhibits a reduced tendency. The Bayin River is located in the hinterland of plateau, ice and snow melt water of mountainous area increased, and also the water supply increased, it is advantageous of the water resources recycling in a short period. Due to the unique plateau geographical environment on the Tibetan Plateau, water in the circulation process of runoff and evaporation dissipation occupied an important part, eventually leading to the lake water level continued to decline, the area of lakes atrophic degeneration.

Secondly, the lake area reduction is related to the increasing of production and living water supply in the region in the same period. Large-scale irrigation district construction has carried out in the Bayin River Basin in the 1950's and 1960's of the 20th century, when the largest cultivated area reached 13,300 hm^2. It is limited to the economic development, the development of agricultural science and technology level and level of investment in water conservancy, caused a series of ecological problems at that time, including soil secondary salinization, soil erosion, desertification and other problems. In 1990's of 20th century, the level of economic development is raised, and the water conservancy construction, the extent of lake's area atrophy has been obviously influenced by the production activities of human life, the specific impact remains to be further in-depth analysis and study.

4 Lakes and wetlands, the ecological water demand analysis

4.1 Ecological water requirement calculated

According to the calculation method of ecological water demand for water balance, the loss of lake water is that the total amount of lake water discharged out the lake minus that of water entered into. According to the analyzing results above, the water quantity for maintaining lakes basic ecological functions of the Keluke Lake and the Tuosu Lake is that needed for maintaining basic function area and wetland area of the lakes. In order to maintain the basic functions of ecological environment system of the lakes, water balance in the lakes required does not change too much, to set that the underground water is under the conditions maintained dynamic balance, and ecological water demand of the lakes is mainly used to maintain water balance of the lakes, i. e to supplement the net loss of water surface evaporation, the calculation formula is as follows.

$$WW = \sum A(E - P) \times 10^3$$

where WW denotes the water demand for surface evaporation consumption, m^3/a; A is measured water surface area of evaporation, m^2; E the corresponding evaporative capacity on water surface, mm; P the corresponding precipitation on water surface, mm.

Here yearly and monthly rainfall data measured by the Zelingou and the Delingha hydrometric stations and the Delingha Meteorological Station are selected, the annual precipitation series of the three station are synchronous to the years of 1956 ~ 2006 according to the regularities of rainfall distribution in the Basin and the results of "Qinghai water resources investigation and assessment".

4.2 Water evaporation on saltwater lake and Saline Lake

The evaporation of saline lake is closely related to its brine density closely, and the evaporation rate decreases with the brine density increases. The evaporation of brine of different concentration are all lower than that of the fresh water, with the concentration is increased and the evaporation ratio decreased gradually, so the brine concentration is the main factor to effect on the brine evaporation rate. Due to the effect of salinity, actual evaporation capacity of saline lakes is small than that of freshwater. For the salt water lakes and brackish water lakes, a two-year evaporation test has been carried out by the Lanzhou branch of Chinese Academy of Sciences in the months from July to August in 1989 ~ 1990 at the near lake zone of the Xiashe Hydrometric Station on south bank of the Qinghai Lake. The evaporation observation on lake water and spring water is carried out respectively, the results show that lake water evaporation in the 20 cm-evaporating dish is about 15% less than that of spring water. This is because the saltwater reduces the saturation vapor pressure to inhibit the water surface evaporation. In order to fully study the relationship between saltwater and freshwater evaporation of the lakes, a measuring station had been established from May 1, 1988 to September 31 on Haixin Mountain in the center of Qinghai Lake by the Hydrological Bureau of Qinghai province, to carry out 5 months of daily evaporation measurement from water surface, finally get the E601- pan daily saltwater evaporation, and the 20 cm pan saltwater and freshwater daily evaporation measured data. Through correlation analysis, the 20cm- pan daily evaporation measured from freshwater and saltwater exists the best correlation, and the correlation coefficient is reached to 0.98. The years average evaporation (1965 ~ 2002) of the Qinghai Lake is 930.9 mm, and the freshwater evaporation observed at several near around hydrometric stations is 945.5 mm on average, so when the lake salinity is 14‰ (i. e. salinity of the Qinghai Lake), the evaporation rate is 98‰. According to the experimental results of the Qinghai Institute of Saline Lake of the Chinese Academy of Sciences, it is shown that on different concentrations, saturated brine evaporation and evaporation ratio of saturated brine and freshwater are all decreased with the brine specific gravity increased, when the brine specific gravity is 1.21, the ratio evaporation value is 74%. Water resources is divided according to the salinity, salinity between 0‰ to 1‰ belongs to the freshwater lake, that between 5‰ ~ 8‰ the brackish water lake, larger than 30‰ the Saline Lake. According to the measured data in October of 2008, the salinity of Tuosu Lake is 30‰, accordingly, the Tuosu Lake and Gahai Lake belong to saline lakes. According to the analysis above, a preliminary estimation evaporation of the Keluke lake, Tuosu lake and Gahai Lake are 1,193.3 mm, 954.6 mm and 895 mm respectively. The calculated evaporation of the Keluke lake, Tuosu lake and Gahai Lake is shown in Tab. 1.

Tab. 1 Years average monthly and annual evaporation of the Keluke, Tuosu and Gahai Lakes

(Unit: mm)

Item	1	2	3	4	5	6	7	8	9	10	11	12	All the year
Keluke Lake	22.7	34.6	74.6	123.2	164.1	163.1	174.5	163.1	124.2	80.9	41.8	26.5	1,193.3
Tuosu Lake	18.2	27.7	59.7	98.6	131.3	130.5	139.6	130.5	99.3	64.8	33.4	21.2	954.8
Gahai Lake	17	25.9	55.9	92.4	123	122.3	130.9	122.3	93.1	60.7	31.4	19.9	894.8

4.3 Eco-water demand of the downstream terminal lakes and wetlands

The surrounding area of the Keluke Lake and Tuosu Lake is low mountains, hills and piedmont alluvial fans to gentle incline to the Keluke Lake and Tuosu Lake where are as the center of gentle inclined plain with low precipitation and large evaporation. The dominant factor for the plant community growth and spatial distribution pattern is not the atmospheric temperature and precipitation, but the underground water level and soil salt content as well as the soil properties and eolian deposit those controlled by the groundwater. Reference to domestic related research, water resources investigation and assessment as well as the results of "Study on Reasonable Disposition and the Load Carrying Ability of Water Resources in Western Area", the shrub forest evapotranspiration in Qaidam Basin is 340 mm, transpiration of riparian scrub and meadow plants is 340 mm.

According to the evaporation (average annual evaporation in 1990 ~ 2006) and rainfall data

(average annual rainfall in 1990 ~ 2006), in accordance with the maintenance of the basic function required area (i. e. the downstream wetland riparian scrub area is 110 km^2, wetland water area 40 km^2; the Keluke Lake's water area 59. 6 km^2, the Gahai Lake's water area 32. 5 km^2, the Tuosu Lake's water area 135 km^2) of wetlands that connected lakes and lakes on the lower Bayin River coccyx, the ecological water demand is calculated as follows; The eco-water demand for riparian scrub and meadow is 16,601,000 m^3, wetland 41,841,000m^3, Keluke Lake 59,851,000 m^3, Tuosu Lake 103,344,000 m^3, Gahai Lake 22,942,000 m^3, the total water demand is 244,579,000 m^3, see Tab. 2. The calculation results of years eco-water demand of terminal lakes and wetlands on the lower Bayin River (riparian scrub and meadow area is fixed, lake's water area in accordance with the remote sensing image results of 1990 – 2006 are calculated) are shown in Tab. 3.

Tab. 2 The eco-water demand for maintaining terminal lakes and wetlands of the lower Bayin River coccyx

Area: km^2 Eco-water demand: $\times 10^4$ m^3

Item		Area (km^2)	Evaporation (mm)	Preciptation (mm)	Plant transpiration (mm)	Water demand ($\times 10^4$ m^3)
Riparian scrub and meadow		110. 0		189. 1	340	1,660. 1
Water area	Wetland water supplement	40. 0	1,235. 1	189. 1		4,184. 1
	Keluke Lake	59. 6	1,193. 3	189. 1		5,985. 1
	Tuosu Lake	135	954. 6	189. 1		1,0334. 4
	Gahai Lake	32. 5	895	189. 1		2,294. 2
Total						2,4457. 9

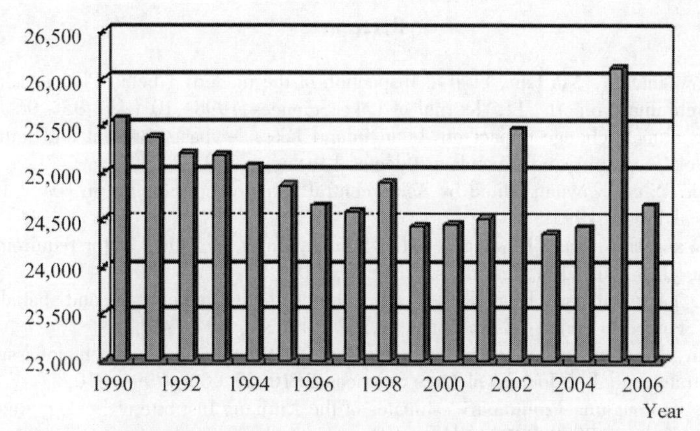

Fig. 3 Years eco-water demand of terminal lakes and wetlands on lower Bayin River coccyx((1990 ~ 2006)

The calculation results of years eco-water demand of terminal lakes and wetlands on the lower Bayin River are shown in Fig. 3. It is shown from Fig. 5 that the requirements of ecological function area (Keluke Lake's area is 59. 6 km^2, Tuosu Lake's 135 km^2, and Gahai Lake's 32. 5 km^2) is to maintain the area of terminal lakes and wetlands on the lower Bayin River coccyx in 2000, i. e. the requirement is to maintain the eco-water demand of 244,579,000 m^3, According to the comprehensive analysis and evaluation on Lake area change and the calculation results of the said lakes, the eco-water damand of 244,579,000 m^3 of terminal lakes and wetlands on the lower river coccyx in 2000 can be minimum ecological water demand for maintaining the basic function of lakes and wetlands on the lower Bayin River coccyx.

5 Conclusions

(1) The area of the Tuosu Lake, a terminal lake, of the lower Bayin River basin is shrinked 1.9 km^2 from 1972 to1990 in the years from 1972 to 1990, reduced by 0.106 km^2 in each year on average. From 1990 to 2000, the lake water area is decreased from 149.8 km^2 to 135 km^2, by 14.8 km^2, which is 1.35 km^2 of annual atrophy on average. In recent years, the degree of atrophy has increased. Under the background of global climate anomalies, in addition to local water resources circulation to affect the lakes atrophy in a period, the development of national economy and the exploitation and utilization of water resources have also been the important factors for influencing the lakes atrophy.

(2) Elevation of the Keluke Lake and Tuosu Lake is basically in the lowest points of the Delingha basin, the Delingha basin is the closed lake basin area, so does not consider the effect on water quantity flowing out of the lake via leakage and underground runoff, the basic eco-water demand of the lakes on the lower Bayin River is mainly consider to supplement the water evaporative water loss of the lakes. The Keluke Lake is a freshwater lake, Tuosu Lake and Gahai Lake belongs to saline lake, in accordance with the theory of brine evaporation is lower than that of fresh water evaporation, the water evaporation of Keluke Lake, Tuosu Lake and Gahai Lake has been calculated.

(3) From the calculation results of comprehensive analysis and evaluation on years eco-water demand and Lake area change, the ecological water requirement of 244,579,000 m^3 of the downstream termanal lakes and wetlands in 2000 can be the minimum ecological water demand to maintain the basic function of termanal lakes and wetlands in the lower Bayin River Basin coccyx. It is completely necessary for deep and comprehensive study on the eco-water demand of the lakes in the Bayin River Basin for the future, and also for the ecological system protection in the alpine, arid and half arid area, and for rational allocation of water resources.

References

Li Shijie, Li Wanchun, Xia Lan, et al.. Inspection of the modern Tibetan Plateau lakes, changes in the preliminary report [J]. Journal of Lake Sciences, 1998, 10 (4): 95 – 96.

Qin Boqiang. climate change impact on Asian inland lakes – past, present and future. Nanjing Institute of Geography and Limnology (doctoral thesis), 1993.

Yang Chuande. Shao Xinyuan edited by Asian central lakes recent changes in [M]. Beijing; Meteorological Press, 1993.

Xu Zhixia, Wang Hao, Tang Kewang, et al.. Stimulation of ecological water requirement of lakes minimum [J]. Journal of Resources, Science, 2005, 27 (3): 140 – 143

Ma Tonghua, Yang Guishan, Duan Tao, et al.. Chinese lake number, size and spatial distribution of [J]. Science in China: Earth Sciences, 2011, 41(3): 394 – 401.

Yang Guishan, Ma Ronghua, Zhang lu. Chinese lake status quo and facing major issues and protection strategies [J]. Journal of Lake Sciences, 2010, 22 (6): 799 – 810.

Jia Baoquan, Ci Longjun. Preliminary estimates of the Xinjiang Institute of Ecology water consumption [J]. Ecology, 2000, 20(2): 243 – 250.

Xu Zhixia, Wang Hao, Dong Zengchuan, et al.. Nasihu lake district of ecological water requirement minimum [J]. Journal of Hydraulic Engineering, 2006, 37 (7): 784 – 788.

Lixin Hu, Song Yudong, Li Yuetan. Lakes minimum ecological water requirement of the [J]. Arid Land Resources and Environment, 2007, 21(2): 114 – 117.

Tan Xiaoming. Analysis of Dongting Lake area minimum ecologica' water demand [J]. Yangtze River, 2009, 40(14): 30 – 37.

Tang Kewang, Wang Hao, Liu Chang. Northern Shaanxi Red alkali. Nao lakes change and the ecological water demand [J]. 2003, 18(3): 304 – 309.

Li Shijie, Li Wanchun, Xia Lan. Tibetan Plateau lakes change and study the preliminary report

[J]. Journal of Lake Sciences,1998,10(4) :95 – 96.

Wang Qingchun, Zhang Guosheng, Li lin. Tsaidam Basin, nearly 40 years of climate change and its impacts on agriculture[J]. Of Arid Meteorology, 2004,22(4) :29 – 33.

Liu Yanhua. Tsaidam Basin, rational utilization of water resources and ecological environment protection [M] Beijing: Science Press,2000:13 – 37.

Hong Jialian, Lu Ruizhi. North of China sea area of the four salt brine evaporation calculation of its distribution [J]. Geography,1988,7(2):17 – 26.

Li Wanchun, Li Shijie, Pu Pei. China. Plateau lagoons water surface evaporation estimates – to Zige Tangco example [J]. Journal of Lake Sciences, 2001,13(3) :227 – 232.

Loose in l, rise, SALT. Chaka Salt Lake hydrology, water chemistry, and resources to research and development [J]. Of Salt Lake Research, 2005,13(3) :10 – 16.

The Preliminary Research on Water – related Ecological Compensation Framework in Yellow River Basin

Han Yanli[1] , ***Huang Jinhui***[1] , ***Wang Wanmin***[2] ,***Ge Lei***[1] ,***Lou Guangyan***[1] and ***Huang Wenhai***[1]

1. Yellow River Water Resources Protection Institute, Zhengzhou, 450004,China
2. Yellow River Conservancy Commission, Zhengzhou, 450003,China

Abstract: After the eco – environment, the eco – protection need and the eco – compensation status were analyzed, the major field of eco – compensation was listed in Yellow River watershed. The ecological compensation key area was preliminary study; finally we give advice of compensation for watershed ecological to strengthen basic research and key areas of case study.

Key words: Yellow River watershed,eco – compensation,the key areas,frame system

Ecological compensation has drawn increasing attention from various sectors of the community as an important measure of effective resource and environment management. NPC deputies and CPPCC members have proposed for many times to call for establishment of related mechanism and policies. And basin ecological compensation thereafter becomes an important area of widespread concern. A series of related researches have been conducted in academic community; and experiments and demonstrations have been conducted by central and local governments actively to find out methods and measures to carry out basin ecological compensation. Ecological compensation in Yellow River basin has just entered into the exploratory stage. Based on national policies, each province of the basin has carried out related practices of basin ecological compensation, conducted preliminary study on aspect of principles, methods and policies of ecological compensation. However, several problems on study and practice still exist mainly due to non – uniform understanding to concept and content of basin ecological compensation, lack of scientific basis for standard setting, not yet formed effective participation mechanism of stakeholders of compensation policies and not deep enough study on feasibility of fund management, utilization and effective supervision mechanism and effective implementation methods. Therefore, based on ecological environment and compensation status quo of Yellow River basin, referring to relevant theories and achievements on ecological compensation at home and abroad, this paper carried out preliminary study on Yellow River basin ecological compensation types, proposed recommendations for contents of different ecological compensation types and safeguard system of basin ecological compensation, and all of these formed the exploratory research for the establishment of Yellow River basin ecological compensation mechanism.

1 Overview of Yellow River Basin

Yellow River is the second largest river of China, with main river channel of 5,464 km and drainage area of 7.95×10^5 thousand km^2, it originates from Yueguzonglie Basin on north of Bayan Har Mountains of Qinghai – Tibet Plateau, runs through Qinghai, Sichuan, Gansu, Ningxia, Inner Mongolia, Shaanxi, Shanxi, Henan and Shangdong and into Bohai Sea in Kenli County of Shandong Province. According to natural geographical features, it is divided into 3 reaches: from riverhead to Hekou Town of Tuoketuo Inner Mongolia forms the upper reach, then from Hekou Town to Xingyang County of Henan Province forms the middle reach, and the left forms the lower reach. The upper reach, with large number of canyons and abundant water resources, holds many existing and in construction water conservancy and hydroelectricity projects, the middle reach constitutes the main source area of Yellow River floods and sediment; entered into plains area, with gentle flow, severe siltation and frequent river swing, the lower reach becomes the world known "Aboveground River" because its beach surface elevation inside the banks is 3 ~ 10 m higher than outside ground.

Major rivers of Yellow River basin are Huangshui River, Fenhe River, Weihe River, Yiluo River, Qinhe River, Dawen River, etc. Huangshui River is originated from southern slope of west Daban in Qilian Mountain with catchment area of 32,863 km^2; Fenhe River is originated from

Ningwu county Shanxi Province with catchment of 39,471 km^2; Weihe River is originated from Weiyuan County of Gansu Province with catchment of 134,766 km^2; Yiluo River is originated from Lantian County of Shaanxi Province with catchment of 18,881 km^2; and Qinhe River is originated from Qinyuan County of Shanxi Province with catchment of 13,532 km^2. Dawen River is originated from Yiyuan County of Shandong Province with catchment area of 9,098 m^2. And Weihe River, Fenhe River, Huangshui River, Yiluo River and Qinhe River are major water inflow area of Yellow River, Huangshui River, Fen River, Weihe River, Luohe River and Qinhe River basins hold concentrated population and rapid and agricultural development which severely polluted water and influence more on water quality of Yellow River mainstream.

Uneven spatial and temporal distribution of water resource in Yellow River basin holds incoordinate water sand relationship of less water and more sand. The average sand content in the basin reaches 31.3 kg/m^3, while 62% runoff come from upper reach and 93% sand come from middle reach. The total water resources amount to 6.47×10^{10} m^3, the available water resources amount to 3.149×10^{10} m^3 among which surface water resources amount to 4.77×10^{10} m^3 while the supply quantity of surface water reaches 3.667×10^{10} m^3, indicating utilization level of 88% which means excessive development. There are rich of mineral resources and 114 kinds of mineral resources have been proven, and coal resource accounts for 46.5% of total nationwide reserve.

2 Water ecological environment issues in Yellow River basin

As the development of economy and society, water consumption in Yellow River increases continuously, and the expanding water demand make the incoordinate water - sand relationship even more severe subsequently. Major ecological environment issues are significantly presented by: the decline of water conservation due to shrinkage of wetlands of lakes and swamps of the Yellow River source area and grassland degradation; excessive utilization of water resources, ecological water demand is diverted which leading to zero flow, excessive extraction of groundwater in some areas causing groundwater funnel and land subsidence, and pollution of groundwater threatening safety water supply at extraction area. Lack of water for ecological environment makes the water - sand relationship even worse, leading to siltation and shrinkage of watercourse, severe flood disaster and degradation of the estuarine ecological functions. Water quantity into the sea is largely reduced, wetland area of estuarine delta is shrinking, and biodiversity is getting lost. Soil erosion in Loess Plateau is still large, man - made erosion occurs at times, and ecological deterioration has not yet been effectively controlled. Increasing artificialization of river ecological system blocked the contact between upper and lower reaches and waterways, reduced the ecological diversity of river, impacted the functions of river ecological system and deteriorated water ecological system. Unreasonable human development and construction activities damaged the basin ecological system severely, in addition to increasing water body pollution and soil erosion, leading to broken of ecological environment and reduction of biodiversity. Decline of service function of water ecological system has threatened the development of social economy and restrict its sustainable development.

3 Ecological compensation status quo and types in Yellow River basin

3.1 Ecological compensation status quo in Yellow River basin

Yellow River is short of resources, and the contradiction between water supply and demand has become increasingly prominent as the rapid development of social economy. In order to use water resources reasonably, Yellow River Conservancy Commission adopts water use permits total control management for mainstreams and main tributaries. In 2003, it conducted the pilot of water right conversion was conducted in Inner Mongolia and Ningxia Province, set up implementation method for management of Yellow River water right conversion, and made clear stipulations for approval procedures and permissions of water right conversion, preparation of technical documents, and period, cost, organization, implementation, supervision and management of water right conversion. In 2010, Shaanxi Province carried out implementation plan for the compensation of water pollution in

Weihe River basin, set pollution compensation amount according to the monthly average concentration of chemical oxygen demand, opened special account for pollution compensation fund which would be spent on establishment and operation of sewage treatment plant, recovery of ecological environment, prevention and treatment of water pollution, etc.

Mineral resources in Yellow River basin is rich, however, ecological environment protection was ignored in the regional development process leading to soil erosion, ground surface settlement and other environmental problems. Related departments have issued some department rules and regulations and measures, for instance, collection and utilization and management method for compensation of soil erosion in exploitation of coal, oil and natural gas in Shaanxi Province, method for the management of security deposit for ecological recovery due to mineral exploitation in Shanxi Province, method for management of compensation of ecological environment, and collection, utilization and management method for compensation and treatment of soil erosion in Shanxi Province, etc. Implement the regulations on protection of oil field ecological environment in Gansu Province, Shanxi Province's implementation method of security deposit system on prevention and treatment of geological disaster in mines, coal sustainable development fund, security deposit for mine recovery and treatment, regulations on prevention and treatment of geological disaster in Shanxi Province, implementation method of security deposit system for prevention and treatment of geological disaster in mines. Mine exploitation unit should pay certain environment compensation or mining security deposit to treat and recover regional ecological environment. In order to prevent ecological environment deterioration due to excessive consumption of natural forest resources, the state carried out protection project for natural forest resources, grassland restoration project, forest ecological benefit compensation fund system and so on, derated agricultural tax for forest restoration, provided seedlings and afforestation fees subsidies for household who abandoned their farmland for forest restoration and other measures. Currently, these measures have achieved significant results.

Generally speaking, ecological compensation in Yellow River basin is mainly carried out for ecological compensation of some regions, and almost nothing has been done on basin level. And problems exist on mainly five aspects. Currently, ecological compensation are mostly carried out by some industries or departments issuing departmental or industrial ecological compensation plan or method based on their own consideration due to lack of special institute and organization on ecological compensation. There must be one institute to coordinate and organize various industries and departments involved to carry out ecological compensation. The second comes to the subject of compensation responsibility. It is difficult to define accountabilities for upper and lower reaches in the basin ecological compensation mechanism, and this is a common issue. The third will be the compensation standard. There still no scientific achievements on ecological compensation standard. Generally speaking, value assessment on basin ecological services is the base to establish basin ecological compensation standard, but it is difficult to implement in practice due to its large calculating results. The fourth is the consultation of stakeholders. The compensation plan can not reflect interest of stakeholders due to lack of common understanding between upper and lower reaches on ecological compensation. The fifth is that there still has no overall framework of ecological compensation on basin level relating to water, and in lack of common understanding on major regions and areas.

3.2 Type and region of ecological compensation in the Yellow River basin

It is determined by the different geographic topography, climate characteristics and resources of upper and lower reaches of Yellow River basin that there have significantly different ecological functions and issues in upper and lower reaches. With rich wetland resource of important water conservation function, the upper reach holds the important agricultural production base Hetao Plain with main ecological function of agricultural products supply. In the middle reach, soil erosion is severe, the ecological environment is weak, while Fen – Wei basin constitutes the important agricultural production zone; in the lower reach, serious river siltation and flood control situations exist, there are various species resources with high biodiversity in wetland within embankments on both sides, and the lower reach alluvial plain is an important grain production base, too.

Ecological compensation concerning water involves in various levels and scales, different subjects in different types of economic society and ecological construction activities impact different objects, different stakeholders and different responsibilities, rights and interests in the ecological compensation concerning water. Divide the ecological compensation concerning water in Yellow River basin into three categories of water resource development and construction, water resources protection and recovery and others according to influence features of ecological environment concerning water. Water resource development and construction category includes water energy development, mineral resources and water sources development; water resources protection and recovery category includes restriction and prohibition of development, and recovery and treatment area of important water ecology; restriction and prohibition area of development includes Three - River Headwaters water conservancy and protection area, windbreak and sand fixation area, centralized drinking water source area, prevention and protection area of soil erosion, and others; water ecology recovery and treatment area includes important river ecology recovery area, soil erosion treatment area, groundwater system protection and recovery area, etc. Energy development includes mineral development projects of coal, oil, natural gas and so on; water energy development mainly refers to construction of major water conservancy project; and water source development refers to large water source development projects such as interbasin water transfer. Others category refers to general production and living water discharging issues. See Tab. 1 for specific ecological compensation types.

Tab. 1 Ecological compensation types in Yellow River basin concerning water

Form of activity	Compensation type	Compensation range
Protection and recovery category	Restricted and prohibited area for development	Three - River Headwaters water conservancy and protection area
		Windbreak and sand - fixation area
		Prevention and protection area of soil erosion
		Centralized drinking water source area
		Flood storage and detention
	Important water ecology recovery and treatment area	Important river ecology recovery area
		Major treatment area of soil erosion
		Groundwater protection and recovery area
Development and construction category	Development of mineral resources / energy	Development and construction of coal, oil and energy chemical industry
	Water energy development	Construction of major water conservancy projects
	Water source development	Construction of interbasin water transfer project
Others	General production and living sewage discharge	Inter - provincial water pollution control

Ecological compensation in restricted and prohibited development regions. The national main function regions planning indicate that it is prohibited to develop important ecological functions. The scope of restricted and prohibited development regions in the Yellow River basin mainly includes water conservation and reserve areas, windbreak and sand - fixation areas, prevention and protection areas for soil erosion, centralized drinking water source areas and flood storage and detention areas. Water conservation and reserve areas in the Yellow River basin are Three - River Headwaters Yellow River headwater region, Gannan Yellow River water conservation and reserve area and Ruoergai Grassland wetland water conservation and reserve area, which are respectively located in Yushu County of Qinghai Province, Gannan Tibetan Autonomous Prefecture of Gansu Province and Ruoergai County of Sichuan. These regions are the main water resource areas of the Yellow River basin, the ecological and environmental protection in the source region are crucial to conserving and regulating safety of water resources in the middle and lower reaches. Many ecological and environmental problems existed in source areas currently not only affect development of local areas, but also affect water resources and ecological environment in lower reach, creating a great impact on socio - economic development of the Yellow River basin. Though the national investment for implementation of ecological protection and project construction initially goes into effect, there is still a large gap between input of eco - environmental protection and demands for it, for which the ecological compensation is urgent to be established so as to form a stable and standard long - term mechanism of benefit compensation and encourage ecological environment construction and protection activities. Most of the drinking water sources in the Yellow River basin are formed through construction of reservoirs, which are mainly located in Taiyuan of Shanxi, Zhengzhou and Kaifeng of Henan, Jinan and Qingdao of Shandong Province, and Baotou and Hohhot of Inner Mongolia. In order to ensure safety of the drinking water source, ecological protection measures, water environmental protection measures and administrative measures have been taken in water resource areas, which restricted economic development in these areas so that the ecological compensation mechanism in important drinking water resource areas is necessary to be established. The flood storage and detention areas of the Yellow River include flood storage and detention areas of Dongpinhu Lake and flood storage and detention areas of Beijin Bank, which are respectively located in Dongping County, Liangshan County and Wenshang County of Shandong Province, Puyang City of Henan Province and Liaocheng of Shandong Province. In order to ensure safe operation in flood storage and detention areas, the development and construction activities are prohibited in these areas, which restricts their socio - economic development so that the study on ecological compensation mechanism in flood storage and detention areas is necessary to be conducted.

Ecological compensation in areas for water ecological restoration and management of important rivers and lakes. Water ecological restoration and management in the Yellow River basin include ecological and environmental problems such as shrinkage of wetlands, dry lakes, ground subsidence, water pollution and soil erosion, due to over - exploitation of water resources, which need to be restored and improved through water ecological environment engineering as well as non - engineering measures, and study on ecological compensation in water ecological restoration and management areas also shall be carried out, Which include protection and restoration of the groundwater system, key control area for soil erosion, and ecological protection and restoration of important rivers and lakes. Groundwater exploitation is concentrated in the Yinchuan Plain, Taiyuan Basin, Yuncheng Basin, Guanzhong Plain and North China Plain, and the largest over - exploited area is in Shanxi Province. Key control areas for soil erosion are located in ravine area of the Yellow River Plateau and that of the Loess Hill, including control area for coarse sediment in Helong Section, control area at the upper reaches of Jinghe River and Beiluo River, control area at the upper reaches of Zuli River and Weihe River, control area at the middle and lower reaches of Huangshui River and Taohe River as well as control area of Sanmenxia Reservoir of Yiluo River. Control area of important rivers and lakes include main stream of Wuliangsu Sea, estuary of the Yellow River, the Weihe River, Fenhe River and Huangshui River etc.

Ecological compensation for development of mineral resources (energy). There are rich mineral resources in the Yellow River basin, coal resources are mainly concentrated in Inner Mongolia, Shaanxi, Shanxi and Gansu etc., the reserves of oil and gas respectively account for 40% and 9% of the state's total geological reserves with large scale of Xishan Coalfield, Huoxi Coalfield in Taiyuan of Shanxi, Weibei Coalfield in Shaanxi, Dongsheng coalfield in Inner Mongolia, Yuxi Coalfield, Zhongyuan Oilfield, Shengli Oilfield in Henan, Yanchang Oilfield, Changqing Oilfield in Shaanxi etc. Development of these resources, on the one hand, caused pollution to surface water and excessive extraction of groundwater, which had an impact on local water resources, on the other hand, it caused shortage of water resources in the Northwest drought area resulting in fighting for water resources between agricultural irrigation and ecological environment protection, causing damage to vegetation on the land surface and having a serious impact on ecological environment, so that study on ecological compensation mechanism shall be carried out to restore ecological environment and promote sustainable utilization of water resources.

Ecological compensation for hydropower development. Hydropower development in the Yellow River Basin is concentrated in the main stream, particularly at reach from Longyangxia to Qingtongxia, which is intensive reach for cascade development of the Yellow River, some large - scale water control project have been built at the middle and lower reaches of the Yellow River, while plateau aboriginal fish species are distributed at the reach from Maqu to Qingtongxia, the Yellow River aboriginal and economic fish species are distributed at reach from Hekou to Longmen and reach from Tongguan to Xiaolangdi, national and provincial protection areas for wetland are also distributed in these areas, so that it needs to analyze the positive and negative effects of project construction on regulating runoff as well as the impact on river ecosystem, the study on compensation methods and channels shall be conducted to restore water ecological environment.

Ecological compensation for development of water resources. Water resources development projects in the Yellow River basin include inter – basin water transfer project and important water control project, the projects such as water transfer from Yellow River to Tianjin, the west route of South – to – North Water Transfer Project and water transfer from the Yellow River to Qingdao will re – distribute regional or watershed water resources and have an impact on water environment and water ecosystem in water resources areas, Xiaolangdi Water Control Project, Sanmenxia Water Control Project cause an impact on household water – consumption as well as industrial and agricultural production, producing long – term beneficial or adverse effects on the regional or watershed ecological environment. Study on ecological compensation shall be carried out to assess its influence, adjust relationship of the stakeholders so as to promote regional or watershed development and achieve coordination between economic construction and ecological protection.

Ecological compensation for general production and domestic sewage drainage Water resources of the Yellow River basin are in severe shortage with conspicuous contradiction between supply and demand as well as serious water pollution situation. Wastewater drainage will have a great impact on water environment of the regional reaches, further influence water utilization at lower reach of the river. The upper and lower reaches of the Yellow River basin involve in different administrative sectors and regions, which may have an impact on different stakeholders. Study on ecological compensation shall be carried out to adjust relationship of the stakeholders, to coordinate the contradiction between the socio – economic development and environmental protection, and to maintain sustainable socio – economic development.

4 Main content of ecological compensation in the Yellow River basin

4.1 Ecological compensation in restricted and prohibited development areas

Restricted and prohibited development areas of the Yellow River basin feature ecological functions such as water conservation, soil conservation, fragile and sensitive ecological environment, which are related with national or larger range of ecological safety areas, development of these regions will have an impact on the local water resources development and utilization, further affect socio – economic development of the regions or river basin, so that compensation in these regions takes country as the subject, the compensation objects are residents, enterprises and institutions as well as local government affected by restricted and prohibited development areas.

Compensation mode. Beneficiaries of ecological environment in restricted and prohibited development areas are usually in a large scale with uncertain subject, compensation by government has become main mode for such ecological compensation which can be paid by central financial transfer and compensated in the form of premium, remission or drawback of tax, subsidies and reclamation fees etc.

Source of compensation funds. Referring to laws and regulations that have already been issued by the country to collect ecological environment tax and ecological compensation of water resources, incorporate it into water price, extract a certain proportion fees from water price or create watershed ecological compensation funds to use for construction of the region's ecological protection and flood insurance fund in flood storage and detention areas.

Compensation projects include returning farmland to forests and animal breeding grounds to pastures, natural forest protection, water saving and pollution control at upper reaches, pollution control projects in water source areas, isolation and protection works as well as ecological migration engineering, education and training of residents, control projects for cleaning up small river basins, ecological protection and construction projects for public environmental protection facilities in regions restricted and prohibited for development.

4.2 Ecological compensation in regions for water ecological restoration and management

According to division of ecosystem services, ecological compensation in regions for water ecological restoration and management of important rivers and lakes in the Yellow River basin include ecological compensation for control of soil erosion, protection and restoration of groundwater system as well as control and restoration of important rivers and lakes.

The subject and object of compensation. In case that the main beneficiaries are difficult to be determined, the state and local governments shall be the compensation subject, in case that the main beneficiaries are clear, the residents, enterprises and institutions or the government benefited shall be compensation subject. Compensation objects are mainly social and economic entities paying lots of cost for protection of ecological environment, as well as the entities whose interests are impaired due to excessive utilization of water resources, environmental pollution and ecological destruction.

Compensation modes include supporting local ecological protection and construction projects through taxes, preferential policies of loans, compensating food and fertilizer to losses caused by returning farmland to forests and animal breeding grounds to pastures, making local people master certain skills through technical training, such as skills for house feeding and conservation tillage. Directly give funds to ecological protection builders to conduct soil erosion control, protection and restoration of groundwater system and control of important rivers and lakes, or the beneficiaries pay to government departments then the government transfers payments to compensation object in accordance with certain standards, or allocate funds through planning construction projects to carry out soil and water conservation construction and management and restoration projects for ecological environment.

Sources of compensation funds include fiscal transfer payment of the state and local area, investment in water and soil conservation program, compensation fee for conservation facilities, fees for prevention and control of water and soil erosion, resource development taxes, sustainable development fund for coal development, margin for restoration of mine environment, margin for restoration of ecological environment, groundwater protection fund, groundwater ecological compensation fees and social contributions.

Compensation projects include control project as well as development and construction project in small watershed, soil erosion control and ecological migration project etc. Subsidies for closing down, suspending, merging and transferring of enterprises reaching discharge standards in regions with important rivers and lakes as well as urban domestic sewage and solid waste disposal facilities in these regions; funds for soil and water conservation as well as protection and management of forests, and water amount in areas with over - exploitation of groundwater, construction of water quality monitoring system, water - saving project, groundwater recharging project and project for water transferred out.

4.3 Ecological compensation for development of hydropower (water source)

Subject and object of compensation. The development and construction as well as operation of hydropower (water source) will have beneficial and adverse effects in water resources and ecological environment of regions or river basin, and have a significant impact on stakeholders, so that the beneficiaries of development and utilization of hydropower (water source) not only shall protect the ecological environment, but also shall bear liability of compensation. Therefore, the compensation subjects are residents, enterprises and institutions as well as local government benefiting from hydropower (water source). Compensation objects are residents, enterprises and institutions or government affected by construction of hydropower (water source) project.

Compensation modes include compensation by corporate and resources stakeholders, supporting by compensation from government, form of corporate compensation includes cash compensation as well as restoration and management, government compensation can be realized by way of establishing fund for restoration and management of ecological environment or sharing fund for ecological compensation, as well as ecological migration subsidies, off - site development project, water rights transaction.

Source of compensation funds. Source of compensation funds includes government fiscal transfer system as well as collection of ecological compensation fees, preferential credit policies, welfare contributions at home and abroad. Margin system of ecological compensation. The development and construction project shall pay a certain number of ecological compensation margin then can obtain permission for the project's development and construction, the margin shall be collected according to the cost demanded for annual ecological damage and shall meet the full cost required for control, or establish ecological compensation fund for hydropower (water source) development project, collect ecological compensation tax of river basin, establish water rights transaction market, water receiving area shall pay a certain amount of cost to water source areas for purchasing water.

Compensation projects include prevention and control of soil erosion, water ecological environment protection and restoration project and water rights transaction of hydropower (water source) development project.

4.4 Ecological compensation for construction of mineral Resources project (energy & chemical)

Compensation subject and object. Compensation subjects are beneficiaries of mineral resources / energy development and local government. Compensation objects are residents, businesses or government affected by project.

Compensation modes. Prepare project concerning pollution control and environmental protection planning in accordance with the requirements, protect and plan direct compensation according to approval of ecological and environmental protection departments, or transfer payments to ecological environment compensation object by a third party, usually the governmental environmental protection agency, in accordance with certain standards.

Sources of compensation funds. Establish margin system for restoration and management of mineral resources, the development and construction project shall pay certain amount of ecological compensation margin to obtain permission for construction, open special margin accounts for ecological restoration, the payment shall be supervised by the governmental ecological environment protection agency, or collect mineral resources tax.

Compensation projects include treatment to regional ecological damage as well as restoration and reconstruction of regional ecological environment in old abandoned mine fields during construction and operation of mineral resources (energy project), and contradiction between industrial and agricultural water utilization in areas short of water resources is conspicuous, agricultural water - saving measures shall be implemented to save water resources for industrial projects. And projects of cash compensation for residents whose lands are damaged, personnel placement, environmental and economic compensation for mine fields with industrial structure transformation as well as techniques recovery compensation. etc.

4.5 Ecological compensation for general production and living sewage drainage

General production and living sewage drainage refers to the ecological compensation for the consumption and pollution discharge of end water consumption households, which is closely related to the water quality, water quantity and hydrological process between the upstream and downstream of river, with obvious regional influence. For the upstream and downstream of Yellow River Basin, it focuses on the water quantity and water quality of different regional interfaces of the river. If one of them can not meet the requirement, bad influences would be made on the economic and social development of downstream area. Otherwise, if some effective measures are taken by the upstream area to provide water resources with enough quantity and quality for the downstream area, without obtaining the corresponding compensation, the enthusiasm of the upstream area to ensure the water quality and water quantity would also suffer serious setback. Therefore, there is a problem of ecological compensation between the upstream and downstream of river basin due to water pollution.

Compensation subject and object. The compensation subjects are all levels of local people's governments. The compensation objects are local residents, enterprises and institutions, and local government, who suffer the influence. Compensation mode. Through making clear ownership of water right and pollution discharge right, realize the water right and pollution discharge right trading, establish distribution system and trading mechanism of water right (water resource use right) and right of pollution discharge into river, and the higher level government shall be responsible for the distribution and trading of water right and pollution discharge right of river basin cross – administrative areas.

Compensation mode. In order to safeguard the safety of production and living water consumption of residents of upstream and downstream in the river basin, and safeguard the sustainable utilization of water resources of river basin, according to the water quality and water quantity of the upstream and downstream water body, the ecological compensation mode of the general production and living water consumption can be divided into pollution discharge trade, water right trade, eco – label system and enterprise compensation and so on.

Source of compensation fund. All levels of local governments set up special fund for ecological compensation, and the corresponding level financial department opens a special account for management. The fund mainly comes from the fixed investment of the corresponding level financial department, such as the charge of water resources and pollution discharge fee.

Compensation projects include the water right trading project, which refers to that the downstream area pays the use fee to the upstream area. In order to use the redundant water quantity of upstream area, which is saved by the upstream area through taking effective measures of saving water resources, the downstream area pays the use fee to the upstream area through the water right trading platform established by the higher level government. The pollution discharge right trading project refers to that, the downstream area applies to the higher level government for purchasing the main pollutants, and then the agency in charge signs the pollutant discharge right trading contract with the demander and delivers it to the upstream area as ecological compensation. The compensation project of the upstream area to the down stream area refers to that, if the upstream area causes the excessive use of original water right and excessive quantity of pollutants discharged, for not taking active and effective measures for water resources management and water pollution prevention and control, which makes bad influence on economic and social development of downstream area, the higher level government shall order upstream area makes the compensation to downstream area.

5 Conclusions

The study on ecological compensation is a subject combining social science and natural science, involving the knowledge of many disciplines, such as water resources, economy, ecology, law and management, which is complex. At present, the basis of ecological compensation in the Yellow River basin is poor. There are many problems needed to be researched and studied to establish the complete and systematical basin ecological compensation mechanism. It is suggested that, for the ecological compensation of Yellow River basin, firstly, some key basic studies shall be made, for instance, the study on the service function value of typical ecological system of Yellow River basin, the study on classification of ecological compensation of Yellow River basin, the study on the ecological compensation of Yellow River basin, the ecological compensation standard, the study on ecological compensation mode, the study on evaluating the implementation effect of ecological compensation, special ecological compensation standard study. Currently, there is no scientific and uniform method. And the method of ecological compensation standard, the measurement, calculation and determination of ecological compensation standard shall be the key points of study. In addition, make the case study and implementation on the key field the most urgent, with a certain basis for building ecological compensation, mutually support, complement and promote with basic study, deeply discuss how to establish long – term effective basin ecological compensation mechanism, strengthen the basin resources management, form the long - term effective mechanism, promote the coordinating development of society, economy and ecological environmental protection.

References

Zhuang Guotai, Wang Jinnan. International Symposium on Ecological Compensation Mechanism and Policy Design[M]. Beijing: China Environmental Science Press, 2006.

Yang Daobo. Some Legal Problems Regarding Ecological Compensation in River Basins[J]. Environmental Science and Technology, 2006, 29(9):57 - 59.

Wang Jinlong, Ma Weimin. Discussion on Issues of Ecological Compensation of Basin[J]. Journal of Soil and Water Conservation, 2002, 16(6):82 - 83.

Zhou Dajie, Dong Wenjuan, Sun Liying, et al. Ecological Compensation in Management of Water Resources of the River Basins[J]. Journal of Beijing Normal University (Social Science Edition), 2005(4):131 - 135.

Wu Liqiang, He Junshi. Research on the Ecological Compensation in the Suzihe Basin[J]. China Rural Water and Hydropower, 2009(8):46 - 47.

Guo Pengheng, Li Yuexun. Ecological Compensation Mechanism of Chemu River Water Protected Areas[J]. Journal of Henan University of Urban Construction, 2008, 17(3):40 - 41.

Chen Zhaokai, Shi Guoqing, Yang Tao. Study on Ecologic Compensation of Water Resources of the Yellow River Basin[J]. Yellow River, 2008, 30(2):40.

Li Xiaobing. Thinking on Building an Ecological Compensation Mechanism in Jinsha River Watershed in China[J]. Journal of Yunnan University of Finance and Economics, 2009,136(2): 135 - 137.

Zhang Yu. Ecological Compensation Pattern for Wellhead of Inter-basin Transfer Project in China [J]. Journal of Northeast Normal University, 2008, 234(4):23 - 26.

Zhao Xu, Yang Zhifeng, Xu Linyu. Study and Application on the Payment for Ecological Services in Drinking Water Source Reserve[J]. Acta Ecological Sinica, 2008,28 (7):3152 - 3156.

Guo Yijun, Wu Huixuan, Lin Zhoufeng. Study on Ecological Compensation Mechanism of Reservoir Water Source Protection Areas[J]. The Science Education Article Collects, 2007:197 - 198.

Task Force on Eco - compensation Mechanisms and Policies. Eco - compensation Mechanisms and Policies in China[M]. Beijing: Science Press, 2007: 95 - 110.

Wang Bentai, Zou Shoumin, Li Yuan, et al. The Practical Ecological Compensation - case Analysis and Exploration[M]. Beijing: China Environmental Science Press, 2008.

Li Guoying. Exploration and Pactice of Water Right Trading in Yellow River Basin[J]. China Water Resources, 2007(9):30 - 31.

Liu Xiaoyan, Wang Jianzhong, Yu Songlin,et al. Building Water Market of the Yellow River and Optimizing Allocation of Water Resources[J]. Yellow River, 2002,24(2):25.

Zhang Qinghua. On Ecological Benefit Compensation System of Water Resource[J]. A Dissertation for Law Master Degree of Sichuan University, 2005, 10.

Yang Yongfang. Analysis on Key Issues of Ecological Compensation of Different Sections of the Yellow River Basin[M]. Nanjing, Symposium on National Environment and Resources Law, 2008:443 - 445.

Fan Yi. Study on Environment Compensation Mechanism of Water and Soil Conservation in Shanxi Province[J]. Shanxi Water Resources, 2008(10):29.

Zhao Chengzhang, Jia Lianghong. Ecological Performance and Sustainable Problems with the Grazing Forbidden Project in the Resource Regions of the Yellow River[J]. Journal of Lanzhou University, 2009(1):38 - 41.

Lv Jin. Research on Ecological Compensation Mechanism of Water Source Conservation Areas Abroad[J]. China Environmental Protection Industry, 2009(1):65 - 67.

Guan Yanzhu. Suggestions on Building Ecological Compensation for Drinking - water Source Protection Area[J]. Xiamen Science & Technology, 2007(6):17 - 18.

Wang Jinlong, Ma Weimin. Discussion on Issues of Ecological Compensation of Basin[J]. Journal of Soil and Water Conservation, 2002(16):82 – 83.

Cao Mingde, Wang Fengyuan. Discussion on Legal Issues of Ecological Compensation for Interbasin Water transfer[J]. Academic Journal Graduate School Chinese Academy of Social Sciences, 2009(2):7 – 11.

Guo Yijun, Wu Huixuan, Lin Zhoufeng. Discussion on Ecological Compensation Mechanism of Reservoir Water Source Protection Areas[J]. The Science Education Article Collects, 2007(6):197 – 198.

Ouyang Hui. Suggestions on Improving the Resources and Ecological Environment Economic Compensation System for oil Resources Development[J]. Review of Economic Research, 2007(17):21 – 22.

Qin Lijie, Qiu Hong. Study on Water Resources Compensation in Songliao River Basin[J]. Journal of Natural Resources, 2005(1):17 – 20.

Zhang Qinling. Thought on Building the Compensation Mechanism of Soil and Water Conservation of Headwater Region in Middle Route of South North Water Transfer Project[J]. Soil and Water Conservation in China, 2008(6):4.

Cheng Linlin, Hu Zhenqi, Song Lei. The Policy Design on the Eco – compensation Mechanism of Mineral Resources in China[J]. China Mining Magazine, 2007(4):12 – 13.

Song Lei, Li Feng, Yan Lili. A Study of the Definition of Ecological Compensation for Mineral Resources[J]. Journal of Guangdong Institute of Business Administration, 2006(12):24 – 25.

Exploration and Practice of the Environment Protection Measures in River Engineering

Sun Hongbin, ***Dong Yu*** and ***Sun Meng***

The Authority of Water Way Works to the Sea of Huaihe River Jiangsu Province, Huaian, 223200, China

Abstract: Huaihe River Sea-entering Channel is a strategic backbone project, which expands the Huaihe River flood way out, improves the Hongze Lake flood control standard, ensures the safety of 20×10^7 people and 30×10^7 mu arable land. Environmental impact assessment system is carried out in engineering construction, and the "three simultaneous" management system - simultaneous design, construction, and use, is taken for environmental protection. Optimal scheme is confirmed in the early planning stage; in the construction stage, environmental protection problems are managed strictly by construction management department, construction units and supervising unit; in trial operation phase, levee and beach farmland ecological environment have been restored and soil conservation measures are implemented; land utilization is controlled strictly by using non-cultivated land and accounting for less fertile farmland during the resettlement. 4.117×10^9 yuan has been invested recently in Huaihe River Sea-entering Channel project, and 2.17×10^8 yuan, accounting for 5.4% of the total investment, is invested in environmental protection. The project passed the water acceptance and was put into use in May of 2003. In 2006 it passed the final acceptance of construction. The project has discharged 7.7×10^8 m^3 during the experience of two Huaihe River flood in 2003 and 2007, which plays a key role in valley security and regional development. The project with a novel architectural style, a beautiful environment and convenient transportation is becoming a water conservancy tourism projects consisted of engineering landscape, natural scenery, garden art, popular science training and entertainment, according to the purpose of green water conservancy and ecological water conservancy during the construction and operation management.

Key words: river channel, construction, environmental protection, practice

With the social and economic development, increasingly importance has been attached to strengthen the construction of harmonious society and achieve the harmonious relationship between man and nature. Water is the source of life, and is also the source of its development. Human survival and development cannot be separated from water. Effectively solving the water problems (such as flood, drought, soil erosion, water pollution etc.) in the process of promoting the building of harmonious society is the inevitable requirement for socio-economic development, is the top priority of people in harmony with nature, and is also the core concept of implementing the scientific development concept. Following the above philosophy, Huaihe River Sea-entering Channel project construction and management were focused on building an environment-friendly project. It is not only a new sublimation in thought, but also a useful exploration in practice.

1 Outline

Huaihe River Sea-entering Channel project, which can expand Huaihe River downstream flooding way out, enhance the Hongze Lake flood control standard and ensure the security of 2.0×10^7 population and 3.0×10^7 mu cultivated land. In the downstream area, is a strategic backbone projects. Its main river channel span 163.5 km and the designed current capacity was 2,270 m^3/s. It originates from the Hongze two floodgate in the west. Eastward along the north side of Irrigation Canal, flow into the Yellow Sea at the Biandan port. Its north and south levees are apart from 580 m. The construction of the project include: Dike River project, Five hubs involving two rivers, Huai'an, Huaifu, Binhai and seaport, five acrossing river road bridges, 29 buildings which pass through the levee and northern levee drainage and irrigation impact treatment project.

According to the requirements of the Environmental Protection Law of the People's Republic of China, Huaihe River Sea-entering Channel project environmental impact assessment report was submitted to the State Environmental Protection Administration in 2001 and in the same year National Environmental Protect Bureau made a reply. Its main contents include: ① To strengthen the protection of ecosystems of Red-crowned Crane Nature Reserve of Yancheng and Haikou. In addition, appropriate preventive measures must be implemented according to the monitoring results. ② Minimize permanent and temporary occupation of farmland, avoid or mitigate the impact of capital construction on farmland and farmland tillage layer and reduce the destruction of vegetation, disposal areas and mud field should be set to avoid the larger villages and towns, as well as surface runoff. Recultivation or afforestation of spoil areas should be smoothed to prevent new soil erosion. ③ Strengthen the environmental management of construction. The waste water like pit water and mud field backwater should be handled to prevent pollution of surface water; mud field backwater shall not be discharged directly into fish ponds, so as not to cause fish kills; drainage sediment should be properly disposed to prevent secondary pollution. ④ According to "Huaihe River Basin Water Pollution Prevention Plan" and estuarial environmental function zoning, the comprehensive treatment of the water environment of "Qingan River" and "Zhangjia River" and "Women River" should be strenghened to completely control south base sewage. ⑤ The immigration placement must save the land resource and the residential area layout should be planed reasonably.

According to the above requirements, in project construction, the " three stimulaneity " approach that has been adopted in the environmental protection construction with the "simultaneous design, construction and use" of the main work, has gradually embarked on a road of innovation. 4.117×10^9 yuan was invested in Huaihe River Sea-entering Channel recent project, of which about 2.17×10^9 yuan was invested in environmental protection, accounting for 5.4% of the total investment. It mainly includes the 4.281×10^7 yuan cost of ernvironmental protection measures during the construction period, the conservation of water and soil investment of 1.247×10^8 yuan, the flood discharge environmental protection investment of 2.899×10^7 yuan in 2003, the related environmental protection invested $2.065,5 \times 10^7$ yuan in the immigration placement and so forth.

2 The implementation of environmental protection measures

2.1 Planning and design stage

Based on the direction of river channel, designers has selected and compared three lines. The first is the abandoned Yellow River line. Its length is 219 km. The earth volume is enormous. The project is difficult to implement and can't be combined with the drainage on the north of channel. The second is the two river – Huaishu River – Xinyi River line. The length is 205 km. Its work is huge. In case the Huai River and the Xinyi suffered, the flood discharge capability will be severely constrained, which threatens the downstream safety. The third is the drainage northern line, which opens up new river along the north side of the canal and forms two rivers and three levee banks with canal. Its outstanding advantages are that the engineering can be combined with the distance and the line is short and low-lying. This line is the most flexible and proactive if the bad situation like "the Changjiang River and Huaihe River rise at the same time or the Huaihe River and Xinyi encounter" appear.

Optimized design scheme: ①The design of floodplains flood was replaced by the design of Hung combined with beach. Huaihe River Sea-entering Channel project continuously used the design of the embankment beam of water and floodplain flood in the early 1980s. In the project, the width of the dike beach was 2.5 km which covered an area of more than 3,000,000 mu migrated 200,000 populations. It's a tremendous project. However, according to the actual situation of the socio-economic development along the route, the design of excavation of deep base and Hung combined with beach was adopteed under the premise of guaranteeing the function and benefits of the project. It narrowed the distance of the dike down to 750 m and covered only an area of 100,000 mu. It saved the land of 200,000 mu and had a huge economic and social benefits. ② The project of North-South hong excavation close to the embankment was replaced by the project of north base south. The cross-section design of flood channel in our country always was adopted to the project of two Hong separation and digging hong to build embankment close to the dike. In the cross-section design of Sea-entering Channel project, north base south replaced the traditional methods, which formed a reasonable pattern of high water level and low water level separated emissions and clean water and wastewater diversion. It simultaneously had protected the farmland of 38,000 mu and reduced more than 160 crossing bridges. This innovative design may not only reduce the environmental impact to the lowest degree but also create a new beautiful natural landscapes.

2.2 Bidding stage

Firstly, those formal construction units which have good qualification, actual strength, reputation and experience will be taken into account and selected. Secondly, bidding units are required to make a commitment to protect the environment according to the bidding documents. While the bidding units are compiling construction design documents, a plan for environment of construction and living area needs to be writed out, which includes the utilization and stacking of construction waste slag, protecting measures of soil erosion and drinking water, control measures of noise, dust, waste gas and oil during constructing, hygienic facilities of construction and living area, treatment measures of dejection and garbage and site clearing after construction.

2.3 Construction stage

Following points must be focused on: ① including protection measures of Red-crowned Crane Nature Reserve; ② engineering measures of water and soil conservasion; ③water quality during construction period, air quality, water environment, construction workers' health and resettlement environment, engineering measures of soil and water conservation, measures of construction land occupation and second plowing and management measures after engineering handover operation.

3 Environmental investigation

3.1 The organisation, scope and methods of investigation

China Institute of Water Resources and Hydropower Research has undertaken the investigation task of Huaihe River waterway to sea engineering environmental construction. A field study of main engineering site which consist of waterway dike, hub buildings, office places of engineering management, main residue field, yard, mud field and main buildings accoss dikes. By having an informal discussion with construction units and administrative departments, the composition, properties and expropriation occupied of engineering, management of environmental institutions and management measures and operational characteristics of environment protection are known by the investors. It is seen that sea-entering channel affect ecological environment by having an informal discussion with relevant government department leaders of 2 cities and 4 counties along the project and doing questionnaires. Protection measures of environment of each setting point, living quality of migrants, recovery of beach farmland and the impact of 2003 flood channels are investigated by discussing with the masses and doing different kinds of questionnaires. The implementation of environmental measures of 11 construction units and 3 supervision units are grasped by questionnaires. Water quality of sea-entering channel, main blowdown channels along and estuary areas are monitored by commissioning the local environmental monitoring department of Huai'an and Binhai. Huaihe River waterway to sea the recent completion of the project environmental acceptance of the investigation report have been completed on the basis of the above work.

3.2 Review of investigation

The optimum scheme has been made about channels and dikes in the early planning stage. Construction management department, construction units and supervision units strictly administer environmental issues during construction period. The requirement from the EIA report and approved views are implemented and environmental pollution and destruction is controlled effectively in several aspects, such as control of land occupation and soil erosion, disposal of various types of waste water and solid waste, prevention of air pollution and noise, protection of ecological environment, population health and beach farmland, environmental monitoring , etc.

In the trial run, dikes and beach farmland ecological environment have been restored. Water and soil conservation measures are implemented. The appearance of the slag field and the yard can't be seen and overall it is in a good condition. Treatment measures of sewage and waste are taken. Management places of project management system have done the beautification greening. It does not have a bad influence on Red-crowned Crane Nature Reserve in the project. In 2003 flood channels affected the beach farmland, estuary ecological system, salt pan production and estuary deposition obviously. Beach farmland productivity has recovered due to the taked measures. Currently estuary ecological environment and salt pan production have been restored. Ecological recovery requirements and measures from the EIA report and approved views have been implemented. What's more, it has a good effect.

In the resettlement, land is under strictly control. Measures are taken to make full use of non-cultivated land and account for fertile land as little as possible. The streets and houses are unified planning and layout to make rational utilization of the limited land resources. Environment of migrants resettlement is good and there is no big pollution phenomenon. Migrants of the most sites have a large degree rise in housing, water, electricity, transportation, and other aspects of the life level. Their living conditions have improved. Now most migrants are satisfied with the living conditions. Measures of environment protection about migrants from the EIA report and approved views have been implemented. Besides, a good effect has been received.

4 Evaluation of environmental measures

From the living environment perspective, Huaihe River Basin had happened the heaviest rainfall and the most serious flood in 2003 since 1991. The water level of Hongze Lake rised quickly. Then Huaihe River waterway to sea started to drain flood on the sixth day after the whole water line acceptance. It beginned from July 4 to August 6, lasting 33 d and the flood discharge reached 4.4×10^{10} m^3. Huaihe River waterway to sea drained flood safety to lower the water level of Hongze Lake, reduce the flood control pressure of Hongze Lake dike and avoid the detention loss surrounding Hongze Lake polder. It played a significant role in saving more than 20 million people and 30 million hectares of arable land of the lower Huaihe valley.

During 2005, the Qubei area continued raining, luckly no flood. The locals said, due to the water the Qubei area was poor in the past. The irrigation canal was digged to improve global water resources condition. "A basin of water on the head (Hongze Lake), the eight legs cut down (drainage channels)". There were a difficulty in irrigation in some part of the Qubei area, leading to low production and instability. Nowadays earth worm diversion can help irrigatation and the pump station can help drain. "The living environment changed greatly after digging Huaihe River Waterway to sea."

From the residential environment perspective, there are more than 17,000 households, 63,000 people along the 163.5 km long Huaihe River sea-entering channel. Such a large resettlement activity is the first time in the history of Jiangsu Water Conservancy Construction. The new buildings and tile-roofed houses appear rapidly in Huaibei Plains, Jiangsu, like bamboo shoots after a rain. 78 central villages where more than 12,000 households migrants, more than 50,000 people have settled, except some migrants moving out and dispersing outside have been built along Huaihe River waterway to sea. In central villages, the area and quality of housing, health conditions and living environment of migrants are much better than before. A villa area is added to meet the requirements of wealthy farmers in Beishan, Funin. The brightly colored villa buildings appear in the poor Canal North rural areas. At present, it is a beautiful and picturesque place with fresh air and superior condition. Seeing this, a folk "poet" can't help humming out a graceful poem. A new type of Chinese characteristics socialistic new countryside is presenting to the world.

From the clean-up shunt, it is pioneering that water conservancy implement scientific development view and construct environment-friendly projects. The sewage from Huai'an was discharged into the canal through Qing'an River leading to complaint from lower canal. Presently the sewage from Huai'an was discharged into the Nanhong through Qing'an River, making sure to protect the water of the canal and the Beihong from polluting. Furthermore, the water quality of Yancheng has been improved a lot.

From ecological protection situation, although the sea-entering channel is far away from Red-crowned Crane Nature Reserve, measures are still taken to protect the ecological environment and try to reduce the noise, dust, waste gas and oil, while construction. At the same time, engineering measures of the environmental ecological recovery are strengthened. Tree planting and construction of beach ecological system are focused on solving to improve the vegetation coverage, create the ecological levee and protect the wetland environment. The sea-entering channel will be made into a good ecological environment of "green corridor", "wetland corridor" and "passage corridor" and finally become a wildlife habitat of "paradise".

In recent years, Huaihe River Waterway to sea is a new and large-scale water conservancy projects in Jiangsu, which passed the whole line acceptance and is put into use after handing over to the management unit in May, 2003 and passed the completion acceptance in 2006. From the 2003 to 2007, Huaihe River Waterway to sea project which experienced two Huaihe River floods, accumulating total discharge 7.7×10^9 m^3 plays a key role in maintaining river basin security and regional development. In the construction and operation management with the aim of green water conservancy and ecological water conservancy, the novel engineering construction style, a beautiful environment and convenient transportation are becoming a water tourism project which includes engineering landscape, natural scenery, botanical garden art, popular science training and leisure entertainment.

References

Water Resources Department of Jiangsu Provience. Huaihe River Waterway to Sea Recent Project Completion Acceptance of Engineering And Construction Management Work Report [R]. Nanjing: 2006.

Huaihe River Basin Water Pollution Prevention Plan [R]. 1999.

Chinese Institution of Water Resources and Hydropower. Huaihe River Waterway to Sea the Recent Completion of the Project Environmental Acceptance of the Investigation Report [R]. 2005.

Sun Hongbin, Song Honghua. Exploration and Practice Project on Saline Land Soil and Water Conservation [J]. Jiangsu Water Conservancy, 2006 (7): 34 - 35.

Sun Hongbin, Zhou Heping, Dong Yu, et al.. Research on the Impact of Operating the Beihong Overtopped Brake on the Drainage State of the Qubei Area of Huaihe River Waterway to the Sea [J]. China Rural Water and Hydropower, 2010 (5): 99 - 102.

Risk Analysis of Living Environmental Hazard for Floodplain Residents of Lower Yellow River

Huang Jiantong[1] and ***Yin Yue***[2]

1. Yellow River Institute of Hydraulic Research of YRCC, Zhengzhou, 450003, China
2. School of Environmental and Municipal Engineering of Qingdao Technological University, Qingdao, 266520, China

Abstract: Inundation risk frequency of lower Yellow River residents is analyzed according to inundation risk of lower Yellow River residents during different times since 1949, and combined with new regulation principles and continuous improvement of future engineering and non-engineering measures in upper and middle Yellow River. The analyzed result shows that the floodplain inundation frequency of lower Yellow River is the same as operating frequency of state detention areas. Therefore, technically speaking, it is possible for residents in the lower floodplain to receive state compensation when they become victims of flood inundation, just like residents in state flood detention areas.
Key words: floodplain residents, environmental hazard, risk, lower Yellow River

Lower river channel of Yellow River is intensely deposited; it easily deposits, easily rechannels and easily breaks due to the imbalance between little water and much sediment. The flood problem is serious. To ensure a proper flood discharge capacity, the lower river channel has been mainly treated with river channel broadening and embankment strengthening of all time, resulting in a floodplain with area of 3,154 km^2 within the 4,860 km^2 river channel, which affects 15 cities, 43 counties (districts), 1,928 villages, 1.895×10^6 people, 0.227×10^6 hm^2 farmland and $32,066.67 \times 10^6$ million mu forest land in Henan and Shandong provinces.

1 Hazards hitting floodplain residents of lower yellow river in history

As shown by incomplete statistics, floodplain residents of lower Yellow River have been hit by various levels of flood for 31 times since 1949, the accumulated flood-hit population is $8.871,6 \times 10^6$ people, the flooded farmland is 1.746×10^6 hm^2 and there are $1.535,2 \times 10^6$ flooded houses. The details are shown in Tab. 1: Statistics of Hazards Hitting Floodplain Residents of Lower Yellow River in History.

As suggested by Tab. 1, flooding occurred in 13 years out of the 25 years from 1949 to 1974 (1974 is not included, this is the period before the third bank reconstruction for Yellow River), the minimum overbank flow is 5,890 m^3/s (happened on Sep 3rd 1973), the flood probability is 52%, i.e. the flood almost happened every two years; flooding occurred in 9 years out of the 12 years from 1974 to 1985 (the third bank reconstruction period for Yellow River, a high flow period as well), the minimum overbank flow is 5,640 m^3/s (happened on Sep 20th 1978), the flood probability is 75%, i.e. the flood happened three times every 4 years; flooding occurred in 7 years out of the 14 years from 1986 to 1999 (the period after Longyangxia Reservoir in the upper reaches of Yellow River was put into use and before the operation of Xiaolangdi Reservoir), the minimum overbank flow is 3,860 m^3/s, the flood probability is 50%, i.e. the flood happened every two years; overbank flooding occurred in 2 years out of the 12 years from 2000 to 2011 (the period after the operation of Xiaolangdi Reservoir), the flood probability is 16.7%, i.e. the flood happened every six years.

Tab. 1 Statistics of Hazards Hitting Floodplain Residents of Lower Yellow River in History

Year	Max. flow of Huayuankou (m^3/s)	Number of flooded villages	Population ($\times 10^4$ people)	Farmland ($\times 10^4$ hm^2)	Number of flooded houses ($\times 10^4$)
1949	12,300	275	21.43	2.984	0.77
1950	7,250	145	6.90	0.933	0.03
1951	9,220	167	7.32	1.679	0.09
1953	10,700	422	25.20	4.664	0.32
1954	15,000	585	34.61	5.116	0.46
1955	6,800	13	0.99	0.237	0.24
1956	8,360	229	13.48	1.811	0.09
1957	13,000	1065	61.86	13.146	6.07
1958	22,300	1708	74.08	20.319	29.53
1961	6,300	155	9.32	1.653	0.26
1964	9,430	320	12.80	4.820	0.32
1967	7,280	45	2.00	2.000	0.30
1973	5,890	155	12.20	3.860	0.26
1975	7,580	1289	41.80	7.607	13.00
1976	9,210	1639	103.60	5.000	30.80
1977	10,800	543	42.85	5.585	0.29
1978	5,640	117	5.90	70.500	0.18
1981	8,060	636	45.82	10.185	2.27
1982	15,300	1297	90.72	14.496	40.08
1983	8,180	219	11.22	2.848	0.13
1984	6,990	94	4.38	2.535	0.02
1985	8,260	141	10.89	15.60	1.41
1988	7,000	100	26.69	102.41	0.04
1992	6,430	14	0.85	95.09	
1993	4,300	28	19.28	75.28	0.02
1994	6,300	20	10.44	68.82	
1996	7,860	1374	118.80	247.60	26.54
1997	3,860	53	10.52	33.03	
1998	4,700	427	66.61	92.20	
2002	3,160	196	12.0	29.25	
2003	2,580		14.02	37.80	0.36
Total		13,275	918.58	2,627.34	153.52

It can be concluded from the flood conditions of the four above mentioned periods that flood probability for floodplain residents of lower Yellow River is about every 2 ~ 6 years.

Agriculture is the main pillar of economy for floodplain of lower Yellow River and there's almost no industry due to the poor natural environment and sand, the influence of flood, restriction of production environment and related state regulations. Floodplain residents of lower Yellow River thus have low income and suffer from great loss, economy of the floodplain in Henan Province and Shandong Province lags far behind and a typical poor area has come into being. There're four national and five provincial poverty stricken counties with a poverty stricken population of 800,000 in the floodplain upstream of Taochengpu at lower Yellow River.

2 Analysis of flood probability in floodplain of lower Yellow River after the operation of Xiaolangdi Reservoir

According to the Short-term Flood Regulation Scheme for Middle and Lower Yellow River approved by the central government in 2005

2.1 When the forecasted peak flow is smaller than 4,000 m^3/s

Xiaolangdi Reservoir shall conduct water and sediment regulation in due time, and shall conduct flood discharge under the principle that the flow of Huayuankou is no larger than the bankfull flow of main lower river channel.

2.2 When the forecasted peak flow is 4,000 ~ 8,000 m^3/s

(1) If it is reported to have heavy rain in a medium term in middle Yellow River or a flood with sediment content of no less than 200 kg/m^3 happens at Tongguan, Xiaolangdi Reservoir shall be operated under balance of inflow and outflow.

(2) If there's no heavy rain in a medium term in middle Yellow River and the flood happens at Tongguan has sediment content of less than 200 kg/m^3, and the peak flow of Xiaohuajian is smaller than bankfull flow of main lower river channel, Xiaolangdi Reservoir shall be operated under the principle that the flow of Huayuankou is no larger than the bankfull flow of main lower river channel.

It means that the operation mode of Xiaolangdi Reservoir for medium scale flood becomes partial regulation now instead of doing nothing according to preliminary design.

According to the newly approved regulation principles, actual flood series after the operation of Longliu Reservoir is selected for analysis of flood probability of floodplain in lower Yellow River.

Statistics of flood with flow larger than 4,000 m^3/s and sediment content of Huayuankou since 1986 are shown in Tab. 2.

Tab. 2 Statistics of flood with flow larger than 4,000 m^3/s and sediment content of huayuankou since 1986

Year	Station Name	Flood Peak		Sediment Peak		Remarks
		Flow (m^3/s)	Time	Sediment content	Time (m - DT)	
1986	HYK	4,130	07 - 12 16:00	50	07 - 15 18:00	According to the new regulation principles, flood flow of HYK shall be no larger than 4,000 m^3/s
	HSG	76	07 - 13 16:00	1	07 - 13 8:00	
	WZ					
	TG	3890	07 - 11 20:00	55	07 - 11 0:00	
1987	HYK	5,450	08 - 27 13:40	151	08 - 29 4:00	According to the new regulation principles, flood flow of HYK shall be no larger than 4,000 m^3/s
	HSG					
	WZ	40	08 - 30 4:00	07	08 - 31 10:00	
	TG	4600	08 - 29 3:00	92	08 - 30 8:48	
1988	HYK	7,000	08 - 21 2:00	194	08 - 12 15:12	Flood process of TG was from 8:00 of Aug 4th to 20:00 of Aug. 5th. Floodplain of lower Yellow River was flooded
	HSG	1,430	08 - 11 4:00	29	08 - 7 8:35	
	WZ	1,030	08 - 16 14:00	14	08 - 15 12:00	
	TG	8,260	08 - 07 4:00	363	08 - 09 14:00	

Continued Tab. 2

Year	Station Name	Flood Peak		Sediment Peak		
		Flow (m^3/s)	Time	Sediment content	Time (m – DT)	
1989 (1)	HYK	6,100	07 – 25 8:00	188	07 – 25 8:00	Floodplain residents of lower Yellow River became victims of flood
	HSG	1,230	07 – 12 6:00	14	07 – 11 14:06	
	WZ	100	07 – 19 8:00	8	07 – 20 8:00	
	TG	7,280	07 – 23 16:00	434	07 – 29 13:25	
1989(2)	HYK	5,140	08 – 21 2:00	60	08 – 23 8:00	According to the new regulation principles, flood flow of HYK shall be no larger than 4,000 m^3/s
	HSG	812	08 – 20 9:28	9	08 – 18 12:00	
	WZ	323	08 – 18 5:30	17	08 – 18 7:48	
	TG	4,940	08 – 20 14:00	70	08 – 20 2:00	
1990	HYK	4,440	07 – 9 20:00	79	07 – 10 2:00	According to the new regulation principles, flood flow of HYK shall be no larger than 4,000 m^3/s
	HSG	343	06 – 22 2:00			
	WZ	56	06 – 29 2:00	2	07 – 5 20:00	
	TG	4,430	07 – 8 8:00	134	07 – 7 7:00	
1992	HYK	6,430	08 – 16 19:00	454	08 – 16 20:00	Floodplain residents of lower Yellow River became victims of flood
	HSG	160	08 – 14 22:00	3	08 – 14 8:00	
	WZ	358	08 – 13 20:00	15	08 – 13 9:12	
	TG	4,040	08 – 15 0:00	297	08 – 14 10:20	
1993	HYK	4,300	08 – 7 8:36	154	08 – 7 16:00	According to the new regulation principles, flood flow of HYK shall be no larger than 4,00 m^3/s
	HSG	241	08 – 2 19:30	2	08 – 16 8:00	
	WZ	624	08 – 6 22:36	28	08 – 6 16:00	
	TG	4,440	08 – 6 5:36	162	08 – 7 8:00	
1994 (1)	HYK	5,170	07 – 10 20:00	150	07 – 12 11:36	Floodplain residents of lower Yellow River became victims of flood
	HSG	702	07 – 13 22:00	13	07 – 12 22:18	
	WZ	52	07 – 14 14:00	4	07 – 15 8:00	
	TG	4,890	07 – 9 7:43	425	07 – 10 2:00	
1994 (2)	HYK	6,300	08 – 8 11:20	241	08 – 8 6:30	Floodplain residents of lower Yellow River became victims of flood
	HSG					
	WZ					
	TG	7,360	08 – 6 17:18	351	08 – 13 13:25	
1996 (1)	HYK	7,860	08 – 5 15:30	353	07 – 31 8:00	Floodplain residents of lower Yellow River became victims of flood
	HSG	1,980	08 – 4 9:00	21	08 – 3 22:00	
	WZ	1,420	08 – 5 22:36	19	08 – 14:00	
	TG	4,230	08 – 3 0:04	468	07 – 29 8:00	
1996 (2)	HYK	5,560	08 – 13 3:30	155	08 – 14 20:00	Floodplain residents of lower Yellow River became victims of flood
	HSG	420	08 – 10 18:00	5	08 – 10 20:00	
	WZ	785	08 – 10 1:00	10	08 – 10 0:00	
	TG	7,400	08 – 11 6:00	263	08 – 12 2:00	
1998	HYK	4,660	07 – 16 16:00	161	07 – 16 11:36	Floodplain residents of lower Yellow River became victims of flood
	HSG	502	07 – 18 20:00	6	07 – 17 2:00	
	WZ	394	07 – 11 9:00	12	07 – 11 14:36	
	TG	6,500	07 – 14 17:39	227	07 – 14 12:00	

Note: HYK – Huayuankou; HSG – Heishiguan; WZ – Wuzhi; TG – Tongguan

According to the 26 years of hydrological data from 1986 to 2011, flood with flow larger than 4,000 m^3/s happened in 11 years at Huayuankou. In respect of the regulation principles approved in 2005, flood with flow larger than 4,000 m^3/s (that's larger than bankfull flow) happened in 6 years at Huayuankou, i. e. 1988, 1989, 1992, 1994, 1996 and 1998, which means the flood happens about every 4.3 years and the flood probability is 23%.

Flood probability of floodplain of lower Yellow River is 23% according to the flood regulation scheme for middle and lower Yellow River approved by the central government in 2005. The chance of flood with flow larger than overbank flow is reduced due to improving flood discharge capacity of the lower river channel of Yellow River brought by the water and sediment regulation of Xiaolangdi Reservoir, and the improving impoundment rate of Xiaolangdi Reservoir for medium scale flood brought by increasing industrial and agricultural water consumption and developing water and soil conservation projects in middle Yellow River.

3 Conclusion

Based on the above analysis, flood probability for floodplain residents of lower Yellow River is every 4.3 years, safety standard of the floodplain is 20 – year – return flood, which are consistent with national standard of "5 – year – return flood for land acquisition of reservoir, 20 – year – return flood for resettlement". 50% of the flood discharge and detention areas in Huaihe river basin have an operating frequency of 3 ~ 5 – year – return flood, the other 50% have 5 ~ 8 – year – return flood. Floodplain residents of lower Yellow River generally live a normal life and work on the farm, and they move out temporarily when flood menaces. The floodplain is the location of flood discharging and storing and the place of residents' livelihood. It is in nature a special river channel that serves for flood storage and flood detention.

Therefore, it is possible for floodplain residents of lower Yellow River to receive state compensation when they become victims of flood inundation, just like residents in state flood discharge and detention areas.

Study on Ecological Project of Zhaojunfen Flood Storage and Detention Area

Yan Dapeng*, *Ma Zhuoluo*, *He Zhiyin* and *Ha Jia

Yellow River Engineering Consulting Co. , Ltd. , Zhengzhou, 450003, China

Abstract: Zhaojunfen flood storage and detention area is one of the six ice-prevention emergency flood storage and detention areas planned to be constructed in Inner Mongolia section of the Yellow River, and featured with a relatively high frequency of utilization. In order to solve a series of social problems and long-term development problems caused by the construction of Zhaojunfen flood storage and detention area, the planning and management objectives in the paper are adjusted to the construction of an ecological remediation type flood storage and detention area to explore the construction of an ecological culture landscape project mainly based on ice flood diversion project, ecological construction and regional culture excavation and development, and the formulation of reasonable regulation and operation method, i. e. scientifically regulating according to the regulation and storage capacity of Zhaojunfen flood storage and detention area under the premise that the safety of ice prevention is ensured, and therefore the flood storage and detention area can play an important role in flood resource utilization and ecological protection. In the study, we abide by the principle of nature respecting, people-oriented and harmonious human and nature, and change the Zhaojunfen flood storage and detention area from a single social benefit mode to a comprehensive, coordinated and uniform mode of social, economic and eco-environment benefits, so as to ensure the sustainable development as well as provide a new reference experience for the constriction and management of the flood storage and detention area.

Key words: Zhaojunfen, flood storage and detention area, ecological project

1 Introduction

According to the overall arrangements for the ice-prevention emergency flood diversion project of Yellow River in Inner Mongolia, Zhaojunfen flood storage and detention area is one of the six ice-prevention emergency flood storage and detention areas planned to be constructed in Inner Mongolia section of the Yellow River, and located upstream the junction to theYellow Rive of Xiliugou and in the Dalad Zhaojun Town of Ordos in Inner Mongolia in the right bank of Yellow River, wherein the range of the flood storage and detention area is as follows: the east part is bounded by the stem dike of Yellow River, the south part is bounded by Ergouwan highland, Zanzibar road, Chengguaizi and Sierkaiwan highlands, the area is reached to the east part for the connection of Xinjiandui to the Sierkaiwan in the west and the old Yellow Rriver dike in the north. In the area, the physiognomy is relatively flat, i. e. the around areas are high and the centre area is low, and the area have been repeatedly inundated in ice flood period. In the ice flood period, the water level of Yellow River is high, the water can flow into the flood storage and detention area, and the water also can be discharged from the flood storage and detention area into the Yellow River with the reduction of ice water level, so that the water storage capacity of river channel in Inner Mongolia section of the Yellow River can be effective reduced, the pressure for intensively discharging the water storage of thawing channel of the Yellow River to the ice prevention part can be reduced, and the disaster of ice flood can be reduced. The designed capacity of Zhaojunfen flood storage and detention area is 32.96×10^6 m^3, the impoundment area is 18.65 km^2, the total length of the dike is 19.155 km, wherein the new dike is 14.815 km, and the old dike used therein is 4.34km. The flood diversion point of Zhaojunfen flood storage and detention area is located in the point having a Yellow River flood-prevention dike number of 301 + 040, wherein the flood-diversion sluice is shown in a two-way energy dissipation mode, i. e. both the water inlet and outlet can be controlled by the sluice. At present, the first-stage project of Zhaojunfen flood storage and detention area has been completed, and the design for the second-stage project have been completed.

In the flood storage and detention area, the current plant species is single and the ecological system is relatively fragile due to the serious salinization of soil. At the same time, the application of flood detention area will aggravate the salinization trend of the soil, and constraint the activity of farmer on the long-term land investment, therefore large areas of land will be in a barren trend, and the development of the local agricultural economy and the stability of rural society will be seriously affected. Thus, the construction of Zhaojunfen flood storage and detention area will bring a series of social problems while bringing social benefits. By considering the relevant policies of the country for the use and management of flood detention area and combining with the specific circumstances of Zhaojunfen flood storage and detention area, the local government departments provide the point that the land within the flood storage and detention area will be applied to ecological restoration and cultural tourism landscape construction, so that the social and ecological problems caused by the construction and application of flood storage and detention area can be solved, the adjustment of industrial structure can be promoted, the regional economic development in a new form can be promoted; on the other hand, a beautiful environment for ecological landscape can be created, the history and national culture can be developed, the development for the ecological tourism culture industry area in Zhaojun city (which is a scenic spot of grade 3A and adjacent to the flood storage and detention area), good economic and ecological environmental benefits can be carried out by means of a relatively low project investment, the principle of nature respecting, people-oriented and harmonious human and nature can be really carried out, and the Zhaojunfen flood storage and detention area can be changed from a single social benefit mode to a comprehensive, coordinated and uniform mode of social, economic and eco-environment benefits.

Therefore, the construction of ecological project, the combination for the requirement of comprehensive development, and the improvement of project benefit are the urgent requirements for ensuring the sustainable development of social, economic, ecological environment in Zhaojunfen flood storage and detention area under the condition that the original ice prevention and flood diversion functions are maintained. The frequency of utilization of Zhaojunfen flood storage and detention area is relatively high, specifically the interval for the probability of use is 5.5 years, so that the planning and management objectives of the area shall be adjusted to the construction of an ecological remediation type flood storage and detention area to realize the unattended mode of flood storage water area and play flood storage, ecological environment protection, wetland restoration, biological diversity, tourism and other functions of the flood storage and detention area. Thus, the development tasks thereof are identified as ice flood diversion, ecological construction and regional culture excavation and development, while giving attention to aquaculture, tourism, training and education.

2 Study on overall project layout

The ecological project of Zhaojunfen flood storage and detention area is designed on the main basis of original project of flood storage and detention area, wherein the corresponding ecological services supporting projects will be carried out by considering ecological construction, water surface and wetland landscape creation, tourism development, aquaculture and other comprehensive utilizations, and the projects mainly comprise six parts, i. e. the project for the increase of dike and the widening of green belt in flood storage and detention area, the project of lake, the project of road, the project of bridge, the project of communication to the water system of Zhaojun city, and the project of ecological culture landscape.

2.1 Project of dike and green belt widening

In the original design, the dike of northwest side is started from the Yellow River dike number K299 + 950 and reached to Yuquankui turning to the Sierkaiwan highland, and the length thereof is 12.21 km, wherein the implementation length of the first stage is 4.1 km, and the rest part of 8.11 km is the content of the second stage. In the design, the area of the flood storage and detention area is expanded with 0.78 km^2 in the west side, i. e. the area of the flood storage area is expanded west to the place close to the west side, thereby increasing a new dike of 2.55 km long and removing the west dike of about 1 km in the original design, and the rest dike of the second stage will be implemented according to the requirements of original design. The grade of the new dike is 4, the elevation of the dike is 1,010.03 m, the top width designed is 6 m, and both the slope ratio of the upstream and downstream slopes are 1:3.

In order to meet the requirements for the balance of earth excavation and backfilling and ecological landscape, the dike within the flood storage and detention area is widened and provided with a landscape green belt, wherein the widening is carried out inside, the increased width is 14 m (west)/24 m (east), and the elevation of the increase width is consistent that of the original dike. Due to the high physiognomy, the ridge sections of Ergouwan and Zhaojun Town did not be provided with a dike, and the root of the ridge is provided with a green belt of 14 m (west)/24 m (east) wide by backfilling, in which the height of backfilling is the same with the elevation of dike top. As the Zhaojun dike is located in the entrance of the landscape, it is widened west with 34 m along the dike top of 6 m wide according to the requirements of landscape layout, wherein the elevation for the top of widened green belt is the same with that of the dike top. In order to meet the requirements of landscape design, the coefficient of all the widened slopes of the dike is 1: 5, specifically the east slope of Zhaojun dike is slowed from the current 1: 3 to 1: 55.

2.2 Project of lake

In order to improve the natural ecological environment, create a good landscape effect, and meet the requirements of aquaculture, navigation and etc, a project for the construction of an ecological landscape lake will be carried out, wherein the excavation area of the lake is 4.95 km^2, and the earthwork excavation volume is 7.03×10^6 m^3.

The whole lake is mainly composed of four artificial lakes, specifically the west side of Zhaojun dike will be provided with a lake (Jingning Lake) with the area about 3.3 km^2, the west side of old Yellow River dike will be provided with a lake (Haoyue Lake) with the area about 0.73 km^2, the space between the east side and south and north dikes of Zhaojun dike will be provided with a lake (Naoerpan Lake) with the area about 0.34 km^2 through the Naoerpan gully, and the westmost side of the flood storage and detention area will be provided with a lake (Yinma Quan) with the area about 0.1 km^2.

In order to meet the requirements of fishery culture and navigation, the a strip-shaped rive (Xiangxi River) with the length of about 5 km, the average width of 100 m and the depth of 2.9 m will be excavated between the ice division sluice of northeastern part and the Jingning Lake to communicate the Haoyue Lake and Jingning Lake; moreover, a trip-shaped rive (Yanzhi River) with the length of about 2.5 km, the average width of 30 m and the depth of 2.2 m will be excavated between the Jingning Lake and the Yinma Quan to communicate the whole water system, wherein the total area of the Xiangxi River and Yanzhi River is about 0.48 km^2.

The layout of islands within the lake is mainly divided into three regions which are respectively located in north and south sides of the Haoyue Lake, the north side of the Naoerpan Landscape Lake, and the north and south sides of the Jingning Lake, and the elevation of top part is between 1,005.50 m and 1,008.00 m.

2.3 Project of Road

Taking into account the transportation of the scenic spot, a road around the lake shall be constructed, wherein the designed road will use a combined dike and road, the appropriate position of the widened dike will be provided with landscape roads, the elevation of the road is the same with that of the dike top, the whole length for the road on the dike top is 26.58 km, the road will be constructed according to the standard of two-lane road of grade 3, the width of road bed is 8 m, the width of the pavement is 2 × 3.5 m, the width of the road shoulder is 2 × 0.5 m, the cross slope of the pavement is 1.5%, and the cross slope of the road shoulder is 2.5%.

The space between the road of dike top and the external road is provided with two upper dike connection lines, the total length is 245 m, and the form of the cross section and the pavement structure are the same with that of the road on the dike top.

2.4 Project of bridge

Wang Bridge, located in the point with the Zhaojun dike number ZD1 +237 to ZD1 +337, is one of the main buildings in the planning area, and the construction thereof plays an important role for the enrichment of landscape effect in the scenic spot and the improvement of tourism value. The selection of bridge type is mainly based on the landscape requirements, the multi-span arch bridge with beautiful structure, good landscape effect, simple structure, great carrying potential and low maintenance cost is used, wherein the length of the bridge is 95.65 m, the cross section of the bridge is provided with a sidewalk (1.5 m), a lane (7.0 m) and a sidewalk (1.5 m), the total width of the cross section is 10 m, the lower part of the bridge is shown in a span changing multiple arch structure, the lower part of the arch pier is provided with a pile, and the pile foundation under the pile is formed by pouring the hole.

As the flow section and capacity for the bridge of Zhaojun dike are more than that of the ice flood division sluice, the setting of Zhaojun dike and bridge located in the central area of flood storage and detention area will be adversely affect the flood division in the flood detention area.

2.5 Project for Communication to Water System of Zhaojun city

Zhaojun city is a national scenic spot of Level 3A, adjacent to the east side of the flood storage and detention area facing to Zhaojun city through the dike, and the Haoyue Lake in the flood storage and detention area will be communicated with the Xiangxi Lake in Zhaojun city according to the planning. Because of the point that both the elevations for the bottom and water level of Xiangxi Lake in Zhaojun city are higher than that of the Haoyue Lake, the communication channel will be arranged in the area with a lower physiognomy after passing through the dike of Yellow River with pipe culvert, and the water reached to the side of the Xiangxi Lake can be discharged into the lake by pumping or digging the bottom of the lake. The diversion canal in the side of Haoyue Lake is 120 m, the front of the culvert pipe passing through the dike is equipped with a box culvert of which the size (length × width × height) is 3.25 m × 5.5 m × 2.9 m; the box culvert is internally equipped with a flat steel gate with the size of 5.5 m × 2.1 m; the culvert pipe passing through the dike is composed of two concentrate pipes with the diameter of 1.5m and the total length of 168 m; and the length of diversion canal in the side of Xiangxi Lake is about 700 m.

2.6 Project of Ecological Culture Landscape

Zhaojun flood storage and detention area is tightly adjacent to the tourist area of Zhaojun city, the ecological culture landscape can be created by depending on Zhaojun culture and combining with prairie culture and Yellow River culture. Firstly, take the Zhaojun culture as a main culture line, i.e. providing a main line from scenery of Zhaojun hometown, Zhaojun entering to the palace, Zhaojun going out of the frontier, grassland life of Zhaojun and other historical events to penetrate the whole design, then integrating Zhaojun allusions, the story of Zhaojun going out of the frontier and Zhaojun poetries into the design as important cultural contents to express the deeds and contributions for the lifetime of Wang Zhaojun. In the design, prairie culture and the Yellow River culture are interspersed, some places are provided with a prairie culture leisure area to reflect the local culture by means of local folklore activities, characteristic restaurants and etc., and the multiculture is integrated by means of cross design. Secondly, take the wetlands, trestles, lakes and cultivations as the main line of water resource development, i.e. setting water facilities, integrating the culture into water, and jointly constructing a tourist area of ecological culture landscape.

Planned structure of ecological culture landscape is as follows: a dike, two lakes and four areas.

2.6.1 A dike: Zhaojun dike

Zhaojun dike, which is a separation dike in the centre of the flood storage and detention area, aims at creating a scenic belt with the style of Northern Jiangnan along the lake according to the theme of the culture of Zhaojun going out of the frontier and the scenery of lakes. The dike is similar to the Su dike of West Lake.

2.6.2 Two lakes: Haoyue Lake (East Zhaojun Lake), Jingning Lake (West Zhaojun Lake)

Haoyue Lake is located in the east side of the flood storage and detention area, tightly close to the Yellow River dike, and named with the petname Haoyue of Zhaojun; Jingning Lake is located in the centre of the flood storage and detention area and the west side of Zhaojun dike, and named with "Jingning" which is the reign title of Emperor Yuan of Han Dynasty.

2.6.3 Four areas: a boutique tourist area of Haoyue Lake, a boutique tourist area of Jingning Lake, a natural wetland protection area, and a prairie culture leisure area

The total green area of Zhaojunfen flood storage and detention area is 2,082.11 hm^2, the total area of the road and square is 28.96 hm^2, and the area of public building and services building is 1.14 hm^2.

3 Study on operation mode of project design

When the land within the flood storage and detention area is change to the land of ecological landscape, the water shall be introduced for one time per year in the ice flood period, so as to reduce the loss of flood damage in the ice flood period and ensure the water of the ecological landscape. So the application mode of the flood division area is considered in the paper: when the flood inundation level in the ice flood period of each year is more than the elevation of 1,005.6 m for the bottom plate of the ice flood division sluice, the ice flood shall be divided and the water shall be conducted until the level of the flood division area is1,005.6 m, and then the sluice shall be closed, wherein the reserved capacity of ice flood division is 38.8×10^6 m^3; when the water level of Yellow River is more than 1,008.29 m, the sluice shall be reopened to divide the water, and then the sluice can be closed to stop the operation of flood division when the water level of Yellow River in the flood division point is dropped to 1,007.86 m or the water level within the flood division area is rose to 1,007.73 m; the sluice shall be closed to store water when the water level in the mainstream of the Yellow River is dropped, and opened to discharge the water when the water of Yellow River is completely attributed to the main channel; after discharging the water, the water level within the flood division area shall be 1,005.6 m, the area of the water surface shall be 12.60 km^2, the requirements of aquaculture, shipping, wetland landscape and the like comprehensive utilizations shall be met while evaporating and leaking, the water level and surface shall be gradually reduced, the water level before the ice flood division shall be dropped to 1,003.50 m, and the area of the water surface shall be shrunk to 4.66 km^2; in the next year, the operation mode is the same with that of last year when the water division conditions are met in the ice flood period; if the water division did not be presented in the ice flood period of the next year, the water level shall be controlled at 1,003.50 m by supplementing from the existing driven well within the flood storage and detention area to meet the requirements of water depth in evaporation, leakage, aquaculture, shipping, wetland landscape and the like comprehensive utilizations.

With the construction of large-scale reservoirs in the mainstream of Yellow River, the natural water and sediment conditions are changed greatly, the river siltation condition in Inner Mongolia section is serious year by year, and the water level of Zhaojunfen in the ice flood period is rose year by year, which is unfavorable for the ice prevention and beneficial to the water division in Zhaojunfen flood storage and detention area as the water division probability of Zhaojunfen flood storage and detention area is increased. The analysis for the series information Zhaojunfen hydrometric station shows that the maximum water level in average ice flood period of many years is 1,006.38 m in the 13 years (from 1998 on which the reservoir of Wanjiazhai is put into use to the present), in which the water conducting level is more than 1,005.6 m in 10 years, accounting for 77%.

According to the analysis process for the daily water level of typical year and the estimation based on the actual ice prevention experience of the local flood control experts, the duration for high water level of Zhaojunfen flood storage and detention area in the ice flood period is about 10 d. According to the inlet water level and flow curve designed according to the Zhaojunfen flood storage and detention area, the water level of the flood storage and detention area can be reached to 1,005.6m and above when the water level of Yellow River will be lasted for 10 d in 1,005.8 m.

In the 13 years (from 1998 on which the reservoir of Wanjiazhai is put into use to the present), water conducting level is lower thanl,005.6 m in 3 years, accounting for 23%. If the water cannot be divided at the end of March, the water level of the lake shall bel,003.50 m, the area of the water surface shall be 4.66 km^2, and the remaining capacity is 1.06×10^6 m^3 only. In the condition that the water cannot be divided continuously, the normal landscape water level can be maintained by pumping groundwater.

Under the circumstance that the water resources in the Yellow River Basin is short and the aquatic ecosystem is serious deteriorated, the work of actively regulating and storing flood resources through the flood storage and detention area is very necessary. According to the application mode above, the flood storage and detention area can play an important role in flood resources utilization and ecological protection through the scientific regulation based on the regulation and storage capacity of Zhaojunfen flood storage and detention area under the premise that the safety of ice prevention is ensured.

4 Analysis for impacts of project on Yellow River water and ice prevention

4.1 Impacts of project on Yellow River water

According to the application mode of the design, the water level before the ice flood division is 1,003.50 m and the capacity is 1.06×10^6 m^3 in normal circumstances; the water level after subsiding is 1,005.6 m and the capacity is 14.49×10^6 m^3, so that the water quantity detained in the flood storage and detention area is13.43×10^6 m^3. Due to annual water division, the water quantity annually pumped from the Yellow River is 13.43×10^6 m^3, accounting for 0.69% of the water resources only, so that the impact on the water resources of Yellow River is small. The water quality of Zhaojunfen flood storage and detention area did not be affected, and the water quality of Yellow River will not be affected by subsiding water.

4.2 Impacts of Project on Ice Prevention

The ice prevention capacity in the original design of the flood storage and detention area is 32.96×10^6 m^3. Due to the increase of water storage area of 0.78 km^2 and the earthwork excavation of 7.03×10^6 m^3 in the ecological project, the available maximum ice flood division capacity can be increased to 52.23×10^6 m^3, the quantity of subsiding water (38.8×10^6 m^3) can be reached to 74% of the flood division quantity and the remaining water can meet the requirements of comprehensive development and utilization under the condition that the maximum water level 1,007.73 m (the storage capacity of 53.29×10^6 m^3) is maintained. Therefore, the design of the project achieves comprehensive development and utilization and improves the comprehensive benefit without affecting ice prevention.

5 Benefit analysis

5.1 Tourism benefit

The ecological project of Zhaojunfen flood storage and detention area is composed of dike, widened green belt, landscape square, lakes, water system, Zhaojun dike, landscape bridge and etc., wherein the various landscape nodes are designed according to the overall idea, the ecological landscape is connected with the surrounding natural environment and ecological habitat areas, the ancillary landscape facilities suitable for viewing and playing are constructed according to local conditions, and therefore the local ecological environment will be improved greatly, the city quality will be improved, more tourists will be attracted, the development of local tourism industry will be promoted, and the income of tourism economy can be improved. In 2010, the number of tourist in Ordos City is 4.904×10^6; when the project is completed, the estimated annual increase of the tourist is about 100,000 according to the consideration for the economic and social development and the improvement of ecological environment in Ordos City, and the annual economic benefit can be reached to 20×10^6 Yuan based on the amortization quota per capita consumption of 200 Yuan.

5.2 Ecological benefit

From the ecosystem of project construction structure, a large area of lake type wetland with great ecological service value will be formed when the project is completed, wherein the service value is mainly reflected as follows: climate regulation, water conservation, air purification, protection of biodiversity and etc.

The study of Costanza and etc. (Global Ecosystem Service Value and Natural Capital, published in "Nature" journal in 1997) revealed the principle and method of ecosystem value assessment from the scientific sense, and is considered as the most influential research results of ecology field in recent years. Xie Gaodi and etc. of the Chinese Academy of Sciences revised and formulated a coefficient table of terrestrial ecosystem service value per unit area in China (see Tab. 1) on the basis for the study of Costanza and etc. According to the table, it can be seen that the ecological value of wetlands or water bodies is much higher than that of other lands, the areas of increased water bodies and wetlands after the implementation of ecological project in Zhaojunfen flood storage and detention area are 4.95 km^2 and 7.65 km^2 respectively, and annual ecosystem service value generated by ecological project in Zhaojunfen flood storage and detention area is about 62.61×10^6 Yuan.

6 Conclusions

The study explores the change of flood storage and detention area from a single social benefit mode to a uniform mode of social, economic and eco-environment benefits through the construction of ecological culture landscape project, to ensure the new way of sustainable development and provide a new reference experience for the constriction and management of the flood storage and detention area.

Tab. 1 Different terrestrial ecosystem service value per unit area in China (Unit: Yuan/hm^2 · a)

Type	Forest	Lawn	Farmland	Wetland	Water Body	Desert
Gas regulation	3,097.0	707.9	442.4	1,592.7	0	0
Climate regulation	2,389.1	796.4	787.5	15,130.9	407.0	0
Water conservation	2,831.5	707.9	530.9	13,715.2	18,033.2	26.5
Soil formation and protection	3,450.9	1,725.5	1,291.9	1,513.1	8.8	17.7
Waste disposal	1,159.2	1,159.2	1,451.2	16,086.6	16,086.6	8.8
Biodiversity conservation	2,884.6	964.5	628.2	2,212.2	2,203.3	300.8
Food production	88.5	265.5	884.9	265.5	88.5	8.8
Raw material	2,300.6	44.2	88.5	61.9	8.8	0
Entertainment and culture	1,132.6	35.4	8.8	4,910.9	3,840.2	8.8
Total	19,334.0	6,406.5	6,114.3	55,489.0	40,676.4	371.4

References

Li Luogang, Zhou Jingjing, Yi Jiye. Discussion on Sustainable Development of Flood Storage and Detention Area [J]. Haihe Water Resources, 2007(6).

Shen Yan. Discussion on Management Mode and Development of Flood Storage and Detention Area [J]. Project and Construction, 2009(5).

Li Chuanqi. Ecosystem Service Value Assessment of Flood Storage and Detention Area [J]. Yangtze River, 2008(1).

Xie Gaodi, Lu Chunxia, Leng Yunfa, et al.. Ecological Assets Valuation of Qinghai-Tibet Plateau [J]. Journal of Natural Resources, 2003(3).

Research on Ecological Water Demand in River Based on River Health

Qiao Mingye, *Cai Ming*, *Liu Meng* and *Chu Li*

Yellow River Engineering Consulting Co., Ltd., Zhengzhou, 450003, China

Abstract: River health indicates that the natural function and social function of the river can work properly under current state. As the demand of the human on water resource increases, the ecological water demand in river cannot be satisfied and the ecological environment degraded gradually such that the river health is seriously affected. This article has stated contents of ecological water requirement in river and determining principle, analyzed principle and applicable condition of various applicable calculation methods (hydrology method, hydraulics method, habitat assessment method, holistic analysis method and the like), based on it, proposed to divide ecological water demand in river for maintaining river ecological system structure, function and morphology into water demand for sediment transport in flood season and ecological basic flow out of flood season. Take Ying River of Huai River Basin for example, the research on determining ecological water demand in river has been conducted by selecting section of a completed hydrologic station and adopting Montana method and typical annual minimum monthly runoff process from hydrological calculation methods.

Key words: river health, water demand for sediment transport in flood season, ecological basic flow out of flood season, Ying River

1 Ecological water demand in river and the river health

The river has been always closely associated with the human civilization progress, as development of human society requires more and more water resource, the river ecological environment degrades so as to seriously affect the river health. How to keep the river healthy and achieve harmonious relationship between people and water has attracted more and more attentions.

Currently, there is no strict definition to the concept of river health, from the actual condition in our country, the content of river health can be summarized as a dynamic process of constantly adjusting proportion of river ecological water consumption and social & economic water consumption through appropriate development and protection to achieve a water consumption demand satisfying socioeconomic development without seriously affecting health of river ecological system. Ecological water demand in river is an important guarantee for keeping river healthy, the basis of rational configuration of water resource, the key technical index for conciliating conflict between social economic water consumption and river ecological water consumption and significant meaning for maintaining structure, function and morphology of the river ecological system. Thus, the basic requirement of river health is to satisfy demand on ecological water in river.

2 Contents of ecological water demand in river and determining principle

2.1 Contents of ecological water demand in river

Ecological water demand in river is an organic integrity consisting of polytomized variables, it indicates the minimum quantity of natural water which must be stored and consumed for keeping basic ecological environment function of river or improving ecological environment quality under special objective of ecological and environmental protection, it is also called basic ecological flow. It has prior satisfiability, other water consumption requirement can be considered only after this part of water demand has been satisfied. Research on ecological water demand in river is beneficial to prevent ecological environment degradation caused by river cutout and flow reduction, keep river healthy and achieve sustainable development of the river ecological system.

2.2 Principle of determining ecological water demand in river

Ecological water demand in river is closely related to the river characteristics, river section location and river function, ecological water demand in river should be determined in compliance with following principles:

(1) Subsection consideration principle.

Different sections of river are very distinct in aspects of river gradient, river pattern and river regime, and different sections also have different water demand, therefore ecological water demand in river corresponding to different river sections should be considered, respectively.

(2) Principle of prior satisfiability of primary function.

In different time and different sections, the degree of priority of all river ecological functions is different. When various functions have conflict, the water demand of leading primary function will be satisfied with priority.

(3) Principle of integrated optimization of all river sections.

Based on the discussion of subsection, ecological water demand in river of the whole river sections should be further analyzed comprehensively to research ecological water demand in river of all river sections.

3 Method for determining ecological water demand in river

As consensus of research on ecological water demand in river has been not reached yet, and the theory and method are still under improvement. Currently, the calculation methods of widest application include hydrology method, hydraulics method, habitat assessment method, holistic analysis method and the like. Calculation principle and applicable condition of various methods are shown as Tab. 1.

Tab. 1 Calculation principle and applicable condition of calculation method of ecological water demand in river

Method name	Calculation theory and method introduction	Analysis on applicable condition
Hydrology method	The ecological water demand can be estimated depending on hydrological data, like historic river flow, including Montana method, flow duration curve method, 90% guarantee rate most - dry month average flow method and typical annual minimum month flow method	It is suitable for the case of low requirement to calculation result precision and lack of biological data, like the planned project
Hydraulics method	Hydraulic data can be applied for analyzing relationship between river flow and fish habitat indicative factor, it is necessary to collect biological data, like river flow data, river cross section data and hydraulic characteristics and preference of target species, including wet perimeter method, R2CROSS method, simplified water gauge analysis method, WSP hydraulic simulation method and the like	It is always applied in ecological endangered area, like fish spawning area, it may provide hydraulic basis for other methods
Habitat assessment method	The methods for determining ecological water demand by assessing requirement of aquatic organisms to hydraulic condition include natural habitat simulation system (PHABSIM), instream flow incremental methodology, usable width (UW) method, weighted usable width (WUW) method and the like	In the case that the ecological protected target in river is fixed species and habitat, it is complicated to implement specifically, large amount of manpower and material resources is required to be not suitable for quick application
Holistic analysis method	Take the river basin as a unit, from the integral river ecological system, a comprehensive research on relationship between flow, sediment transport and river bed shape and communities in riparian zone has been taken as per expert opinion, and overall analysis on river ecological water demand has been made, including building block method (BBM) of South Africa, holistic method of Australia and the like	It is necessary to form an expert team, including ecologist, geographer, hydraulician and hydrologist, with high resource consumption, long time and complicated result, it is suitable for assessment of integral ecological water demand of the river basin, but not for quick application

4 Analysis on ecological water demand in the Ying River

The Ying River consisting of numerous tributaries has shallow and narrow river bed, in which the mountain occupies 29.6% of the whole basin area, hilly region occupies 18.5%, plain terrain occupies 51.7%, lake occupies 0.2%, the total head is 665 m, and gradient is 1.20‰.

4.1 Representativeness of the hydrometric station section to the river section

The hydrometric station should be set in the river section with smooth and straight river bed, stable and concentrated flow, no pit in low water part and convenience for layout of test facilities, the length of the smooth and straight section should be always not less than 3 ~ 5 times of the width of the main channel under flood condition, and the test river section should avoid unfavorable factors, like fluctuating back water, rapid silting change, distributary, oblique flow, serious floodplain and the like. From the view of selecting condition of the hydrometric station section, the cross section of the hydrometric station has representativeness to the cross section shape of smooth and straight river bed; as water depth in the river curve is greater than that in the smooth and straight section, in dry season, the curve is safer than the smooth and straight section. Therefore, the cross section of the hydrometric station has good representativeness to the whole river section, it is safe for living beings to apply the minimum ecological flow ascertained in accordance with the cross section data of the ascertain.

4.2 Calculation of ecological water demand in river

In accordance with characteristics of many tributaries and large sediment in Ying River, the water demand for sediment transport in flooding season should be satisfied to prevent the river section accumulating sediment, keep the basic river section shape, in consideration of all gathered materials, Montana method and typical annual minimum monthly runoff process are adopted from hydrological calculation methods for calculating the ecological water demand in river out of flooding season, the water demand for sediment transport is adopted as the ecological water demand in river in flooding season.

4.2.1 Montana Method (Montana Method, Tennant Method)

Montana method was proposed by Tennant, D. L and so on in 1976, the entire year is divided into two calculation periods, April to September are defined as water sufficient period, and October to March are defined as water shortage period, the flow percentage in different periods is different, annual average flow percentage is adopted as ecological flow. See Tab. 2 for different flow percentages in river and corresponding ecological environment status.

Tab. 2 Relationship between flow and fish in river, wildlife, entertainment and relevant environment resources

Qualitative description of habitat	Recommended flow standard (annual average flow percentage, %)	
	Normal water consumption period (October to March)	Fish spawning and nursing period (April to September)
Maximum	200	200
Optimal flow	60 ~ 100	60 ~ 100
Excellent	40	60
Very good	30	50
Good	20	40
Begin to degrade	10	30
Poor or minimum	10	10
Very poor	<10	<10

In accordance with Tennant method, the calculation formula of ecological water demand for maintaining certain function of river is shown as follows:

$$W_R = 24 \times 3,600 \times \sum_{i=1}^{12} M_i \times Q_i \times P_i$$

In the formula, W_R (m^3) indicates the ecological water demand for maintaining certain function of river under average condition of many years, M_i (d) indicates number of days in the ith month, Q_i (m^3/s) indicates the average flow of many years in the ith month, and P_i indicates the ecological water demand percentage in the ith month.

Flow data from 1993 to 2002 day by day of cross section of the completed hydrometric station of Ying River was analyzed, as per standard of Tennant method, 10% of the average flow of many years is the critical condition of river degradation, the average flow percentage from October to March is 10%, the average flow percentage from April to September is 30%, the calculation result of water demand for maintaining certain function of river is shown as Tab. 3.

Tab. 3 Calculation sheet of ecological water demand for maintaining certain function of river

Year	Month											
	Jan.	Feb.	Mar.	Apr.	May	Jun.	July	Aug.	Sep.	Oct.	Nov.	Dec.
1993	0.39	0.27	0.33	2.67	1.48	0.42	0.43	0.37	0.12	0.00	0.20	0.30
1994	0.32	0.23	0.18	0.55	0.10	1.39	8.77	1.57	0.71	0.47	0.36	0.39
1995	0.34	0.16	0.07	0.08	0.02	0.29	2.84	1.09	0.30	0.39	0.28	0.31
1996	0.27	0.10	0.03	0.04	0.05	0.00	5.03	34.16	4.70	2.03	3.24	1.89
1997	1.80	1.10	1.42	1.20	0.93	0.49	1.01	0.09	0.00	0.00	0.15	0.32
1998	0.32	0.27	0.37	0.86	2.28	1.60	1.09	3.01	0.91	0.31	0.38	0.38
1999	0.25	0.12	0.27	0.32	0.49	0.13	0.47	0.36	0.36	0.64	0.58	0.45
2000	0.37	0.35	0.29	0.02	0.00	0.00	9.38	11.16	2.02	2.10	1.45	1.49
2001	1.44	1.31	0.90	0.29	0.07	0.32	0.47	0.28	0.00	0.06	0.00	0.25
2002	0.37	0.23	0.19	0.00	0.58	0.17	1.16	1.23	0.41	0.34	0.24	0.41
Pi	0.1	0.1	0.1	0.3	0.3	0.3	0.3	0.3	0.3	0.1	0.1	0.1
Number M_i of days in each month	31	28	31	30	31	30	31	31	30	31	30	31
Average flow Q_i of many years	0.59	0.41	0.40	0.60	0.60	0.48	3.07	5.33	0.95	0.63	0.69	0.62

$$W_R = 24 \times 3,600 \times \sum_{i=1}^{12} M_i \times Q_{ix} P = 9,701,200.00 \text{ m}^3$$

4.2.2 Typical annual minimum monthly runoff process

The year which may satisfy certain function of river and has no cutout and no serious ecological environment problem is selected as the typical year, the minimum monthly flow or monthly runoff in the typical year is adopted as the average flow or monthly runoff for satisfying annual ecological water demand.

$$W_{Eb} = 365 \times 24 \times 3,600 \times Q_{sm}$$

In the formula: W_{Eb} indicates ecological water demand in river, m^3; Q_{sm} indicates the minimum average flow in typical year, m^3/s.

In accordance with the flow data of cross section of the completed hydrometric station of Ying River from 1993 to 2002, 1998 is selected as the typical year, the minimum monthly average flow of the typical year is 0.27 m^3/s. After calculation, the ecological water demand in river W_{Eb} = 8,514,700.00 m^3.

4.2.3 Water demand for sediment transport

Water demand for river sediment transport indicates the water demand for keeping river water flow and sediment silting balanced, and it is related to river runoff and sediment transportation condition, sand grain, river type and river shape. For the heavily silt-carrying river in northern, the river sediment transport is concentrated in the flooding season, the water flow has high sediment content, the water demand for sediment transport can be calculated in accordance with water demand for unit sediment transport in the flooding season. The calculation formula of water demand for sediment transport is as follows:

$$W_s = S_l \times \frac{1}{S_{cw}}$$

In the formula: W_s indicates the annual water demand for sediment transport, m^3; S_l indicates average sediment transport of many years, kg; S_{cw} indicates average sediment content of many years in flooding period, kg/m^3.

Statistics of sediment data of the completed hydrometric station of Ying River of 45 years from 1956 to 2000 was conducted, the average sediment transport of many years of completed cross section of Ying River is 323,000 t, the average sediment content of many years in flooding season is 9.25 kg/m^3, after calculation, the annual water demand for sediment transport is 34,919,000.00 m^3.

4.2.4 Analysis on calculation result

The water demand for sediment transport 34,919,000.00 m^3 is adopted as ecological water demand in river in flooding season, the basic ecological flow out of flooding season is the maximum value of the calculation result of hydrology method, which indicates that the calculation result of Montana method is 9,701,200.00 m^3 (see Tab. 4).

Tab. 4 Comparison of Calculation Results of Water Demand with Different Calculation Methods

Calculation method	Water demand ($\times 10^4$ m^3)
Montana method	970.12
Typical annual minimum monthly runoff process	851.47
Water demand for sediment transport	3,491.9

5 Conclusions

The basic requirement of river water resource management is to satisfy ecological water demand in river and support river healthy life. The article has analyzed contents and determining principle of ecological water demand in river, theory and applicable condition of various calculation methods, and Ying River system of Huai River basin is taken as an example to analyze the representativeness of cross section of the hydrometric station to the entire river section, the cross section of completed hydrometric station is selected and Montana method and typical annual minimum monthly runoff process are adopted from hydrologic calculation methods to calculate, the maximum value of the result is taken as the ecological water demand in river out of flooding season; the water demand for sediment transport is adopted as the ecological water demand in river in flooding season to satisfy the water demand for sediment transport and the water demand for keeping basic river function. The ecological water demand in river calculated by the method may satisfy the water demand for sediment transport of heavily silt - carrying river in flooding season, and also satisfy the requirement of maintaining the basic natural function and social function of the river out of flooding season, which is beneficial to keep the river healthy and provides reference for rational decision of basic ecological flow for water resource management of same river type in the basin, and it has certain realistic significance.

References

Li Guoying. Maintaining the Healthy Life of the Yellow River [M]. Zhengzhou: Yellow River Conservancy Press, 2005.

Fan Zili. Resource Environment and Sustainable Development of Tarim River Basin [M]. Beijing: Science Press, 1998.

Male J W. Tradeoffs in Water Quality Management [J]. Journal of Water Res. Plan. Manage. ASCE, 1984,110(4):434 -444.

Ni Jinren, Cui Shubin, Li Tianhong, et al.. Discussion on Water Demand of River Ecosystem[J]. Journal of Hydraulic Engineering, 2001, (9):22 -28.

Wang Xiqin, Liu Changming, Yang Zhifeng. Research Advance in Ecological Water Demand and Environmental Water Demand[J]. Advances in Water Science, 2002, 13(4):507 -514.

Feng Huali, Wang Chao, Li Jianchao. Research Advance In Ecological Water Demand and Environmental Water Demand of River [J]. Journal of Hohai University (Natural Science Edition), 2002, 30(3):19 -23.

Xu Zhixia, Chen Minjian, Dong Zengchuan. Research on Methods of Minimum Ecological Water Requirements in River Based On Analysis of Ecosystem [J]. Water Resources and Hydropower Engineering, 2004,(12).

Li Lijuan, Deng Hongxing. Calculation of River System Eco - environmental Water Demand in Haihe and Luanhe Basin[J]. Acta Geographica Sinica, 2000, 55(4):495 -500.

Ni Jinren, Cui Shubin, Li Tianhong, et al.. Discussion on Water Demand of River Ecosystem [J]. Journal of Hydraulic Engineering, 2001 (9):22 -28.

Dong Zheren. Connotation of River Health [J]. China Water Resources, 2005, (4):16 -18.

Willis K G, Garrod G D. Angling and Recreation Values of Low - flow Alleviation in Rivers [J]. Journal of Environmental Management, 1999, 57(2):71 -83.

Study on Ecological Compensation Mechanism of River Bank Resources —Taking the Yellow River in Ningxia for Instance

Su Yanmei*, *Hou Xiaoming*, *Cai Ming* and *Ren Jinliang

Yellow River Engineering Consulting Co. , Ltd. , Zhengzhou, 45000, China

Abstract: The paper was taking the Yellow River in Ningxia for instance, utilized ecological economics theory and research method to assess the ecological service value of the river bank resources; and based on the theory of ecological service value, ecological externality, ecological asset and public goods, and the framework of Ecological Compensation Research, set the ecological compensation mechanism of river bank resources utilization of the Yellow River in Ningxia. The results revealed that the total annual value of indirect service functions of the river bank resources far greater than the direct service functions. River bank resources utilization should be based on a functional regionalization, and it was strictly prohibited in the forbidden zone. The river management department should set up ecological protection fund for the riverbank resources and collect ecological compensation fee from the individual and unit, who utilize the river bank resources, to restore the river bank ecosystem in which the service function has been degraded, and to the environment damage from human activities, and realized the sustainable utilization of river bank resources.

Key words: river bank resources, ecological service, ecological compensation, the Yellow River in Ningxia

Zonal regions within certain range of land and water borders at two sides (periphery) of rivers (lakes) are called as mudflat waterfront resource. Waterfront resource is a kind of important and special land resource, because it not only has the functional attributes of flood discharge, adjusting stream flow and maintaining health nature and ecological environment of rivers (lakes), but also has the land resource property capable of being developing under certain situation. Development and protection for waterfront resource both play significant roles on sustainable developments of economy and society, guaranteeing the flood (storage) discharge capacities of rivers (lakes), preserving positive circle of ecosystem and river health.

1 Study area summarization

The Yellow River flows through 397 km in Ningxia, including 50,000 km^2 drainage area and ten county territories distributed along both shores of Yellow River, occupying 43% land size of the whole region, gathering 64% population of the whole region and creating more than 90% of GDP and 94% financial revenue. In other words, Ningxia exists, develops, and is prosperous because of the Yellow River.

In recent years, the Ningxia Hui Autonomous Region created the Yellow River golden bank, built standardizing dikes and further promoted the target for developing urban belt along the Yellow River as building economic zone along the Yellow River. In order to persistently enable the Yellow River to play the comprehensive functions of flood prevention, water supply, ecology, landscape and so on, the Autonomous Region put forward the concept of the Ningxia Yellow River comprehensive treatment. Comprehensive treatment for the Yellow River channel provides beneficial conditions for developing resources of waterfront mudflat and the Yellow River waterfront accepts a new upsurge of economic development. As the link for connecting economic development to ecological construction, ecological compensation mechanism ensures the sustainable development of the Yellow River waterfront resource.

2 Definitions and theoretic oundation of ecological service compensation

2.1 Definitions of ecological xervice compensation

Ecological compensation is a service function for preserving ecological environment, sustaining, improving or recovering the ecosystem and it is applied to regulating distribution relation between environmental benefit and economic benefit produced by the activities for protecting or destroying ecological environment of stakeholder and internalizing external cost or external income caused by relevant activities so as to become an institutional arrangement and running mode with economic incentives or restrict function. This institutional arrangement not only involves collecting tax (charging) to the destroyer and beneficial owner of ecological environment or providing economic compensation for the behaviors for protecting ecological environment, but also includes establishing restraint mechanism and encouraging mechanism good for preserving, recovering and constructing ecological environment. It also means ecological compensation mechanism is an economic system and focuses on establishing corresponding law and regulation frames, solving the stock and increment problems of ecological environment resources during economic development in one region via the manners of economy, policy and market and so on and easing the unbalance development problem during different regions to gradually reach and embody balance and coordinated development among regions. Thus, people's enthusiasm for protecting and constructing ecological environment is aroused to enrich ecological assets ecological environment and sustainably utilize resources.

Ecological compensation has become an important effective measure to coordinate economic development and ecological conservation. Foreign scholars consider that "ecological (environment) compensation" mainly refers to compensate the decrease or destroy for current ecosystem function or quality caused by economic development or economic construction through easing the ecosystem condition of the destroyed region or establishing a new habitat with equivalent ecosystem function or quality to keep the stability of ecosystem, which approaches the ecological compensation in ecology meaning of our country, i. e. "compensations for the damages on ecological function and quality occurred during development, the purpose of those compensations is to improve the environmental quality of damaged region or create a new region with similar ecological function and environmental quality".

2.2 Theoretical foundation of ecological service compensation

Ecological economic axiology considers that natural ecological asset and social economic assets have essential differences, but they can be converted mutually. Those assets all are elements of social total assets and resource environment has useful value all the time no matter the analysis based on labor value theory or analysis of marginal utility theory. Furthermore, the effective use value and surplus existing value thereof decrease following aggravation of pollution and ecological damage, while the labor and consumption cost supplied by human for sustaining regenerative resource environment grow increasingly. In terms of sustainable development, balance allocation between economic growth and ecological conservation is required, as well as overall equalization between economic yield and ecological yield, wherein the precondition is that ecological yield (natural ecological conservation) plays an important role on regional development and has value just lick industrial products. Besides, parts of ecological products identically have regional service functions for exchanging and trading.

From present researches, theoretical foundation for establishing ecological compensation mechanism includes public goods property of resource, paid application theory of ecological environment resource, internalization theory of external cost and theories on efficiency and fairness.

Public goods property of resources results in externality of ecological construction investment and decides the necessarily of ecological compensation and it is the root cause of ecological compensation problem. Public goods property of resources and compensation theory for use consider: "ecological environment belongs to public goods, so human always use public goods excessively due to noncompetitive during production consumption and cause tragedy of the commons at last. Tragedy of the commons can be avoided if the beneficial owners are required to pay and or use with fee via establishing a rational system which is ecological compensation system so as to promote regional sustainable development."

Externality theory requires internalizing external effect via taxation and subsidy and realizing consistence of private optimization and social optimization. External effect theory has been applied in the field of ecological conservation. Therefore, internalization theory of external cost has become the standard for computing ecological compensation and the theoretical foundation for promising relevant implementation.

The theories on efficiency and fairness provides basis for realizing ecological construction and establishing ecological compensation mechanism by governments. At present, the problem how to help the poor via ecological compensation is in the process of research. Economic compensations are supplied to the regions and units who pay the price and contribute for protecting and recovering ecological environment and functions thereof so as to embody the fairness of development.

2.3 Study status of ecological srvice compensation

Researches on ecological compensation in foreign countries started early and much profitable experiences capable of being referenced were formed. Foreign researches on ecological compensation began around the ecological service of forest ecosystem. Its realizing mechanism of ecological compensation mostly research marketing mechanism implementation of ecological compensation apart from governmental transfer payment and emphasizes the behaviors and selection of micro main body in ecological compensation, economic reason of ecological compensation, implementing way of marketized compensation, concrete mechanism for compensation and other aspects. Overseas practices on ecological compensation contain: government purchase, special fund, resource tax, private transaction, open market trade, tax preference and donation of international organizations and so on.

The research scope on ecological compensation problem in China mainly includes: quantitatively measure and calculate various service functions of ecosystems throughout the country by referencing international research thread on ecosystem service function; and describe the significance for implementing ecological compensation and provide theoretical basis for the researches on measuring and calculating ecological compensation standards. On the view of ecological compensation practices, China began to make energetic explorations on the ways and measures for carrying out ecological compensation since 1980s. shanbei shelter forest project and Qingchengshan forest compensation of Sichuan and others all are the achievements on ecological compensation at early stage of China.

3 Evaluation of ecological service function for river shoreline

3.1 Current situation of development

Overall mudflat area of Ningxia Yellow River reaches 46,373.33 hm^2 (do not include the mudflats of Shapotou Reservoir and Qingtongxia Reservoir), including 41,980 hm^2 marginal bank (23,260 hm^2 at the left beach and 18,720 hm^2 at the right beach) occupying 90.53% and 4,393.33 hm^2 river islands occupying 9.47%. Development of river beach resources mainly comprises agricultural cultivation (including cultivation), waterfront landscape construction, sand collection work, traffic occupation for crossing river and other manners, wherein cultivation or breeding takes 71.52% of river beach resource and is the major application mode of river beach at present.

Due to excellent water conservation and natural conditions, Ningxia Segment of the Yellow River forms rich natural wetland resources in the river channel. Currently, the patterns for protecting river beach resources mainly are reflected as the wetland protection carried through along the river by combining with ecological conservation. In recent years, the major wetlands of Ningxia along the Yellow River have been brought into protection include Weining Plain Wetland, Wuzhong Yellow River Wetland, Yinchuan Plain Wetland, Huangsha Gudu Yellow River Wetland, Pingluo Tianhe Bay Yellow River Wetland and Huinong Wetland.

3.2 Valuation of ecological service function

3.2.1 Ecological service function

Ecosystem service function is the natural environmental conditions and effectiveness ecosystem that people live by and are formed and sustained by the ecological process. Ecosystem service value is an important method for quantizing and evaluating the effect of ecological service function . Service value of ecosystem is realized during the interaction with human and largely affected by human activities. If human do not understand or ignore the service value of ecosystem, unreasonable human activities will cause less material benefits to human provided by ecosystem, decrease ecosystem service value and directly threat the ecological foundation of human sustainable development. On the basis of correctly learning the ration between human activities and ecosystem service value, recovery of the damaged ecosystem through drawing up reasonable policies and implementing effective human control can help relive and improve the service value of ecosystem . Thus, analyzing the influence on ecosystem service value caused by human activities has great theoretical and practical significances on promoting the sustainable developments of human society and economy and ecological environment.

3.2.2 Valuation of ecological service function

Presently, the studies on ecological service function have been deep into the studies on types of ecosystem service function, service mechanism and evaluation method and other aspects from initial discussions on concept and connotation. Ecological service function in functional zone of river shoreline can be concluded as product supply (aquatic product, agricultural products and so on), adjusting function (climate control, water and soil conservation, conservation of water sources, degradation of pollutants and so on), biodiversity conservation (habitat supply) and information function (scientific investigation, landscape, entertainment and so on). Combining with mercerization degree, product supply and information function are marketizing ecological service functions; the adjusting function is direct marketizing ecological service function, while biodiversity is non-market ecological service function.

Valuation method of ecological service function mainly includes market price method, travelling expense method, shadow project method, asset value method and the like. The market price method is suitable for check computation of environmental value effect without expenditure but with market value. The travelling expense method is ordinarily used for evaluating those natural sights without market prices or the value of environmental resource. The tourism value is estimated depending on the expenses paid for those consuming those environmental goods or services. Shadow project method is a special mode of replacement cost method and it is a measuring method for building a new engineering to replace original ecological environment system by supposing after the environment is damaged via an artificial method firstly and estimating the economic loss caused by environment damage (or pollution) with the expense for building the new engineering. Asset value method is to estimate the loss caused by environment pollution or the benefit by improving the environmental quality depending on the variation of asset value following the change of environmental quality .

Wetland locates at the transitional zone between water area and land and can generate many special functions playing vital roles to human, such as ecological function, resource function and service function. Due to versatility of ecosystem function and service, ecosystem service function of wetland has many values to bring huge economic and social benefits to human . Total economic value of wetland ecological function can be divided into direct use value and indirect use value based on benefit evaluation.

3.2.2.1 Direct use value

Direct use value of waterfront resource mainly refers to the product supply function. The ability of supplying product is evaluated by the direct market price method. Major products provided by waterfront resource include fish and rice and so on. Planting area of rice is about 26,666.67 hm^2 and the area for breeding fish reaches 993.33 hm^2 The product function provided by shoreline resource of the Ningxia Yellow River is 0.89×10^9 Yuan/a approximately calculating based on corresponding mu production value of 2010.

3.2.2.2 Indirect use value

Indirect use value of waterfront resource involved in the Ningxia Yellow River mainly include biodiversity conservation, climate regulation, water and soil conservation, water conservation, degradation of pollutant, scientific investigation and landscape entertainment, tourism and so on. Waterfront area provides habitat for inhabiting and spawning to various wild lives and is a key ecological service function of waterfront resource. Regulation of part climate mainly depends on thermal capacity of large water body to effectively ease urban heat island effect and improve the air humidity by combining with bush and trees. The metabolical processes and physiochemical processes of river bank, aquatic plants and soil of river bottom hold back part organic and inorganic dissolved substances and suspended matters in rain pollutant or river water body to convert many toxic harmful compounds into nontoxic or useful substance. So, the water body is cleared and water quality is improved to reach the effects of purifying and beautifying the environment.

According to relevant research achievements, waterfront resource statistics of the Ningxia Yellow River is stated as the Tab. 1.

Tab. 1 Waterfront resource statistical table of the Ningxia Yellow River (Unit: hm^2)

Forest land	Landscape construction	Farmland	Wetland	Desert
100,013.33	840	30,880	474,893.33	13.33

Costanza and other persons clearly defined the principle and method for valuation of ecosystem from scientific meaning, which have been regarded as the most influential scientific achievements in the field of ecology during recent years. Xie Gaoding and other persons served in the Chinese Academy of Sciences revised the ecological service value factor table of Chinese land ecosystem under unit area (see the Tab. 2) on the basis of referencing the researches made by Costanza and other people.

Tab. 2 Service values of chinese different land ecosystems under unitaArea

(Unit: Yuan/(hm^2/a))

Type	Forest	Grassland	Farmland	Wetland	Water body	Desert
Gas regulation	3,097.0	707.9	442.4	1,592.7	0.0	0.0
Climate control	2,389.1	796.4	787.5	15,130.9	407.0	0.0
Conservation of water source	2,831.5	707.9	530.9	13,715.2	18,033.2	26.5
Soil formation and protection	3,450.9	1,725.5	1,291.9	1,513.1	8.8	17.7
Waste disposal	1,159.2	1,159.2	1,451.2	16,086.6	16,086.6	8.8
Biodiversity conservation	2,884.6	964.5	628.2	2,212.2	2,203.3	300.8
Food production	88.5	265.5	884.9	265.5	88.5	8.8
Raw material	2,300.6	44.2	88.5	61.9	8.8	0.0
Entertainment culture	1,132.6	35.4	8.8	4,910.9	3,840.2	8.8
Total	19,334.0	6,406.5	6,114.3	55,489.0	40,676.4	371.4

According to the ecological service value factor table of Chinese land ecosystem under unit area mentioned in the Tab. 2, please reference the ecological service value factor of forest for the forest land of waterfront resource involved in the Ningxia Yellow River. Landscape construction mainly is greening plant according to the calculation of Grassland. And the supply of raw material has been considered in direct use value, so the indirect use value for waterfront resource of the Ningxia Yellow River is about 4×10^9 Yuan RMB in total and details are shown as the Tab 3.

Tab. 3 Indirect use value for ecological service function of waterfront resource involved in the Yellow River (Unit: $\times10^4$ Yuan)

Project	Forest land	Land construction	Farmland	Wetland	Desert	Total
Gas regulation	465.17	8.92	204.92	11,345.44	0.00	12,024.45
Climate control	358.84	10.03	364.77	107,783.45	0.08	108,517.18
Conservation of water source	425.29	8.92	245.91	97,698.86	3.61	98,382.59
Soil formation and protection	518.33	21.74	598.41	10,778.42	0.00	11,916.89
Waste disposal	174.11	14.61	672.20	114,591.29	3.22	115,455.42
Biodiversity conservation	433.27	12.15	290.98	15,758.39	0.44	16,495.23
Food production	13.29	3.35	409.89	1,891.26	0.02	2,317.80
Entertainment culture	170.12	0.45	4.08	34,982.31	0.77	35,157.71
Total	2,558.42	80.16	2,791.15	394,829.40	8.13	400,267.27

4 Ecological compensation mechanism of waterfront resource development

4.1 Basic elements of ecological compensation mechanism

According to the research achievements on theoretical framework of ecological compensation, the basic elements of ecological compensation mechanism mainly include: subject and object of compensation, compensation standard and compensation manner and the like. The three points are the difficult points for implementing ecological compensation and form the core content of ecological compensation mechanism together. New questions will be encountered on above three aspects when establishing the ecological compensation mechanism and constructing an institutional environment, reasonable organization arrangement and perfect diverse ecological compensation manner which are used for ensuring valid operation of the mechanism to form new difficult points during implementation. The problems how to clear define the compensation responsibility, how to decide the quantity of compensation, and which means should be adopted for realizing corresponding compensation and others should be solved by ecological compensation mechanism.

4.2 Basic principle of ecological compensation

The basic principle of ecological compensation should have instructional significance on the decision and implementation for the whole ecological compensation and it runs through every link of ecological compensation, which must be carried through and followed by ecological compensation. Premier Wen Jiabao proposed the principle of "protection provided by developer, recovery provided by the destroyer, compensation provided by the beneficial owner, payment provided by the party for discharging pollution" in the sixth national environmental protection conference. The basic principle of ecological compensation must be peculiar to ecological compensation, should embody the essence and target of ecological compensation and the operating principle of ecological compensation and should be the norm capable of reflecting the foundational property and essentiality of national basic policies related to ecological compensation and environment protection. Fairness, efficiency, sustainability and payment of beneficial owner are four basic principles which must be carried out and followed during ecological compensation activities. In China, the construction of national ecological compensation system must make full use of the government and markets to carry forward sequentially.

4.3 Operating mechanism of ecological compensation

4.3.1 Subject, object and target for compensation

According to the principle of "compensation provided by the beneficial owner", the subject of ecological compensation shall be the beneficial owner of ecological environment protection in theory. Due to the public goods feature of environment, all people may be the beneficial owners of environment protection behaviors, but not all ecological beneficial owners are the subjects of ecological compensation. Hence, the subject of compensation cannot be defined too general during implementation, or else the system will be out of operability. During developing waterfront resource, the subject of compensation should be the individuals or units who engage in developing waterfront resource in the river channels.

Object of ecological compensation means the object directed to rights and obligations among objects. Specific to law relations of ecological compensation, the object of compensation is the compensation activities implemented around the construction of ecological interest. That is, the object of this ecological compensation is the construction activities for recovering the degeneration of river shoreline ecosystem service function and damage of ecological environment caused by human activities.

Object of ecological compensation refers to the social organizations, regions and individuals enjoying material, technical and capital compensation or tax preference and other allowances based on legal rules or contract items because their normal living and working conditions or property utilization or economic developed are adversely affected caused by providing ecological service and ecological products to the society, developing ecological environment construction, adopting green environmental friendly technology or protecting ecological environment. The paper considers: ecological compensation for development of waterfront resource is the general term of serial compensations, recovery, comprehensive treatment and others for natural subject which is ecological environment self, the damages on ecosystem and natural resources and the pollution on environment caused by the economic activities of human. So, the target of ecological compensation shall be natural ecological environment of waterfront resource.

4.3.2 Compensation means

According to the subject of compensation, forms of ecological compensation can be classified as governmental compensation and market compensation. Presently, governmental compensation mechanism is the major form of ecological compensation due to the advantages of strong directivity of policy, goal-oriented and easy to initiate and others, but this form also has the disadvantages of inflexible system, difficult to decide the standards, high operating cost, and pressured finance and so on. The compensation means of market compensation system is flexible and can be regarded as a supplementary form for governmental compensation system, along with low operating cost and wide applicable scope. But, market compensation system has the disadvantages of asymmetric information, high transaction cost, blindness, boundedness and short-term. Wang Jinnan pointed out that "Means of market compensation involve government payment, one-to-one transaction, market trade and eco-mark, while the major means of governmental compensation include financial transfer payment, ecological friendly tax policy, and implementation of preferential policies for ecological conservation, construction projects and regional development."

For compensation means, foreign countries mainly adopt the mean for establishing ecological compensation fund like National Forest Fund of France. The Treasury Book of United State has "Restoration and Control (Reclamation) Fund for Mines Wasteland" and every region also has relevant Restoration and Control Fund for Mines Wasteland. Costa Rica sets up national forestry compensation fund as the compensation fund of private forest owner. In Germany, special fund for reclamation is reserved for newly developed mineral area. America enables security deposit system for restoration and control; Canada fetches certain proportion of compensation expense from the gate receipts of tourism sector subject to forests such as forest parks, botanical gardens and natural preservation zones to the forest cultivation department. European Union carries out carbon tax to achieve ecological benefit compensation and America collects graze tax in state-owned forestry area. Columbia charges to the discharger and beneficial owner.

Waterfront scope belongs to river management, so the management of waterfront resource belongs to river management department. The design establishes a waterfront ecological conservation fund in the river management department to collect ecological compensation from the individuals and units developing waterfront resource in the river channel for restoring the degeneration of waterfront ecosystem service function and damages on ecological environment caused by human activities to realize sustainability of waterfront resource application.

5 Conclusions and discussions

Establishment of ecological compensation mechanism based on evaluation of ecological service function is the research focus of ecological compensation mechanism. By taking Ningxia Segment of the Yellow River for example and adopting the theories and research methods of ecologic economics, the paper evaluates the value of waterfront ecological service function in this segment and establishes the ecological compensation mechanism of development on the Ningxia Yellow River waterfront resource on the basis of relevant researches on ecological compensation framework according to the foundational theories of ecological service compensation, including value theory of ecological service function, environmental externality theory, ecological assets theory and public articles theories.

The indirect use value for ecological service function of Ningxia Segment of the Yellow River waterfront is 4×10^9 Yuan/a and the direct use value is 0.89×10^9 Yuan/a approximately. So, the ecological value for protecting waterfront resource is far greater than the value of direct development. In the processes of developing the Ningxia Yellow River-based economic zone and urban expansion, sustainability of ecological service function value must be valued and various ecosystem service functions within the region shall be considered regularly to scientifically, rationally protect and utilize waterfront resources of the Yellow River from the angle of entire ecosystem. Waterfront development shall be based on function zoning it is strictly forbidden to carry through development activities in the zone where is forbidden to develop. A waterfront ecological conservation fund is set up in the river management department to collect ecological compensation from the individuals and units developing waterfront resource in the river channel for restoring the degeneration of waterfront ecosystem service function and damages on ecological environment caused by human activities to realize sustainability of waterfront resource application. Furthermore, a policy guarantee and supervision system shall be established for ensuring the implementation of ecological compensation mechanism.

References

Cao Weidong, Cao Yuhong, Cao Youhui, et al.. Evaluation and Developmental Research on Yangtze River Waterfront Resource of Anhui Wuwei County [J]. Journal of Anhui Normal University (Natural Science), 2006, 29(6): 586 - 590.

Hydroelectric Power Planning and Design Institute of MWR. Technical Rules of Management Planning for National Rivers (Lakes) Waterfront Utilization, 2008.

Wang Jinna, Zhuang Guotai. Ecological Compensation Mechanism and Policy Design [M]. Beijing: China Environmental Science Press, 2000: 12 - 23.

Cuperus R., Canters K. J., Piepers A. G., Ecological Compensation of the Impacts of a Road [J]. Ecological Engineering, 1996(7): 327 - 349.

Ruud C, Marco M G J, Bakermans H A, et al.. Ecological Compensation in Dutch Highway Planning[J]. Environmental Management, 2001, 27(1).

Costanza R, d Arge R, De Groot R, et al.. The Values of the World's Ecosystem Services and Natural Capital[J]. Nature, 1997(387): 253 - 260.

Daily G C. Nature's Services: Societal Dependence on Natural Ecosystem [M]. Washington D C: Island Press, 1997.

Ouyang Zhiyun, Wang Rusong,Zhao Jingzhu. Ecosystem Service Function and Yalue Evaluation of Ecological Economy Thereof [J]. Chinese Journal Applied Ecology. 1999, 10(5): 635 - 640.

Millennium Ecosystem Assessment. Ecosystems and Human Well - being: Biodiversity Synthesis [R]. World Resources Institute, Washington, DC. 2005.

Zhang Zhiqiang, Xu Zhongmin. Value Evaluations for Ecosystem Service and Natural Assets [J]. Chinese Journal of Ecology, 2001, 21(11): 1918 - 1926.

Xie Gaodi, Lu Chunxia, Leng Yunfa at al. Value Evaluation on Ecological Asset of Tibetan Plateaus [J]. Journal of Natural Resources, 2003, 18(2): 189 - 196.

Zheng Hua, Ouyang Zhiyun, Zhao Tongqian,et al. Influences on Ecosystem Service Function from Human Actibities[J]. Journal of Natural Resources, 2003, 18(1): 118 - 126.

Zhang Lubiao, Zheng Haixia. Research Progress and Formation Mechanism of Basin Ecological Service Market [J]. Natural ecological conservation, 2004, 12: 38 - 43.

Zhuang Dachang, Ding Dengshan, Dong Minghui. Benefit and Loss Evaluation for Ecological Economy of Degenerated Wetland Resource in Dongting Lake [J]. Geographical Science, 2003, 23(6): 680 - 685.

RICHARD T W, WUI Y S. The Economic Value of Wetland Services: a Meta - analysis[J]. Ecological Economics, 2001(37): 257 - 270.

Ouyang Zhiyun, Wang Rusong, Zhao Jingzhu. Value Evaluation on Ecosystem Service Function and Ecological Economy Thereof [J]. Chinese Journal of Applied Ecology, 1999(5): 636 - 640.

Xin Kun, Xiao Duning. Estimation of Wetland Ecosystem Service Value in Panjing Area [J]. Chinese Journal of Ecology, 2002(8): 1345 - 1349.

Zhang Zhiqiang, Xu Zhongmin, Cheng Guodong. Value Evaluation on Ecosystem Service and Natural Assets [J]. Chinese Journal of Ecology, 2001(11): 1918 - 1926.

Li Wenhua. Study on Ecosystem Service is the Core of Ecosystem Evaluation[J]. Resource Science, 2006, 28(4): 4.

Research Status on Environmental Flow of Yellow River and Key Topics

Wang Ruiling, *Huang Jinhui*, *Wang Xingong*, *Ge Lei*, *Lou Guangyan* and *Feng Huijuan*

Yellow River Water Resources Protection Institute, Zhengzhou, 450003, China

Abstract: This paper systematically introduces the current situation, obtained results and application & practice of the research on environmental flow of Yellow River, deeply explodes the key bottle - neck problems in the current research and the direction urgent to be researched, analyzes the development course and the solutions to key technology of oversea research on environmental flow. Based on the oversea research experience, methods and the requirements of development and protection of Yellow River regulation in new period, this paper proposes the thoughts and research direction, key topics and expected objectives urgent to be resolved in recent future about the research on environmental flow of Yellow River and preliminarily builds the framework of research on environmental flow of Yellow River.

Key words: environmental flow of Yellow River, research status, key topics

1 Introduction

Environmental flow of Yellow River means the river runoff conditions required for well maintenance of riverbed, water quality and ecology of rivers on condition that the natural functions and social functions of rivers are all kept in balance. Clause 21 in "Water Law of the People's Republic of China" clearly specifies that "the exploration and utilization of water resources in semi - arid and arid regions shall fully consider the water demands of the ecological environment", and Clause 30 specifies that "The formulation of exploration & utilization plan of water resources and the deployment of water resources shall maintain the proper flow of rivers and the proper water level of lakes, reservoirs and groundwater as well as the natural purification capacity of waters". In 2009, our country proposed to "implement the strictest water resources management system", to build up three "red lines" of water resources management as well as to "speed up the transfer from water supply management to water demand management". Therefore, it is the important content of implementation of "Water Law" and "the strictest water resources management system" to properly determine the water demand for the ecological environment of rivers, maintain the health of rivers and promote the sustainable utilization of water resources.

Yellow River is one of the rivers in China that water shortage and water conflicts are most prominent. The unbalance problem of ecology caused by water shortage and water pollution of Yellow River is extremely serious, which has had a severe effect on the sustainable utilization of water resources and social and economic sustainable development in Yellow River basin. For this reason, the research on environmental flow of Yellow River is demanded urgently. This research, aiming at the key issues urgently to be resolved for water resources management in Yellow River basin, is to deeply study the water demands by Yellow River itself and those by human economy and society in the macro, strategic and overall level, is to accurately master the bearing capacity of the water resources and water environment in this basin, formulate rational and scientific practice proposals for the environmental flow of Yellow River and provide technical support for achievements of the functional continuous flow of Yellow River, and is of great significance to maintain the health of Yellow River and support the economic and social sustainable development with the sustainable utilization of water resources.

2 Current situation and issues of research on Environmental Flow of Yellow River

2.1 Current situation of the research

Most area in the Yellow River Basin is within semi – arid and arid region. Austere problems of water resources and water ecology bring more and more attention to the research on environmental flow of Yellow River. Our country, especially the research institutes of Yellow River Conservancy Commission, has conducted lots of exploratory research, providing sound scientific basis and technical support for the united dispatching of the Yellow River water and the improvement of river ecology.

Research on environmental flow of Yellow River began with the concern for water for sediment transport. The key "Research on the Rational Allocation and Optimal Dispatching of Water Resources in Yellow River Basin", one of scientific and technological project of "the Eighth Five – Year Plan", analyzed the water for sediment transport in river courses in lower reach of Yellow River in flood period and non – flood period. And also, "Research on Water Protection for the Sanmenxia downstream of Yellow River", a sub – project of the key scientific and technological project of "the Ninth Five – Year Plan", explored the minimum water demanded by ecological environment in river courses and estuary lower than Sanmenxia reach of Yellow River by Tennant method and proposed that the water amount demanded by ecology in non – flood season shall be not less than 5×10^9 m^3. And the key project for Yellow River regulation, "Research on Water Demand by Ecological Environment of Mainstream of Yellow River", proposed the water demand amount of the ecological environment at 10 key hydrological cross – sections of the mainstream with Tennant method, the driest month flow with 90% assurance factor, etc.. Another research, "Research on the Key Technology for Health Restoration of Yellow River", the technology support project of "the 11th Five – Year Plan" of the nation, proposed the proper environmental flow of 8 cross – sections of Yellow River with the same methods mentioned above. Besides, the cooperation program of China and Holland, "Research on Water Demand for Ecology at the Estuary of Yellow River", proposed the water demand for the environment with a certain scale of fresh water wetland at the estuary of Yellow River with the comprehensive methods, such as hydraulics, landscape ecology, etc. The "Research on Flow Indexes of Key Reaches of the Mainstream and Branches of Yellow River", one of public welfare special projects currently conducted by Ministry of Water Resources, will research on environmental flow to keep a sound river function in the key branches, such as Huangshui, Weihe River, etc. when deeply researching the river flow conditions required by the aquatic species in the mainstream of Yellow River. At the same time, many colleges and scientific research institutes also did exploratory researches on the water demand by the ecological environment of Yellow River and achieved important results.

Compared with those of other rivers in China, the researches relevant to the environmental flow of Yellow River achieved breakthroughs and innovation regarding to the introduction of new foreign concepts, thoughts and technical methods, the use of multiple disciplines including ecology, hydrology, hydraulics and other as well as the participation of stakeholders, and obtained a better ecological effect in term of the application of the management and practice of Yellow River. For instance, the technology support project of "the national 11th Five – Year Plan" scientifically defines the meaning of environmental flow, which emphasizes that the research on water demand by river ecology shall be done within the spectrum of society – economy – nature. It pays more attention to the combination of ecological water flow and the practice of river regulation and development. It had held a meeting with the presence of shareholders and sought for their suggestions prior to the first ecological dispatching at the estuary in 2008. The cooperation program of China and Holland, "Research on Water Demand by Ecological Environment at the Estuary of Yellow River", for the first time, did exploratory research to introduce public participation into the research on ecological water demand, built up the evaluating system for the ecological water demand by freshwater wetland at the estuary and the effect analysis with ecological hydrology method and proposed the ecological water demand for a certain protection objective and scale, achieving significant research achievements which is well applied to the practice of river regulation.

Based on above research work, Yellow River Conservancy Committee also did exploratory researches to apply environmental flow to the regulation, development and protection of Yellow River, actively conducted the ecological dispatching of Yellow River, and did ecological dispatching at estuarine delta based on water and sediment regulation prior floods successively in 2008, 2009 and 2010, which achieved significant ecological effect.

2.2 Key issues in the current research

Although previous researches on the environmental flow provides forceful support for the regulation, development and management, Yellow River is a river which is the most complicated to regulate and is of the most arduous regulation task in China and even in the world. Compared with other rivers, the special water and sand relation, austere situation of water resources, complicate social and economic background in the basin and serious water ecology issue of Yellow River make its environmental flow research more difficult and challenging. Meanwhile, the implementation of the strictest water resources management system and maintenance of healthy life of Yellow River ask for a higher requirement for the regulation, development and management of Yellow River. Therefore, many new theories, techniques and methods met in the research on the environmental flow of Yellow River and its implementation and management in this new era are to be researched and discussed further, which are obviously shown in the following aspects.

(1) The environmental flow of Yellow River is still in need of solid base research and systematical ecological monitoring data as support.

Correct understanding on the response relation between the evolution rule of ecosystems of Yellow River, the changes of river flow, exploration and utilization of water resources, etc. is the basis for the research on the environmental flow of Yellow River. But the current research related to this is not enough or even blank in scientific research on Yellow River. In addition, the research on the environmental flow of Yellow River lacks of long - term and systematical ecological monitoring data. At present, the deep research and conduction of many subjects are restricted due to unclear ecosystem history, unclear current situation and insufficient data support. This has become a serious bottle - neck in the research on the environmental flow of Yellow River.

(2) The research on the environmental flow of Yellow River is requiring to build up the research methods and technical system suitable for the reality of Yellow River.

Currently, most of the research methods for the environmental flow of Yellow River are still hydrological and hydraulic. They seldom use the systematical methods, such as comprehensive analysis, expert experience, ecological hydrology, etc. and have not found out research methods and technical system suitable and feasible for the Yellow River according to the water demand characteristics of ecological environment of rivers.

(3) The research on the environmental flow of Yellow River requires to further systematically analyze the action mechanism of water - ecology.

Flow required by the river itself and by the indicating animals and plants, coupling relationship between water and ecology, etc. are still not identified or resolved scientifically, and this make the current conclusions of research on ecological water demand lack of systematical theory basis but only restricted to the proposal of flow. Lacking of the concern for "natural flow regime" of rivers make the research has a long way to go to reach the requirements of Yellow River regulation and exploration. It is another serious bottle - neck problem in the research on the environmental flow of Yellow River in this stage.

(4) Ecological protection objectives of the environmental flow of Yellow River are required to be specified and the ecological effect is required to be checked scientifically.

The current research on the environmental flow of Yellow River lacks of the concern for rational ecological protection objective, quantified protection scale and ecological benefit of water supplement, thus it requires to build up the relation between protection objective and ecological water flow and between ecological water flow and ecological benefit. And this restricts the application of research achievements to the regulation and practice of Yellow River in a certain extent.

(5) The research on the environmental flow of Yellow River is to be ungraded to management

and policy level urgently, thus to provide a forceful technical support for the flow dispatching and allocation for a sound function of the Yellow River.

Restricted by the interior complexity and the lack of basic theory and basic data, etc. , the concepts, thoughts, methods and technology of the research on the environmental flow of Yellow River have not yet comprehensively considered the unique requirements and characteristics of the ecological environment in the basin and the regulation and exploration practice of Yellow river. This makes the research related to ecological water demand always on the theoretical level. So it still has a long way to go to the actual water resources allocation and has not been raised to management and policy level completely.

3 Research in the world

The ecological water demand study in the world began in the 1940s when the biologists studied water course internal flow. The 1950s and 1960s belong to the embryonic stage of eco – environmental water demand theory. Between 1970s and 1990s, during which the ecological water demand theory developed fast, people started to pay attention to the relationship between biological factors and hydrological factors, and many ecological water demand calculation methods were also developed at that time, such as the Tennant method, channel flow incremental method. Since the 1990s, people began to consider ecological flow demand maintaining the river ecosystem and even the basin ecosystem integrity. Compared with other countries in the world, the research on the environmental flow of South Africa and Australia currently draw more attention and recognition of the public. Holistic method has become their mainstream method to determine the environmental flow, meanwhile stakeholders were involved in the research to reasonably identify river ecosystem protection objective, and determine the rationality of environmental flow outcome.

Compared with the domestic study, the foreign environmental flow study has the following significant features.

(1) more complete monitoring and evaluation system for river ecosystem and data – sharing mechanism, the long – term data available as a powerful foundation support for research and application of environmental flows.

(2) Emphasized on research on the relation between hydrological processes and ecological processes, and emphasized on the application of hydraulic models, habitat models, landscape ecology model to build direct relation between the form of flow and species and groups, endowing it with a solid ecological and hydrology basis.

(3) stressing on both flow and water amount, including "what", "when", "where" and "how".

(4) Formed a more mature system and systematic research procedures, including contents of rivers asset evaluation, conservation goals identification, environmental flow, environmental flow implementation, the assessment of implementation effect, etc.

(5) Stressed that the environmental flow is an integral part of modern watershed management, attached importance to the adaptive management of environmental flows. It is believed that to identify and develop a trade – off programme which balances the interests of all parties is the core in establishing and implementing environmental flow. Many research outputs so far have been put into the basin management and ecological dispatching practice as an important control means in the routing work.

Overall, compared with domestic related research area, the developed countries has formed a more complete evaluation system for river ecosystem monitoring and and data – sharing mechanism. The environmental flow study has a solid ecology and hydrology basis, and forms a more mature approach system and systematic research procedure, attaches importance to the implementation and application of environmental flows taking it as an important part of integrated river basin management.

4 Framework and key topics of research for the environmental flows of Yellow River

Aiming at the prominent issues existing in the Yellow River environmental flows, some thoughts and research direction in urgent need to carry out have been proposed based on the experiences gained in the world in the field, and the new tasks and new requirements proposed in the new era of maintenance healthy life of Yellow River and realization of functional continuous flow of Yellow River.

4.1 Thoughts for the Research on the Environmental Flow of Yellow River

The Yellow River environmental flow is not just the issue of water quantity, it should also answer the questions like "What is", "What time", "Where" and "How", while take into consideration the hydrological processes, suitable ecological objectives, the reasonable protection scale, implementation and management, monitoring and evaluation etc. such a range of issues. At the same time, the environmental flow is not only an ecological and environmental issue, it is also a social issue. The research on the Yellow River environmental flows should fully integrate water resources characteristics, eco - environmental characteristics and basin socio - economic background of the Yellow River, take full consideration of the views of the public and stakeholders, the knowledge and experience of experts.

Based on the points above, the research on the environmental flow of Yellow River shall include the following principle steps:

(1) Reasonably determine the protection objective and protection scale.

(2) Identify the interaction mechanism of water demand law and water - ecological system.

(3) Calculation and analysis of environmental flow.

(4) Implementation and management of environmental flow.

(5) Monitoring and evaluation of the effect of environmental flow implementation.

4.2 Key topics of the research on the environmental flow of Yellow River

In line with the proposal of strictest water resources management and the vision of Yellow River Flow management for a sound river function, based on the previous research on the environmental flow of Yellow River, aiming at significant problem in urgent need to be solved, it is proposed to start with main tributaries and reaches with prominent issues to deploy environmental flow research in spectrum of flow account, implementation, management and evaluation. Researches are to analyze eco - system and function in the target reaches, to identify rational protection target and ecological objective, to study ecological indicators and water demand in the wetlands, to figure out responding relationships between indicator species and river flow regime, between wetlands and flow regime. Then, using the systematic approach of integral method, expert experience method etc., and integrating the multi - demands of society and economy and natural ecosystem and reaches function, environmental flows of important sections of the Yellow River and the sections of main branches that inflow Yellow River are raised. By combining with the water dispatching of Yellow River, the research proposed the scheme and program for management, implementation, monitoring and evaluation for environmental flow of important sections of Yellow River and the sections of main branches that inflow Yellow River, so as to provide technical support for the functional continuous flow dispatching of Yellow River.

(1) Asset assessment and reasonable protection goal determination of Yellow River ecosystem.

Analyze ecosystem structure and functional characteristics of Yellow River, determine important rivers assets of the various reaches and make evaluation, identify important ecological conservation objectives for each reach, the protection order and the main threat factor, and determine the ecological function positioning of each reach and the ecological water demand composition.

(2) Research on river habitat changes and water demand law of typical reaches of Yellow River.

Research the growth law of representative types of fish and the demand mechanism of hydraulic conditions and water environmental factors, couple hydrodynamic model, the fish habitat model, the water environment model etc., and establish response relation between target species and habitat and hydraulic conditions and water environmental factors, simulate the quantitative relationship between the of different flow conditions and habitat quality, to determine the appropriate runoff conditions for thrive of representative fishes of typical reaches.

(3) Research on water - ecology action mechanism and water demand law of wetland along the typical reaches of Yellow River.

Analyze floodplain wetland evolution and hydrological regime changes of typical reaches of Yellow River and the response relation between the evolution of the river regime, take in - depth research on interaction mechanism of wetland hydrological factors and ecological factors and indicator species water demand law, couple hydrodynamic model, groundwater model, ecological landscape decision - making model, to establish a comprehensive evaluation system of ecological water demand of wetlands, to simulate the along river floodplain wetland succession law in different flow processes, and to propose water demand process to maintain different sizes (or range) of wetland along rivers.

(4) Coupling research on the environmental flow of important section of Yellow River.

On the basis of the above research, use the integral method and expert experience method, couple fish water demand, wetland water demand, river self - purification water demand, sediment transmission water demand, take full account of the functional location and needs of the river, establish the environmental flow evaluation and analysis system of the hydrology - ecology - the stakeholders of the Yellow River, integrate the multi - demands of society and economy and natural ecosystems, consider the possibility of achieving water resources allocation of Yellow River, propose the environmental flows of typical reaches of Yellow river to meet the ecosystem protection and human needs.

(5) Research on the implementation and management of environmental flow of Yellow River based on the reservoirs dispatching of Yellow River.

Couple water use of the socio - economic system, determine the dynamic management objectives of environmental flow, combined the Yellow River's past yearly, monthly, ten - day water dispatching programs, develop implementation plans of environment flow of Yellow River based on its reservoir dispatching, establish assessment feedback system reservoir dispatching effects, apply the scenario analysis technology, through the process of simulation - evaluation - adjustment - simulation, use multi - objective decision analysis to find the optimized program, develop the adaptive management system of environmental flows of Yellow River based on its reservoir dispatching.

(6) Research on monitoring and evaluation system of implementation effect of environmental flow of Yellow River.

Establish monitoring and evaluation system of implementation effect of environmental flow of the typical reaches of Yellow River, tracking monitor and evaluate implementation situation, the effectiveness and environmental risk of the environmental flow of Yellow River, make revisions and adjustments of the environmental flow implementation program based on monitoring and assessment results.

4.3 Expected objectives

On the basis of the above research, systematically identify the ecosystem water - eco - coupling mechanism of Yellow River, determine the water demand law of ecosystem indicator species in Yellow River, and quantitatively establish dynamic ecological goals and management objectives, determine the environmental flow that complies with the river ecosystem protection and basin social and economic development, propose the indicator of maneuverability of functional continuous flow of Yellow River, develop the implementation, monitoring and evaluation program of Yellow River environmental flows, initially establish Yellow River reservoir multi - objective real - time dispatching technology system that taking into account of ecological protection, provide technical support for the deepening of Yellow River ecological dispatching, its functional continuous flow, and maintenance of the healthy life of Yellow River.

5 Conclusion and suggestion

This paper systematically described the domestic and international status and the results achieved of Yellow River environmental flows, analyzed the problems and major bottlenecks in the development of this work. According to the requirements of the new era in development and protection of Yellow River, and learning from the foreign research experience, proposed the basic research ideas and research projects and expected goals that is in urgent need to implementation of the environment flow of Yellow River, initially set up the overall framework for future research on the environmental flow of Yellow River.

Given that the research on environmental flow is involved in the disciplines of water resources, ecology, eco – hydrology, ichthyology, hydrology, hydraulics, water environment etc. and natural, economic, social and engineering systems, being associated to multi – sectoral of water conservancy, agriculture, forestry, environmental protection, government and others, it need multi – disciplinary, multi – angle, multi – sectoral, long – term tireless research and efforts to solve the problem, propose the following suggestion for environmental flow research of the Yellow River: ① establish long – term mechanism of the Yellow River environmental flows and research platform, to increase basic research projects of the Yellow River environmental flow; ② to establish a relatively complete river ecosystem monitoring system and network; ③ the mechanism for public participation and expert knowledge systems into environmental flow of the Yellow River system; ④ environmental flows into the integrated watershed management system to enhance the practicality of the Yellow River environmental flow of research results; ⑤ increase input and personnel training of the Yellow River environmental flow research. Establish research mechanism and the security system for the environmental flow of Yellow River, provide a strong theoretical, technical and institutional support for new leap of the realization of the functional continuous flow dispatching of Yellow River and the maintenance of healthy life of Yellow River.

References

Liu Xiaoyan, Lian Yu, Huang Jinhui, et al.. Yellow River Environmental Flow Study [J]. Science & Technology Review, 2008(17): 24 – 30.

Chang Bingyan, Xue Songgui, Zhang Huiyan, et al.. Study of Rational Allocation and Optimal Scheduling of Yellow River Basin Water Resources [M]. Zhengzhou: Yellow River Conservancy Press, 1998.

Cui Shibin, Song Shixia. Study of Yellow River below Sanmenxia Segment Water Environmental Protection [R]. 2002.

Huo Chuanhe, Zhang Xinhai, et al.. Study on Ecological Water Use and Controlling Indicators of Yellow River Basin [R]. 2003.

Hao Fuqin, Huang Jinhui, Li Yun. Study on Eco – Environmental Water Demand for Main Stream of Yellow River[M]. Zhengzhou: Yellow River Conservancy Press, 2005.

Huang Jinhui, Hao Fuqin, Gao Chuangde, et al.. Study Review of Yellow River Mainstream Ecology and Environmental Water Demand [J]. Yellow River, 2004, 26 (4): 26 – 27, 32.

Liu Xiaoyan. Yellow River Environmental Flow [M]. Zhengzhou: Yellow River Conservancy Press, 2009.

Liu Xiaoyan, Shen Guangqing, Li Xiaoping, et al.. Water Demand to Maintain the Lower Yellow River Main Channel Bankfull Flow of 4,000 m^3/ s[J]. Journal of Hydraulic Engineering, 2007(9).

Liu Xiaoyan, Lian Yu, Ke Sujuan. Yellow River Estuary Ecological Water Demand Analysis[J]. Journal of Hydraulic Engineering, 2009, 40 (8): 956 – 968.

Lian Yu, Wang Xingong, Huang Chong, et al.. Evaluation of Ecological Water Demand of Yellow River Estuary Wetlands Based on Eco – hydrology [J]. Journal of Geographical Science, 2008, 63(5): 451 – 461.

Lian Yu, Wang Xingong, Liu Gaohuan, et al.. Study of Environmental Water Demand and Evaluation of Yellow River Mouth Wetland Based on Eco - hydrology [C]. Yellow River Water Conservancy Committee. Collected Papers of the Yellow River International Forum - Sustainable Use of Basin Resources and Benign Maintenance of River Deltas Ecosystems. Zhengzhou: Yellow River Water Conservancy Press, 2007.

Wang Ruiling, Lian Yun, Huang Jinhui, et al.. Eco - efficiency Evaluation Yellow River Delta Wetland Replenishment. Yellow River, 2011, 33 (2): 78 - 83.

Bas Peroli, Marcel Marchand, Michiel Van Eupen (Eds.). Yellow River Delta Environmental Flow Study [R]. 2007.

Ni Jinren, Jin Ling, Zhao Ye'an, et al.. Initial Study of Lower Yellow River Minimum Eco - environmental Water Demand [J]. Journal of Hydraulic Engineering, 2002 (10): 1 - 7.

Yang Zhifeng, Liu Jingling, Sun Tao, et al.. Basin Ecological Water Demand Law [M]. Beijing: Science Press, 2006.

Luo Huaming, Li Tianhong, Ni Jinren, et al.. Study of Ecological Environment Water Demand Characteristics of Sediment - laden River [J]. Science China Technological Sciences, 2004 (34) (Supp. I): 155 - 164.

Shi Wei, Wang Guangqian. Study of Minimum Ecological Flow of the Lower Yellow River [J]. Journal of Geographical Science, 2002, 57(5): 595 - 601.

Tang Yuan, Wang Hao, Chen Minjian. Lower Yellow River Ecological Water Demand and Estimation [J]. Journal of Soil and Water Conservation, 2004, 18 (3): 171 - 174.

Zhao Xinsheng, Cui Baoshan, Yang Zhifeng. Study of Ecological and Environmental Eater Demand of Typical Wetlands in Yellow River Basin [J]. Environmental Science, 2005, 25 (3): 567 - 572.

Zhao Weihua. China River Zobenthos Macro Pattern and Lower Yellow River Ecological Water Demand [D]. Beijing: Chinese Academy of Sciences, 2010.

Zhang Wenge, Huang Qiang, Jiang Xiaohui. Study of Instream Ecological Flow Based on Physical Habitat Simulation [J]. Advances in Water Science, 2008, 19(2): 192 - 197.

Yellow River Conservancy Commission. Comprehensive Planning for Yellow River Basin [R]. 2010.

Yellow River Engineering Consulting Co., Ltd. Comprehensive Management Plan of Yellow River Estuary [R]. 2009.

Yan Denghua, Wang Hao, Wang Fang, et al.. Primary Exploration of China's Ecological Water Demand System and Key Topics [J]. Journal of Hydraulic Engineering, 2007, 38 (30): 267 - 273.

Dayson M., Bergkamp G., Scanlon J.. Environmental Flows - Essential for Rivers. Zhang Guofang et al.. Trans. [M]. Zhengzhou: Yellow River Conservancy Press, 2006.

Study on Evaluation of Effect of Ecological Water Regulated in the Yellow River Delta

Wang Xingong[1], *Huang Jinhui*[1], *Wang Ruiling*[1], *Liu Bo*[2], *Wang Ding*[3], *Huang Wenhai*[1] and *San Kai*[4]

1. Yellow River Water Resources Protection Institute, Zhengzhou, 450004, China
2. Hohai University, Nanjing, 210098, China
3. Henan Institute of Education, Zhengzhou, 450003, China
4. Yellow River Delta National Nature Reserve Administration Bureau, Shandong, Dongying, 257091, China

Abstract: Diaokou River is one of the old courses of the Yellow River into the sea before 1976 and its estuarine delta is an important part of the Yellow River Delta National Nature Reserve. To recover the increasingly degenerated ecology of the wetland in the estuarine delta of Diaokou River, the eco – water supplement to the wetland during the water and sediment regulation period was decided by the Yellow River Conservancy Commission in 2010. In this study, taking Diaokou River as one of case study, the effects of the Yellow River ecological water regulation were evaluated, and the follow – up assessments and analyses on the factors such as vegetation, wetland, soil salt content, ground water, etc. before and after the eco – water supplement are made with the methods including the real – time monitoring, remote sensing image interpretation and ground water model simulation concerned. The result shows that the eco – efficiency of the eco – water supplement to Diaokou River is remarkable, which not only preliminarily impedes the developing trend of the invasion of salty groundwater into the degenerated wetland area of the river, but also alleviates the soil salinization within the region to a certain extent, i. e. the damaged sensitive habitat and the wetland vegetation structure are restored along with the preliminary formation of the reed marsh and reed meadow—the suitable habitat for the aquatic birds and the obvious increase of the biodiversity of the aquatic birds therein.

Key words: Yellow River Delta, ecological water regulated, evaluate

Before 1976, Diaokou River was the old way of Yellow River into sea. Since Yellow River changed to the current flow way through Qingshuigou River, due to change of water and sand, ocean dynamics and influence of human activities, topography of Diaokou River's former watercourse, the spare flow way of Yellow River, as well as natural patterns and the ecological environment, has been largely changed. At the same time, severe deterioration of wetland ecology and damage to biodiversity largely impacted its ecological function as an important component of Yellow River Delta national nature reserve, threatening the stabilization of Yellow River Delta ecological system and sustainable development of regional economy. In July 2009, Yellow River Conservancy Commission decided to "use flow way of Diaokou River to implement ecological water diversion", seek to use flow ways of Diaokou River and Qingshuigou River alternatively to restore the water and sand delivery function, extend the period entering into the sea, further improve the recovery of deteriorated Yellow River Delta wetland and achieve the well maintenance of ecosystem at estuary area. In June 2010, Yellow River Conservancy Commission carried out experiments of Yellow River Delta ecological water diversion and recovery of water flowing through Diaokou River flow way during "Water and Sediment Regulation" period. Flowing through Diaokou River and ecological water diversion in reserve area will improve regional ecological environment. And implementation of assessment studies on effectiveness of Diaokou River ecological water diversion with correct results will provide not only technical support for specific water demand process in ecological water diversion, but also important technical bases and support for following optimization of Yellow River ecological regulation, maintenance of river health and further improvement of Yellow River Delta ecological environment.

1 Overview of research area

Yellow River Delta generally refers to the fan – shaped zone at the estuary of Yellow River formed due to many years of siltation extension, swing, diversions and precipitation. This continental weak tidal strong accumulation estuary is located between Laizhou Bay and Bohai Bay in north Shangdong Province with range of approximately from east longitude 118°10′ to 119°15′ and from north latitude 37°15′ to 38°10′. In modern times, Yellow River Delta refers to the area taking Ninghai as the vertex, with Taoer estuary in the north and fan – shaped zone at Zhimai River mouth in the south, forming a 135° angle, it covers more than 6,000 km^2 with coastline of 350 km. Yellow River Delta, covering most of Dongying City and some of Binzhou City of Shangdong Province administratively, holds the second largest oil field in our country. Yellow River estuary belongs to weak tidal continental category. When large quantity of sand entering estuary area, the sediment carrying capacity of water flow at river – sea cross reduces suddenly, in addition to ocean dynamics insufficient to deliver such a huge amount of sediment, thus forms the natural evolution principle of siltation, extension, swing and diversions at the tail which leading to constant change of estuary, outward shift of coastline and expansion of estuary delta area. Since 1985, Yellow River has been 6 large changes of flow way. Due to special geological environment, Yellow River Delta, where storm surge easily occur, is one of the severely afflicted area of storm surge, and tide invasion constitutes one of the major reasons for salinization of soil in this area.

Due to heavy landform changes and large quantity of land batture, slopes, large depressions and so on at here and there of Yellow River Delta, it forms unique micro landform of hill, slopes and depressions in alternate arrangement. Its land zone with unique ecotype formed in tail swing of Yellow River expands increasingly, and large area of shallow shoals and wetlands exist at river – sea cross. It becomes an important "transfer station" and wintering & breeding place of birds migration in inland of northeast Asia and west circum – Pacific, the largest delta in nationwide, and the widest, completest and youngest native wetland ecosystem in warm temperate zone in our country. In order to protect Yellow River wetland ecosystem and birds' habitat environment, China has set up Yellow River Delta national nature reserve (Fig. 1) which is located newly silted area on both sides of Yellow River estuary and divided into two major parts: the north part is located at former flow way of Diaokou River and the south part is located on both sides of current flow way of Yellow River. With protection objectives of new wetland ecosystem and rare and endangered birds, the reserve covers a total area of 15.3×10^4 hm^2 among which the core area is 5.8×10^4 hm^2, buffer area 1.3×10^4 hm^2, and experimental area 8.2×10^4 hm^2.

Diaokou River (old course of Yellow River) has the total length of about 55 km. The water level of Diaokou River course rose quickly due to ice slush at the estuary congested the course. The dyke break and flood diversion were made at Luojiawuzi in January, 1964, and water was formed from Diaokou River into the sea, which was the ninth course into sea on Yellow River Delta, and the period of water flowing is 12 years and 5 months. Yellow River Diaokou River tail reserve is an important part of Yellow River Delta National Reserve Area, and an important habitat of rare birds, such as swan and great bustard, having significant protection value. In May, 1976, according to flood protection task at that time, the water level of west estuary approached to the course changing standard of 10m, Shandong provincial government organized to make the artificial course migration at Luojiawuzi, therefore, Yellow River changed its course from Diaokou River course to Qingshuigou course. Up to now, Diaokou River course has stopped for 35 years. Yellow River changing its course to Qingshui stopped the supply of freshwater resource and sediment for this area, made the Diaokou River lose the leading element of maintaining the hydrological and ecological balance of this area. The straightforward succession of ecological system suffered serious interference, even reversion, the area of freshwater wetland shrank seriously, the biological diversity was damaged seriously, which needed the freshwater supply urgently.

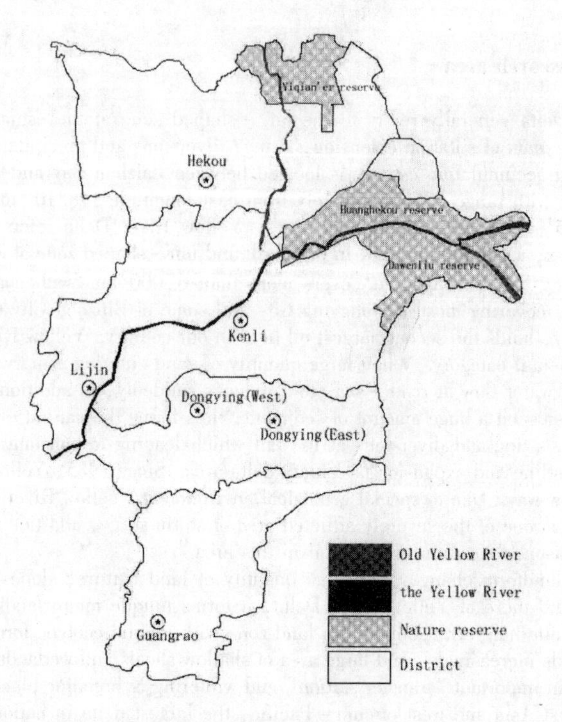

Fig. 1 Yellow River Delta (Dongying City) and nature reserve

2 Research method

Applying technologies like remote sensing, geographic information systems, GPS and groundwater model, this study uses the ecological investigation methods and technologies to investigate and monitor the ecological water diversion of Yellow River Delta, the test background of Diaokou River course returning to flow water, and the ecological situation after water flowing. By using the groundwater model, this study simulates the groundwater regime, accurately grasp the ecological dynamic changes of Diaokou River course before and after the ecological water diversion, and comprehensively evaluate the effect of water diversion on aspects of groundwater regime, wetland landscape, vegetation evolution, and the change of birds and their habitats. Change of groundwater and its influence evaluation are based on the groundwater level monitoring of Diaokou River and tail wetland water supply area before and after water diversion, and the groundwater numerical simulation system of Yellow River Delta is established for ecological water diversion. This study applies the MODFLOW model to make the quantitative analysis on the groundwater change law of Diaokou River and its banks, water supply area of tail reserve area and the involving area before and after water diversion. By appling remote sensing and quadrat investigation, combining the method of spatial sere substituting for time sere, the study evaluates the wetland landscape and vegetation change. By appling the on - site birds observation, combining the remote sensing interpretation, the change of birds and habitats is analyzed and evaluated.

3 Evaluation of effect of ecological water diversion

3.1 Situation of ecological water diversion in 2010

From June 19, 2010 to July 8, 2010, the Office of Flood Control, YRCC made the tenth

"Water and Sediment Regulation" and ecological water diversion, based on the integrated dispatching of reservoirs in the middle reaches of Yellow River. It experienced 48 days, and realized the sediment ejection of reservoir, silt reduction of courses, water supply of wetland at estuary and the test of the course into sea of Diaokou River returning to flow water, through the integrated dispatching of reservoirs of Wanjiazhai, Sanmenxia and Xiaolangdi. At 9:00, on June 24, the test of ecological water diversion of Yellow River Delta and Diaokou River returning to flow water was started formally in Shangdong Dongying Yellow River Cuijia Control Works. Then, with the gradual increasing flow of Yellow River Lijin Station, Diaokou River course flows water through Cuijia regulating sluice and Luojiawuzi gate. On July 13, Beipian natural reserve area of Diaokou River course met the requirements of water diversion, and the water flowed from dike breach to Beipian wetland. On July 17, the Cuijia pumping ship was started to pump water to Diaokou River and was stopped at 8:00, on August 5. Up to 0:00, on August 5, the flow through Diaokou River course was amounted to 36,084,700 m^3, sediment, 208,160 t, water quantity into sea, 3,240,000 m^3, and water supply quantity of Diaokou River wetland restoration area, 8,050,000 m^3.

3.2 Influence of ecological water diversion on groundwater

3.2.1 Establishment of groundwater model

The basic thought on the model establishment is that, through collecting and arranging the information related to research area, establish the conceptual model suitable to the area, and then take it as basis to establish the corresponding mathematical model, use Visual modflow software to simulate it, analyze the groundwater change before and after ecological water diversion and the influence range of ecological water diversion; according to the water level distribution of groundwater output by the groundwater model, obtain the buried depth distribution law of groundwater, at the same time, investigate the vegetation cover and growth situation in the research area before and after ecological water diversion, analyze the relationship between the water level of groundwater and vegetation distribution, and evaluate the ecological effect brought by water supply in natural reserve area. The model is established with the prerequisite that Diaokou River flows water, and the simulation area is the current course area of Yellow River Delta, and the area of Diaokou River course, 70 km long from east to west and 71 km from north to south.

3.2.2 Analysis of simulation results

3.2.2.1 Groundwater

The numerical simulation results show that the overall flow direction of groundwater flow in simulated region is from southwest to northeast, water inflow from the upper reach of Yellow River to its estuary until into the sea along the direction of Yellow River. Due to poor permeability of soil in this region, lateral flow of water is not pronounced, at the same time, the flow in the simulation range has a number of local small - scale groundwater flow system, such as surrounding area of reservoirs, ponds and solar salt fields, so the groundwater level distribution has a certain spatial differences. Specifically, because of water supply from Yellow River, the groundwater level in the current flow path and near the Diaokou River is high with dense isolines, river water supplied for groundwater leak into both banks of the river course, which further influences the groundwater level near the river course. After ecological water supplement, surface water bodies were formed in ecological protection zone in tail channels of Diaokou River and the south bank of the current flow path with water depth of about 1.5 m, the groundwater level in protection zone was generally high forming a larger hydraulic gradient with surrounding area, which slowly supplied groundwater to surrounding area making groundwater level of these areas correspondingly rise.

3.2.2.2 Impact of water supplement on groundwater along Diaokou River

The simulation results show that daily average leakage recharge capacity of precipitation to aquifer is 110.3×10^4 m^3, being the main supplying item to aquifer and accounting for 90.7% of the total recharge capacity; and the second is recharge capacity of river water leakage to groundwater after river flowing through with daily average leakage capacity of 10.2×10^4 m^3, accounting for 8.85% of total recharge capacity. Due to strong evaporation in summer, phreatic evaporation becomes a main discharging item in the simulated region with daily average evaporation of 419.2 × 10^4 m^3, accounting for 98.1% of total discharging items. Precipitation becomes the largest supplying item of this region. Flowing through Diaokou River then immediately stabilize leakage to aquifer, river leakage into aquifer becomes the second largest supplying item in this region, then the groundwater level along the Diaokou River rises. However, because river supply is in linear distribution in space and permeability coefficient of sediments at the bottom of the river bed is small, the leakage recharge capacity of river to the aquifer is still limited, accounting for less than 10% of the total recharge capacity.

In order to better analyze the effect extent and range of flowing through Diaokou River to the abandoned channel of and areas along the Yellow River, compare calculated average water level at positions with different distance from the same cross section of the Diaokou River to the river with and without flowing through Diaokou River, the position of cross section away from the upper reach of water supplement area of the nature reserve is about 6 km. The groundwater level distributions of the right river bank under different conditions are shown in Fig. 2.

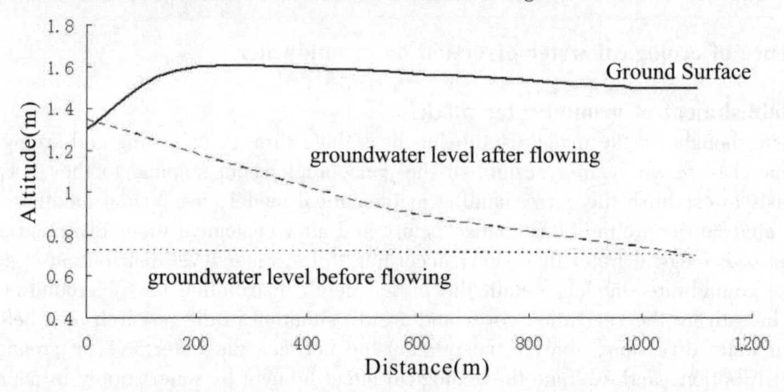

Fig. 2 Comparison of waterlevel at a cross section under the condition of flowing through Diaokou River and actual situation

From Fig. 2, it is known that compared with condition of without flowing through Diaokou River, flowing through Diaokou River can supply to groundwater along the river with apparent rising and restoration of groundwater level. With continuous flowing of the river, the groundwater along the river is successively supplied, the maximum rising rate of groundwater level around Diaokou River is 65 cm. At the same time, the restored flowing through Diaokou River has certain impact on groundwater along the river. The simulation analysis shows that groundwater in range of 550 m away from Diaokou River apparently rose with water level rising more than 15 cm, but subsequent water level rise rapidly weakened. After restoration of flowing for two months, the groundwater level had no obvious differences with and without actual flowing through in range of 1,100 m away from Diaokou River. In general, effected range of restoration of flowing through Diaokou River on groundwater along the river is 1,100 m, of which the groundwater level apparently rose in range of 550 m along the river.

3.2.2.3 Impact of water supplement on groundwater of wetland water supplement area of Diaokou River

The simulation results show that daily average leakage recharge capacity of precipitation to aquifer is 3.6 × 10^4 m^3, being the main supplying item to aquifer, accounting for 58% of the total recharge capacity. The second is recharge capacity of river water leakage to groundwater after river flowing through with daily average leakage capacity of 2.6 × 10^4 m^3, accounting for 41.7% of total recharge capacity. Due to strong evaporation in summer, phreatic evaporation becomes a main discharging item in the simulated region with daily average evaporation of 13.9 × 10^4 m^3, accounting for 98.1% of total discharging items. In case of wetland water supplement, the proportion of water supplement to aquifer from wetland is bigger and bigger as time goes by. Wetland water supplement will become the primary supplying item in the seventh and eighth week of simulation period. Evaporation is always the largest discharging item in this region, because the simulation lasts from the end of June until mid – summer, the evaporation will continue to increase as time goes by.

Wetland reserve in tail channels of Diaokou River leak to aquifer making groundwater level rises under the condition of water supplement. In order to better analyze the effect extent and range of ecological wetland water supplement in tail channels of Diaokou River on surrounding area, points around the wetland with different distance to wetland shall be taken and analyzed. Comparison of groundwater level at points with different distance away from the wetland of the last 1 d in simulation period under the condition of before and after water supplement derived from ecological protected zone in tail channel of Diaokou River without water supplement and status quo model is shown in Fig. 3.

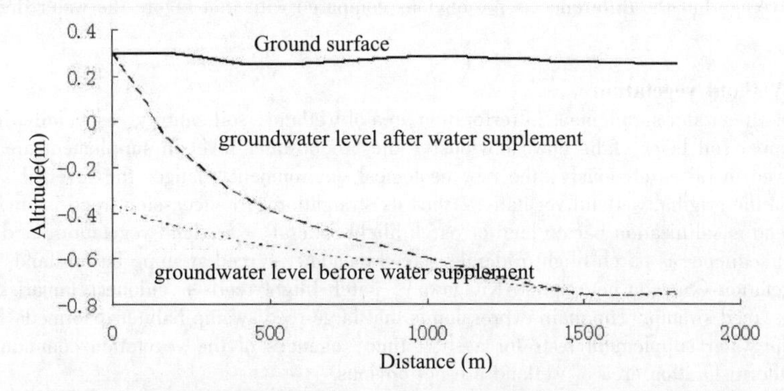

Fig. 3 Comparison of groundwater level at surrounding area with and without water supplement of ecological wetland in tail channel of Diaokou River

From Fig. 3, it can be seen that in the last 1 d of simulation period, compared with situation of no water supplement, water level at the edge of protected zone increases to 70 cm, water level at area of 100 m away from the protected zone increases by 45 cm, which indicates that the farther away from the protection zone, the smaller the effected extent. At the same time, wetland water supplement has a certain impact on groundwater of surrounding area. In range of 500 m away from restoration area, rise and restoration of groundwater level is more obvious with level rising more than 15 cm, while the subsequent rise of water level rapidly weakened, after restoration of flowing through measures for two months, groundwater level in range of 1,500 m away from the protected zone has little differences with that under the condition without water supplement, and the groundwater level in this region does not rise significantly. In general, the effected range of water supplement in the protected zone of tail channel of Diaokou River to groundwater in surrounding area is about 1,500 m, of which it has an obvious rising effect on groundwater in range of 500 m away from the restoration area.

3.3 Influence of ecological water diversion on soil and vegetation in wetland

3.3.1 Soil salinity and water content

In Diaokou River region, the water content of upper soil layer is commonly higher, fluctuating with an extent of about 0.2 g/g, and the difference in succession phase of different communities is obvious ($p < 0.05$). The water content of upper soil layer in chionese tamarisk phase is the lowest and followed by that of grass phase, and that in Suaeda heteroptera Kitagawa phase is the highest. Upon ecological water diversion, the diversion of much fresh water will greatly rise up the groundwater level in restoration area and the surrounding area and further affect the water content of upper soil layer there. Compared the background survey with the ecological survey after water diversion, and upon analysis of variance, it is founded that the difference of water content of upper soil layer in different community phases before and after the ecological water diversion is obvious ($p < 0.01$), and after the ecological diversion, the water content raises obvious. While researches indicate that water content of upper soil layer is in positive correlation with the soil salinity at the 10 cm and 30 cm layer, and both can be better fitted with exponential function model.

With the ecological diversion, much freshwater supply relieves the soil salinization in wetland to some extent. The survey results before and after the ecological water diversion indicate that the soil salinity in different plant communities upon ecological water diversion takes on downtrend, and the difference of the soil salinity in the 0 ~ 30 cm layer before and after the water diversion is obvious ($p < 0.05$), in which the soil salinity in the 10 cm layer fell down by 55% on the average and that in the 30 cm layer fell down by 41% on average, while that in deep soil actually fell down by 0. 16 on average but the difference is not obvious compared with that before the water diversion.

3. 3. 2 Wetland vegetation

Since after water supplement in restoration area of wetland, soil salinity, especially the soil salinity in upper soil layer, falls down obviously, the groundwater level in supplement area and the area involved in raises obviously, the new ecological environment changes the reversal succession direction of the original wetland vegetation, thus its straightforward succession trend is obvious. The general trend is salinization barren land or parch blight ruderal – wetland vegetation, and the main succession sequence is parch blight ruderal – reed meadow + reed swamp, barren land – annual saline vegetation (Suaeda heteroptera Kitagawa), parch blight reed + chionese tamarisk – reed meadow + reed swamp. The main expression is that large reed swamp habitat is formed. However, because the water supplement lasts for a short time, changes of the vegetation communities surrounding the restoration area of wetland are not obvious.

The restoring succession of the damaged Diaokou River wetland is a long – term succession. It can be respected that with the deep development of the recovery of water flowing in Diaokou River and the ecological water supplement in wetland, the straightforward succession trend in the restoration area in wetland will be more obvious, and finally form two succession sequences, namely, succession sequence of halophilic vegetation and succession sequence of wetland vegetation. Compared the background survey with the ecological survey after water diversion, plant species actually increase upon the water diversion from the former number of 13 to the current of 17. This indicates that the water diversion obviously improved the soil situation in the restoration area, making some species re – settle down here. But in general, the plant species in the restoration area of Diaokou River are still comparatively less.

3. 4 Influence of ecological water diversion on birds in wetland

The recovery of water flowing in Diaokou River and the ecological water supplement in wetland have an obvious influence on the landscape pattern in wetlands of the water supplement area. It mainly finds expression in the basic change of wetland pattern there. Before the ecological water diversion, it was mainly ecological patches, and the wetland deterioration was serious, mainly including such patches as regressive tidal flat, parch blight ruderal meadow, shrub, holt, etc. , in which only the low area has a little water patch (inwelling). After the ecological water diversion, most of restoration area became ecological wetland patches, forming the wetland covering an area of 513. 3 hm^2, accounting for 62. 1% of the total restoration area. In addition, the habitat type in the wetland restoration area before and after water supplement changes greatly. It finds mainly expression in the large reed swamp and water habitat that are formed. The improvement of the ecology type of wetland provides a good habitat for the birds in the protected area, making birds quantity there increase greatly. According to the monitoring results of waterfowl in bird nomadic phase in 2009 and 2010, there was 18 species with 4,403 water fowls in 2009; and this number was up to 20 species with 10,800 water fowls, with an increase of 1. 45 times compared with that in 2009. Quantity change of waterfowl in bird nomadic phase before and after the ecological water diversion takes on the following characteristics: ①the number of rare birds under special protection increases obviously. For instance: the number of red - crowned crane increases by 11, Oriental White Stork by 18, black stork by 14, gray stork by 25, big swan by 80, small swan by 50; ②quantity of birds increases greatly. For instance: sea crow increases by 2,220, spot - billed duck by 1,850, and mallard by 900. The quantity increase of birds, especially rare birds, indicates, to a great extent, that Diaokou River ecological water diversion exerts an obvious and better effect on ecological improvement.

4 Conclusions and suggestions

The evaluation results of this study suggested that the experiments of Yellow River Delta ecological water diversion and restoration of water flowing through Diaokou River flow way carried out in 2010 initially showed comparatively significant ecological effect, the Yellow River Delta tail channel wetland water supplement had positive effects of restoration for the degenerated wetland, which are mainly as follows: ①Contained the impedes the developing trend of the invasion of salty groundwater into the degenerated wetland area of Diaokou River. The ground water simulation and monitoring suggested that groundwater level in range of 1,100 m along the flowing through Diaokou River was rose and groundwater level in range of 550 m obviously rose, with a maximum rising range of 65 cm; effect range of The Diaokou River tail channel wetland water supplement to groundwater of surrounding area was about 1,500 m, groundwater level in area in range of 500 m away from the wetland supplement area was significantly rose, with maximum rising range of 45 cm. ②The rapid developing trend of soil salinization got under control. The soil salinity along the Diaokou river decreased significantly, especially the soil salt content in the 0 ~ 30 cm layer is significantly reduced, laid the foundation for the growth and development of freshwater wetlands aquatic and hygrophilous vegetation. ③Manual water supplement preliminarily contained the reverse succession trend of vegetation in Diaokou River area, repaired the damaged sensitive habitats and wetlands vegetation structure, and preliminarily formed the reed marsh and reed meadow, which are the main suitable habitats for aquatic birds. ④The vegetation change of birds habitats lead to the significant improvement of wetland habitats. On the basis of the original suitable habitat of protected species of swan, large - scale limicoline protected birds habitats got significantly restored, and the number of aquatic birds increased significantly, particularly in the Diaokou River the inhabitation and breeding of Oriental White Stork, which prefers freshwater wetland and has exacting demands for breeding habitat, highlighting the improvement effect of habitat after the Diaokou River wetland ecological water supplement.

As succession and development of ecosystem, the restoration and succession of degenerated wetlands is a comparatively longer period, the ecological benefits and ecological effects of restoration of the Diaokou River water diversion will progressively appear and develop due to the long - term implementation of the water transfer and retardation of ecosystem succession. Therefore, systematic and long - term monitoring and research mechanism of response relationship between water resources and ecological succession should be established in the future to carry out long - term tracking and monitoring, surveys and studies of implementation effectiveness of water transfer and succession of sensitive habitats of representativeness, to get comprehensive, timely, accurate knowledge concerning dynamic ecological changes of Diaokou River course before and after the ecological water diversion, and to provide strong technical support for optimization of the water dispatching of the Yellow River, for maintenance of beneficial cycle of estuarine ecosystem, and for promotion of smooth construction implementation of efficient ecological economic zone of the Yellow River Delta.

References

Zhao Yanmao, Song Chaoshu. Scientific Survey of Yellow River Delta Nature Reserve[M]. Beijing: Chinese Forestry Press, 1995.

Wang Xingong, Xu Zhixiu, Huang Jinhui, et al. Study on Ecologic Water Demand of Fresh Water Wetland of the Yellow River Estuary[J]. Yellow River, 2007, 29 (7): 33 - 35.

General Editorial Room of Yellow River Journal of Yellow River Conservancy Commission. Summary of the Yellow River Basin[M]. Zhengzhou: People's Publishing Company of Henan Province, 1998.

Li Guoying. Ponderation and Practice of the Yellow River Control [M]. Zhengzhou: Yellow River Conservancy Press, 2003.

Yellow River Conservancy Commission. Summary of the Yellow River Downstream Ecological Dispatching Work [R]. Department of Water resource Management and Regulation of Yellow River Conservancy Commission, 2010.

Preliminary Studies on Test Evaluation System of Pollution Management of Water Function Zone

Yan Li[1] , ***Zhang Junfeng***[1] , ***Xu Ziyao***[2] , ***Ma Xiumei***[1] and ***Hao Yanbin***[1]

1. Yellow River Water Resource Protection Institute, Zhengzhou, 450004, China
2. School of Civil and Environmental Engineering, Cornell University, Ithaca, New York, 14853, USA

Abstract: In this paper, in order to advance tough water management measures, the framework of test evaluation system including target layer, system layer, item level, index level was initially built based on the responsibility according to the national laws and regulations, and the current professional work of water faction zone management. This paper discusses how to determine all objects and choose examination index in application of the test evaluation system.

Key words: water function zone, evaluation system

1 Statement of the problem and study significance

1.1 Statement of the problem

Water function area is the basic unit of water resources management and protection by water conservancy departments as per laws, management of water function area has run through each link of water resources protection system. In 2009, our country proposed to carry out the strictest water resources management system, planed three "Red Lines" of water resources management, one of which is to clarify red line to restrict pollutants in water function area. After that, No. 1 Central Document in 2011 made the important decision on establishment of a pollutant restriction system in water function areas and establishment of a water resources management responsibility and assessment system. In January, 2012, the State Council issued [2012] No. 3 "Opinion of the State Council on Implementing the Strictest Water Resources Management System", further put forward the measures of implementing the strictest water resources management system, and clarified to incorporate main indicators of water resources protection into comprehensive assessment system of local economic and social development. Therefore, under the new situation and new requirements of water resources management, it has become an urgent task to fulfill the strictest water resources management system to establish water quality and carrying capacity management & assessment in water function area to enable specific measures of water resources protection to be implemented by rigid means of assessment.

1.2 Orientation of study and main thoughts

Ensuring the fulfillment of the strictest water resources management system is the key point to realize efficient water resources utilization and effective protection, while establishment of management responsibilities and assessment system of water resources is the important means to ensure the fulfillment of the strictest water resources management system. This paper presents that, at present, under the conditions that the strictest water resources management system is far from well establishment, and management means is not perfect, gradual perfection of water resources management responsibilities and assessment system, and promotion of fulfillment of the strictest water resources management system through assessment means are the first and most important task necessary to achieve for construction of assessment system framework; the second level of task is to arrange and design assessment content concerning water resources protection through construction of assessment system framework; the third level is to research and determine specific problems about technology, such as assessment methods and evaluation methods. This paper mainly orients to discuss the first two levels of problems, aiming to promote fulfillment through assessment, taking responsibilities of water administrative departments entitled by relevant laws, regulations and policies as the basis, takes managing current business in water function area as the principal line, has primarily established a framework of assessment and evaluation system of water quality and pollutant carrying management.

2 Principles of designing framework of assessment system

2.1 Principle of depending on jurisdiction

This paper takes carrying capacity assessment, formulation of opinions on pollutants restriction, water quality monitoring, management of pollution discharge outlets and others which are authorized by laws and regulations and determined by relevant policies and other responsibilities and authorities of water resources protection as bases to set up assessment indicators and its system. Main bases include "Water Law of the People's Republic of China", "Law of the People's Republic of China on Prevention and Control of Water Pollution", "Management Procedures of Outlets into the Yellow River", "Management Procedures of Water Function Area", "Decision of the Central Committee of CPC and the State Council on Accelerating the Water Conservancy Reform and Development" (No. 1 Central Document in 2011), "Opinion of the State Council on Implementing the Strictest Water Resources Management System" (GF[2012] No. 3), "Notice on Strengthening Water Resources Protection and Management in Provincial Boundary Buffer Zone" (BZY[2006] No. 131) and so on.

2.2 Principle of being comprehensive and systematic

During the current period and in the future, monitoring and management of pollutant restriction in water function areas still focus on assessing carrying capacity, putting forward opinions on restricting total amount of pollutant discharge, water quality monitoring and monitoring and management of pollution discharge outlets. According to work practice and needs of development of water resources protection, this paper arranges the above water resources protection in function areas overall and systematically, and fulfills and incorporates the important work into framework of assessment system.

2.3 Principle of operability

The evaluation indicator system of this framework tries to have clear levels, be simple and brief, and connect with existing indicator system of water resources protection. Selected indicators are relatively representative and independent, and easy to be obtained and calculated. When determining assessment scope, assessment content and assessment depth, fully consider the current situation of establishment of water resources protection capacity and possibility of realization, so as to make assessment system operable

3 Detailed design of framework of assessment system

3.1 Overall structure

According to design basis, principle and overall plan of the above assessment system framework, the study proposes assessment system framework of water quality and pollutant carrying capacity in water function area, combining quantitative assessment of indicators and qualitative assessment of system establishment and other situations. This assessment system framework has four levels: target level, system level, project level and indicator level. Among them, system level proposes the relatively independent system content in five aspects, from the perspectives of responsibilities of water resources protection departments endowed by laws, regulations and rules, and fulfillment of the strictest water resources management system; project level determines specific projects for assessment, according to specific management works in function area concerning system level; indicator level refines and decomposes assessment projects of project level to each assessment indicator or assessment content. See Fig. 1 for logical relations of this assessment system framework.

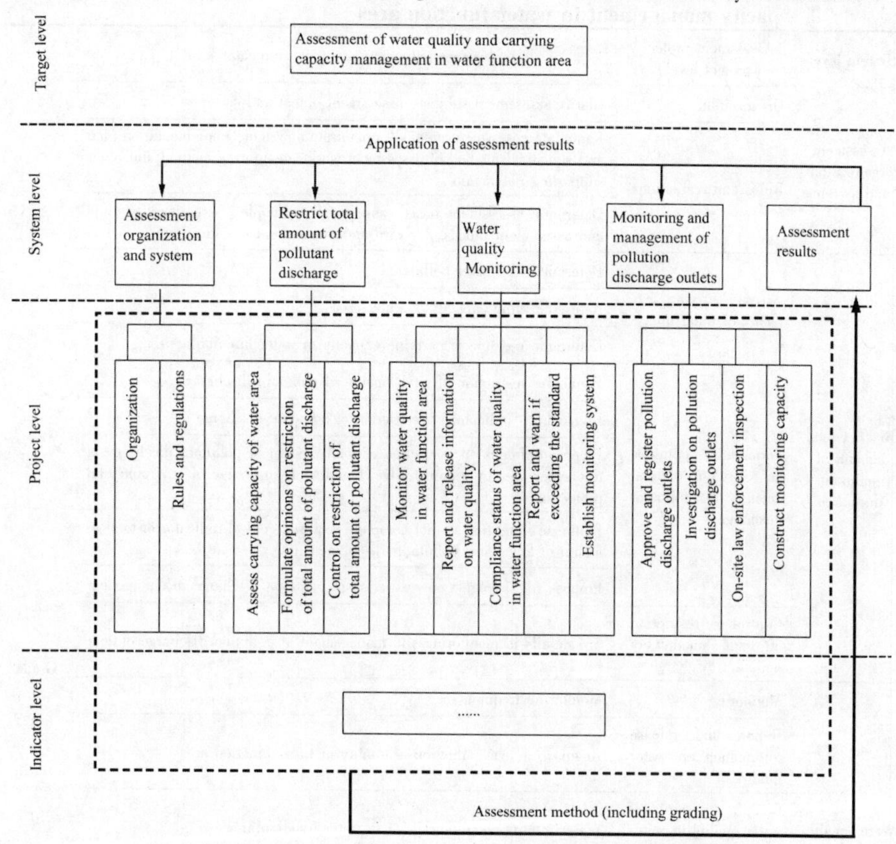

Fig. 1 Logical constitutional diagram of assessment system framework of water quality and carrying capacity management in water function area

3.2 Core content

System level and project level are the core of this assessment system framework. System level includes five systems: assessment organization and system, restriction of total amount of pollutant discharge, water quality monitoring, monitoring and management of pollution discharge outlets and application of assessment results. Among them, establishment of responsibilities and assessment system of water resources protection is the premise and base of ensuring performance of assessment work, therefore, situation of system establishment is the important part of this assessment system; system of restriction of total amount of pollutant discharge mainly assesses whether county-level and above water administrative departments or basin organizations perform their responsibilities of assessing carrying capacity and putting forward opinions on restriction of pollutant discharge; water quality monitoring is the responsibility of water conservancy departments and the determinative factor in determining whether assessment data is reliable, therefore, it is a necessary part of assessment system framework; monitoring and management of pollution discharge outlets is the important system and direct means of water administrative departments restricting carrying capacity of river and protecting water function area, and shall be assessed as an important system. According to main business concerning the above system, specific assessment projects and assessment indicators are proposed, see Tab. 1 for details.

Tab. 1 Composition of assessment system framework of water quality and carrying capacity management in water function area

System level	Assessment project (project level)	Assessment indicator / content (indicator level)	
Assessment organization and system	Organization	Clarify assessment subject, assessment object and so on	Grade
	Rules and regulations	Formulate assessment method, organization system, implementation plan and statistical method of assessment work, monitoring method and other supporting documents	
		Determine assessment basis, assessment principle, assessment method and assessment process, reward and punishment measure, etc.	
Restrict total amount of pollutant discharge	Assess carrying capacity of water area	Determine key water pollutants	
		Assess carrying capacity of water area	
		Determine red line of carrying capacity in water function area	
	Formulate opinions on restriction of total amount of pollutant discharge	Formulate restriction of total amount of pollutant discharge	
		Break down indicator of restriction of pollutant discharge	
		Propose opinions on restriction of total amount of pollutant discharge of this water area to administrative departments in charge of environmental protection	
		Publicize units that do not complete required control indicator of total amount of key water pollutants discharge	
		Propose requirements on restriction of pollutant discharge in emergency	
	Control on restriction of total amount of pollutant discharge	See details in monitoring and management of pollution discharge outlets	
Water quality monitoring	Monitoring	Monitoring frequency	
	Report and release information on water quality	Regularly release situation of quality in water function area	
	Compliance status of water quality in water function area	Water quality compliance rate in water function area	
	Report and warn if exceeding the standard	Report to related People's Government to take measures and warn to administrative departments in charge of environmental protection	
	Establish monitoring system	Coverage rate and capacity establishment of water quality monitoring	
Monitoring and management of pollution discharge outlets	Approval of pollution discharge outlets	Divide according to jurisdiction and approve according to laws	
	Registration of pollution discharge outlet	Registration of units set at pollution discharge outlet	
	Investigation on pollution discharge outlets	On - site investigation frequency, investigation scope etc.	
		Require pollution discharge unit to report information and statements related to pollution discharge outlet and statistics of pollution discharge of pollution discharge outlet of the last year	
	On - site law enforcement inspection	Inspection frequency, inspection scope, release and warning of results	
	Establishment of monitoring capacity	Establishment of measuring and monitoring facilities on pollution discharge outlets	
Application of assessment results			

4 Description of application of assessment system

4.1 Determination of assessment objects

Assessment system is the management system to evaluate, diagnose and continuously improve assessment objects, there are different assessment contents for different assessment objects, therefore, when applying assessment system, assessment subjects and assessment objects shall be determined clearly. According to related regulations of "Water Law of the People's Republic of China" on jurisdiction and responsibility of water resources management, this assessment system framework correspondingly divides assessment objects into two classes:

The first is that, assess basin management organizations for items directly charged by basin management organizations, and assessment content are mainly results of their control and management, including responsibilities of basin organizations, measures taken as per laws and so on.

The second is that, assess county-level and above local government and its water administrative department according to division of authority and responsibility, and assessment content are mainly their water resources protection measures taken in the local administrative region and implementation results.

"Opinion of the State Council on Implementing the Strictest Water Resources Management System" clearly points out that the State Council assesses the fulfillment of water resources protection indicators of each province, autonomous region and municipality directly under the Central Government. Referring to management of administrative region, the applicable assessment subjects and objects for this framework is shown as Tab. 2.

Tab. 2 Applicable assessment objects of assessment system of water quality and carrying capacity management in water function area

Assessment subject	Assessment object
State Council	The local People's Government of provinces, autonomous regions and municipality directly under the Central Government
Water administrative department of the State Council	Basin management organization
The local People's Government of provinces, autonomous regions and municipality directly under the Central Government	City level local People's Government
City level local People's Government	County level local People's Government

4.2 Assessment system needs further improvement

Establishment of responsibilities and assessment system of water resources protection is the premise and base of ensuring performance of assessment work. At present, there is no complete assessment and accountability system for water resources protection, and system basis and basic principles as guidance are in lack. Promotion and fulfillment of "red line of carrying capacity" needs urgently to formulate and implement strict assessment and accountability system, including the assessment method of the strictest water resources protection management system, implementation plan and statistical method of assessment work, monitoring method and other supporting documents, and pays attention to perfecting the mechanism for public participation, adopts many ways to listen to various opinions to further improve transparency of decision making.

4.3 On selection of assessment indicator

According to design principle of being comprehensive and systematic, this assessment system framework involves many assessment contents and assessment indicators, in the practical application of assessment in water function area at present, it may be not feasible for some assessment objects to list all indicators in the framework as assessment contents once. Therefore, it is suggested that applying departments can select some important indicators from this framework for assessment, according to assessment object's governed scope, specific responsibilities, requirements of strategic objectives, current water resources protection works and situation of capacity establishment, and if assessment indicators involve many factors, they can be clarified in additional factor level of indicator level. In the future, with the development of management in water function area, additional assessment indicators can be added correspondingly to gradually perfect assessment system. .

4.4 On application of assessment results

Application of assessment results is foothold of assessment system, correct application of assessment results can inspire, guide and constrain the actual work, which is objective of assessment. Therefore, in the relevant system, it shall clarify application form of assessment results through reward and punishment measures, so that assessment results can promote actual work in turn and assessment system becomes a complete and closed system.

5 Conclusion

Currently, in the early stage of implementing the strictest water resources management system, this paper primarily explores assessment and evaluation system of water quality and pollutant carrying management in water function area, proposes assessment system framework, and the application of this framework has positive effects on promotion of implementation of the strictest water resources management system. If complete assessment system is established in the following stage, it needs to further select assessment indicator, determine methods of assessment and evaluation, and continuously strengthen establishment of monitoring capacity and scientific and technological support.

Acknowledgements

This work was supported by the National Department Public Benefit Research Foundation (201001011).

References

Shang Chunjing, Zhang Zhihui, Li Xiaodong. Design of Evaluating Indicator System of Circular Economy for Construction Enterprises. 2009 (9) : 110-113.

Wang Chuanyu, Zhang Yamin, FengXirong, et al. Evaluation Index System Design of Port Enterprises' Energy-saving and Pollutants Reduction. 2010 (11): 45-47.

Techniques for High – efficient Utilization of Water Resources in Gully Areas of the Loess Plateau and Existing Problems

Yan Xiaoling[1], *Liu Haiyan*[1], *Song Jing*[2] and *Duan Bailin*[1]

1. Xifeng Supervision Bureau for Yellow River Water and Soil Conservation, Qingyang ,745000,China
2. Upper and Middle Yellow River Bureau, YRCC, Xi'an, 710021,China

Abstract: Presently it is an urgent task to scientifically and reasonably utilize water resources for the purpose of developing rural economy, improving the living standard of rural and urban people, maintaining the sustainable development of cities, controlling soil erosion, and improving the ecological environment of the gully areas on the Loess Plateau. The present situation of local water resources utilization is investigated and the techniques for high – efficient utilization of rural and urban water resources are introduced from the aspects of control and utilization of the runoff collected from plateau surface, slope surface and gullies, as well as the pilot project of rainwater collection for irrigation in Xincheng District of Xifeng City. Furthermore, suggestions are made, i. e. strengthening the construction of rainwater collection facilities to alleviate the water shortage in rural and urban areas, taking engineering and biological measures for groundwater recharge, attaching importance on water saving and pollution control to ensure water supply for industry and agriculture, and finding new ways to settle the water demand of petroleum production.

Key words: gully area of the Loess Plateau, high – efficient utilization of water resources, runoff resources, utilization of water resources for agriculture, utilization of urban water resources

Being a main landform area of the Loess Plateau, the gully area enjoys better natural conditions than other areas of the plateau. However, because of both natural and human reasons, the gully area of the Loess Plateau has been suffering from serious water loss and soil erosion for a long time, accompanied with frail ecological environment. This area is facing not only arduous task of development, but also serious problems of environment control, all of which require large quantity of water. Water loss and soil erosion affect the reasonable and effective utilization of water resources. Exploitation and utilization of water resources need water saving and protection, which are of the same importance for sustainable utilization of water resources. The gully area of the Loess Plateau is struggling against deficient water resources, serious water loss and soil erosion, and low agricultural productivity. Reasonable and high – efficient utilization of water resources are becoming more and more necessary, in order to help the peasants cast off poverty and become better off and to promote the improvement of ecological environment.

1 Present situation of water resources utilization in the gully area of the Loess Plateau

1.1 Shortage of water resources

The gully area of the Loess Plateau is located in the hinterland of the Loess Plateau in west China, where there is a serious lack of water resources. Standing in the central part of the gully area, Xifeng District has annual mean water resources of 2.29×10^8 m^3 and utilizable water resources of 0.80×10^8 m^3. Water resources per capita is only 260 m^3/a, which is one – ninth of national level and one – sixth of the provincial level. Xifeng District has low groundwater table, making water utilization difficult and expensive. In Xifeng District, the industrial and agricultural production and people's daily life mainly relies on groundwater. However, local groundwater table is declining continuously and water funnels have appeared in some regions because of overexploitation of groundwater for quite a long time. Local annual mean rainfall is 561 mm, which represents the precious water resources. However, affected by the monsoon, the rainfall shows spatiotemporal misdistribution. More than 60% of annual rainfall comes in form of rainstorm or continuous rainfall mostly in the period from July to September, which makes there drought in the beginning and the end of a year and excessive rain in the middle of the year. Frequency appearance and rotation of drought and excessive rain badly endangers the industrial and agricultural production. Meanwhile, Xifeng District is suffering from serious water loss and soil erosion. Rain wash and flood erosion lead to forward movement of gully head, expansion of gully banks, downcut of gully, and year – by – year decrease of plateau surface. The process of urbanization and industrialization is speeding up, so large quantity of water is required for urban industry, construction, tree and grass planting, and urban beautification. Existing water resources are far away from meeting the demand of production and daily life. Water shortage makes agricultural production depend on rainwater, which restricts the production activities and daily life of local peasants and the development of local economy. Plateau surface is covered by dry soil due to water shortage, making trees and grass difficult to survive and grow, and ecological environment difficult to get recovered naturally. Therefore, shortage of water resource is the bottleneck problem for the social and economic development in the gully area of the Loess Plateau.

1.2 Sharp contradiction between supply and demand, and difficulties of development and utilization of water resources

Along with the urbanization progress, there emerges a great upsurge in construction of various development zones, together with large – scale infrastructural construction and real estate development. Besides, population growth and industrialization progress, especially large – scale exploitation of petroleum resources in the gully area of the Loess Plateau, require increasing quantity of water. It brings heavier negative influence on efficient utilization of resources, ecological environment and water, and makes the imbalance between water supply and water demand become more serious. Groundwater, which is essential to human existence, is being overexploited and most of the water resources easy to be exploited and utilized have already been put into service. Furthermore, the management system of water resources is in a decentralized state. All the abovementioned jointly produce many negative factors for efficient utilization and management of water resources, and make further development and utilization more difficult.

2 Techniques for water resources utilization

2.1 Techniques for utilization of agricultural water resources

In the gully area of the Loess Plateau, soil water is a kind of important agricultural water resources. It possesses the basic characteristics of water, but differs from the gravity water. Soil water can not be dispatched and exploited. It can only be absorbed by plant in – situ and directly goes back to the atmosphere through evaporation. Basically soil water is naturally utilized, not requiring large construction investment. Soil water is undoubtedly an important or even the only kind of agricultural water resources in the regions lacking of surface water and groundwater. In the north China, soil water makes up 60% ~ 70% of the rainfall resources, generally 360 ~ 420 mm. In the growth period of wheat, soil water consumption can amount to one – third of the total water consumption (Zhao Ancheng, Li Huaiyou, et al, 2006).

2.1.1 Techniques for control and utilization of runoff resources

The rainwater runoff in Xifeng District shows discontinuous and unstable spatiotemporal distribution, and large quantity of rainwater runoff is wasted in vain. In order to alleviate the deficient water supply for agriculture and ensure the water source for agricultural production in arid regions, the rainwater runoff is partially collected and utilized by three steps, i. e., runoff collection – accumulation and storage – high – efficient utilization. By such means, the intermittent and discrete rainwater runoff is transformed into a stable system capable of continuous water supply, and rainwater can be collected from a large area and then used for a small area. Application of this technique not only effectively controls water loss and soil erosion, but also alleviates water shortage for agriculture, which is favorable for agricultural environment improvement and high - efficient agriculture development. Through collection, storage, and high - efficient utilization of rainwater runoff, artificial control and redistribution of rainwater resources are realized to attain the purposes of actively controlling water loss and soil erosion, improving the rainfall utilization rate, and alleviating water shortage for agriculture.

2.1.1.1 Control and utilization of water resources on plateau surface

Through hardening road surface, roofs and courtyards, rainwater is collected and stored in water cellars and ponds, serving for developing courtyard economy. The gully area of the Loess Plateau is the most favorable place for culture of high - grade apple and the main producing area of apple. However, shortage of water resources restricts both the yield and the quantity of the apple in the area. It is encouraged to apply water - saving techniques such as drop irrigation, micro irrigation, and sprinkler irrigation to the orchards in the whole area, aside from popularizing the techniques of film mulch, grass mulch, interplant, and sward, etc.

The water collection and storage system in courtyard consists of three parts, i. e. rainwater collection from courtyard and roofs, rainwater harvesting, and runoff storage. Rainwater is mainly collected from the water harvesting surface such as roofs, courtyard ground, and surrounding road surface. On the one hand, because of the scattered distribution of local residents, the land around the courtyards can be rebuilt to improve the efficiency of rainwater collection; on the other hand, the materials of water harvesting surface can also be improved to achieve better rainwater collection performance.

In the gully area of the Loess Plateau, the rural residence always has the roofs covered by concrete, machine - made tiles, cement tiles or grey tiles. The roofs stay away from human pollution, so the rainwater collected from the roofs has low content of various noxious substances, basically satisfying the national standards for potable water (Gao Jianling, Zhao Ancheng, et al, 2010). However, the rainwater collected from courtyard ground always mixes up with other pollutants, resulting in rainwater polluted. Such water is undrinkable but available for irrigating the orchards and vegetable greenhouses in courtyards, thereby making contributions to courtyard economy.

It is an engineering measure to construct the mini hydraulic works which specially serve for collection, storage, control and utilization of rainwater. Through the procedures of 'collection → storage → delivery → irrigation or potable water for people and livestock', rainwater runoff is transformed into the water for living and production in the rural areas, making it possible to develop courtyard economy and provide water - saving irrigation for crops, trees and grass, as well as to properly develop livestock breeding and processing industry. The rainwater hydraulic works mainly consist of rainwater collection works, water storage works, water supply works, and irrigation facilities.

There are mainly six modes for high - efficient utilization of the runoff in the gully area of the Loess Plateau:

(1) Road in village → gravel pavement → rainwater collection from road surface → water storage in water cellar and pond → low - pressure pipe irrigation system.

(2) Hardening treatment of courtyard → rainwater collection → water storage in water cellar → drop irrigation system in orchard

(3) Hardening treatment of courtyard → rainwater collection → water storage in water cellar → livestock in warm shed → biogas generating pit → drop irrigation system in orchard

(4) Hardening treatment of courtyard → rainwater collection → water storage in water cellar → livestock in warm shed → biogas generating pit

(5) Alley way diversion → alley way block - up → runoff storage area (low - lying land, waste pumping wells, reclaimed land from original alley ways) → constructing ponds at gully head, and ridge along gully → path of collected water → avoid runoff flowing into gully from gully head.

(6) Catch canals on plateau surface, or gently - rolling land → level terrace → planting

yellow day lily or grass at terraces → tillage for water and soil conservation.

2.1.1.2 Control and utilization of water resources on slope surface

Along with the implementation of Western Development, Project of Returning Farmland to Forestry (Grass Planting), and Ecological Project of the Yellow River for Water and Soil Conservation, many of the original barren mountains and hills become green. In the gully area of the Loess Plateau, where the rainfall is 500 ~ 700 mm, large area of the original barren mountains and hills is covered by a great variety of trees and shrubs, economy trees and high - grade pastures. Water - collecting techniques are taken for forestation and grass planting. Under the principle of planting proper trees at the places with different landforms, a number of three - dimensional structures are set up by taking the most favorable techniques on forestation, involving mixed forests of arbors and shrubs, broadleaf trees and coniferous trees, heliophilous species and skiophilous species, and planting forage grass in the blank areas of woodland. Water harvesting slopes, level ditches, horizontal stages, and narrow terrace which can impound the water, are also built, and trees and grass can grow within such retaining works. The plants such as trees and grass effectively impound rainwater, slow down runoffs and winds, and retain the surface soil. These retaining works not only maximize the collection and utilization of rainwater, but also improve the utilization rate of water resources and the survival rate of trees and grass.

The growing leaf canopies of tress can prevent raindrops from directly hitting land surface, which not only weakens splash erosion to soil caused by raindrops, but also traps part of rainfall. Rotten litter of woodland forms a layer with loose structure, which has fine water - absorbing capacity and water permeability. Root system of trees can fix soil, alleviate shallow landslide, and avoid collapse of riverbank or reservoir shore caused by wave impact. Large area of woodland or grassland can condition micro climate by increasing air humidity, regulating air temperature, and reducing soil evaporation, etc. Besides the function of safeguarding, forest trees can make economic contributions by producing a certain quantity of wood, fuel, fertilizer, forage and oil plants.

To transform cultivated hillside field into terraced field is a key technique aiming at high - efficient utilization of rainfall for water and soil conservation and ecological environment construction. After transformation from cultivated hillside field to terraced field, change of original micro landform takes place, making the original inclined field surface turn to be flat. The field surface with gentle gradient has better infiltration velocity than that of hillside field, so raindrops directly infiltrate into the soil and become soil water instead of forming runoff on field surface easily, thereby reducing soil erosion. Meanwhile, field ridges blocks the hillside surface, making the slope length shorter and gathered runoff less. Furthermore, soil layer becomes thicker after the transformation, which improves the water holding capacity of soil.

By building orchards in hillside fields and taking the techniques of film mulch, grass mulch, interplant, and drop irrigation, utilization of water resources is maximized to increase income, prevent water loss and soil erosion, and obtain satisfactory social, economic and ecological benefits.

2.1.1.3 Control and utilization of water resources in gullies

Techniques for control and development of gullies and for optimization and construction of dam systems: Control and development of gullies is the key mode for utilizing the rainwater runoff on the Loess Plateau, and it is an essential part of the ecological environment construction for water and soil conservation on the Loess Plateau in the 21st century. Some scientific findings and rich experience have been achieved in this field in the past many years. Medium and small - size warping dams are built to control gullies section by section, and finally gullies are sealed off by key dams. In the early stage, the impounded water is used for irrigation, and in the later stage, the silt storage can serve for crop cultivation.

Modes for utilization of water resources in gullies include "spring water or the water collected by the key gully - erosion control works - water pumping by small - size high - efficient facilities - water storage by water tower on plateau surface - water delivery through pipelines - potable water for people and livestock", "spring water or the water collected by the key gully - erosion control works - water pumping by small - size high - efficient facilities - water storage by water tower on plateau surface – water delivery through pipelines – water – saving irrigation system for orchards", "check dam – warping dam – small reservoir – fish culture or lifting water up to plateau surface for vegetable plot irrigation or for the water – saving irrigation system of orchards", "check dam – warping dam – key dam – protection forest on gully banks – protection forest in gully bed" for sediment trapping, "laterite debris surface – land preparation by fish – scale pit – scrub forest (mainly seabuckthorn)", "mesa – level terrace – portable irrigation equipment – potable water for people and livestock – planting melons and vegetable by plastic film mulching" for economic development, "spring water or stream runoff – water pumping by small – size high – efficient facilities – water storage by head tank at the halfway of hillside – water delivery through pipelines – potable water for people and livestock – water – saving irrigation system for orchards in hillside fields or for cash crops" (Gao Jianling, Zhao Ancheng, et al., 2010).

2.1.2 Techniques for development of agricultural water resources

Through regulating soil water reservoir reasonably and high – efficiently, runoff resources are utilized to cover the shortage between water demand of crops and water supply potential of soil, so as to improve the utilization efficiency of rainfall. The systematized techniques for overall development of water resources consist of four parts, which are techniques for high – efficient utilization of rainwater, techniques for maintenance of soil water reservoir, techniques for rainwater collection and runoff development, and techniques for supplementary irrigation, respectively.

The systematized techniques for overall development of water resources tend to solve three problems in water resources utilization, i. e. "cropping accommodating to rainfall and avoiding drought" to improve rainfall utilization (to settle the contradiction between the spatiotemporal misdistribution of water supply and the water demand of crops), "maintenance of soil water reservoir" to improve water supply potential of soil, "techniques for rainwater collection and runoff impoundment" to develop runoff resources, and "techniques for water – saving supplementary irrigation" to improve the utilization efficiency of water resources. Based on studying the relationship between 'the spatiotemporal distribution of local rainfall, runoff, and soil resources' and 'the growth of crops, trees and grass', the systematized techniques for comprehensive utilization of rainfall, soil water, and runoff (secondary water resources) are put forward in order to coordinate and efficiently utilize these three kinds of water. Detailed measures include cropping accommodating to rainfall and avoiding drought by rearranging the layout structure of crop production; protecting and regulating the continuous water supply potential of soil by taking biological measures and engineering measures; developing and utilizing secondary water resources; and improving the performance of water resources utilization and further promoting the ecological, social and economic development in the control area.

It is of great significance to popularize rainwater collection and water – saving irrigation works to develop courtyard economy in the gully area of the Loess Plateau. To build rainwater cellars under the financial and technical support from the government and technology departments makes great contributions not only to potable water of the people and livestock in dry and rainless regions, but also to the agricultural production irrigated by the rainwater collected in water cellars. Studies show that annually 6.27×10^8 m^3 of runoff is collectable from part of the land used for residence, mining area and traffic on the Loess Plateau in Shaanxi Province. It is preliminarily estimated that the collectable runoff on the Loess Plateau amounts to 28.8×10^8 m^3. If half of it is used for grain production, there would be an increase of 28.8×10^8 kg in yield, which can make the total grain yield, 50×10^8 kg, on the Loess Plateau increase by around 30%.

2.1.3 Techniques for water – saving irrigation

Water – saving irrigation is to minimize the water loss in the course of "water source – water delivery – water distribution – irrigation and water consumption of crops", and to maximize the crop yield and production value contributed by per unit of water consumption through effectively utilizing natural rainfall and irrigation water and taking necessary measures regarding water resources, engineering, agriculture and management. Water – saving agriculture generally involves four aspects, including reasonable development and utilization of water resources; saving water during water delivery, water distribution, and field irrigation; techniques for water saving and yield increase; and management of water utilization and water - saving. Water - saving irrigation refers to flood irrigation, furrow irrigation, border irrigation, sprinkler irrigation, and drip irrigation, etc. Border irrigation consumes water 15% ~20% less than flood irrigation, sprinkler irrigation consumes water 60% less than canal irrigation (furrow irrigation and border irrigation), and drip irrigation consumes water 30% less than border irrigation and making a yield increase of 30% ~50%. Drip irrigation makes the overall effective utilization of the delivered water come to 95% ~97%, which is 30% ~40% higher than that of canal irrigation and requires floor area can reduced by 1% ~3%.

2.2 Collection and storage of rainwater in urban area

While the new urban area of Xifeng District, which is located in the gully area of the Loess Plateau, was being constructed in 2003, the successful experience of rainwater utilization in rural area was further applied to this urban area in order to fully utilize natural rainfall, alleviate the load of flood control, inhibit water loss and soil erosion on the plateau surface, improve regional ecological environment, and promote the sustainable economic and social development in the gully area of the Loess Plateau as well. To collect and store natural rainfall is an important approach for settling water resources shortage, which transforms rainstorms and flood disasters into water resources. The Tianhu Project, a pilot water project against drought by rainwater collection, began to be built in the new urban area of Xifeng City. After completion of the project construction, three reservoir lakes in different size were built, having gross water storage capacity of 1.328×10^5 m^3 and year - to - year capacity of 4×10^5 m^3. This project brings "four transforming" to utilization and management of rainwater and floodwater in urban area, i. e. transforming flood control to flood utilization; transforming discharge of rainwater and floodwater to storage of rainwater and floodwater, changing the wastes into valuables; transforming scattered small - size hydraulic works for rainwater utilization in rural area to the concentrated large - scale hydraulic works in urban area; and transforming hillside control, gully control and gully head protection for water and soil conservation to water origin control. It brings about remarkable social, economic and ecological benefits. Meanwhile, the project of collecting and storing rainwater and floodwater for ecological environment protection was constructed in Xifeng District

The project is of great importance and significance for collection of surface rainwater and floodwater in the urban area of Xifeng District, as well as for protection of Dongzhiyuan area and urban flood control, comprehensive utilization of water resources, water and soil conservation in urban area, improvement of ecological environment, and irrigation for agricultural production. Dongjiaohu Lake, Xihu Lake, Qingyanghu Lake, and Beihu Lake were built at the proper places on the four directions of Xifeng City, and Donghu Park was rebuilt. These five lakes were connected up by an artificial canal around the city, thereby forming an integrated system for storage and utilization of water resources on the plateau surface.

After completion of the project construction, the total water area comes to 131 hm^2, with total storage capacity of 2.87×10^6 m^3 and year - to - year storage capacity of 6.85×10^6 m^3, equivalent to the original water consumption for two years in urban area. With the help of high - efficient and water - saving irrigation measures taken in suburbs, the water required by irrigation of 6,667 hm^2 of farmland can be fulfilled. The price of the water lifted up to Xifeng form Bajiazui Reservoir, which is 15km away from the urban area, through nine stages of pumping stations is 1.06 Yuan/m^3, so 6.85×10^6 m^3 of water can save 7×10^6 Yuan. Moreover, the project can prevent 1.18×10^6 m^3 of sediment from flowing downstream by erosion, which is equivalent to the sediment retained by 100 medium - size warping dams in a year. Furthermore, the water area of the city becomes larger, which changes the regional climate and reduces the costs for flood discharging works in urban area. Therefore, rainwater and floodwater are collected and utilized, which brings about considerable benefits and promotes the sustainable social, economic, and ecological development in Xifeng District. Meanwhile, the project becomes a part of the urban and rural water source system involving collection and storage of rainwater and floodwater and power irrigation, thereby realizing high - efficient utilization of urban rainwater and floodwater.

3 Existing problems about water resources utilization in the gully area of the Loess Plateau

3.1 Serious damage of water resources and heavy task for its control

There are two main rivers in Xifeng, i. e. Puhe River and Malianhe River. Along with the rapid industrial development and population growth, discharge of wastewater and sewage are increasing greatly year by year. Presently the water of Malianhe River already can not be used for irrigation and drinking. Puhe River is also suffering from pollution, making some water quality indexes exceed the limit. The groundwater, which is essential to people life, is facing the potential threats of the pollution caused by petroleum exploitation. It is urgent to take control measures as soon as possible, otherwise water pollution will not only directly damage local water resources and affect the sustainable development of water resources and local economy and society, but also affect people's health and living standard improvement. Therefore, there is a hard long way to go.

3.2 Serious water loss and soil erosion threatening the city safety

Because of the less quantity of vegetation, high density of buildings, and large area of impermeable underlying surface made of cement or asphalt in urban area, rainfall quickly transforms into surface runoff which gathers and flows away in vain, not only resulting in the waste of large quantity of resources, but also producing strong destructive power. The Huoxianggou Gully, which is in the northeast of Xifeng City, is the discharge port of the original urban district. Long - term centralized water discharge induces the gully head moving forward and gully banks expanding rapidly, which directly threatens the safety of the city and the lives and properties of local people.

3.3 Overexploitation making continuous decline of groundwater table

Groundwater recharge decreases sharply because more and more urban area is hardened, and moreover, the domestic water of urban population also depends on groundwater, making urban groundwater table have a decline of around 0.7 m each year. Water funnels have appeared in some regions. The ecological environment is damaged seriously, which endangers the living conditions of local people.

3.4 Weak awareness of water saving and serious waste of water

Local people do not have well understanding of water resource. They just think groundwater is inexhaustible in supply and always available for use. In addition, recycle and treatment of sewage and wastewater has not been popularized in urban area. Most of the people have weak awareness of water saving and waste of water is commonly seen in every industry and trade, making water shortage aggravated by human factors.

3.5 Difficult to make unified management of water resources

In the management system of water resources, the classification of rights and responsibilities is not clear between the administrative authorities and the undertaker of development and utilization. It is difficult to harmonize the management of surface water and groundwater. Water supply, water consumption and water discharge need to be coordinated in management. It is difficult to put the management system for the following items into effect, i. e. exploitation and utilization of water resources, water pollution control, and water resources protection. These factors restrict unified and reasonable development and utilization of water resources, and affect the sustainable development of water resources.

4 Suggestions

4.1 Strengthening facility construction for rainwater collection and storage to alleviate water shortage in urban and rural areas

It is necessary to mobilize the whole society to actively construct the facilities for collection and storage of rainwater. The construction shall be under unified planning and in accordance with local conditions. The integrated network consisting of the facilities with various size and scattered distribution shall be built for flood control and water storage. Rainwater shall be retained and utilized in – situ. By means of rainwater collection and storage, it becomes possible to adjust the spatiotemporal misdistribution of rainwater and transform rainwater into utilizable resources. This is a new way for development and utilization of water resources. It helps to setting up the water – saving society, featured by harmonious relationship between people and water and cyclic utilization of resources, to alleviate the contradiction between demand and supply of water in the gully area of the Loess Plateau.

4.2 Taking engineering measures and biological measures for groundwater recharge

Firstly, pollutant discharge system and water drainage system shall be built separately during urbanization construction and highway construction. The urban network for discharge or impoundment shall be built in accordance with urban geographic features in order to accumulate and utilize rainwater as much as possible. Secondly, the area of green vegetation and permeable ground shall be enlarged to encourage rainwater infiltration. Finally, deep wells and infiltration basins shall be built at proper places of the urban area in order to effectively recharge the groundwater by increasing rainwater infiltration capacity.

4.3 Attaching importance to water saving and pollution abatement, and solving the problems of industrial and agricultural water resources

Besides water – saving propaganda, it is necessary to call all the people to save water by making supporting policies, rules and regulations. Under the principle of "water saving foremost and pollution abatement fundamental", the development scheme for water – saving economy and society need to be worked out as soon as possible with the purpose of improving the utilization rate of water resources and alleviating the shortage of industrial and agricultural water resources.

4.4 Finding new ways to settle the water demand of petroleum production

There are rich mineral resources, including large quantity of petroleum and coal reserves, in the gully area of the Loess Plateau. In recent years, it is true that the large – scale exploitation of petroleum and coal resources drive the development of local economy. However, it is incontestable that water loss and soil erosion and environmental disruption come along with recourses exploitation.

Regional economy has improved a lot since Xifeng Oilfield went into production in 2003. Presently, more than 800 oil wells are drilled in a range of about 200 km^2 in Xifeng, annually producing 0.5×10^6 t of crude oil. The petroleum production contributes 1×10^8 yuan of tax revenue to Xifeng annually. Meanwhile, 15 thousands of workers move into Xifeng, which induces the development of local service sector (the third industry). Investigations show that to build an oil well consumes 3,000 m^3 of water, and to product 1t of crude oil requires 5 m^3 of water. Present water consumption in the oilfield is 0.9×10^6 m^3/a for building new oil wells and 2.5×10^6 m^3/a for producing crude oil, with annual water consumption If 3.4×10^6 m^3. Water supply depends on the loess phreatic water. The total loess phreatic water in Xifeng petroleum area amounts to about 2×10^6 m^3/a, of which, 1.5×10^6 m^3/a is allocated for urban water supply. If 3.4×10^6 m^3 is required by petroleum production, it is obvious that such a water demand exceeds the exploitable water resources. The fact is that petroleum production is scrambling for the water resources against urban water supply, which not only seriously affects the urban water supply, but also brings potential pollution to groundwater. Recently, the oilfield has begun to draw deep confined water to satisfy the water demand. In addition, each year 200 to 300 oil wells are built in Xifeng Oilfield, and petroleum production increases by 0.2×10^6 t annually. Such rapid development means a great challenge to the water resources in the arid regions. Thus, it is imperative that new ways must be found to settle the water demand of petroleum production.

Along with the rapid development of urbanization, continuous growth of agriculture and economy, and progress of new countryside construction, the problems about water resources becomes serious day by day. Presently, under the condition of water shortage, it is an urgent task to scientifically and reasonably utilize natural water, surface water, and groundwater to maintain the sustainable development of urban construction and industrial and agricultural production.

Acknowledgments

This study was funded by S&T Promotion Program by Ministry of Water Resources, China (No. TG1017).

References

Zhao Ancheng, Li Huaiyou, et al. Study on Technologies for Control and Utilization of Water Resources in Gullied Rolling Loess Area [M]. Zhengzhou: Yellow River Conservancy Press, 2006.

Gao Jianling, Zhao Ancheng, et al. Study on Comprehensive Control Mode for Gullied Rolling Loess Area Based on Regulation and Utilization of Runoff [M]. Beijing: China Water Power Press, 2010.

Study on Optimizing the Import Water Project of Diversion Project to Alleviate Eco-environmental Influence: a Case Study of Water Source Project in Guyuan Region of Ningxia Province

Yang Yuxia[1], *Huang Jinhui*[1], *Hao Fuqin*[2], *Ma Xiumei*[1], *Yan Li*[1], *Hao Yanbin*[1], *Ceng Wei*[1] and *Zhang Shikun*[1]

1. Yellow River Water Resources Protection Institute, Zhengzhou, 450004, China
2. Yellow River Basin Water Resources Protection Bureau, Zhengzhou, 450004, China

Abstract: Urban and rural drinking water safety project in Guyuan region of Ningxia province involves headwater region of Jinghe river, and the Provincial Scenery District in Jinghe river headwater, and Liu panshan Natural Protection Region. The ecological environment in this region is very fragile. The quantity of water diversion is 3.98×10^7 m^3 of the multi-year average. This project will require larger amount of water in dry period of the long time average, the ratio of water diversion volume in dry season is 30.7% of total water quantity. There are nine diversion dams and the amount is too many. Eco-environmental water demand of Jinghe River cannot be reach to 10% of the long time average runoff, especially in dry season. This project will take great influence on the regional eco-environment. By optimizing the Import Water Project, it is put forward increase 50% of the water diversion volume in flood season & decrease 30% of the water diversion volume in dry season of Jinghe River and the Cedihe River respectively, reduce quantity of diversion dams obviously, ensure eco-environmental water demand, This study alleviates unfavorable influence of the project on eco-environment of headwater region of Jinghe River, Especially to channel Habitats in dry season.

Key words: eco-environmental influence, optimize, import water project, Guyuan region of Ningxia province, drinking water safety project

1 Outline of project

Water source project in Guyuan region of Ningxia province was located in Middle South Ningxia Province, which in Guyuan city. This project draws water from Jinghe River mainstream & the branches of Cedihe River, Xiehe River and Nuanshuihe River. To central south Arid Area of Ningxia in order to solve the problem of Urban and Rural Drinking Water. Water source project in Guyuan region of Ningxia province involves headwater region of Jinghe River, and the Provincial Scenery District in Jinghe River headwater, and Liupanshan Natural Protection Region. The ecological environment in this region is very fragile.

This project is composed of Longtan Reservoir、Zhongzhuang Reservoir、Nuanshuihe Reservoir, and three booster pumping stations, five diversion dams and ten Tunnels. Water Supply population of 2009 was 1.108×10^8 and population of 2025 for Water Supply was more than 1.31×10^8. Water diversion of 2025 is 3.98×10^7 m^3 of the multi-year average.

2 The original import water project and problems

2.1 The original import water project

2.1.1 The diversion process:

In the case of typical year, the monthly cited water of design average year 2025 is shown in Tab. 1, and the monthly diversion process is listed in Fig. 1.

Tab. 1 **The monthly diversion of the original Import Water Project of 2025 at the different design frequency**

Design frequency	The monthly water diversion($\times 10^4$ m^3)												
	Jan	Feb	Mar	Apr	May	June	July	Aug	Sept	Oct	Nov	Dec	Total
the multi-year average	225	188	220	299	357	339	430	464	487	381	340	250	3,980
$P = 20\%$	170	145	266	385	441	397	591	667	355	262	488	442	4,609
$P = 50\%$	264	181	187	243	309	251	464	653	388	340	357	265	3,901
$P = 75\%$	180	152	191	228	362	373	584	477	128	108	221	146	3,150
$P = 95\%$	174	156	152	163	207	250	209	191	83	88	178	140	1,991

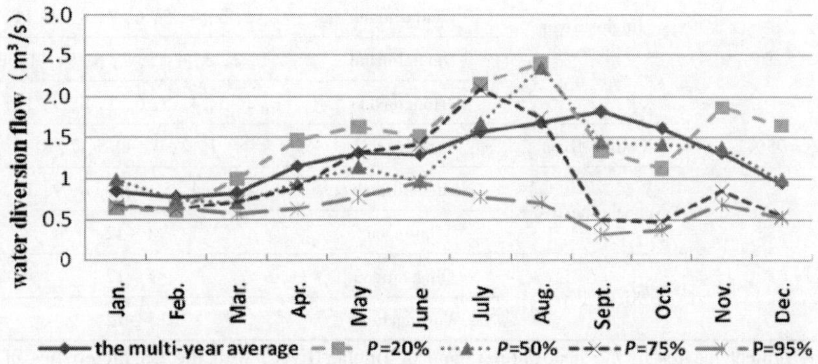

Fig. 1 The monthly diversion process of the project at the different design frequency

2.1.2 Eco-environmental water demand

Tab. 2 Analysised Satisfaction degree of the monthly eco-environmental water demand at the different design frequency

Design frequency	River	Diversion sections	Unsatisfiable months
The multi-year average	Jinghe River mainstream	Longtan Reservoir	1, 2, 3
		Landazhuang	1, 2, 3, 4, 12
		Huanglinzhai	1, 2, 3, 12
		Hongjiaxia	1, 2, 3, 12
	Cedihe River	Shizuizi	2, 3
	Xiehe river	Qingshuigou	1, 2, 3, 5, 12
		Woyangchuan	1, 2, 3, 5, 6, 11, 12

Continued Tab. 2

P = 20%	Jinghe river mainstream	Longtan Reservoir	1, 2, 3, 5, 6, 11, 12
		Landazhuang	1, 2, 3, 5, 6, 11, 12
		Huanglinzhai	1, 2, 3, 4, 5, 12
		Hongjiaxia	1, 2, 3, 4, 5, 12
	Cedihe River	Shizuizi	1, 2, 3, 5, 6
	Nuanshuihe River	Nuanshuihe river	1, 2, 4
		Baijiagou	1, 2
	Xiehe River	Qingshuigou	1, 2, 4
		Woyangchuan	1,2,4
P = 95%	Jinghe River mainstream	Longtan Reservoir	1, 2, 3, 4, 5, 7, 8, 11, 12
		Landazhuang	1, 2, 3, 4, 5, 6, 7, 8, 9, 11, 12
		Huanglinzhai	1, 2, 3, 4, 5, 7, 8, 9, 11, 12
		Hongjiaxia	1, 2, 3, 4, 5, 6, 7, 8, 9, 11, 12
	Cedihe River	Shizuizi	1, 2, 3, 4, 5, 7, 8
	Nuanshuihe River	Nuanshuihe river	1, 2, 3, 4, 5, 6, 7, 8, 9, 11, 12
		Baijiagou	1 ~ 12
	Xiehe River	Qingshuigou	1 ~ 12
		Woyangchuan	1 ~ 12

The project locates in the headwater region of Jinghe River, with the Characteristics of small flow and low sediment concentrations et al. So the eco-environmental water demand is considered mainly about the ecological basic flow to maintain ecological functions of the river. According to the characteristics and hydrological datas of the river, Eco-environmental water demand was calculated by Tennant method. Thus, according to eco-environmental water demand determined by this paper and the original Regulating Calculation process, Satisfaction degree of the monthly eco-environmental water demand at different design frequency was analysised. The results are shown in Tab. 2. Eco-environmental water demand could not met mainly in dry season. The number of unsatisfiable months is increased significantly at 95% frequency.

2.1.3 Diversion dams layout:

The project is mainly composed of diversion dams and Storage Project. The original Import Water Project has 9 diversion dams and showed in Tab. 3 (which contains no Longtan Reservoir and Nuan shuihe Reservoir), Zero flow from Upstream diversion dams to Nuanshuihe Reservoir will be Occurred because cuts branches in Upstream Nuanshuihe Reservoir in dry period, and will Products greater unfavorable Influence to eco-environment.

2.2 Problems of the original import water project

From Tab. 1 and Fig. 1, we can see that the peak appeared not only in the flood season, but also in the dry season of Nov. and Dec., The original Import Water Project was required larger amount of water in dry period of the long time average. Comparing The quantity of Water diversion of of Jan. and Feb at the different design frequency, The quantity of Water diversion at $P = 95\%$ was slightly larger than $P = 20\%$. The ratio of water diversion volume at $P = 95\%$ was bigger. In Tab.

2, The monthly eco-environmental water demand of Jinghe River cannot be ensured at different level. The number of months, which eco-environmental water demand could not be met, was increased significantly. The quantity of diversion dams was too much. The original Import Water Project will have great effect on the headwater region eci-environment of Jinghe river.

Tab. 3 The diversion sections and related water system

Serial number	Diversion sections	Related water system
1st	Landazhuang	The first branch of Jinghe River
2nd	Huanglinzhai	The third branch of Jinghe River
3 rd	Hongjiaxia	The Second branch of Jinghe River
4 th	Shizuizi	Cedihe River mainstream
5 th	Toudaohe River (including Toudaohe River、Chagou and Xihe River)	The third branch of Nuanshuihe River
6 th	Baijiahe River	The first branch of Nuanshuihe River
7 th	Baijiagou	The first branch of Nuanshuihe River
8 th	Qingshuigou	The first branch of Xiehe River
9 th	Woyangchuan	Xiehe River mainstream

3 Import water project optimization

Facing the problems of the original design, the Import Water Project was optimized following three principles: ①the principle of water diversion according with runoff quantity. ②Priority should be given to eco-environmental water demand, and then considering cited water. ③optimizing the layout of diversion dams, reducing the number of diversion dams.

3.1 Diversion process

According to the above principles to optimize the Import Water Project. The monthly water diversion is showed in Tab. 4 and Fig. 2. Which increase 50% of the water diversion in flood season & decrease 30% of the water diversion in dry season of Jinghe river and the Cedihe River respectively. Diversion process is more reasonable after optimized.

Tab. 4 The monthly water diversion of 2025 at the design frequency after optimizing

Design frequency	The monthly water diversion ($\times 10^4$ m^3)												
	Jan.	Feb.	Mar.	Apr.	May	June	July	Aug.	Sept.	Oct.	Nov.	Dec.	Total.
The multi-year average	175	144	172	266	341	323	495	544	560	401	330	229	3,980
$P=20\%$	116	98	218	354	412	348	678	802	915	518	459	414	5,332
$P=50\%$	204	125	125	184	265	201	598	757	782	602	384	266	4,494
$P=75\%$	118	97	133	171	327	334	793	450	254	201	221	126	3,225
$P=95\%$	101	89	81	92	139	184	141	121	117	272	155	108	1,600

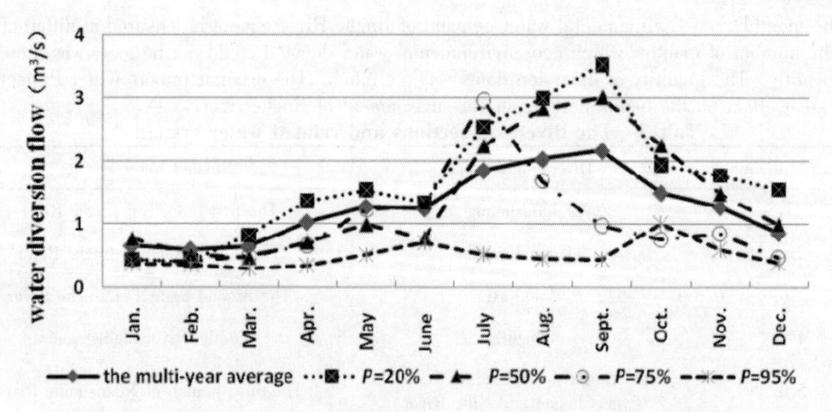

Fig. 2 Diversion process at the design frequency after optimizing

3.2 Eco-environmental water demand

Eco-environmental water demand was calculated by Tennant method and was equal to 10% of the long time average runoff. Priority should be given to eco-environmental water demand, and then considering cited water. Eco-environmental water demand can be completely satisfied.

3.3 Diversion dams layout

Diversion dams with smaller water diversion were cancelled, including Landazhuang, Huanglinzhai, Toudaohe River and Baijiahe River. The number of Diversion dams is five afer optimizing. water diversion of cancelled Diversion dams of Jinghe river was cited from Longtan reservoir, and water diversion of cancelled Diversion dams of Nuanshuihe river was cited from Nuanshuihe reservoir Which is decreased the influence scope of channel Habitats.

4 Concluding remarks

The project is located in the headwater region of Jinghe river. The ecological environment in this region is very fragile. The original Import Water Project was only Considered Project Investment and ignored eci-environment problems lower of diversion sections of Jinghe River basin. After optimized the import water project, eco-environmental water demand is ensured, diversion dams is reduced, water diversion is increased in flood season and water diversion is decreased in dry season. This study alleviates unfavorable influence of the project on eco-environment of headwater region of Jinghe River, Especially to channel Habitats in dry season.

Implement the Ecology – oriented Strategy in Xiaokaihe Irrigated Area and Strive to Create Ecological Water Conservancy

Wang Jingyuan, *Fu Jianguo*, *Pang Qihang*, *Liu Cuili* and *Bao Jianping*

Binzhou Xiaokaihe Yellow River Irrigation Bureau, Binzhou, 256600, China

Abstract: The administration in the Xiaokaihe Yellow River – diversion Irrigated Area has gradually explored and accumulated a set of effective utilization and treatment measures of sediment in practices, always adhering to the ecological oriented strategy, transforming from passive defense to active action, thus established a integrated utilization system of sediment from the Yellow River, solved the environmental degradation problem caused by the Yellow River sediment, and fully utilized sediment resources to benefit local economy by the following measures. The administration in the irrigated area focused on water conservation and storage, enhances soil and water conservation, and has been involved in protection and improvement of ecological environment.

Key words: water saving, water storage, greening, beautifying, sediment utilization

The irrigated area closely integrates natural ecosystems and artificial ecosystems; the administration in the Xiaokaihe Irrigated Area has developed a technology – oriented, resource – saving and ecological protection modernized irrigated area, by changing the time and spatial distribution of water resources in the irrigated area through engineering measures and taking water resource as a carrier, to balance relations between local people and natural environment, between local people and water recourse, thus successfully utilizes sediment from the Yellow River. The measures has achieved a comprehensive, coordinated and sustainable development in the Irrigated area, established a water – saving, harmonious, green ecological – oriented irrigation project, thus to build a complex ecological system with harmonious relations between local people, water recourse, and natural environment.

Xiaokaihe Yellow River – diversion Irrigated Area is located on the left bank of the Yellow River downstream, only 60 km away from the Lijin Hydrometric Station, the last hydrometric station downwards in the Yellow River, Xiaokaihe Yellow River – diversion Irrigated Area is located in the Yellow River Delta hinterland, runs through central region of Binzhou City, Shandong Province, between north of the Yellow River levee and south bank of Dehuixinhe River in Wudi County, covering five counties (districts) such as Bincheng, Huimin, Yangxin Zhanhua, and Wudi, 23 towns, 667 villages, with a controlling land area of 2,247,300 mu, including 1,600,000 mu of arable land, 1,100,000 mu of designed irrigated area, 1,234,000 mu of actual irrigated area, which designed diversion flow is 60 m^3/s. The Irrigated Area has been completed and put into operation since November 1998, the total length of main canal is 91.5 km, the main canal is composed of 51.3 km of sediment transportation canal, 4.16 km of settling basin, and 36.04 km of water transportation canal, under total investment of 2.30×10^8 yuan. Xiaokaihe Project was upgraded to the provincial key project in 1998; awarded as a "high quality project" by the Ministry of Water Resources in 2002, and known as the "the demonstrative irrigation project in construction & management of Yellow River – diversion irrigated areas of Shandong Province"; its canal greening project was awarded as "the national youth green demonstration project" by the China Central Committee of Communist Youth League and the National Forestry Bureau; the area was also named as "National Water Conservancy Scenic Area" by the Ministry of Water Resources in December 2010.

Up to now, the irrigated area has diverted nearly 3×10^9 m^3 of water from the Yellow River, to solve the problem of brackish water for generations of 420,000 people, which effective irrigated area is 110 mu, with substantial social, economic and ecological benefits, and is highly praised by local people.

1 Ecological protection and gradual improvement of environment

Local administration focuses on ecological protection from early construction phase of the irrigated area; the research achievement "Construction and Management in Xiaokaihe Yellow River – diversion Ecological Irrigated Area", completed in 2006, was awarded as the First Prize of Water Conservancy Science and Technology Progress in Shandong Province, and also known as the "Yellow River – diversion Ecological Irrigated Area."

1.1 Water saving, efficient utilization of water resources

The water resource has been fully saved by multi – channel, multi – measures under the principles of balance between supply and demand, overall balance, macro control, and high economization, and the area strives to develop into a water – saving irrigated area.

1.1.1 Water – saving by projects

The first measure targets at water – saving materials for the project. 51.3 km sediment transportation main canal was fully provided with polyethylene plastic film, composite geomembrane, and concrete prefabricated board, thus achieving a impermeable water – saving transformation of the whole section, reducing the loss during transportation, the annual saving water volume is 4.5×10^7 m^3, 10% of the branch canal lining project has been completed; under – branch – canal works are being in a tract development and scale operation by utilizing national land development projects, roads, canals, ditches, and forests are being in integrated planning and full implementation; part of branch canals has been hardened for water saving, and conveyed water to the field, the completed development area has been up to 100,000 mu.

1.1.2 Water – saving by effective management

The second measure is effective management. The main canal has 55 branch canal intake gates; on one hand, these gates have been seriously damaged with less than 30% of intact rate, thus making great adverse influence on water volume regulation; on the other hand, local people often switch on the diversion gates without permission to consume excessive water, resulting in waste. Irrigated area administration has closed all branch canal intake gates, made anti – theft transformation, thereby reducing water waste.

1.1.3 Water – saving by adjusting planting structure

The last measure is adjustment in the planting structure. duty water rate is relatively low in Wudi County in the most downstream, cotton has been planted there; Zhanhua has been actively involved in development of jujube industry. The required irrigation water volume for both the above crops is much lower than that of wheat, thus saving water.

1.2 Water reservation by river and reservoirs to effectively improve ecology

The Yellow River water should be stored in flood season and used in dry season to the question that it is available but not reliable. To achieve these goals, some measures ought to be taken, as follows: Firstly, five plain reservoirs with the storage capacity of more than 1.0×10^7 m^3 has been constructed in the irrigated area to solve the drink problem of 150,000 rural people and livestock. Therefore, every household is able to drink tap water and urbanization of drinking water in rural has been realized. The West Sea Reservoir, becoming one of five ecological reservoirs which are surrounding Binzhou City in the several corners, has provided water for newly – constructed districts in Binzhou City. Moreover, there are 12 small – size reservoirs built by towns and villages, which are in a canal – joining – reservoirs pattern. Secondly, the settling basin in the middle of irrigation area has the total area of 2,400 mu, with the water area of 1,500 mu. It boasts pure water glistening and clusters of swamp with luxuriant grass growing all the year round. There are patches of well – arranged wild willows, a separate aged one independent into good view and thickly blossomed cattails only in the growth of fresh water in the settling basin. For preventing criminals catching birds with the net, the staffs from irrigation and forestry sectors are not only on patrol irregularly, but installed four high – definition camera for supervising. It is typically ecological wetlands possessing ducks, pheasants, egrets and other birds to live, to grow and to multiply and owning all kinds of natural plants of the Yellow River Delta such as wild soybeans in slightly higher ground. Thirdly, river sluices are cross the downstream of 11 large and medium – sized rivers covering from east to west in the irrigated area, which have been put up for impounding rain floods and saving storage and irrigation tail water. In total, the irrigation area cumulatively has 8.0×10^7 m^3 of water. Besides other 9 new wetland nature reserves, Wudi wetland nature reserve in the irrigation area has been established under the help of State Forestry Administration and the World Wide Fund for nature (WWF). Now it has obtained 1×10^6 mu of reed wetlands. Furthermore, it has been an appropriate habitat for rare birds such as swans, white storks and gulls thanks to the perennial water reservoirs and wetlands, and a comfortable paradise for people harmoniously combine with nature of the clean water and lovely fish.

1.3 Landscaping and improving the environment step by step in stations

The Binzhou Xiaokaihe Yellow River Irrigation Bureau (BXYRIB) consists of seven control stations, and it has conducted a master plan for them based on their own locations, soils and adjusted measures to local conditions, in order to improve the working and living conditions of staff embodying with the people – oriented philosophy. Consequently, they have achieved the great achievement every stations boasts special scenery following the principle emphasizing both on virescence and economic benefits. For example, the ornamental value plants have been planted such as Nesaea Pedicellata, Sophora, gingko, Chinese Pagoda Tree and Crape Myrtle, while the economical vegetables and fruits have been cultivated just like South Korea Housi, Zhanhua Brumal Jujube and Wudi Golden – silk Jujube as well. The BXYRIB has been striving to be an ecological unit with green landscapes throughout the year, flowers blossoming annually and harvests expecting always.

1.4 Intensive development of lotus pools along both sides of main canal

Sediment transporting canal is 51.3 km in length, and its upstream is over ground. At the constructing time, it shaped a dozen meters to dozens of meters, or even hundreds of meters borrow pits which are deserted and waterlogged after digging. With continuous attempt, groups of lotus pools have been developed, with the area of nearly 3,000 mu, enjoying the guaranteed water in both quality and quantity owing to the locations are close to main canal and irrigated by the Yellow River water. The formed brand, Xiaokaihe white lotus root, has received widespread praise. During the scorching summer, visitors from everywhere who are fond of the fragrant flowers, green lotus leaves and jumping fish come here to take a picture of lotus flowers, play with their beloved and go fishing. Not only can they supply the attracting scenery, but also they are eco – wetlands and have considerable incomes.

1.5 Greening canals extended in depth and gradually

From 1999 to 2001, total 150,000 fast – growing poplar saplings have been planted, which are suitable for local environment; local administration erected 5 – meter – wide green belts on both sides of the 91.5 km of main canal, planted four rows of trees that are currently growing well and look promising. The sediment transportation main canal is constructed on the ground, thus weeds over – grew in the borrow pits due to embankment excavation, but the border of canals has been clarified and confirmed. Any contractor must accept the three – in – one bundled contract project including project management, channel greening, land and water resources development, and bids in open tender, total of 101 contractors have signed contracts for 20 years, thus the measure pioneers a project management pattern integrating administrative control and open contraction, strengthening a close relationship between the authority and local people, achieve the harmonious development between the irrigated area and the masses. In 2002, National Mother River Protection & Greening Action on – the – spot meeting was held here. On July 22, 2003, the CCTV focus chat show made a special report on the area in the evening of the 15th congress opening of Communist Youth League.

Greening works in the branch canals are being in a depth extension with the implementation of continuously constructed infrastructures in the irrigated area, Greening works in the lateral canals are also being in a depth extension with the construction of small – sized rural water conservancy key counties, especially "the four – in – one implementation of road network, water network, crop filed network, and forest network".

2 Long – distance transportation and comprehensive utilization of sediment

It is inevitable to divert sediment while the water from the Yellow River is being diverted due to the Yellow River is a highly – sandy river; generally, the settling basin in Yellow River – diversion irrigated areas is built in the head of canals, in consideration that Xiaokaihe irrigated area is characterized by fertile land throughout the entire upstream and densely – populated villages, the settling basin to be built on canal head not only requires immigration due to relocation of villages, and also causes desertification in canal head, thus seriously affecting the ecological environment, and the harmony between local people, society and natural environment. To this end, Xiaokaihe administration built the settlting basin on the midstream of the irrigated area where is 51.3 km away by adjusting the main canal slope, to concentrate deposition of sediment from the Yellow River, under its innovative and practical spirit, based on the model test by China Institute of Water Resources and Hydropower Research, thus creating a precedent of long – distance sediment transportation. Here the land is vast, low – lying, saline – alkaline, desolate and uninhabited, not even a blade of grass grows, and the ecological environment is more severe.

The settling basin covers an area of 2,400 mu, ensures its annual dredging volume of 300,000 ~ 400,000 m^3 by digging operation; the smaller Yellow River sediment contains a variety of fertility that can improve the saline – alkali soil, the original barren land has become a cotton – rich fertile land; and the abandoned land has became highly – productive crops land, the land for annual dredging does not require any compensation cost of any relocation, the masses strive to be the first to plant crops here. General dredging operation is conducted in fall and winter seasons when the cotton has been harvested, it is shown from statistics that local administration has settled and improved total 1,500 mu of land; the annual benefit has reached 1.2×10^6 yuan. five new materials plants, which were constructed closely to the settling basin where contains coarse sediments, adopt new technologies, new processes, fire sand lime bricks by utilizing sediment to replace clay bricks, with 5.6×10^7 pieces of brick of annual output, the used sediment volume is about 200,000 m^3, and its annual output value is more than 2.0×10^7 yuan, not only consuming sediment, but also saving 150 mu of arable land each year, turning waste into treasure, and answering multiple purpose.

Over the past 13 years since establishment of the irrigated area, both erosion and deposition are basically balance in the 51.3 km of sediment transporting canal where the water can be normally conveyed, any large – scaled dredging operation has not been made, thus creating a precedent for long – distance sediment transportation; firstly, via sediment canal, 45.4% of the total sediment volume is conveyed to the settling basin where is 51 km far away from the downstream, thus to achieve the long – distance transportation in the sediment line. 50.6% of the total sediment volume is conveyed into the branch canals sediment detained reservoirs and fields, thus to achieve the long distance transportation on the broad area. 60% of the sediment from branch canals stays in the sediment detention reservoirs and widely applied into construction as building materials, most of this part is used for various purposes via settling in branch canals; 40% of this part flows into the fields; 4.0% of the total sediment stays in the sediment transporting canal to maintain a balance between erosion and deposition; the research *Long – distance transportation and allocation optimization project in the Xiaokaihe Irrigated Area* has been awarded as the First Prize of Science and Technology Progress in Shandong Province in 2011.

The BXYRIB has gradually explored and accumulated a set of effective utilization measures of sediment resources in practices, always adhering to the ecological oriented strategy, transforming from passive defense to active action, thus established a integrated utilization system of sediment from the Yellow River, solved the environmental degradation problem caused by the Yellow River sediment, and fully utilized sediment resources to benefit local economy by series of measures such as engineering design, construction, and management.

The Causes and Strategies for the Deterioration of Ecological Environment of Baiyang Lake

Zhang Jun*, *Luo Yang*, *Zhou Xushen*, *Xing Haiyan*, *Xu Wei*, *Wang Hongcui* and *Lin Chao

Haihe River Water Conservancy Commission, Ministry of Water Resources, Tianjin, 300170, China

Abstract: The Baiyang Lake is one of the most important wetlands in the plain in North China. In recent decades, the wetland is facing more and more stern test. Baiyang Lake was chosen as one of the pilots of river and lake health program in the Haihe River Basin. Through the survey in 2011, the results indicated that there are four reasons for the deterioration of ecology environment in Baiyang Lake as follow:

(1) the basic reason for ecological environment issue is the decrease of the amount of inflow and the decline of water level, which resulted in the pollutants concentration, water self - purification ability drops and vicious circle of the ecological environment of Baiyang Lake.

(2) The scale of aquaculture in Baiyang Lake is huge. The aquaculture feed and feces directly go into water and deposit at the bottom of the Lake.

(3) There are numerous residents surrounding Baiyang Lake. Moreover, there is rapid rourism development in Baiyang Lake upstream. Therefore, the large direct discharge of sanitary sewage in Baiyang Lake produces huge environmental pressure.

(4) The economy develops well around upstream area, with population growth and increase of industrial and agricultural water consumption. That is the main reason that the water quality of all upstream rivers of Baiyang Lake is not good. Most rivers mainly receive upstream industrial and sanitary sewage, which actually become a river for waste water. Based on the analysis of ecological environment, the authors put forward some strategies for the protection and remediation of ecological environment as follow: ①inter-basin water transfer; ② optimal culture model; ③strengthening the control of pollution sources; ④ecological restoration technology.

Key words: Baiyang Lake, ecological environment, river and lake health, the Haihe River Basin

1 Introduction

Baiyang Lake is located in triangle centre of Beijing, Tianjin, Shijiazhuang, which is in the administrative district of Anxin, Rongcheng, Xiongxian, Gaoyang of Baoding City and Renqiu of Cangzhou City. Baiyang Lake belongs to the Daqing River downstream in the Haihe River Basin, collecting runoff and waste water from 9 rivers of the Juma, Fu, Tang, Bao, Cao, Xiaoyi, Ping, Baigou channel and Zhu long in the south, west and north, with the catchment area of 3.0×10^4 km^2 area. The flood will go into Duliujian River and into Bohai (seen in Fig. 1). It is the largest freshwater lake in North China with wide water surface and marsh wetland, which is called "the kidney of the North China". Baiyang Lake has many ecological functions such as flood storage, dry climate mitigation, regulation of rainfall, water purification, groundwater recharge and maintaining biological diversity and protection of rare species resources .

2 Current water quality of Baiyang Lake

In recent years, the water quality of Baiyang Lake is worsening year after year, and water ecological environment is being seriously threatened. In 2011, the result of Baiyang Lake health survey showed that the water quality of upstream rivers and Baiyang Lake was as follows:

(1) For the upstream river, the water quality of Fu river (Anzhou Section) was inferior to Class V. The main pollutants exceeding the standard request were ammonia nitrogen, chemical oxy-

gen demand and the biochemical oxygen demand in five days. The water quality of Xiaoyi River (Gaoyang Section) was Class Ⅴ. The main pollutants exceeding the standard request were biochemical oxygen demand in five days. The other upstream rivers were always dry status in the river state.

(2) The water quality of Guancheng, Liutong and Guolikou Section in Baiyang Lake was relatively good, which belonged to class Ⅲ. The water quality of Quantou, Caiputai and Zaolinzhuang was Class Ⅳ. The main pollutant exceeding the standard request was biochemical oxygen demand in five days. The water quality of Anxinqiao, Dazhangzhuang and Wangjiazhaiputai and Zaolinzhuang was inferior to Class V. The main pollutant exceeding the standard request was ammonia nitrogen. In addition, the investigation results indicated that more than 80% area in Baiyang Lake appeared eutrophic status.

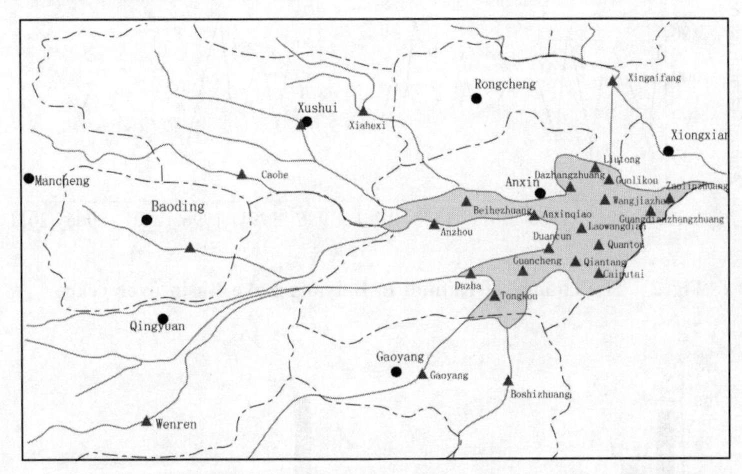

Fig. 1 The Sketch map for upstream river and Baiyang Lake

3 The biodiversity of Baiyang Lake

The wetland biodiversity conservation is one of the important functions of Baiyang Lake. However, Baiyang Lake biodiversity and ecological function of wetland has been severely damaged as the lake is frequently dry and shrinking. Through the investigation for biological status, the results are as follows.

(1) 99 species phytoplankton were found, including17 Cyanophyta, 2 pyrrhophyta, 4 Chrysophyta, 3 Xanthophyta, 24 Bacillariophyta, 9 Euglenophyta, 36 Chlorophyta, 4 Cryptophyta.

(2) 37 species Zooplankton belonging to 19 Genus and 9 family were found, including 19 Cladocera, 14 Copepoda, 4 Rotifera.

(3) 17 species Macrobenthos belonging to 3 Phylum and 7 Order were found, including 9 Mollusca, 4 Hirudinea and Oligochaeta, 4 aquatic insect.

Compared with historical data, it is found that the biodiversity in Baiyang Lake greatly decrease and the species composition has changed. The result indicates that the biodiversity of Baiyang Lake is heavily damaged.

4 Causes of deterioration of Baiyang Lake ecological environment

4.1 Water quantity of runoff to Baiyang Lake

Baiyang Lake is in the middle reache of Daqinghe River Basin, receiving many tributaries flows of Taihang Mountain. The catchment area is 31.2×10^3 km^2, and the average annual runoff is

3.57×10^9 m^3. However, the water quantity into Baiyang Lake gradually dropped down since the 60s of the 20th century because many large and medium - size reservoirs were built in the upper reaches of Baiyang Lake. Especially in the 80 s of the 20th century, the Baiyang Lake had been dry for five times. The rainfall and runoff over years in Baiyang Lake are shown in Fig. 2 and Fig. 3. The water quantity of runoff in Baiyang Lake reached 2.55×10^9 m^3 in 1996, but natural runoff into Baiyang Lake is almost zero since 2000. The Baiyang Lake was often supplied by other water resources such as Yuecheng reservoir and the Yellow River. Without enough water, self purification ability of the lake is weak and water quality is getting worse. Therefore, not enough water for Baiyang Lake is one of the biggest pressures of ecological environment, which is the main reason for Baiyang Lake health.

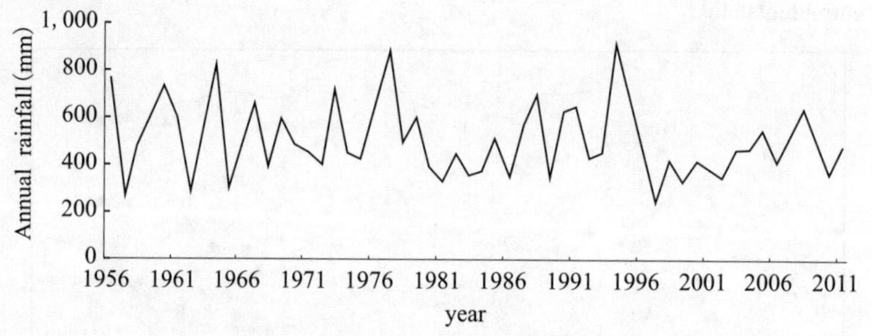

Fig. 2 The change of rainfall in Baiyang Lake Basin over years

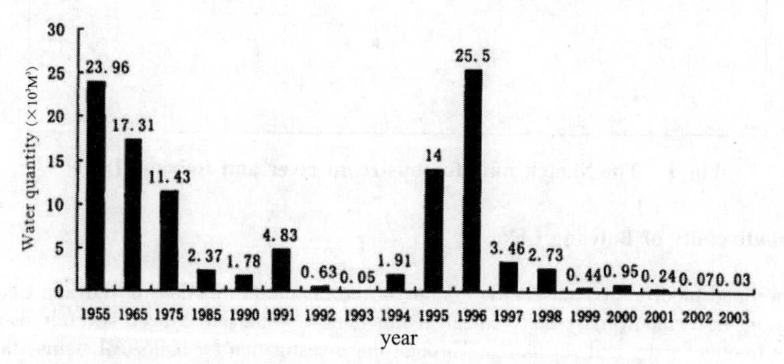

Fig. 3 The water quantity of runoff for Baiyang Lake over years

4.2 Aquaculture and poultry farming

The area of cage culture and captive crab with dike is getting larger in order to develop the economy for local people. The breeding density increases year after year. At present, there are 80,000 Mu water surface in and around the lake, with more than 50,000 t aquatic products including 24 cultured species and 35 wild species. Baiyang Lake has become an important base of aquatic products in North China, and offers plenty of aquatic products for Beijing, Tianjin, Shijiazhuang. However, only little part of the bait was eaten by fishes. Most bait sinks to the bottom of the lake. In addition, the massive excrement from high density fish cage also directly deposits in water. It is no doubt that it causes environmental pollution problem no matter bait and excrement of fishes. It is also reported that cultivation scale greatly exceeded the bearing capacity of Baiyang Lake. In addition, not only aquaculture is large scale, the scale of poultry farming is also big, especially for duck. The species of duck includes Kangbeile, Jinding, Yingtaogu, Beijingbai besides spotted duck. At present, there are about 1×10^6 ducks in Baiyang Lake. The waste and manure of ducks directly or indirectly go into the lake, greatly aggravated eutrophication of Baiyang Lake.

4.3 Local residents and visitors

There are 10 townships, 39 water villages, 134 semi – water villages in Baiyang Lake. The population in the lake area is about 100,000 people, and there is 243,000 people around the lake. Therefore, it is about 343,000 people in total. In addition, with the rapid development of Baiyang Lake tourism industry, number of floating population is increasing in Baiyang Lake, especially in the peak tourist season. Therefore, it produces large amount of sewage and garbage in Baiyang Lake and the surrounding area every day. At present, due to the lack of effective treatment measures, most of the domestic sewage directly discharged into the lake. The rubbish is piled casually, and goes into lake with rainfall. It causes the water pollution of Baiyang Lake.

4.4 The development of upstream economy

The development of upstream economy is one of important cuases for the deterioration of the ecological environment of Baiyang Lake. First, the increase of population in Baiyangdian river basin is one of the main factors which result in the increase of water consumption. From 1974 to 2007, the population of Baiyang Lake Basin increased from 9.7×10^6 to 1.4×10^7, rised by 44.3%. Along with the increase of population, the demand for water resources is also increasing. The water consumption of upstream area result in decrease of water resources of downstream river, so the runoff for Baiyang Lake is decreasing. Second, the grain output of Baiyang Lake Basin increased from 2.6×10^9 kg in 1970 to 6.6×10^9 kg in 2007. With the grain yield growth, the amount of water consumption of agriculture and non – point source pollution also increase. Water consumption in agricultural is the major water use. In 2006, the amount of water consumption of agriculture was up to 78.8%. In addition, a large number of farmland were reclaimed in the lake area and surrounding area, so the non – point source pollution such as pesticide, chemical fertilizer is also one of the important sources of water pollution in Baiyang Lake. Third, sewage discharge along with the economic development increases year after year. From the middle of 1960s of the 20th century, there was no natural flow in Fu River, which became the discharge channel of Baoding city sewage and industrial wastewater. The main pollutants are ammonia nitrogen, chemical oxygen demand, biochemical oxygen demand in five days, volatile phenol, fecal coliform, etc. Based on investigation result, annual sewage discharge of Baoding city is more than 2.00×10^8 t. However, the sewage treatment capacity of Baoding city cannot meet this requirement, thus a part of untreated sewage directly goes into Baiyang Lake through Fu river. For other upstream rivers, the sewage discharged into the river in non – flood season, cannot go directly into Baiyang Lake due to little water quantity. However, these pollutants deposited in the river will enter Baiyang Lake with rainfall runoff during the flood season.

5 Strategies for ecological environmental problem of Baiyang Lake

5.1 Interbasin water transfer

The national wetlands protection project was officially launched in May 25, 2006. Baiyang Lake is included as one of the important wetland ecological water supplement demonstration projects. In fact, the water transfer work had already been started in Baiyang Lake before the demonstration projects. In recent years, the water of Baiyang Lake was supplied many times from the Xidayang reservior, Wangkuai reservoir, Angezhuang reservoir, Yuecheng reservoir and Yellow river, seen in Tab. 1. Baiyang Lake received water about 1.149×10^6 m^3. The water alleviated to some extent ecological crisis of Baiyang Lake wetland, which improved ecological environment and played an important role in maintaining the ecological balance in Baiyang Lake wetland.

Tab. 1 The diversion water amount of Baiyang Lake from 2000 to 2009

($\times 10^6$ m^3)

Year	2000	2001	2002	2003	2004	2005	2006	2007	2008	2009	2010	2011	Total
Quantity	41	67	86	116	160	43	157	0	158	128	100	93	1,149

5.2 Optimal cultivation model

According to the shallow macrophytic Lake features in Baiyang Lake, a kind of the biologic chain organic cultivation mode for Baiyang Lake should be established. Fish feeding and poultry raising should be in an ecological way. At the same time, it is important to strengthen aquaculture management, control breeding scale and capacity, and make a plan for aquaculture development. These efforts will reduce the aquaculture pollution and control water eutrophication for Baiyang Lake.

5.3 Strengthening the control of pollution sources

The main idea of pollution control is source control. Therefore, the treatment of pollution in Baiyang Lake must strictly control pollution sources, including the surrounding agricultural non - point sources and urban sewage outlet in the upstream. First, more large - scale sewage treatment plants should be built; enhance the research of sewage treatment technology in order to meet discharge standards. Moreover, we should try to use more recycled water and improve the utilization efficiency of water resources. Second, it is necessary to improve the efficiency of agricultural water usage and farming methods, reduce usage of pesticide and fertilizer, reduce pollution discharge and save water resources. Third, it is also important to strengthen propaganda and education, improve the environmental protection consciousness of the local people, and do punishment when any companies and individuals act against environmental laws and regulations.

5.4 Ecological restoration technology

Ecological restoration technology should be strengthened in order to control and reduce aquaculture pollution of Baiyang Lake, improve the ecological environment of the lake. On one hand, some aquatic plants can be introduced in Baiyang Lake such as Lantana camara, potamogeton crispus and water chestnut. The growth of aquatic plants can consume N, P and other nutrients, so it could be used to control eutrophication and purify water body. On the other hand, the study for the local indigenous microorganism should be carried out. The ecological environment in cultivation area could be restored through microbial degradation of organic pollutants.

Acknowledgement

The authors would like to thank Gao Yuepeng, Xu Mingxia for assistance in field work and sample pretreatment. This paper is supported by the Commonweal Projects Specific for Scientific Research of the Ministry of Water Conservancy of China (Grant No. 201101018) and Central divided water resources fee project (Grant No. 1261120411636)

References

Xiao Guohua. The Status of Culture Environment and Ecologic Fish Culture Development in Baiyang Lake[J]. Hebei Fisheries, 2011,214(10): 47 - 49.

Zhang Xiaogui, Liu Shuqing, Dou Tieling, et al. Strategies for Controlling Water Environmental Pollution in the Area of Baiyangdian Lake[J]. Chinese Journal of Eco. Agriculture, 2006,14(2):27 - 29.

Cheng Chaoli, Zhao Junqing, Han Xiaodong. Analysis of the Water Quality and Quantity of Baiyang Lake Wetland in Recent 10 Years[J]. Haihe Water Resources, 2011(3): 10 - 12.
Jin Songdi, Li Yonghan, Ni Caihong. The Potamogeton Crispus Absorption for Nitrogen and Phosphorus in Water[J]. Acta Ecologica Sinica, 1994, 14(2): 168 - 173.
Lin Qiuqi, Wang Chaohui, Si San. Study on the Feasibility of Utilizing Hydordictyon Reticulatum to Treat Eutrophic Waters[J]. Acta Ecologica Sinica, 2001, 21(5): 814 - 819.
Men Sushi, Zhao Fang, Cui Xiuli. Discussion for Qquatic Vascular Plant Comprehensive Utilization and Water Purification of Baiyang Lake[J]. Environmental Science, 1995(1):32 - 34.
Wu Xiaolan, Li wei, Ma Xiaoneng. Application of Micro Ecological Agent in Healthy Aquaculture [J]. Fisheries Science & Technology Information, 2006, 33(3): 130 - 133.

The Analysis of the Livestocks' Carrying and Grazing Capacity in Theory of Source Area of the Three Rivers in Qinghai

Liu Zhengsheng[1], *Cui Changyong*[1] and *Wang Wanmin*[2]

1. Yellow River Engineering Consulting Co., Ltd., Zhengzhou, 450003, China
2. Yellow River Conservancy Commission, Zhengzhou, 450003, China

Abstract: Because of natural conditions restrictions of Source area of the Three Rivers in Qinghai, the way of local animal husbandry production management is traditional and extensive in long – term, in recent years, with the development of economical society, the contradiction among the population, resources and environment here becomes prominent, and the animal husbandry production is in serious overloading state. According to statistics, at present, about 90% of the grasses appear different degrees of degradation in the source area, and 50 years ago, plant yield per unit area fell 30% to 50%. To protect the ecology and improve the people's livelihood, according to the principle "deciding the livestock amount by the grass, keeping the balance between grass and livestock", combined with the comprehensive construction of well – off society and the requirement of improving the local residents living conditions, it should be done to use the "3S" technology, have the survey and analysis on all kinds of grassland of the source area, and put forward the reasonable carrying capacity to provide the technical support to realize "keep the balance between grass and livestock" of Source area of the Three Rivers and protect the eco – environment.

Key words: Source area of the Three Rivers in Qinghai province, "3S", the carrying and grazing capacity in theory, keep the balance between grass and livestock

1 Put forward the problem

On April 6, 2011, the Premier of General Office of the State Council of the People's Republic of China, Wen Jiabao, in the executive meeting of the State Council of research on promoting the deployment and pastoral areas and rapid development of the policy, pointed out that, in 2020, the level grazing area will fully realize balance between grass and grazing, grassland's ecology will enter in a benign circulation, grazing area in further will be optimized in economic structure, herdsmen conditions of production and living have an overall improvement, basically realize the goal of the comprehensive construction of well – off society. Source area of the Three Rivers is one of the six grazing areas in China as the important supply base in animal products. The vast grassland is ecological barrier, plays a vital role in all above. In recent years, global warming, human unreasonable intervention on the ecological system, such as overgrazing, excessive mining, abuse of digging, the mouse insect damage, lead to the grassland degenerate eco – environmental deterioration. Therefore, it is necessary to fully consider grasslands' bearing capacity. On this basis, we can reasonably determine the carrying and grazing capacity in theory.

2 The general situation of Source area of the Three Rivers in Qinghai province

Qinghai province's Three – River Source Area is located in China's western Qinghai – Tibet plateau hinterland, in the south of Qinghai province. Three – River Source Area has an altitude of 4,200 m in average is the birthplace of the Yangtze River, the Yellow River and the Lancang River, which is also known as the "Chinese water tower" and even "Asian water tower". The geographic coordinates is 31°30' ~ 36°04' of northern latitude, 90°28' ~ 102°18' of east longitude. And the total area is 294,700 km^2, including 97,600 km^2 in the Yellow River source area, 159,800 km^2 in the Yangtze River source area, 37,300 km^2 in the Lancang River source area. Administrative regions involved in 16 counties of Yushu, Guo luo, Hainan and Huangnan and Tibetan autonomous prefectures and Tanggula Mountain village in Golmud city.

Historically, Source area of the Three Rivers was grasses, medium - bodied, lakes, wild animal population with various plateau grassland meadows area, known as the ecological "virgin region", also is the area that breed the most big rivers on Eurasian continents, of which all come from the source area Source area of the Three Rivers, among them, the total water in the Yangtze river is 25%, the total water in the Yellow River is 49% and 15% in the Lancang River. Water resources and eco - environmental situation of Source area of the Three Rivers are crucial to the sustainable development of economical socity of the big three basins and to regional even or national ecological safety.

Source area of the Three Rivers is the main living area of minorities, and the graziery has developed for thousands years, which has been a long history. Graziery is always the economic subject of grazing area of Source area of the Three Rivers, which is the survival leading industry of the people of all ethnic groups. The unique natural geographical environment of Source area of the Three Rivers breeds the unique plateau type animal and plant resources, has the most abundant biodiversity at alpine region where there are the main plant species like Bellardatus Kobresia, Stipa capillata Linn, Carex tristachya, Saussurea gossypiphora, Roegneria kamoji Ohwi, Bluegrass, Elymus, Achnatherum splendens etc. Alpine meadow and alpine grassland are the main vegetation types and natural grassland in Source area of the Three Rivers.

Since the 1950s, along with the increase in population of Source area of the Three Rivers and number of cattle, overgrazing exists in grassland, which results in the grass productivity reduction, grassland degradation, and the function reduction of water and water - soil conservation. Especially in recent years, with the development of society and economy, the contradiction among the population, resources and environment become very prominent, the graziery is in overgrazing situation. The more cattle, the more increasingly fierce in grassland eco - destruction, according to the analysis of remote sensing data, there has currently about 90% of the grass appearing different degrees of degradation in Source area of the Three Rivers, to compare with 50 years ago, grass production per unit area fell 30% ~ 50%. According to actual situation's estimated number of cattle in Source area of the Three Rivers is 20.212 million sheep unit, overload 7.61 million sheep unit, the overload rate is reached to 60.4%.

3 The calculation thought of the carrying and grazing capacity

In the grass - livestock balance of "deciding the livestock amount by the grass", it is based on water and soil resources as the constraints, and the algorithm is the reverse calculates of livestock - grass - land. To determine the number of cattle, a key factor is to determine the income level of herdsmen and need sheep unit. In the measurement, the first consideration is the nomads' animal husbandry income ratio, also as the net income of animal husbandry accounts for about the proportion of the net income of the nomads. The second one is to consider the nomads' future income level and country's expectations on improve nomads' income, namely the expected income level, which will be converted into the sheep unit each nomads needs. The third is to consider the sell rate and commodity rate have a positive correlation on nomads' income. The last is to consider to adjust the structure of the nomads' income.

According to the requirement of realizing the balance between grass and livestock in 2020, and we must strive to make natural grassland back to the early 1980s grassland, at that time, compared with the current grass production, the future grass production will increase 30% ~ 40%. At the same time, we can develop some artificial grasslands in the conditional regions, rise grass artificially, increase grass production and its size, according to the estimation of the basic feeding demand in winter. The construction of irrigating forage - land, according to the overall thinking of grazing area's water conservancy development "small construction, big protection", according to the water resources carrying capacity and the need of protecting the grasslands, in good place of water and soil resources condition, we can make a batch of constructions of water - saving irrigation forage - land, increase supply of feeding grass, combine with the means such as the fence, captive breed-

ing, animal species improvement, adjust the structure of animal community, change producing ways of animal husbandry, give fully play to the natural ability of self – repairing, protect and restore the grasslands' ecology, and increase the nomads' incomes.
According to the following formula:

$$Y_1 = \sum_{i=1}^{n} X_i S_i d_i$$

$$Y_2 = \sum_{h=1}^{n} X_h S_h$$

where: Y_1 is natural grassland area plant yield, kg; X_i is all kinds of natural grassland area planning, hm^2; S_i is all kinds of natural grassland plant yield per unit area, kg/hm^2; D_i is natural grassland utilization rate of forage grass; Y_2 is an artificial grassland, straw, feed, and various ways of hours in the ability, kg; X_h is an artificial grassland, straw, feed, and hours area, hm^2; S_h is an artificial grass – and straw, feed, and plant yield per unit area hours, kg/hm^2.

4 The analysis of carrying and grazing capacity in theory in Source area of the Three Rivers

4.1 The analysis of carrying and grazing capacity in theory in natural grassland

4.1.1 The investigation of natural grassland

By use of the "3S" technology, and combined with typical survey, we can determine all kinds of grasslands' area in Source area of the Three Rivers and correctly estimate the region's the bearing capacity of the sustainable development. Through the data analysis of remote sensing survey of the present situation in land use, natural grassland area where is in Source area of the Three Rivers is 17,486 hm^2, accounting for 59.3% of the total area of the source region, in which we can use is 15,294 hm^2, accounting for 87.5% of the area of natural pasture. See Tab. 1.

Tab. 1 The questionnaire of grassland area by remote sensing in present grazing region in Source area of the Three Rivers (unit: khm^2)

Basin	Natural grassland	Available grassland	Available grassland's ratio in natural grassland (%)
The Yellow River	7,189	6,725	93.6
The Yangtze River	7,467	6,082	81.5
The Lancang River	2,830	2,487	87.9
Source area of the Three Rivers	17,486	15,294	87.5

4.1.2 The carrying and grazing capacity in theory of natural grassland of the years in target

According to the survey of the amount of grass production, per hectare products 600 kg to 4,050 kg pasture grass in Source area of the Three Rivers, combined with the survey of the source area's available pasture grass area by remote sensing, we can calculate edible and fresh grass's quantity is 18.271×10^9 kg in the present situation years, in accordance with each sheep's eating fresh grass yield of 4 kg/d, so, the suitable grazing capacity of natural pasture in Source area of the Three Rivers is 12.515×10^6 sheep unit.

Through the protective measures like banning grazing, grazing in rest, grazing take turns, eco-immigration, actively adjusting the animal husbandry ' s industrial structure, forming good eco – industrial chain, improving grassland ' s eco – environment to realize the grass become natural regeneration, compared with present situation ,natural grass production will get a 30% to 40% increased. Hereby forecasting, natural pasture grass can provide fresh grass of 24.58×10^9 kg in 2020 and can carry 16.832×10^6 cattle sheep unit.

4.2 The analysis of carrying and grazing capacity in theory in artificial grassland

Artificial grassland is the grassland that use agricultural technology measures to product forage grass in the cultivated land or natural grassland. Compared with natural grassland, artificial grassland has higher land productivity and labor productivity. According to the survey, artificial grassland ' s in general average production is as 2 ~ 5 times as natural grassland per hectare. According to the statistics, current situation ' s artificial grassland can provide 96×10^6 kg fresh forage in Source area of the Three Rivers, and it can raise 66,000 sheep unit cattle.

At 2020 year level, according to the basic demand in winter, every herdsmen ' s artificial grassland gets to 0.33 hm^2, it estimates that artificial grassland of Source area of the Three Rivers in target years will reach to 55,800 hm^2, which can produce 260×10^6 kg pasture and can carry 177,000 sheep unit.

4.3 The analysis of carrying and grazing capacity in theory in irrigated grassland

Source area of the Three Rivers follows the traditional herding means in long times, it is basically dependent on the nature, and natural factors have big influence on animal husbandry production. Developing water – saving irrigation in forage land is helpful for adjusting pressure of natural grassland in grazing and make the grassland to take a break, which improve productive level of animal husbandry and anti – disaster ability, which promote the change from depending on the nature to feeding in house, intensified direction and provide basic guarantee on sustainable development of grassland animal husbandry. According to the statistics, current situation ' s artificial grassland can provide 0.31×10^6 kg fresh forage in Source area of the Three Rivers, it can raise 21,000 sheep unit cattle.

On the basis of the actual situation in Source area of the Three Rivers, in future, the focus of the development of animal husbandry irrigation is to make the existing irrigation area ' s mating of continuous construction and water – saving transformation better, to improve the level of management, to give full play to the economic benefits of the existing irrigation area. At the same time, in the basis of consolidating the existing irrigation area, combined with water and soil resources, we should use the form of concentration and scatter getting unite and develop the forage land of part water – saving irrigation. According to the water and soil resources of Source area of the Three Rivers, in target years, we will develop 29,000 hm^2 forage land, which can produce 4.60×10^8 kg pasture and carry 317,000 sheep unit.

4.4 The total grass production and the carrying and grazing capacity in theory

According to the above analysis, the future development of animal husbandry of grassland at Source area of the Three Rivers need a shift from extension type to the connotation type, farmers' income will change from mainly increasing on the number of cattle to improve livestocks' species and their structure, to the direction of increasing commodity rate. In accordance with the principle: "let grass to decide the number of animals, keep grass and livestock balance ", in target years, through grazing in rest and grazing take turns, natural grassland ' s adjustment and adaption, which make all kinds of grasslands' carrying and grazing capacity in theory get greatly higher, according to the principle of keeping grass and livestock balance to strictly control number of cattle at 17.326×10^6 sheep unit in target years. See Tab. 2.

Tab. 2 The condition of grass production and carrying and grazing capacity at all kinds of grasslands in Source area of the Three Rivers in different level years

(unit: 10,000hm^2; $\times 10^8$ kg; 10,000 sheep unit)

Basin	Natural grassland			Artificial grassland			Irrigated grassland			Carrying and grazing capacity in theory	
	Acreage	Grass production	Carrying and grazing capacity	Acreage	Grass production	Carrying and grazing capacity	Acreage	Grass production	Carrying and grazing capacity	Acreage	Carrying and grazing capacity
The Yellow River	672. 55	115. 58	791. 60	2. 85	1. 34	9. 20	1. 90	3. 03	20. 80	119. 96	821. 60
The Yangtze River	608. 19	88. 65	607. 20	1. 71	0. 77	5. 30	0. 62	0. 98	6. 70	90. 41	619. 20
The Lancang River	248. 67	41. 51	284. 30	1. 02	0. 47	3. 20	0. 39	0. 62	4. 20	42. 60	291. 80
Source area of the Three Rivers	1,529. 42	245. 75	1,683. 20	5. 58	2. 59	17. 70	2. 90	4. 63	31. 70	252. 96	1732. 60

5 Conclusions

By use of the "3S" technology and combined with typical survey, we can estimate all kinds of grassland area. In accordance with the principle "deciding the livestock amount by the grass, keeping the balance between grass and livestock ", we should consider the requirement of constructing well – off society and improving the local residents' living conditions, increase the effective supply of pastoral areas through developing the irrigation and forage land and planting the pasture artificially, research the reasonable carrying and grazing capacity in theory to ensure that basically balance between grass and livestock in Source area of the Three Rivers, so as to achieve the purpose of protecting the local grassland and the recovery of eco – environment, and provide scientific support to guarantee the national eco – security, promote national unity and the stability in border areas, and promote the regional coordinated development .

References

Qinghai Province. The grassland law of the People's Republic of China[S]. Qinghai Prataculture, March, 2008.

Bureau of Statistics in Qinghai Provincial. Qinghai Statistics Yearbook [M]. Chinese Statistical Press, 2008.

Guo Kezhen, Kang Yue, Zhao Shuyin, et al. The Rules of Grassland in Grazing Area and Drainage Technology sl –2005 [S]. Chinese Water Conservancy and Hydropower Press, Water Conservancy industry standard of the People's Republic of China, 2005.

Ni Wenjin, Chen Zhijun, Chen Quchang, et al. The Plan of Grassland Eco – protection and Water Resources Security in Chinese Grazing area [R]. Rural Water Conservancy Department of the Ministry of Water Resources, Water Conservancy Science Institute of the Grazing Area of Ministry of Water Resources, Chinese Lrrigation and Drainage Water Development Center, July 2003.

Li ping, Xing Yongjun, Duan Shuijiang, et al. The plan of grassland eco – protection and water resources security of Qinghai grazing area[R]. Qinghai Povincial Wter Rsources Breau, April 2003.

Qinghai Povincial Implement of Government Policy Task in Tibet. Tibetan Ntural Gassland Gazing Frbidden implementation Pan(2009 ~ 2020) [R]. Agriculture Animal Husbandry office of Qinghai Province, April 2009.

Sun Guangchun, Gao Zhiyong, Wang Yuanyuan, et al. Water Sving Irigation Dvelopment Panning of grazing area's feedingland in Qinghai province[R]. Water Cnservancy of Qinghai Povince, September 2010.

Jia Youling. Several Teories and Pactical questions about the Blance between Gass and LIestocks. Grassland journal [J]. December 2005, the 13th book, the 4th Sage.

Qinghai Provincial People's government, Qinghai Provincial Development and Reform Commission. Qinghai provincial Pan for Al – round Cnstruction of Wll – off Sciety ({2005}98) [R]. June 24, 2005.

Ministry of Agriculture of the People's Republic of China. Management Mthods of the Blance between Gass and LIestocks[S]. Put in to Pactice in the Frst in March 2005.

Research on the Hydrology Ecological Effect in the Intake Area of Large Water Diversion Project

Wang Liming and *Xu Ning*

Water Resource Conservation Scientific Institute of
Haihe River Water Resources Commission, MWR, Tianjin, 300170, China

Abstract: South-North Water Diversion Project inevitably shall exert great influence on the ecological environment of Haihe River Basin. This paper studies the ecological environment of different hydrologic units such as ground water, streams, lakes and estuaries in the Intake Area in Haihe River Basin under different water diversion conditions. Based on the research achievements in China and abroad, this paper establishes the evaluation index system from the hydrologic, physicochemical and biological perspectives and analyzes the influence of South-North Water Diversion Project on the ecological environment in the Intake Area in Haihe River Basin. The analysis shows that after the launch of South-North Water Diversion Project, the hydrologic characteristics and the river water quality of the main rivers, lakes and wet lands in the Intake Area in Haihe River Basin change significantly. The deep ground water funnel rebounds to some extent. The ecological conditions at the estuary change for the better gradually.

Key words: South-North Water Diversion Project, eco-hydrology, index system

1 Research background

Haihe River Basin is the political and cultural center of China and one of the most developed regions. It is also a region in severe shortage of water resources. The per capita water resources are just 293 m^3, less than 1/7 of the national average and 1/24 of the world average and far less than 1,000 m^3 per capita, the recognized international shortage standard. With the rapid economic and social development of the river basin, the conflict between supply and demand of water resources becomes more and more serious and the excessive exploitation of water resources causes a series of ecological problems. For example, the rivers become dry or cut off, the wet land shrinks, the water bodies are severely polluted, the ground sinks constantly and the ecosystem at the estuaries deteriorates . South-North Water Diversion Project is a key strategic project which can relieve the severe shortage of water resources in North China and has significance in alleviating the water resources and ecological crisis in Haihe River Basin .

2 Water allocation in the intake area in Haihe River Basin

The intake area in Haihe River Basin in South-North Water Diversion Project covers Haihe River Plain off Jing-Guang Railway in the east except Jidong, Luanhe in Hebei and the plain in the east of North Shandong. It covers an area of 10.2×10^4 m^3, 80% of Haihe River Plain. There are sixteen large and medium-sized cities such as Beijing, Tianjin, Baoding, Shijiazhuang, Xingtai, Handan, Langfang, hangzhou, Hengshui, Anyang, Xinxiang, Hebi,Jiaozuo, Puyang, Dezhou and Liaocheng, which are the main water-deficient areas in Haihe River Basin.

According to the Overall Plan for South-North Water Diversion Project, Middle Line Phase I and East Line Phase Ⅰ and Ⅱ shall be completed before 2020 and Middle Line Phase Ⅱ and East Line Phase Ⅲ shall be completed before 2030. In 2020, the water distributed from Changjiang River to Haihe River through the middle line is 6.24×10^9 m^3 and that through the east line is 1.68×10^9 m^3. The total amount is 7.92×10^9 m^3. In 2030 (in Tab. 1), the water distributed from Changjiang River to Haihe River through the middle line is 8.62×10^9 m^3 and that through the east

The Commonweal Projects Specific for Scientific Research of the Ministry of Water Conservancy of China(Grant No. 201101018).

line is 3.13 × 10^9 m^3 (by the water divisions of the main channel). The total amount is 11.75 × 10^9 m^3.

Tab. 1 Table of water distribution in Haihe River Basin in South-North Water Diversion Project Unit: × 10^8 m^3

Provincial-level administrative region	2020			2030		
	Middle line phase Ⅰ	East line phase Ⅱ	Subtotal	Middle line phase Ⅱ	East Line Phase Ⅲ	Subtotal
Beijing	10.5	0	10.5	14.9	0	14.9
Tianjin	8.6	5.0	13.6	8.6	10.0	18.6
Hebei	30.4	7.0	37.4	42.3	10.0	52.3
Henan	12.9	0	12.9	20.4	0	20.4
Shandong	0	4.8	4.8	0	11.3	11.3
Total	62.4	16.8	79.2	86.2	31.3	117.5

3 Analysis of hydrology ecological effect in the intake area

3.1 Grading of hydrology ecological effect

After South-North Water Diversion Project starts to supply water, the water resource distribution structure in Haihe River Basin should also change accordingly, as well as the hydrologic regime and the ecological water demands. The hydrology ecological effect in the intake area in Haihe River Basin can be graded into three levels: primary, secondary, and senior. For the direct influence of the water supplied by South-North Water Diversion Project, the hydrologic regime in different ecological units in Haihe River Basin such as rivers and lakes shall also change accordingly. This is the primary ecological effect caused by South-North Water Diversion Project. The ground water in the intake area shall be forbidden or limited in exploitation because of the supply of water so the decrease of groundwater level slows down. This is the secondary ecological effect. To protect the supplied water resources and improve the local water functional region as well as achieve the water quality target corresponding to the goal of economic and social development, we must make great efforts to save water and reduce emission. This is also known as secondary ecological effect. With the improvement of the hydrologic regime in the rivers and the lakes, the habitats of the living things shall be improved as well as the biological diversity, the fishes, the birds and even the higher mammal. This is known as senior ecological effect(in Fig. 1).

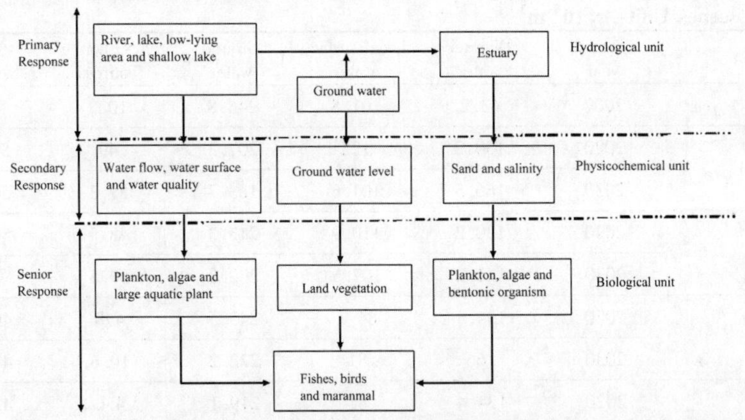

Fig. 1 Hydrology ecological effect in Haihe River Basin after South-North Water Diversion Project starts to supply water

3.2 Set Different Water Diversion Scenes

According to the principle of giving priority to the use of Changjiang water and unconventional water resources, controlling the use of the surface water in the local area and the Yellow River water, and using as little ground water as possible, this paper sets plans for different water diversion scenes.

(1) Target year: Now 2007 and in the future 2020 and 2030.

(2) Hydrologic series: series 1980 ~ 2005, mean precipitation 501 mm; Series 1956 ~ 2000, mean precipitation 535 mm.

(3) Water drawn from Yellow River: 5.12×10^9 m^3.

According to the two hydrologic series, in 2020, East Line Phase Ⅰ and Middle Line Phase Ⅰ of South-North Water Diversion Project start to supply water and achieve the expected effect. The total water supplied to Haihe River Basin is 7.922×10^9 m^3. There are two possibilities in 2030. Firstly, East Line Phase Ⅳ and Middle Line Phase Ⅱ of South-North Water Diversion Project start to supply water and achieve the expected effect. The total water supplied to Haihe River Basin is 11.75×10^9 m^3. Secondly, Phase Ⅱ of South-North Water Diversion Project is not completed as expected. The water transferred in Middle Line Phase Ⅰ shall increase by 20%. The total water supplied to Haihe River Basin is 9.18×10^9 m^3. See Tab. 2 and Tab. 3

Tab. 2 Plan for Setting Different Water Diversion Scenes

Hydrologic series	Progress of water diversion project	Plan	Target Year	Water transferred ($\times 10^8$ m^3)	
				Middle line	East line
1956 ~ 2000	Phase II implemented as expected	F1	2020	62.42	16.8
			2030	86.21	31.3
	Increase the water transferred at Middle Line Phase Ⅰ by 20%	F2	2020	62.42	16.8
			2030	75	31.3
1980 ~ 2005	Phase Ⅱ implemented as expected	F3	2020	62.4	16.8
			2030	86.2	31.3
	Increase the water transferred at middle line phase Ⅰ by 20%	F4	2020	62.4	16.8
			2030	75	16.8

Tab. 3 available water quantity of Haihe River Basin under different water diversion scenes Unit: $\times 10^8$ m^3

Plan	Target year	Diverted water	Local surface water	Ground water	Other sources	Total supply
Reference year	2007	43.2	101.8	246.8	10.2	402
F1	2020	130.2	122	207.3	49	508.5
	2030	166.5	101.6	183.7	57.3	509.1
F2	2020	130.2	110.9	213.3	48.9	503.3
	2030	142.2	107.5	202.8	50.5	503
F3	2020	134.4	84.1	243.5	4.8	466.8
	2030	166	81	222.2	10.6	479.8
F4	2020	134.4	81	219.1	4.6	439.1
	2030	143	91.1	221.4	11.1	466.6

3.3 Evaluation of hydrology ecological effect

This paper analyzes the hydrology ecological effect from four aspects: the hydrologic regime of rivers and lakes, the water quality in the water functional division, the ground water level and the salinity changes at the estuaries.

3.3.1 Changes of the hydrologic regime of rivers and lakes

Research shows that the minimum ecological water requirement of Haihe River Basin is 2.851 $\times 10^9$ m^3 and the suitable ecological water requirement is 7.0 $\times 10^9$ m^3. After South-North Water Diversion Project is launched, some ecological water use which has been diverted and misappropriated before shall be returned to the ecological system of rivers and lakes. Based on ROWAS water resources distribution model, this paper figures out that the river ecological water demands of the rivers in the basin and the plain in 2020 and 2030 is between 3.314 $\times 10^9$ and 7.698 $\times 10^9$ m^3, as shown in Tab. 4. In case of Series 1980 ~ 2005 (Plan F3), the minimum ecological water demand of the rivers in the basin in 2020 and 2030 can be satisfied but the suitable ecological water requirement cannot be satisfied. In case of Series 1956 ~ 2000 (Plan F1), the suitable ecological water requirement at the target years can be satisfied practically. 60% of the rivers in the plain always have water after the restoration measures such as ecological water supplement and replacing water with green. The ecological conditions of the riches of Yongding River, Hutuo River and Zhang River which suffer severe desertification have been improved obviously.

Tab. 4 Ecological water demand of the rivers and storage capacity of lakes and wetland in the intake area of haihe river basin under different scene plans

Target year series	Scheme no.	Target year	Ecological water demand of rivers ($\times 10^8$ m^3)	Water demand of lakes and wetland ($\times 10^8$ m^3)
1956 ~ 2000	F1	2020	67.02	31.47
		2030	76.98	36.32
	F2	2020	74.75	29.42
		2030	74.44	34.50
1980 ~ 2005	F3	2020	54.45	19.70
		2030	63.87	16.63
	F4	2020	52.12	25.32
		2030	33.14	22.25

There are twelve main plain wetlands in Haihe River Basin such as Baiyangdian so the protection and restoration of the river ecosystem there must be given priority. The minimum ecological water demand of the twelve wet lands is 8.77 $\times 10^8$ m^3 and the corresponding ecological water surface area is 816 km^2. After South-North Water Diversion Project has been launched, the annual water retention capacity is between 1.662 m^3 and 3.632 m^3. If the ecological water surface area of the wet land is further increased within a certain period, the ecological benefits shall be improved significantly. Among them, the surface area of the wetlands of Hengshui Lake, Dalangdian Reservoir, Beidagang Reservoir and Datun Reservoir which are the water head areas of South-North Water Diversion Project shall be stabilized and maintained. At the same time, South-North Water Diversion Supporting Project plans to build up 27 medium and small-sized water regulation and storage stations and the newly added storage capacity is 1.58 $\times 10^8$ m^3. A new "Two-way Six-cross" water-net backbone project frame which focuses on delivery arteries of South-North Water Diversion Project and is supplemented with delivery branches shall be formed in the intake area of Haihe River Basin. Along the delivery route a south-north reticular wetland cluster shall be formed.

3.3.2 Changes of Water Quality in the Water Functional Division

The intake area in Haihe River Basin in South-North Water Diversion Project refers to 155 water functional divisions and the total river length is 7,487.3 km. In 2020 and 2030, the dirt holding capacities of main pollutant COD are 1.85×10^5 t/a and 207 thousand t/a respectively (see Tab. 5). According to the principle of "three first and three last" of South-North Water Diversion Project, till the end of 2,007, 121 municipal sewage plants have been built up in the intake area in Haihe River Basin. The annual processing capacity is 3.781×10^9 t and the actual processing amount is 2.486×10^9 t. Now Beijing, Tianjin and Hebei which are all in Haihe River Basin have issued strict sewage discharge standard in succession. The concentration of the main pollutants must be in accordance with Level 1B or Level 1A of Standard for Discharge of Pollutants of Municipal Wastewater Treatment Plant (GB18918 ~ 2002). To satisfy the demand for water body functions to the satisfaction of the economic and social development within the river basin, they adopt the strictest water resources management system and water pollution prevention policies, control the total water consumption, improve the water use efficiency and promote waste water reclamation and reuse. By this way they increase the treatment rate of sewage discharged into the rivers by 90% and basically achieve the goal of water quality in the water functional divisions.

Tab. 5 Compliance with the standard in the water functional divisions in the intake area under different scene plans

Plan	Target year	Number of water functional divisions which are in accordance with the standard	Quantity of COD discharged into rivers ($\times 10^4$ t)	Compliance rate of water functional divisions (%)
F1	2020	75	19.4	97
	2030	155	21.1	98
F2	2020	75	20.5	93
	2030	155	22.4	96
F3	2020	75	19.6	94
	2030	155	21.4	96
F4	2020	75	20.2	97
	2030	155	22.0	99

3.3.3 Changes of Ground Water Level

Under different water resources allocations in 2030, this paper uses Modflow Model to simulate the changes of the ground water level and analyze the ground water recharge, the changes of water discharge and the trend of ground water level in the intake area in Haihe River Basin in South-North Water Diversion Project. (Tab. 6)

(1) Shallow ground water.

In case of Precipitation Series 1980 ~ 2005, the average annual recharge and the discharge predicting outcomes of the shallow ground water under different scenes from 2007 to 2030 are listed in Tab. 6. Results show that from the increase of the water diversion in Phase I of South-North Water Diversion Project (F2) to the Water Diversion Scene in Phase II of South-North Water Diversion Project (F1), with the increase of leakage in water diversion, the surface water recharge in the shallow ground water shall increase; with the decrease of the exploitation of the shallow ground water, the return water of well irrigation shall decrease. From Scene F2 to Scene F1, the exploitation of shallow ground water shall decrease. The deeper it flows, the less it shall discharge. The phreatic water evaporation capacity shall also increase to some extent. The above features in recharging and discharging show that the recharging and discharging structure of shallow ground water are changing for the better. In general, After Project Phase II has been completed, shallow groundwater recharge and discharge change for the better but the shallow water bearing system still presents a negative balance. The excessive exploitation of shallow groundwater remains.

Tab. 6 Prediction of recharge, discharge and storage of shallow ground water in the intake area in Haihe River Basin (2007 ~ 2030)

Plan	Recharge (×10⁸ m³)									Discharge (×10⁸ m³)					Storage
	Precipitation Infiltration Recharge	Mountain Front Side Recharge	Surface Water Recahrge: Diverting Water from the Yellow River Infiltration	Surface Water Recahrge: Water Diversion Infiltration	Surface Water Recahrge: Riverway Infiltration	Surface Water Recahrge: Subtotal	Well Irrigation Return Recharge	Seawater Infiltration	Total	Exploitation	Phreatic Water Evaporation Capacity	Cross Flow Discharge	Discharge into Sea	Total	
F1	126.1	18.4	9.6	12.6	14.1	36.3	10.2	0.2	190.7	135.6	30.4	35.9	0.6	200.4	-9.7
F2	126.1	18.4	9.6	12.2	14.1	35.9	10.4	0.2	190.5	140.9	29.1	36.1	0.5	206.6	-16.1

(2) Deep ground water.

In case of Precipitation Series 1980 ~ 2005, the average annual recharge and the discharge predicting outcomes of the deep ground water under different scenes from 2007 to 2030 are listed in Tab. 7. In general, after Project Phase II has been completed, the exploitation of the deep ground water decreases to some extent. The water head of the deep pressure water is higher than that in Phase II and the cross-flow discharge and recharge between the deep layer and the shallow layer are lower. Results show that when Phase II starts to supply water, the deep water bearing system is in good conditions. Generally it is in a balanced state.

Tab. 7 Prediction of recharge, discharge and storage of deep ground water in the intake area in Haihe River Basin (2007 ~ 2030)

Plan	Recharge (×10⁸ m³)			Discharge (×10⁸ m³)				Storage (×10⁸ m³)
	Cross Flow Recharge	Seawater Infiltration	Total	Exploitation	Cross Flow Discharge	Discharge into Sea	Total	
F1	35.9	0.2	36.1	30.7	5.91	0.35	36.96	-0.86
F2	36.1	0.21	36.31	31.66	6.1	0.37	38.13	-1.82

3.3.4 Changes of salinity at main estuaries

There are two main estuaries such as Luanhe River Estuary, Haihe River Estuary and Zhangweixin River Estuary. The average annual water discharged into sea between 1956 and 2007 is 8.966×10^6 m^3. With the decrease of the precipitation and the improvement of the water resources exploitation and utilization, the water discharged into sea gradually decreases. In the 1950s, the average water discharged into sea is 24.1 billion m^3. In the 1960s, it is 16.1×10^6 m^9 and in the 1980s, it decreases to 2.22×10^9 m^3. The average water discharged into sea in Haihe River Basin between 2001 and 2007 is only 1.676×10^9 m^3. In 2007, the water discharged into sea is 4.9×10^9 m^3. The decrease of the water discharged into sea causes the off-shore salinity increases from 29‰ - 30‰ to 32‰. Estuary sedimentation becomes more serious, which significantly influences the estuary ecology and the biological resource in Bohai Sea.

After South-North Water Diversion Project is implemented, the water discharged into sea in Haihe River Basin in 2020 and 2030 under difference scene plans is between 4.97×10^9 m^3 and 7.14×10^9 m^3. See Tab. 8. The salinity at main estuaries and the off-shore salinity are recovered by 19.1% to 35.8%. The area where the salinity decreases by 10% is 15 km² to 31.6 km². The increase of the water discharged into sea plays a very important role in protecting the ecological environment at the estuaries.

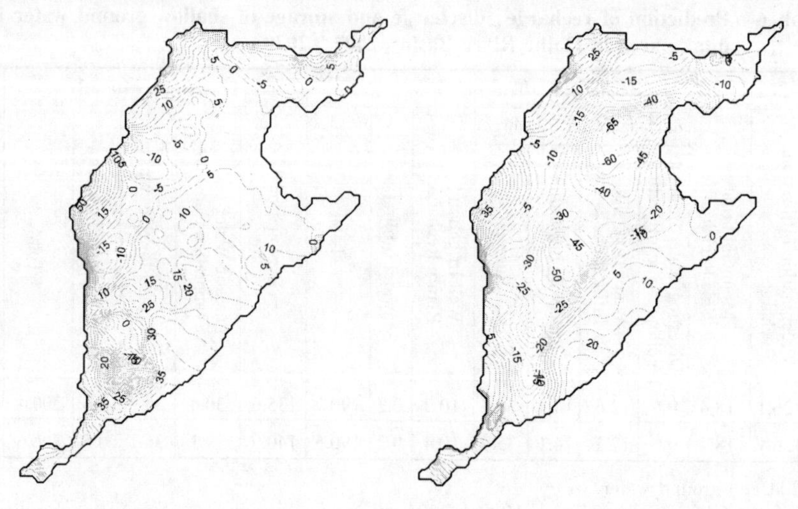

(a) Shallow Ground Water (b) Deep Ground Water

Fig. 2 Ground water isogram in the plain area of haihe river basin at the end of 2030

Tab. 8 Water discharged into sea and the recovery of estuary salinity under different scene plans

Plan	Target year	Water discharged into sea ($\times 10^8$ m^3)	Recovery (%)	Area (km^2)			
				Decrease by 10%	Decrease by 5%	Decrease by 2%	Decrease by 1%
F1	2020	65.5	30.2	26.0	125.9	634.7	1767.5
	2030	68.6	33.0	28.8	139.4	681.7	1927.6
F2	2020	62.1	26.9	22.7	110.2	579.8	1580.8
	2030	71.4	35.8	31.6	152.9	728.8	2087.7
F3	2020	59.2	24.1	19.9	96.7	532.8	1420.7
	2030	51.0	19.5	15.4	74.6	448.1	1167.7
F4	2020	66.5	31.2	26.1	129.5	642.3	1811.3
	2030	49.7	19.1	15.0	72.6	439.7	1146.8

4 Conclusions

According to different water resources allocations in the intake area in Haihe River Basin in South-North Water Diversion Project, supposed that the demand of water resources of Haihe River Basin for the economic and social development is satisfied, this paper analyzes the ecological effect of South-North Water Diversion Project and figures out the plan for optimal ecological benefits through the indicators such as ecological water demand of riverways and wetland, the target rate of the water functional divisions, the ground water level and the water discharged into sea.

In general, when satisfying the demands for water resources of Haihe River Basin for the economic and social development, South-North Water Diversion Project also improves the ecological environment of the rivers and lakes in the Intake Area, promotes the water quality in the water functional divisions and to some extent, alleviates the excessive exploitation of the groundwater and the deterioration of the ecosystem at the estuary. Therefore South-North Water Diversion Project creates obvious ecological benefits. In view of the ecological benefits of South-North Water Diversion Project, the three measures in water conservation, water diversion and waste treatment must be taken strictly at the same time as for Haihe River Basin so that the ecological environment problems caused by resource-based and pollution-based water shortages can be solved effectively and the ecological environment in Haihe River Basin shall change for the better.

References

Ren Xianshao. Complete the Comprehensive Planning and Revision for Haihe River Basin under the Guidance of Scientific Outlook on Development [J]. China Water Resources, 2007(6): 56-58.

Li Changming, Shen Dajun. Influence of South-North Water Diversion Project on Ecological Environment [J]. Discovery of Nature, 1997, 16(2): , 1997,1-6.

Lin Chao, He Shan. Survey of Water Consumption and Calculation of Ecological Water Flow in Haihe River Basin [J]. Water Resources Planning and Design. ,2003(2) 11-18.

Liu Dewen, Yu Hui, Wang Liming. Alanysis of Dirt Holding Capacity and Dirt Discharging Limit of Haihe River Basin [J]. Haihe Water Resrouces. ,2006(6): 4-6.

Haihe River Water Resources Commission of Ministry of Water Resources. Haihe River Basin Ecology and Environment Recovery Water Resource Safeguard and Planning [R]. 2005.

Water Resources Sustainable Development of Dongping Lake, a Flood Storage and Detention Area in China

Yin Junxian[1], *Zhu Xueping*[2,1], *Zhang Lili*[1], *Liang Yun*[3,1] and *Cai Wenjun*[2,1]

1. Department of Water Resources, China Institute of Water Resources and Hydropower Research, Beijing, 100038, China
2. School of Hydraulic Engineering, Dalian University of Technology, Dalian, 116024, China
3. School of Environmental Science and Engineering, Donghua University, Shanghai, 201620, China

Abstract: On the basis of surveying the current situation of water resource utilization and protection and the water ecological tourism resources of the Dongping Lake, which is the most important flood storage and detention area of the lower Yellow River, the challenges and opportunities of Dongping Lake caused by the implementation of the East-Route of South-to-North Water Transfer Project (ERSNWTP) were analyzed. This paper puts forward the ideas of development of the Dongping Lake from the angle of strategy of sustainable development of water resources, and suggests to adjust the old Dongping Lake into comprehensive utilization reservoir, increase surface water, and construct the water ecological protection pattern called "one lake two corridors" and improve the water quality and hydrology monitoring network system.

Key words: water resource, sustainable development, Dongping Lake

1 Introduction

Dongping Lake is located in the bar depressions which caused by the Yellow River and Dawen River intersection, in the low of the Dawen River and the west of Dongping County, with the latitude of 35°30' ~ 36°20' and the longitude of 116°00' ~ 116°30'. The total surface area is 627 km^2 including the new and old part. Dongping Lake pile on the water from Dawen River, connected the canal in the south, communicated with Xiaoqing River and the Yellow River. It is the second largest fresh water lake in Shandong Province, which is rich in water resource. Dongping Lake is the largest flood area in the lower Yellow River. Its main effect is to cut the Yellow River flood peak, regulate the flood of the Yellow River and Dawen River, and control the discharge of the Aishan hydrological station no more than 10,000 m^3/s.

Dongping Lake plays an important role in delaying the flood of the Yellow River and the Dawen River and ensuring the safety of flood control. In the 1950s, the Yellow River Huayuankou stand occurred many times larger floods, such as the 1953 flood (peak flow of 11,200 m^3/s), the flood in 1954 (peak flow of 15,000 m^3/s), the flood in 1957 (peak flow of 13,000 m^3/s), the flood in 1958 (22,300 m^3/s). In all of these floods, Dongping Lake played an role in peaking the flood, bursts Aishan from the disaster. In 1982 August 6, Huayuankou happen floods peak flow of 15,300 m^3/s. The Yellow River water levels above the Dongping Lake are about 1 ~ 2 m higher than the highest water level in 1958, to ensure the lower Yellow River safety, opened the two brake called Linxin, Shilibao, use Dongping Lake peak the flood, the maximum points is 2,400 m^3/s, a total of more than 4.00×10^8 m^3/s of water storage, solve in the lower Yellow River flood threat.

But due to the construction of storage area, the people made a huge sacrifice and the economic development is restricted, and regional economic development is relatively backward too. In 2002, the first stage of ERSNWTP began to implement, as the last level of ERSNWTP, in order to guarantee a safety supply of ERSNWTP, a large number of small processing enterprise surrounding the Dongping Lake has been closed up, and culture in net cage in the lake has been districted too, which make people's loss of original economic income source. The country has issued relevant policy but failed to improve regional people's life level. So, with the opportunity of ERSNWTP implementation, by adjusting measures to local conditions of developmental development strategy, and looking for the sustainable development about Dongping Lake, the local government is trying to make Dongping Lake in flood control, water supply and ensure complete shipping such tasks, at the same time, to make Dongping Lake area people reach common prosperity, and realize the out of poverty. This paper, based on the present situation of water Dongping Lake and the analysis of the challenges and opportunities from the ERSNWTP , is to make full use of regional advantage resources, to realize the sustainable development of water resources and local economy , and to raise the utilization ratio of resource improving the living level.

2 Current situations of water resources

2.1 Situation of water resources utilization

There are three main sources of fresh water resources in Dongping Lake region, including natural precipitation, water of Dawen River and groundwater. In addition, it also contains irregular flood diversion water of the Yellow River. Among them, the key water supply source of Dongping Lake is the incoming water of Dawen River. The Dawen River originated form the Sa Yazi Village located in the southern foot of mountain in Laiwu City, the average water entering into Dongping Lake was about 8.73×10^8 m^3 (1980 ~ 2010) for many years. Moreover, nearly 80.73% (about 7.04×10^8 m^3) of the water concentrates in the flood season, which mostly flows into the Yellow River as disposable water.

To ensure the flood control safety of the lower Yellow River, Dongping Lake plays an important role in controlling the flood as a flood storage area. However the storage function is not obvious, leading to a lower-usage of local surface water resources. According to statistics, the water resources utilization efficiency is only about 5% currently in Dongping Lake region, mainly as a form of groundwater. The usage of groundwater has reached 72.8%, but the surface water resources have a very low utilization, which have greater potential for development. With the economic development in Dongping Lake area, the contradiction between water supply and demand will become increasingly prominent. Meanwhile, we should take the premise of flood safety into consideration, give full play to the regulation and storage, store the incoming water of Dawen River, and improve local water resources utilization and increase rate of water supply.

2.2 Situation of water resource protection

From 1998 to 2004, the water quality of Dongping Lake has a sharp deterioration, which is level V or less every year. To protect the operational safety of water quality for ERSNWTP, a series of measures have been taken to cut off water pollution sources in Dongping Lake. The water quality is significantly improved in lake region and reached the present class Ⅲ ~ Ⅳ, compared with the previous situation—class Ⅴ. However, there are still certain gaps with the goal of the Ⅲ standard of the national surface-water environmental quality as the criterion in the North Water Transfer Project that the State Council on the general Planning of South to North Water Diversion Project approval (the letter [2002] No. 117). Because water quality of inflow rivers is poor in Dongping Lake region, there are heavy pollution treatment in Dawen river and its tributaries.

2.3 Water eco-tourism resources

There are rich aquatic resources, mountains and rivers, trees, a unique ecological landscape in Dongping Lake. The ancient and mysterious historical changes of Dongping Lake accumulated rich cultural connotations. It has prolific tourist resources: ①Water leisure and tourism resources. The age-old water culture and pleasant natural scenery provide a good basis for water area in Dongping. ②Heritage and cultural tourism resources. Dongping has a long history and a special geographical location. A large number of cultural relics and ruins of the Water Margin heroe activities has been left over in this area. The rich natural landscape bears the deep Water Margin cultural, and highlights the market value through the outlaws of the Marsh culture. Dongping Lake has been designated as the major attraction of the mountain East – West Line and tourist areas and the "Water-Margin line" by the provincial government in 1986. ③ Eco – tourism resources. Some staggered landscape of mountains, hills, plains, lakes has formed eco-tourism area, with the advantage of developing eco-tourism resources.

3 Opportunities and challenges of ERSNWTP

The water diversion of the first phase of ERSNWTP brings opportunities and challenges to the water resources sustainable development of Dongping Lake, such as follows.

3.1 Provide stable water supply source for Dongping Lake

The diversion stage of ERSNWTP is from October to the following May. The diversion period is just the dry stage of Dawen River when it appears little runoff about 1.69×10^8 m^3 (the average amount of 1980-01 ~ 2010-12). The first phase of ERSNWTP diverts water into the Dongping Lake with a flow of 100 m^3/s, and out of the lake with a flow of 100 m^3/s including 50 m^3/s diverts through the Yellow River and 50 m^3/s supply for Shandong Province. The project diverts water stably into the lake, which provides stable water supply source for economic and social development of Dongping Lake.

3.2 Maintain the stable water level condition for Dongping old lake

During the first phase of ERSNWTP (start diverting water in 2013), the characteristic water level of Dongping is controlled. ① The water level during July to September is controlled at about 42.0 m and can be raised to 42.5 m during October. ② When the water level of the old lake is below 40.51 m, the project pumps water from the Yangtze River for reservoir storage and the maximum water level is controlled under 41.51 m. ③ When the water level of the old lake is higher than 41.51 m, Dongping Lake makes a water balance of the inflow and outflow of ERSNWTP during the diversion stage. ERSNWTP maintains the stable water level for Dongping operation, which not only provides possibility of landscape water supply but also provides good conditions for developing ecological tourism.

3.3 Offer engineering construction for regional water resources application

The engineering constructions of ERSNWTP water diversion canal helps regional water resources application. The regional government can develop the regional water resources as a water supply source for Jiaodong area based on the Jiping main canal, as well as water supply source for North Shandong Province through Weihe River out lake brake and via the wear Yellow River construction. The measure not only improves the utilization ratio of local water resources, but also improves the regional residents' living standard by collecting water resources fee for increasing the regional residents' income.

3.4 Promote regional economic development

The shipping of Beijing—Hangzhou Grand Canal extends mileage to the south bank of the Yellow River, which will greatly improve the regional transportation structure. The shipping industry promotes the economic development of Dongping Lake.

However, the diversion stage of ERSNWTP may overlap the flood stage of Dawen River, which brings challenges for the flood control function of Dongping Lake to some possible. Shipping will use some water, and water transportation produced water pollutants which influences the surrounding water environment. Shipping will increase the sudden water pollution risk of Dongping Lake, the important main canal of Beijing-Hangzhou Grand Canal.

4 Suggestions for water resources sustainable development

After ERSNWTP diverts water and Beijing-Hangzhou Grand Canal resumes shipping, the storage function of Dongping Lake becomes important since it is the last cascade reservoir of ERSNWTP. Dongping Lake should implement the strategy of water resources sustainable development, make full use of local resources advantage, develop the economic green circle mainly including green tourism and ecological farming, and develop into a holiday tourism economic circle primarily of ecology leisure. The regional transportation and the third industry including wholesale, retail, and catering, housing, will get further development, bolstered by sustained growth in tourism. And then promote the regional economic and social sustainable development. Some suggestions are as follows.

4.1 Adjust the old lake of Dongping into a reservoir of comprehensive utilization

As the last cascade reservoir of ERSNWTP, the storage function of Dongping Lake becomes important after ERSNWTP diverts water and even more important when the third phase of ERSNWTP constructed. The lower Yellow River management of the lake is sure to build flood control constructions, such as standardized embankments, dredged waterway, and flood storage and detention area. Furthermore, Xiaolangdi project decreases the use probability of flood storage and detention area function of Dongping Lake at some extent. Under this situation, we suggest that the function of Dongping Lake will be adjusted from a flood storage area only for flood control to a reservoir of comprehensive utilization which mainly for flood control, but also give consideration to water supply, shipping, ecological farming, ecological tourism and other multi-function using, remaining the flood storage and detention area function as the same.

4.2 Increase the utilization of surface water

Under the precondition of the assuring safety of flood control, people are suggested to make full use of the flood resources. Dongping Lake plays the role of water storage to store the water of Dawen River. The measure improves local surface water utilization efficiency of water, and increases the availability of the Yellow River water resources as well. The regional government adopted a series of pollution treatment measures. Dongping Lake will maintain water quality of the surface water above the Ⅲ class standard, making it a surface water supply source. Therefore, collection of water resources fee by increasing the local water supply for Jiaodong area and North Shandong Province will improve the regional residents' living standard. On the other hand, Dongping Lake water, instead of underground water, can be considered supplying water for the county seat of DongPing.

4.3 Construct the water ecological protection pattern of "One Lake Two Corridors", and improve the network system of water quality and hydrology monitoring

After ERSNWTP diverts water, Dongping Lake mainly accepts water comes from Dawen River and ERSNWTP. Then the pollutants of Dongping old lake mainly come from the Dawen River pollutants, ERSNWTP pollutants, the shipping pollutants and the lake district raising pollutants. Local water ecological protection pattern of "One Lake Two Corridors" is suggested to be constructed to against these pollutants and satisfy the Ⅲ class standard of the surface water quality of ERSNWTP. Strengthen water quality monitoring system and construct ecological protection network will strengthen the emergency response ability of the sudden water pollution incidents.

5 Concluding Remarks

The construction and implementation of the first phase of ERSNWTP and the recovery of the Beijing—Hangzhou Grand Canal firstly bring opportunities for water resources sustainable development of the Dongping Lake. The projects also create conditions for adjusting old Dongping Lake into multi - purpose plain reservoir, increasing the utilized amount of surface water and developing properly eco - tourism and eco - farming industry. However, it brings some challenges for flood control and water quality protection of the lake region, calls for improvements of the network system of water quality and ecological protection to protect water quality safety. Implementation of sustainable development strategy of regional water resources is the key way to improve resources utilization and living standards in Dongping Lake region.

Acknowledgement

We would like to acknowledge the National Critical Patented Projects in the Control and Management of the National Polluted Water Bodies (Grant No. 2012ZX07205005), the National Critical Patented Projects in the Control and Management of the National Polluted Water Bodies (Grant No. 2012ZX07601001), the National Natural Science Foundation of China (Grant No. 51009150), and the Beijing's Project of Science and Technology Plan, China (Grant No. Z111100074511005). The authors are indebted to the editors and reviewers for their valuable comments and suggestions.

References

Wang S. M., Dou H. S., Lakes' Notes in China [M] Beijing: Science Press, 1998.

Dongping Lake Note Editor Committee of the Yellow River WeiShan Construction Bureau in Shandong Province. Dongping Lake's Note [M]. Jinan: Shandong University Press, 1993.

Fan Z. W.. Discussion of Emigrant Poverty and Minimum Living Standard Security of Dongping Lake Region [J]. Yellow River, 2007, 29(7): 73 - 74.

Zhao C. L., Chen H. S., Cao H. S.. Discussion of Developmental Resettlement of Dongping Lake Region [J]. Shandong Water Resources, 2000 (10): 36 - 38.

Lai B.. Research of Tourism Poverty Alleviation for Emigrant of Dongping Lake Region [J]. Yellow River, 2007, 29(6): 76 - 78.

Shi N. Y., Hu P., Zhou J. J., et al. Current Situation and Main Problems of Water Resources Management, Dongping Lake [C]//. The Yellow River Water Resources Management Research Proceedings, 84 - 87.

Chen S. Y., Dong J., Zhang C. Y.. Countermeasure Study on Ecoenvironmental Status of Dongping Lake and Sustainable Development of Its Catchment [J]. Journal of Anhui Agri. Sci., 2007, 35(5): 1436 - 1437.

Study on Effective Utilization of Water Resources of Qiantang Estuary

You Aiju*, *Han Zengcui* and *Lu Xiaoyan

Zhejiang Institute of Hydraulics and Estuary, Hangzhou, 310020, China

Abstract: Runoff is one of the key factors that determine the steady of the environment of Qiantang Estuary. Therefore, effective utilization of water resources of Qiantang Estuary should weigh environmental water requirement and human water requirement. Based on the study of the water resources characteristics, this paper put forward the effective utilization pattern of water resources comprising water resources allocation in macroscopic lay, aiming at restricting water intake volume to guarantee sufficient runoff into the sea; and water resources regulation in microscopic lay, aiming at ensuring periodic water quality safety.

Key words: water resources, effective utilization, water resources allocation, water resources regulation, Qiantang Estuary

1 Introduction

Qiantang River is from Fuchun power plant to Luchaogang cross section, the length is 282 km (Han, Dai and Li, 2003). River reach from Wenjiayan to Ganpu is used to being called estuarine reach, where the dominated motive forces are runoff and tide alternatively. The upstream of the estuarine reach is the fluvial reach where the dominated motive force is runoff, and the downstream, used to being called Hangzhou Bay, is the tidal reach where dominated motive force is tide.

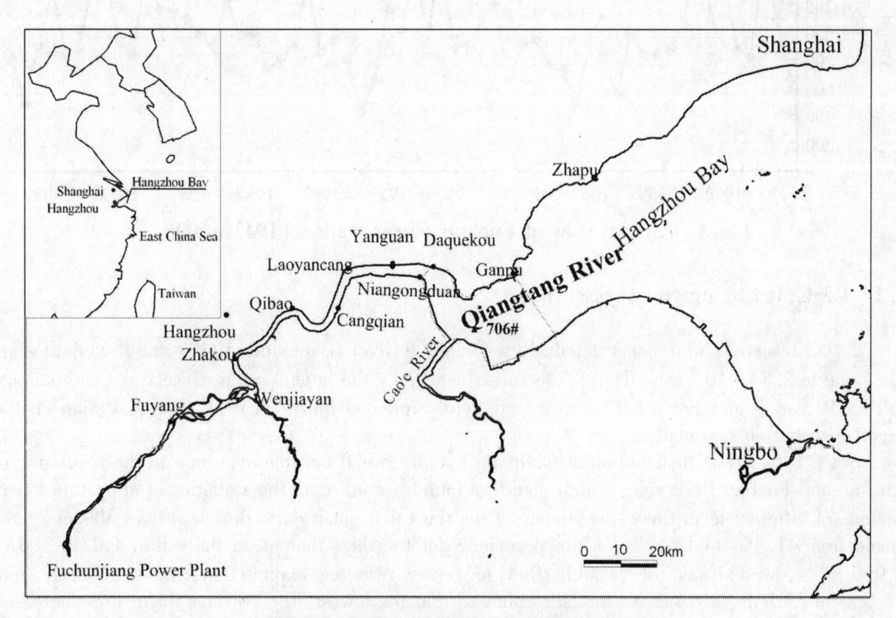

Fig. 1 Layout of Qiantang Estuary

Plain of Qiantang Estuary, which is divided into Hangzhou, Ningbo, Shaoxing, Jiaxing Cities, and occupies 20% land, 35% population, and 55% GDP of Zhejiang province, is the most developed area in Zhejiang province. Since 1990s, due to the population growth, economic development and local water quality pollution, water resource deficiency of Qiantang Estuary plain was becoming more and more serious. Therefore, the demand to divert freshwater from Qiantang estuary was very urgent. However, Qiantang Estuary is a macro tide estuary with big tide range, violently deformed tide wave, abundant sediment from sea and changeable river bed, accordingly, so the environment of the estuary change is very complex. Runoff is one of the key factors determines the steady of the estuarine environment, therefore, effective utilization of water resources of Qiantang Estuary must leave sufficient runoff into the sea, and make sure water quality security at the same time.

2 Water resources characteristics of Qiantang Estuary

2.1 Uneven seasonal and yearly distribution

According to the 1947 ~ 2006 observed flow data, perennial flow of Fuchun power plant is 954 m^3/s, while 70% concentrates in the rainy season from April to July. Meanwhile, the runoff changes largely from year to year, the observed maximum mean annual flow was 1,705 m^3/s in 1954, and the minimum mean annual flow was 411 m^3/s in 1979. Besides seasonal and annual change, the runoff has a 7 ~ 10 a change period, and the difference between the successive wet years and dry years could be 1.5 times, see Fig. 2. The uneven distribution of runoff is the nature restrict of water resources utilization.

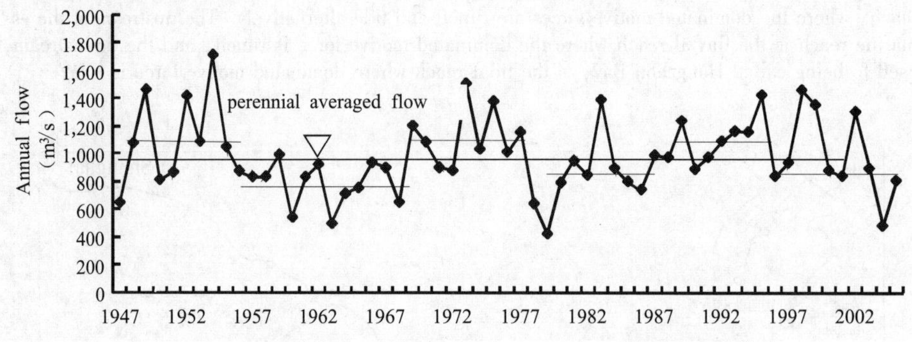

Fig. 2 Annual flow of Fuchun power station (1947 ~ 2006)

2.2 Regulated by upstream reservoirs

2,055 reservoirs were constructed in the Qiantang River Basin since 1949, and the totally storage is up to 2.83×10^{10} m^3. Those reservoirs, especially the large Xin'an Reservoir (with storage 2.16×10^{10} m^3) and Fuchun Reservoir, efficiently reduced the flood peak flow and changed the runoff distribution seasonally.

Since 1960, time distribution of Qiantang Estuary runoff became even due to the regulation of Xin'an and Fuchun Reservoir, which could be found according to the statistics of appearance percentage of different level flow. Percentage of the days that mean daily flow less than 300 m^3/s reduced from 41.8% to 18.3%, while percentage of the days that mean daily flow fall on 300 ~ 1,000 m^3/s, which was the feasible flow for water resource exploitation, was enhanced from 27.9% to 53.9%, see Fig. 3. Statistics showed that the lowest 30 ~ 90 days mean flow increased 260 ~ 240 m^3/s in average year, and 100 ~ 200 m^3/s in 70% ~ 90% guaranteed year. In general, the construction of basin hydraulic engineering could effectively enhance dry flow and resist saltwater intrusion, that is helpful for fresh water develop and use.

2.3 Water demand for balance of sediment transport is large

Cross section of the estuarine reach is the long term dynamic balance of the landward and seaward sediment transportation. Due to shallow water friction effect, the duration of flood tide decreased gradually while trace to upstream, on the contrary, the duration of ebb tide increased gradually. Thus, in dry season, the flood tidal volume equal to ebb tidal volume because runoff occupied a small part in the ebb tide flux, accordingly, flood velocity is higher than ebb velocity and flood sediment concentration is higher than ebb sediment concentration, and result in upstream deposition and downstream erosion. In flood season, with the increasing of flow, ebb tidal volume and ebb velocity increased significantly and sediment carrying capacity enlarged, resulting in upstream erosion and downstream deposition.

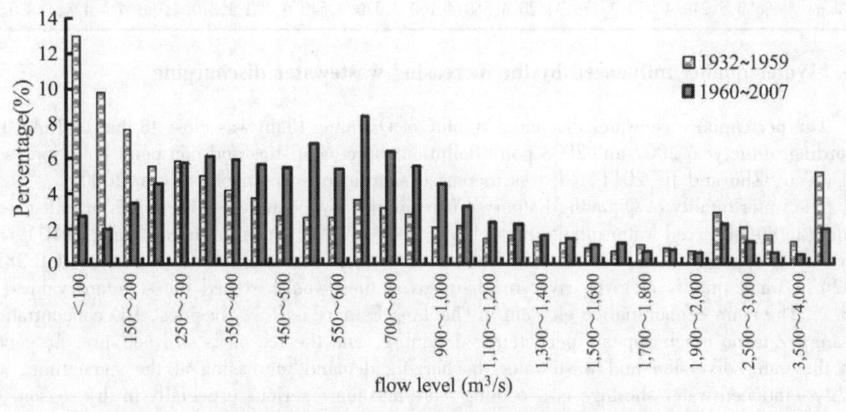

Fig. 3 Appearance day percentage of different flow level

Therefore, runoff is the key factor to maintain the sediment transport balance of the estuary, and the diversion of water resource could cause deposition of the whole reach. Research shows that water demand for balance of sediment transport is no less than 80% of the total water resources (Han, You and Xu, 2006). This is the main restrict factor of water resources use in Qiantang Estuary, typically different from other weak tide estuaries.

2.4 Water quality obviously influenced by saltwater intrusion

The flow discharge to the sea of Qiantang estuary is 3.4% of the Yangtze River, however, the mean tide range is up to 5.6 m, more than 2 times of the Yangtze River. Therefore, saltwater intrusion intension is larger than the Yangtze River. The intensity of saltwater intrusion is proportion to tide range, and inverse to runoff.

Longitudinal distribution of water chlorinity along Qiantang Estuary is declined with the tide trace to upstream, and the tail end of saltwater intrusion site is about Wenyan cross section. Tab. 1 lists the mean chlorinity in 1995 - 2006 of the observational stations along Qiantang Estuary. From Tab. 1, the chlorinity at Ganpu station is high and close to that of sea water, while go upstream, due to the runoff dilution, water chlorinity declined gradually.

Coefficient of Deviation Mean (CDM) is the main index which reflects the variable amplitude of water chlorinity, and it increased gradually when go upstream, and achieved the peak value at Qibao—Zhakou reach. It is because Qibao—Zhakou reach is the zone where runoff and tide act as the driving force alternatively, and result in the frequently change water chlorinity. Therefore, though the mean chlorinity of Qibao—Zhakou reach is relatively low, but for the given period, the water chlorinity could be higher than the criterion for safety (chl ≥ 250 mg/L), and it normally happens in middle of July to November, the dry season with spring tide period, when the irrigation demand is high as well.

Tab. 1 Characteristics variability of chlorinity along Qiantang Estuary

station	Mean annual chlorinity (mg/L)												Mean value	CDM
	1995	1996	1997	1998	1999	2000	2001	2002	2003	2004	2005	2006		
Zhakou	60	40	20	30	20	30	20	20	70	20	20	20	35	0.65
Qibao	330	330	110	170	110	200	140	60	430	170	80	110	207	0.67
Cangqian	850	790	380	390	320	500	380	210	850	840	320	470	560	0.52
Yanguan	2,160	2,620	2,000	1,130	1,330	2,020	1,840	1,160	2,140	2,970	1,490	1,830	1,925	0.38
Ganpu	4,650	5,240	4,970	2,470	3,520	4,580	4,160	3,320	4,540	6,490	3,870	4,790	4,478	0.35

2.5 Water quality influenced by the increasing wastewater discharging

The per capita wastewater discharge volume of Qiantang Plain was close to that of Taihu Basin according to the year 2007 and 2008 point pollution source statistics and non point pollution estimation (You, Zhu and Ji, 2011). It was increased significantly compared to year 2000.

The water quality of Qiantang Estuary upstream of Laoyancang cross section is overall good according to the observed water quality data of year 2001 to 2008 (You, Zhu and Tian, 2010). However, when appraised comply with "Environmental quality standards for surface water (GB 3838—2002)", water quality of given river reach in given time would exceed the standard values, See Tab. 2. The main contamination element in Qiantang Estuary is P, N, besides, DO concentration in Fuyang—Zhakou reach appears persistently declining, and the reason is still not sure. In general, with the water diversion and wastewater discharging demand increasing at the same time, water quality - induced water shortage is becoming more and more serious especially in dry season.

Tab. 2 Water quality class of Qiantang Estuary

Cross section	river	Water quality target	year water quality class							
			2001	2002	2003	2004	2005	2006	2007	2008
Zhaixi		Ⅱ	Ⅲ	Ⅲ	Ⅲ	Ⅳ	Ⅲ	Ⅲ	Ⅲ	Ⅲ
Lushan	Fuchun	Ⅱ	Ⅲ	Ⅲ	Ⅲ	Ⅲ	Ⅲ	Ⅱ	Ⅱ	Ⅲ
Fuyang	River	Ⅲ	Ⅲ	Ⅲ	Ⅲ	Ⅱ	Ⅲ	Ⅱ	Ⅱ	Ⅱ
Yushan		Ⅱ	Ⅲ	Ⅲ	Ⅲ	Ⅲ	Ⅲ	Ⅳ	Ⅳ	Ⅱ
Yuanpu		Ⅱ	Ⅲ	Ⅲ	Ⅲ	Ⅳ	Ⅲ	Ⅳ	Ⅲ	Ⅱ
Zhakou	Qiantang	Ⅱ	Ⅲ	Ⅲ	Ⅲ	Ⅲ	Ⅲ	Ⅳ	Ⅲ	Ⅲ
Qibao	River	Ⅲ	Ⅲ	Ⅲ	Ⅲ	Ⅲ	Ⅲ	Ⅲ	Ⅲ	Ⅲ
Zhetoujiao		Ⅲ	Ⅳ	Ⅳ	Ⅳ	Ⅲ	Ⅳ	Ⅲ	Ⅲ	Ⅲ

3 Effective utilization of water resource

3.1 Framework

Considering the characteristic of water resource of Qiantang Estuary, effective utilization of water resources include not only to control water quantity by water resources allocation in macroscopic lay, but also to guarantee water quality by water resources regulation in microscopic lay, see Fig. 4.

Water resources allocation realize effective water resource utilization by the means of policy, organization system, water management and engineering construction based on overall study on water demand and water supply, relationship between human water demand and ecological water demand, rules of justice and efficiency. Water resources regulation aims at guaranteeing the water quality of water intakes accord with the demanded criterion based on saltwater intrusion and water quality prediction, by means of enlarging discharge volume of upstream water monitoring engineering to adapt to the tide change, and reducing water loss as well.

3.2 Water resources allocation of Qiantang Estuary

3.2.1 Available water resources estimation

Available water resources means the water resources developed on the premise of rational capital investment, feasible technique, and permissible environment capacity, which is unrepeatable, and usually calculated by total water resources minus unavailable water resources and environmental water requirement. The unavailable water resources of Qiantang Estuary is mainly flood water resources, however, it is useful to keep sediment transportation in balance, and is the main component of environmental water requirement. Therefore, available water resources of Qiantang Estuary could be estimated by total water resources minus environmental water requirement. Available water resource of Qiantang Estuary is 20% of the total water resources according to relative researches, close to 6.9×10^9 m^3 (Han, You and Xu, 2006). Available water resource determination is the basic for proceeding water resource allocation.

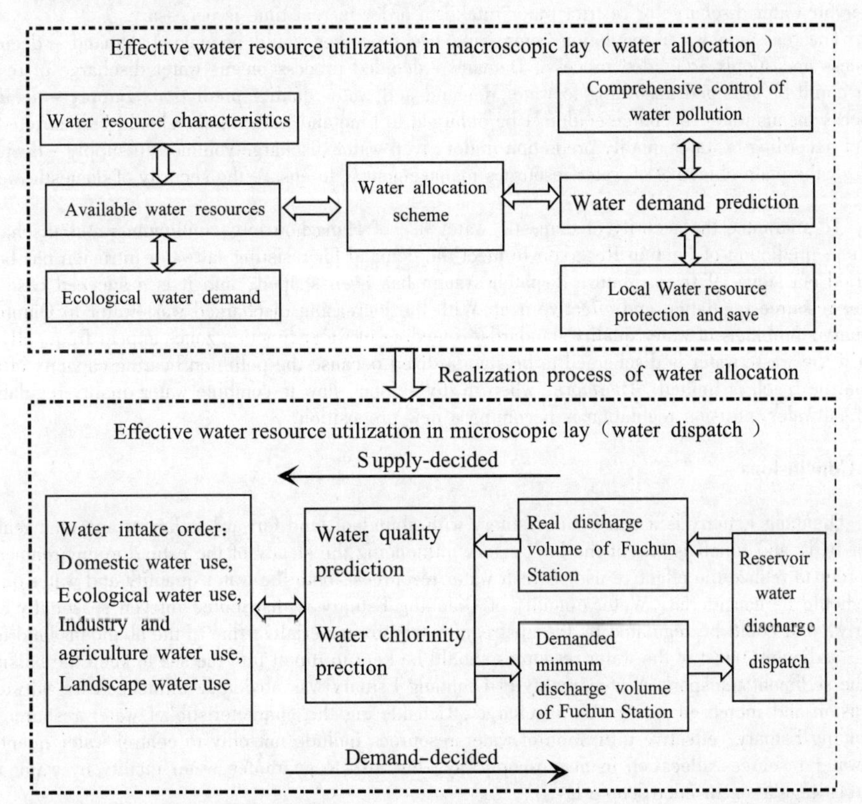

Fig. 4 Framework of effective water resources utilization of Qiantang Estuary

3.2.2 Water resource allocation rules

Firstly, harmonize the water requirement of different region and different industry, former users have priority in water intake, and the total amount of water intake of each user was controlled. Secondly, daily water intake should conform to following priority: domestic water use, industry and agriculture water use, landscape water use and others. Thirdly, lay stress on exerting integrated efficiency of water resources.

3.2.3 Water resources allocation outcome

Annual water resources required to divert from Qiantang Estuary would increase to 4.29×10^9 m^3 in 2020 according to water demand prediction, allocated to domestic water use would be 1.68×10^9 m^3, industry and irrigation water use would be 5.45×10^8 m^3, and landscape water use would be 2.065×10^9 m^3.

3.3 Water resource regulation

Though the mean annual water requirement is less than the mean available water resources, due to the seasonal runoff change, the ratio of water demand over water resources in different season could be very different, and water quality is changeable as well. Therefore, in the dry season when water use is concentrated and runoff is low, to guarantee water quantity and quality by regulating reservoir water discharge or restrict water intake in order in real time is necessary.

The real time water regulation comprising two processes could be named demand - decided process and supply - decided process. Demand - decided process means water discharge of reservoir could be enlarged according to water demand and water quality prediction. Supply - decided process means water discharge couldn't be enlarged in time and water intake should be restricted in order according to water quality prediction under given water discharge volume. In supply - decided cases, the main objective of water resources management is to ensure the security of domestic water use.

To guarantee the security of domestic water use of Hangzhou city, minimum water discharge volume prediction of Fuchun Reservoir to meet the demand for resisting saltwater intrusion has been carried out since 1970's. Mature dispatch system has been shaped, and it is a succeed case for water resource regulation and effective use. With the increasing discharged wastewater to Qiantang Estuary, problems of water quality standards exceeding in water function zones appear frequently even if the waste water is discharged to be standardized because the pollution bearing capacity of the estuarine reach is limited. Therefore, when in dry season, how to combine water quality regulation into saltwater intrusion regulation is becoming a new proposition.

4 Conclusions

Qiantang Estuary is a macro tide estuary with abundant land forward sediment and changeable river bed, and runoff is one of the key factors influencing the steady of the estuarine environment. In order to realize the effective use of fresh water resources, both the water quantity and water quality should be considered. Water quantity of Qiantang Estuary is distributed uneven seasonally and yearly, and could be regulated by the upstream reservoirs materially; due to the abundant land forward sediment, most of the water resources should be kept to runoff into the sea to keep the balance of the sediment transport. Water quality of Qiantang Estuary was obviously influenced by saltwater intrusion and increased wastewater discharge. Considering the characteristic of water resource of Qiantang Estuary, effective utilization of water resources include not only to control water quantity by water resources allocation in macroscopic lay, but also to guarantee water quality by water resources regulation in microscopic lay.

References

Han Zengcui, Dai Zeheng, Li Guangbing. Regulation and Development of Qiantang Estuary[M]. Beijing: China WaterPower Press, 2003.

Han Zengcui, You Aiju, Xu Youcheng. Calculation Methods of Environmental and Ecological Water Demand for Macro – tidal Estuary[J]. Journal of Hydraulic Engineering, 2006(4):16 – 22.

You Aiju, Zhu Junzheng, Ji Shenghua. Estimation Method of Pollution Load in Qiantang Estuary and Its Application[J]. Journal of Zhejiang Hydrotechnics, 2011(1):10 – 13.

You Aiju, Zhu Junzheng, Ji Shenghua. Present Situation and Control Countermeasures on Water Environment of Qiantang Estuary [J]. Environmental Pollution & Control, 2010,32(5):92 – 96.

Study on the Biodiversity Responses to Different Disturbance Gradients in the Wetland of the Yellow River Floodplain Area

Ding Shengyan and *Wang Yuanfei*

College of Environment & Planning, Henan University, Kaifeng, 475004, China

Abstract: Based on the overall vegetation investigation and the usage of the SPSS software, the plant species diversity and community diversity of the wetland in the Yellow River floodplain area (Heigangkou—Liuyuankou) were firstly analyzed in this paper. The landscape heterogeneity of the study area was also quantitatively analyzed by the RS, GIS technology on the basis of the SPOT high - resolution images and the LUCC data. By the comparison of species diversity in different development stages of the wetland (the old floodplain area, the middle floodplain area, the tender floodplain area), and the correlation analysis of the landscape pattern indexes with the wetland community diversity in different spatial scales, the study results showed that the plant species diversity decreased when the human disturbance gradients gradually increased, and the plant community diversities were significantly correlated to the landscape heterogeneity of the wetland.

Key words: wetland, biodiversity, Human - induced disturbance regime, the Yellow River

1 Introduction

Wetland is one of the ecosystems which has very important environmental functions and various biological habitats, and now it is also the most threatened ecosystem on earth (Wang et al., 2003). River wetland is mainly distributed along the rivers and in the plain area of middle - lower reaches, and has a large amount of population, therefore it bears much more pressure from the social and economic development (Liu, 2002). Biodiversity is the foundation of human beings survival and development, however, the increased human activities have caused serious damage to species diversity of wetland landscape ecosystem in recent years, and especially some rare species have lost at the speed higher than ever before (Miska, et al., 2003; Liu et al., 2001; Jiang et al., 1998). The Yellow River watershed is the cradle of the Chinese civilization; the beaches of the Yellow River are not only the flood detention and sand desilting area, but also the land resources for human production and life (Wang et al., 2002; Liu et al., 2008). Under the influences of agricultural development and landscape pattern dynamics, the species diversity of the natural community has changed significantly in the wetland landscapes of the Yellow River (Jiang, 1997; Ding et al., 2004). Therefore, to maintain the sustainable and healthy development of the ecological environment, and to coordinate the relationship between the regional economic development and the nature environment, there is important significance in studying the biodiversity of the wetland and exploring how the biodiversity responds to the human disturbance and landscape development in the Yellow River floodplain area.

2 General situations of the study area

The study area (114°12' ~114°52'E, 34°52' ~35°01'N) is located in the north of Kaifeng city in Henan province, 10 km far from the downtown along the Yellow River embankment, the study area starts from the Bianjing floating bridge of the Shuidao county and to the east of the Yellow River floating bridge of Liuyuankou county (Fig. 1). On Jun 9 th of 1994, the Liuyuankou was approved to establish a natural resources reserve for protecting the wetland ecological environment and winter migratory birds by province government, and it was also put in the list of the World

Wetland Directory. The climate of the study area belongs to the continental monsoonal climate that has a clear distinction between the four seasons, and the precipitation is concentrated in summer. The annual mean temperature in this area is 14.0 ~ 14.2 ℃, the annual frost free days are 207 ~ 220 d, and the annual mean precipitation can get to 627.5 ~ 722.9 mm. The study area has flat terrain, deep and fertile soil that is advantageous for planting of all kinds of crops including economic crops, and the forest coverage of this area is much higher than the mean coverage of forest in the whole country.

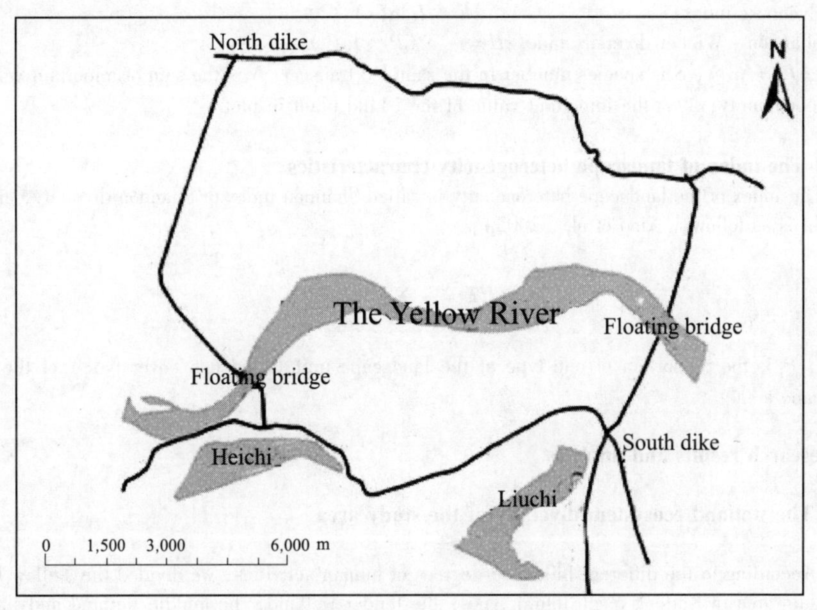

Fig. 1 Illustrative diagram of the study area

3 Materials and methods

3.1 Sample set

By the transect method, the study sites and the plots were selected and set along the south and the north embankments (Liuyuankou—Heigangkou) of the floodplain area of the Yellow River respectively. The investigation contents were the species, quantity, height, abundance and coverage of all occurred weed species. The environmental factors such as geographic coordinates, altitude and slope, and the human disturbance situations, for example, the human disturbance intensities and the surrounding land use status were also recorded during the plant species survey.

3.2 Data sources and processing platform

The data of this paper were the 1:50,000 digital grid topographic map (1975) provided by the national basic geographic information center; the land use status atlas of Henan province (1997); the SPOT remote sensing image of 2004 (resolution of 2.5 m after inosculating); landscape pattern and geographic materials by field investigation and GPS data. And the data processing systems were mainly the ERDAS and ARCGIS.

3.3 Index calculation

3.3.1 Species richness, evenness and diversity indices in different communities

The computing formulas of species richness (R), evenness index (E) and Shannon - Wiener diversity index (H) are as follows:

Richness: $R = S$

Evenness index: $E = H/\ln(s)$

Shannon - Wiener diversity index: $H = -\sum(P_i * \ln P_i)$

where, $P_i = N_i/N$, S is species number in the sampled transect; N is the sum of important value of plant community; N_i is the important value of the i kind plant in plot.

3.3.2 The index of landscape heterogeneity characteristics

The index of the landscape heterogeneity is called Shannon index or Shannon diversity, and its formula is as follows (Xiao et al., 2003):

$$HT = -\sum_{i=1}^{m} P_i \ln P_i$$

where, P_i is the proportion of one type of the landscape unit (land use/cover type) of the total landscape area.

4 Research results and analysis

4.1 The wetland ecosystem diversity of the study area

According to the different influence degrees of human activities, we divided the Yellow River floodplain area in Kaifeng reach into 3 types: the tender wetland, the middle wetland and the old wetland. The old wetland is mainly covered by cultivated plants that are wheat in the summer and bean, corn and rice in the fall. Because both the middle wetland and the tender wetland are always flooded in rainy season, the crop which is mainly wheat just can be planted in winter and spring. Due to the extensive cultivation, the crops usually grow in mix with many grass species in these areas, and they include *Phragmites australis*, *Sesbania cannabina*, *Polygonum amphibium*, *Echinochloa crusgali*, *Polygonum lapathifolium* and others.

4.1.1 The species composition characteristics of the wetland in the tender floodplain area

There are 33 kinds of weed species in the wetland of tender floodplain area. The dominant species of the community are *Aster tataricus* and *Phragmites australis*, their important values are 12.08 and 10.43 respectively; The sub - dominant species are *Paspalum paspaloides*, *Cyperus michelianus* and *Calamagrostis pseudophragmites*, which have the important values all more than 7.0; The associated species in this area are *Eclipta prostrate*, *Cynodon dactylon*, *Xanthium sibiricum*, *Typha minima*, *Polygonum hydropiper* and *Polygonum amphibium*. In general, there are 23 species having relatively lower important values less than 3.0, but accounting for more than 50% of the whole species; and also 12 species' important values are at 3.0 ~ 1.0, the other 11 species' are less than 1.0. They are enriching the plant diversity of the wetland in the tender floodplain area. The analysis of the plant species composition and distribution characteristics suggested that the species number and diversity are influenced by the swing of the Yellow River channel. And some plant species were carried and settled in this area by the influences of the upper reaches of the Yellow River (Tab. 1).

Tab. 1 The community composition and number of natural features of the tenderness floodplain wetlands

Plant species	Abundance (%)	Coverage (%)	Frequence (%)	Height (cm)	Important value
Aster subulatus	16. 42	31. 80	41. 67	97. 70	12. 08
Phragmites australis	10. 99	15. 62	54. 17	118. 54	10. 43
Potentilla supina	17. 09	12. 08	50. 00	3. 67	7. 81
Tamarix chinensis	20. 72	6. 80	33. 33	14. 45	7. 80
Paspalum paspaloides	4. 90	41. 50	8. 33	67. 00	7. 61
Cyperus michelianus	8. 81	16. 77	54. 17	14. 08	6. 70
Calamagrostis	0. 47	3. 60	20. 83	100. 80	4. 59
Eclipta prostrata.	3. 47	4. 50	58. 33	6. 93	4. 18
Cynodon dactylon	3. 79	18. 00	20. 83	14. 60	4. 05
Xanthium sibiricum	2. 21	6. 50	8. 33	57. 50	3. 36
Typha minima	0. 28	10. 00	4. 17	50. 00	2. 78
Juncellus serotinus	0. 16	3. 00	4. 17	70. 00	2. 72
Polygonum hydropiper	0. 09	4. 00	8. 33	53. 50	2. 47
Polygonum amphibium	2. 68	5. 50	25. 00	3. 50	2. 44
Trigonotis peduncularis	1. 80	15. 00	8. 33	4. 00	2. 36
Cyperus rotundus	1. 11	4. 80	20. 83	8. 60	1. 95
Cyperus glomeratus	1. 07	7. 50	8. 33	15. 00	1. 83
Alternanthera sessilis	0. 51	2. 43	29. 17	3. 71	1. 81
Bidens parviflora	0. 19	2. 25	16. 67	12. 75	1. 42
Daucus carota	0. 79	6. 00	8. 33	3. 00	1. 24
Fimbristylis squarrosa	0. 44	5. 00	4. 17	3. 00	0. 86
Cyperus glomeratus	0. 09	2. 00	8. 33	5. 00	0. 75
Equisetum amosissimum	0. 13	2. 00	8. 33	4. 50	0. 74
Digitaria sanguinalis	0. 06	3. 00	4. 17	5. 00	0. 65
Poa aannua	0. 03	2. 00	4. 17	7. 00	0. 61
Rorippa palustris	0. 06	2. 00	4. 17	4. 00	0. 52
Eleusine inidica	0. 06	2. 00	4. 17	3. 00	0. 49
Conyza canadensis	0. 13	2. 00	4. 17	2. 00	0. 47
Plantago asiatica	0. 06	2. 00	4. 17	2. 00	0. 46

4.1.2 The species composition characteristics of the wetland in the middle floodplain area

There are 26 wetland plant species in the middle floodplain area. The dominant species of the community are *Aster subulatus* and *Phragmites austral* and their important values are 18.36 and 13.97 respectively. The sub – dominant species is *Cynodon dactylon*, and the associated species are *Sesbania cannabina*, *Setaria glauca*, *Juncellus serotinus*, *Polygonum lapathifolium*, *Cyperus glomeratus and Cyperus rotundu*. There are obvious seasonal waterlogged characteristics in the middle floodplain area. And the precipitation has an important effect on the abundance and diversity of the species, especially in the transition area from the wetland hydrophyte to the surrounding xerophile. This will also influence the biodiversity and evenness of the plant species in the middle floodplain area (Table 2).

Tab. 2 The community composition and number of natural features of the middle floodplain wetlands

Plant species	Abundance (%)	Coverage (%)	Frequence (%)	Height(cm)	Important value
Aster subulatus	39.00	24.92	70.59	53.01	18.36
Phragmites? australi	23.67	14.70	58.82	97.41	13.97
Cynodon dactylon	28.00	31.67	35.29	16.33	13.37
Sesbania cannabina	0.15	10.00	17.65	109.67	5.60
Setaria glauca.	0.05	10.00	5.88	100.00	4.64
Juncellus serotinus	1.13	17.50	11.76	35.00	4.17
Polygonum lapathifolium	0.05	5.00	5.88	105.00	4.17
Cyperus glomeratus	0.88	12.60	29.41	15.60	3.88
Cyperus michelianus	0.74	12.00	29.41	18.80	3.87
Cyperus rotundus	0.93	10.67	17.65	40.00	3.75
Potentilla supina	2.55	5.20	29.41	5.20	3.07
Polygonum hydropiper	0.10	3.50	11.76	48.50	2.59
Eclipta prostrata	0.39	3.80	29.41	9.80	2.49
Echinochloa colonum	0.98	5.00	5.88	42.00	2.47
Polygonum amphibium	0.15	10.00	5.88	4.00	1.72
Plantago asiatica	0.05	2.00	5.88	30.00	1.50
Bidens parviflora	0.05	2.00	5.88	28.00	1.44
Digitaria ischaemum	0.20	4.00	11.76	7.50	1.42
Digitaria sanguinalis	0.15	5.00	5.88	13.00	1.38
Rorippa palustris	0.15	2.50	11.76	8.00	1.24
Calystegia hederacea	0.05	3.00	5.88	15.00	1.17
Echinochloa crusgali	0.15	1.50	11.76	7.00	1.09
Portulaca oleracea	0.10	1.00	11.76	1.75	0.85
Eleusine inidica	0.25	2.00	5.88	1.00	0.66
Fimbristylis squarrosa	0.05	1.00	5.88	4.00	0.58
Chenopodium glaucum	0.05	1.00	5.88	2.00	0.52

4.1.3 The species composition characteristics of the wetland in the old floodplain area

There are only 7 wetland plant species in this area and the diversity and evenness index are obviously less than the other kinds of wetland communities. The species composition characteristics showed that the dominant species of the kind of community are *Juncellus serotinus*, *Echinochloa crusgali* and *Echinochloa colonum*. And the sub – dominant species are *Polygonum amphibium* and *Muhlenbergia hugelii* (Tab. 3).

Tab. 3 The community composition and number of natural features of the old floodplain wetlands

Plant species	Abundance (%)	Coverage (%)	Frequence (%)	Height (cm)	Important value
Juncellus serotinus	31.90	22.00	75.00	53.33	21.56
Echinochloa crusgali	2.14	90.00	25.00	60.00	20.84
Echinochloa colonum	20.91	27.67	75.00	63.33	20.57
Polygonum amphibium	24.13	20.00	25.00	2.00	10.85
Muhlenbergia hugelii	1.61	1.00	50.00	45.00	8.42
Alternanthera sessilis.	8.04	10.00	25.00	30.00	8.09
Digitaria sanguinalis	10.99	4.50	50.00	2.50	7.20

4.1.4 The comparative analysis of different biological diversity indexes

According to the study results of the community characteristics in the different development stages of the wetland (the tender floodplain area, the middle floodplain area and the old floodplain area), both of the species composition and the quantity characteristics tended to decline with the increasing human disturbance. By using the different index formulas, the species diversity indexes of three types of communities were calculated (Tab. 4). The results showed that the tender floodplain area had the highest species richness and diversity, while the other two kinds of wetland had lower biodiversity; the evenness index of the old wetland was significantly higher than the middle wetland and the tender wetland. These indicated that both of the species biodiversity and composition characteristics were seriously reduced under the human disturbance gradients.

Tab. 4 The biodiversity of different communities

Community types	Richness	Shannon – Wiener diversity	Evenness
Tender wetland	33.000	3.074	0.879
Middle wetland	26.000	2.785	0.855
Old wetland	8.000	1.917	0.922

4.2 The landscape pattern heterogeneity of the wetlands in the study area

4.2.1 The landscape classification system in the study area

According to the physical characteristics such as topography and physiognomy of the Yellow River floodplain area, and the land classification standards of China, and also considering the surrounding environmental conditions and land use situations of the wetland, the landscape in this area can be classified into 8 categories and 23 types (Tab. 5).

Tab. 5 The landscape type classification system in the study area

Code Ⅰ	Types	Code Ⅱ	Types	Code Ⅲ	Types
1	farmland	12	dry land		
2	forest	21	forest land		
		23	afforestation land	231	green belts on the both sides of the embankment
				232	street tree of willow
				233	street tree of pine and cypress
		22	unknown forest land		
3	garden	31	orchard		
4	grassland	41	wild grass land		
		42	weed belts on both sides of the road		
5	water	51	water		
6	building land	61	flood preventing buildings	611	flood control dam
				612	dangerous section building
		63	roads	631	country road (width <5 m)
				632	country road (5 m < width < 10 m)
				633	path (not country road, width <5 m)
				634	blacktop (the coach and van can pass through)
				635	the embankment road of the Yellow River
		64	building land		houses, buildings in the Yellow River scenic area
7	wetland	71	wetland		wetland (excluding the river and swag wetland)
8	unexploited land	81	barren land		
		82	waste land on the embankment platform		

4.2.2 The landscape heterogeneity in different spatial scales

Two sub - regions which have relatively concentrated plots were selected for the analysis of the landscape heterogeneity in different spatial scales. We planned the buffer areas which were far from the center of the sub - regions of 100 m, 200 m, 300 m, and then made the distribution maps of different landscape factors (Fig. 2 and Fig. 3). By using the ArcGIS9.0, the landscape pattern indexes of the regions which put the buffer areas' outside boundaries as perimeters were calculated, and their landscape pattern heterogeneity were also studied.

By calculating the landscape pattern indices of the two sub - regions in three spatial scales, the landscape heterogeneity was firstly increased and then decreased with the increase of spatial scale. For the first sub - region, the landscape heterogeneity was 2.558 in the 100 m spatial scale;

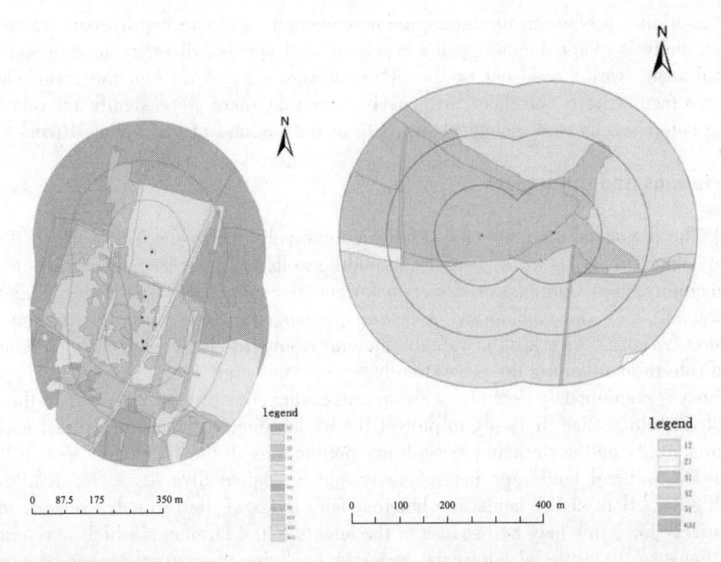

Fig. 2 Distributing of the landscape types of 01 plot

Fig. 3 Distributing of the landscape types of 02 plot

when the spatial scale increased to 200 m, its landscape heterogeneity index slightly increased, but seriously decreased in the 300 m scale. For the second sub – region, its landscape heterogeneity was just 1.229 which was even less than the half value of the first sub – region in the 100 m scale; its landscape heterogeneity index decreased slowly when the spatial scale got to 200 m, while increased steadily in the 300 m scale (Tab. 6). This indicated that the changes of the landscape pattern indices were correlated not only with the spatial scales, but also with the area of the wetland landscape in the sub – regions of the study area.

Tab. 6 Landscape heterogeneity index (LHI) of different scale

	LHI in 100 m scale	LHI in 200 m scale	LHI in 300 m scale
Sub – region 01	2.588	2.749	1.912
Sub – region 02	1.229	1.707	1.378

4.2.3 The correlation analysis of the landscape heterogeneity and community diversity

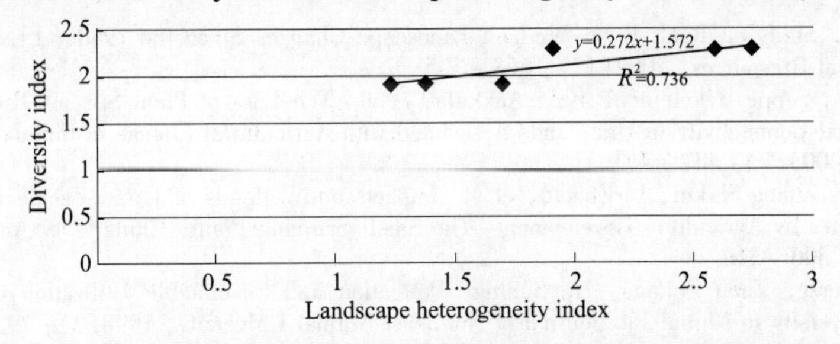

Fig. 4 Wetland landscape heterogeneity index and the diversity index

According to the study results of the landscape heterogeneity in three spatial scales, and also considering the community diversity index of the two sub – regions (2.279 and 1.917 respectively), the relationship between the landscape heterogeneity and species diversity was identified. The correlation analysis of the landscape heterogeneity and species diversity in different study regions and spatial scales was carried out by the SPSS software (Fig. 4). And the result showed that the two analysis factors were correlated with each other that there were significant relationship of the landscape heterogeneity and biological diversity in the wetland of the Yellow River.

5 Conclusions and discussions

(1) The biological diversity of wetland was relatively rich in the Yellow River floodplain area. The main community types in the tender and middle wetlands are: *Aster subulatus – Phragmites australis* community, and *Calamagrostis pseudophragmites* community. The major community types of the old wetland are: the community of *Setaria glauca*, *Juncellus serotinus*, *Cynodon dactylon – Echinochloa crusgali*. As a kind of valuable natural resource, the Yellow River wetland plays a very important role in maintaining the ecosystem balance of the study area. However, in recent years the wetland area is continuously declined with the increasing construction activities in the floodplain area. People should realize that only to protect the wetland ecosystems, we can get and maintain the ecosystem stability and sustainable agricultural productions in the floodplain area in the future.

(2) The wetland landscape heterogeneity and biological diversity were significant correlated with each other. Both of the landscape heterogeneity index of the two sub – regions in 300 m scale were relatively low, this may be because of the effects of the farmlands which were out of the sub – regions. Because the artificial landscape generally excludes the natural biological species, it leads to the decreased species diversity and number. The correlation of the landscape heterogeneity and biological diversity of the two sub – regions can get to 0.736,3, and were positively correlated. This further indicated that keeping landscape heterogeneity of wetland will be an important way to maintain the wetland resources and biodiversity.

(3) Because of the limited time and other reasons, the sampled plots were not very adequate in this paper, therefore it is necessary to do further check and discussion about the relationship between the landscape heterogeneity and diversity, and this is also our focus of work in the future. In addition, according to our study results we suggest changing the single agricultural landscape type in the old floodplain area, maintaining and increasing some natural patches such as wetland and grassland to reduce the disturbance by human beings and to prevent big wind and pest diseases in this area (Forrnan, 1995).

References

Wang Lixue, Li Sen, Dou Xiaopeng, et al. The Meaning of Wetland Protection and Degradation Reasons and Countermeasures of China Wetland[J]. Soil and Water Conservation in China, 2003(7): 8 –9.

Liu Hongyu. Study on Raoli Rvier Wetland Landscape Changes Since the 1980s[J]. Journal of Natural Rresources, 2002(17): 668 –705.

Miska Luoto, Seppe Rekoleinen, Jyrki Aakkula, et al. The Loss of Plant Species Richness and Habitat Connectivity in Grasslands Associated with Agricultural Change in Finland[J]. Ambio, 2003(7): 447 –451.

Liu Hongyu, Zhang Shikui, Li Zhaofu, et al. Impacts on Wetlands of Large – scale Land – use Changes by Agriculture Cevelopment: The Small Sanjiang Plain, China[J]. Ambio, 2004 (6): 306 –310.

Jiang Mingkang, Zhou Zejiang, He Suning. Protection and Sustainable Utilization of Wetland Biodiversity in China[J]. Journal of Northeast Normal University, 1998(2): 79 – 84.

Ding Shengyan, Gu Yanfang, Miao Chen, et al. Comparing Research of Weeds Communities of Different Land Using Types in Typical Area of the Middle – Lower Reach of the Yellow River [J]. Journal of Henan University (Nature Science), 2006(1):75 –78.

Zhang Xuliang, Xiao Zongmin, Xu Zongjun, et al. Biodiversity Characteristics and Protection

Strategy of the Yellow River Delta Coastal Wetland[J]. Wetland Science, 2002(2):125 - 131.

Wang Genxu, Guo Xiaoyin, Cheng Guodong. Dynamic Variations of Landscape Pattern and the Landscape Eco - logical Functions in the Source Area of the Yellow River[J]. Acta Ecologica Sinica, 2002(10):1587 - 1598.

Liang Guofu. Study on the Beach Area Landscape Change along the Yellow River in Henan Province from 1975 to 2000[J]. Henan Science, 2008(2):230 - 234.

Jiang Wenlai. Ecological Environment Problems and Countermeasures During Wetland Resources Development in China[J]. Chinese Land Science, 1997(4): 37 - 40.

Ding Shengyan, Liang Guofu. Evolution of Landscape Pattern Along Yellow River in Henan Province Recent 20 Years[J]. Geography Science, 2004(59): 653 - 7661.

Xiao Duning, Li Xiuzhen, Gao Jun. Landscape ecology[M]. Beijing: Science Press, 2003.

Fornan R T T. Land Mosaics: the Ecology of Landscape and Regions[M]. London: Cambridge University Press, 1995.

Environmental Impact Analysis of Hekoucun Reservoir Project

Huang Haizhen[1], *Dang Yonghong*[1], *Wang Na*[2] and *Gao Xuejun*[1]

1. Yellow River Engineering Consult Co., Ltd., Zhengzhou, 450003, China
2. Zhengzhou Water Investment Holding Co., Ltd., Zhengzhou, 450007, China

Abstract: The Hekoucun reservoir project' s main functions include flood control, water supply and irrigation, which have substantial social benefits. This project will involve the experimental zones of Taihangshan mountains Macaca mulatto chinensis National Nature Reserve in Henan province (below this referred to as the reserve), in which the environment is more sensitive. Some advantages and disadvantages of the environmental impact on Hekoucun reservoir project are analyzed in this paper, and the conclusions are drawn that the project has more positive impacts in general, through putting forward corresponding mitigation measures to avoid, reduce or eliminate some negative effects.
Key words: nature reserve, environment impact, Hekoucun reservoir project

The planned Hekoucun reservoir project is located in the outlet of the last gorge of Qin River, the primary tributary of Yellow River. This reservoir project is in Jiyuan city Henan province, and it is a crucial project for flood and flow control in Qin River, and it is also an important component of flood control system on Yellow River downstream . According to the project design, the environment problem is sensitive because this project construction area is located at the reserve experimental area, and the flood area is located in buffer area and core area (later on these areas become experimental area). This paper analyzes the environmental impact based on investigations of the project area and the project characteristics.

1 The project area environmental status

1.1 The status of the reserve

The most part of project areas are located at the reserve, approved to be established by Henan province government in 1982. This is the largest wild life reserve in northern China. It is located in the south of warm temperature zone, with a various plant species including northern and southern species, most of which are natural secondary forest. The data show that the reserve has about 1,700 kinds of higher plant, and four of them have been considered as the second grade national protected plant, such as Cercidiphyllum japonicum Sieb. Et Zucc, Sinowilsonia henryi Hemsl, TaihangiarupestrisYüetLi and Glycine soja Sieb. et Zucc. Among them there also have some national protect animal, Panthera pardus considered as first grade national protected animal, and naemorhedus goral, Macaca mulatto chinensis and Lutra lutra, these three kinds of animal being considered as the second grade national protected animal. Besides there are 140 kinds of birds, Ciconia ciconia, Ciconia nigra, Aquila chrysaetos and Haliaeetus leucoryphus, and these four species have been considered as first grade national protected birds.

It has a positive effect on rare animal and ecosystem after the reserve established, however, because of the reserve is close to many villages and towns, the core areas forbidden access in this reserve still have a lot of villages, and human activities are frequent. In recent years, the number of rare animals in this reserve, such as Panthera pardus etc. is decreasing. The biodiversity in this reserve is decreasing. The reserve is in a dilemma between human activities and environmental protection.

1.2 Other environmental problem

1.2.1 Threat of flood disaster

In recent years, the flow of Qin River is decreasing, and downstream of Qin River is running dry, and the most river courses drying days can be up to 319 d and continuous river courses time is 240 d.

The problem becomes more and more serious. The river courses caused channel shrinkage, making the Qin River downstream flood level increased and lasted longer time. And the downstream embankment status of the Qin River flood control standard is low, which only meet once in 25 years flood controlling requirements.

1.2.2 Over – exploitation of groundwater

As the main stream has no control project, due to the lower degree of development and utilization of surface water to meet the water demands of rapid economic development, the massive groundwater was exploited, resulting in regional groundwater overexploitation in recent years. Water tables are falling in some areas and the groundwater funnel was formed.

1.2.3 Water pollution

Proposed project reservoir is located in a sparsely poluated area, although the existence of agricultural production causing pollution, the water quality is good. With industrial effluent and municipal sewage draining off into downstream in Wulongkou, the pollution becomes serious. According to the historical data and current status of monitoring the overall water quality in project area was grade Ⅱ, which means the quality is good. The water quality from downstream of Wulongkou to Yellow River estuary was grade Ⅳ, part of the month was inferior grade Ⅴ.

2 The project characteristics

Hekoucun reservoir project has a total capacity of 3.17×10^8 m^3, 11.6 cumulative megawatts installed capacity, 1.272×10^8 m^3 water supply yearly. The dam height is 122.5 m and its normal water level is 275.0 m corresponding capacity of 2.5×10^8 m^3, and backwater length is 20.0 km. The project consists of rock fill dam, spillway, diversion and power generation system and other buildings. The primary mission of this project is flood control, water supply, irrigation and power generation.

3 Environmental impact analysis

3.1 Main advantage effect

After the project implementation, the ability of flood control between the Qin River and Xiaolangdi to Huayuankou of Yellow River control area was improved. It will reduce the threat of flood at Yellow River downstream, relief the pressure of flood control at Yellow River downstream, reduce the industrial and agricultural disaster caused by flood, and ensure a safe environment for the people living in downstream. Meanwhile, this project will maximize its ability of flood control, water supply, irrigation and power generation and effectively promote the economic development of the downstream region.

After the project implementation, it will maintain minimal environmental base flow to 5 m^3/s and change long-term frequent shut-off situation of the Qin River in recent years, effectively maintain and improve the ecosystem of downstream of the Qin River.

After the project implementation, villagers living in the reserve especially in its core area will be resettled, which will reduce the human activities impact and benefit to manage and maintain reserve.

3.2 Environmental impact at construction stage

3.2.1 Environmental impact on water

In the construction period, a large number of construction and management staff will produce amount of sewage. Aggregate processing, concrete mixing, pouring and maintenance will produce washing wastewater. Besides, maintenance of the construction machinery and vehicles can produce oily wastewater. Engineering disturbance the original geomorphologic and destruction of vegetation, engineering accumulation of soil and other materials could easily lead to soil erosion. Construction workers sewage and untreated wastewater and soil erosion will have some impact on the water quality.

Through wastewater and sewage discharge at standard level after processing, taking effective soil conservation measures, strengthen the construction management, the impact of the construction on the water environment is reduced to an acceptable level. Because the effect occurs only during the construction period, the impact is short-term and area-limited.

3.2.2 Environmental impact on atmosphere and noise

The construction machinery and vehicle fuel discharge waste gas and dust, noise in construction area, which will have some impacts on the residents near the construction area, the crops and the animals in reserve. Should strengthen the construction of vehicle maintenance, cover the easy cause dust material and sprinkle water to reduce dust, reduce the speed, set a reasonable construction work time, minimize nighttime construction, and other measures to reduce the atmosphere and noise environmental impact. However, all above impacts only take place in the construction stage, which will disappear after the construction is finished.

3.2.3 Ecologically environmental impact

The plant species near the construction site are widely distributed, and the construction will not affect these plant populations and lead to the extinction of species, and there are no rare and protected plants in construction area.

The species of terrestrial animal in construction area are relatively poor, construction activities have a weak influence on these animal. There is also not Macaca mulatto chinensis activity area in and around construction area, and the impact of construction activities on Macaca mulatto chinensis is weak. However, project will have a direct impact on reptile and amphibian, such as frog, snake etc. Due to the widely distribution in this area, the project activities will not endanger their population size.

Construction activities have a main impact on bird living around river bank, which caused some interference and to some extent destroy their habitat. During the construction period, due to worker stationing on project area and management oversights, harmful activities by human such as poach, illegal logging etc. will be harmful to the reserve.

According to the impacts above, as long as taking effective measures and strengthening management during construction stage, most those impacts can be mitigated and avoided.

3.2.4 Other impacts

Project construction will stimulate the development of neighboring economies, while creating a lot of employment opportunities. Construction personnel consumption will promote local agriculture, food industry and other service industry development, and have a great effect on adjustment of the structure of agricultural industry and the development of tertiary industry. However, project construction will disrupt traffic and normal life of local residents. During the construction period, a large number of workers will station on construction area for a long time, increasing the risk of epidemics. People should strengthen health control and do periodic physical examination and strengthen prevention and monitoring of infectious disease.

3.3 Impacts on hydrological regime during operation stage

After Hekoucun reservoir storage, the current surface water channel will become artificial lake. Water body scale, shape and flow regime etc. will be changed dramatically. The width, depth and volume compared with the status quo will be increased greatly, but the flow velocity will decrease. The water environment in the reservoir area would be changed from river rapid type to a slow reservoir flow type.

The construction of reservoir will change river from natural to artificial control and change the river natural state. It can make a good effect on water environment of downstream of Qin River by discharge 5 m^3/s minimum environmental base flow and increase the downstream channel water quantity in some month in spring, summer and part of the month. It can play an active role in the hydrological regime improvement to downstream of the river. However, the reduction of the downstream water supplies generally will affect the water demand of conservation objectives.

3.4 Impacts on water environment during operation stage

3.4.1 Water temperature

After reservoir reserving water, according to the prediction results presented, the water temperature in reservoir will present vertical gradient distribution. In summer, the temperature of discharge water would be lower than water in natural state. Discharged low-temperature water phenomenon is most obvious in May, and the discharged water temperature is below the natural water temperature of about 4.6 ℃.

3.4.2 Water quality in reservoir

Because of the good quality of water upstream, a big storage capacity of the reservoir and a long water residence time the quality of water released from Hekoucun reservoir is good. According to the prediction, the concentration of COD and ammonia nitrogen come out of reservoir can reach the water quality standard grade II, and it meets the requirement of urban water supply water quality. In the meanwhile, the reservoir water condition regulates water quality in some extent. After mixing water in reservoir, the outlet water quality throughout the year can maintain a stable level. It is a guarantee for the city water supply water quality protection.

3.4.3 Water quality in downstream

Due to the good discharge water quality of Hekoucun reservoir, water from dam site to wulongkou being polluted lighter, and water volume being relatively big, the water quality from dam to wulongkou would be good and the water quality in this section change a little before and after the project construction. There is more and more source of industrial pollution in the section between wulongkou to the estuary of Yellow River, and after the irrigated area using water, water in the reservoir will be reduced. Hence, the section of Qin river estuary to Yellow River water quality is worse compare with wulongkou section, but would be improved a lot compare with before the project construction.

3.5 Impact on the terrestrial ecological environment during operation stage

As the inundated areas are all in the experimental zone of the reserve, aiming at protecting some bushes and some water areas and amphibians, reptiles, birds and small mammals that live in that areas, the influence will be minor. Thus, the reserve function will not change. In addition, the inundated areas villagers' relocation will be conducive to the ecological protection and management of the reserve.

The plants in the inundated areas are mainly Vitex negundo var. heterophylla, Gleditsia heterophylla and other common plants in reservoir area, with Vitex negundo var. heterophylla-Gleditsia heterophylla community the most common one, and not involve rare and endangered species.

Because inundated area is also not Macaca mulatto chinensis regular activity area, the project impact on their habitat will not be big. After the reservoir is impounded, the original Qin River will become the broad water surface reservoir, an ecological barrier for the Macaca mulatto chinensis in each side of the reservoir, hindering the gene exchange between those communities. In order to hinder adverse impact, bridges may be applied at the narrow area that Macaca mulatto chinensis usually pass through; meanwhile, it is necessary to establish management and caring stations, as well as to strengthen the publicity, temporary ambulance and other measures to reduce the engineering construction macaques.

According to the survey, Panthera pardus are seldom seen in the project area. In addition, the activity area of Panthera pardus is extremely large, and they do not directional migrate. It is safe to conclude that the project construction will not significantly affect leopards.

The reservoir will provide excellent living condition for migrating water birds. Also, reservoir area surrounding humidity increases, which is conducive to the growth of a variety of plants, and will also provide a better source of food to the birds.

3.6 Impact on aquatic ecological environment during operation stage

3.6.1 Impact on aquatic ecological environment in reservoir

After reservoir impoundment, the population structure will be selected by the river type transition to the reservoir - lake type. Plankton community structure tends to be more diversified, some of the benthic fauna will migrate to the shallow waters of the tributaries or reservoir tail. Reservoir storage would appear to provide conditions of aquatic vascular plants in coastal and riparian acet.

After reservoir impoundment, the spawning grounds of those fishes that produce demersal eggs or sticky eggs will no longer exist. However, this kind of fish is less demanding on their productive environment, their spawning ground will transfer to shallow water area. The fish bait condition will have greater improvement, and feeding grounds will become more abundant and diverse. Reservoir will provide more wintering sites to let fish live through the winter.

According to the impact of dam block to fish, this project takes the artificial propagation and artificial releasing measures to reduce the negative impact.

3.6.2 Impact on aquatic ecosystem in the downstream

After the completion of the reservoir, the discharged water volume is guaranteed and this is conducive to the breeding of fish. There will form a Living Water District in the downstream, which is beneficial for the survival of aquatic organisms. In spring and summer, the discharged water provides more spawning grounds and breeding conditions, and favorable growth and feeding fish. In the fall, due to human control and the reduction of river water, riverbed width will be narrowed, which will affect the growth of benthic animal and waterweeds.

3.6.3 Impact of low temperature discharge on aquatic ecosystem in the downstream

The low temperature water discharge from reservoir will change the structure of phytoplankton community. Low temperature preferred species will become dominant species in this place, at the same time low temperature water discharge will delay fish spawning, and some fish characteristics such as spawning and feeding will transfer to downstream. This project adopted the multi-level intake structure to reduce the impact of the low-temperature water discharged.

4 Conclusions

During the construction period, wastewater, noise, dust and land occupied by project etc. will impact on local environment and destroy local ecosystem to some extent. These adverse impacts can be eliminated by the environmental mitigation measures. And those effects are temporary, and can be eliminated after project finished.

The Macaca mulatto chinensis gene exchange will be disturbed during operation period, some adverse impacts caused by dam block and low temperature water discharge. These adverse impact above-mentioned can be eliminated, avoided or slow down through taking engineering and ecological compensation measures.

The primary mission of this project is to mitigate flood, guarantee water supply, irrigate and generate power. Villagers living in the reserve relocated will reduce the human activities impact and benefit to manage and maintain reserve. Environmental base flow discharge will protect ecosystem in the downstream. The project has a certain favorable impact and enormous social benefits. Considering the special environment of the project area, based on the effective implementation of environmental protection measures, from an environmental point of view, the project is feasible.

References

Yellow River Conservancy Commission. The Yellow River Recent Focus on Governance Development Planning [M]. Zheng zhou: Yellow River Conservancy Press, 2002.

Taihangshan mountains Macaca Mulatto Chinensis National Nature Reserve Scientific Survey[M]. China Forestry Publishing House, 1996.

GB3838 – 2002 Environmental Quality Standards for Surface Water[S]. China Environmental Science Press, 2002.

Hydrodynamic Experiment on the Polluted Sediment Control with the Application of Ecological Spur – dike

Liu Huaixiang*, *Lu Yongjun*, *Lu Yan* and *Ding Jingjing

State Key Laboratory of Hydrology – Water Resources and Hydraulic engineering, Nanjing Hydraulic Research Institute, Nanjing, 210029, China

Abstract: The ecological effect of hydraulic engineering has been concerned worldwide in recent years. Structures such as spur – dikes are effective in river regulation but also a threat to river ecology. Therefore, in this paper artificial wetland is considered in spur – dike engineering practice for river restoration. In order to study the influence of this ecological project on the discharge of polluted sediment particles, a laboratory experiment was carried out. The flow field and concentration process around the project area was measured under different conditions of discharge, vegetation and sediment concentration input. Results showed that, the reduction of peak concentration was obvious due to velocity variation. The increase of vegetation occupation ratio would lead to the transition of concentration process diagram from "single – peak" to "multi – peak" type. And it's very helpful for the enhancement of water quality and the pollutant decomposition. The pollution reduction amounts and the best parameters of this ecological approach were also calculated.

Key words: spur – dike, ecology, hydrodynamic model, pollution

1 Introduction

Lots of environmental and ecological problems have been caused by hydraulic engineering practice. Generally they threaten the river ecology through the way of changing water – sediment dynamic structure and river boundary conditions. For example, construction of hydraulic structures may increase the turbidity nearby and so that directly influence the photosynthesis of aquatic vegetation. The erosion and siltation of river bed induced by engineering may also destroy the habitat of creatures. In the 1970s and 1980s, the Nanning – Guangzhou navigation channel project in the Pearl River built more than 330 hydraulic structures such as spur – dikes. The destruction of aquatic creatures' habitat was enormous. Therefore, how to avoid the ecological threat of hydraulic engineering and to study the relative envrionmet – friendly technology is a key to the sustainable economic development. Hohmann suggested that, the river regulation should control the stress of human activities and keep the species diversity in order to restore the natural conditions of rivers. Thus the composiation of traditional hydraulic structures and ecological restoration technology is prevailing in both researching and engineering practice. For instance, Lu and Wang studied the application of ecological methods such as artificial fish nest in regulation projects. Among the ecological methods, the artificial wet land is the most common and effective one. It is very helpful in forming aquatic organism communities. The velocity in the dam – field zone of spur – dikes (the bank zone behind dikes) is often slow so that it is very suitable for reconstruction of aquatic vegetation. Then artificial wet land and ecological bank zone will be formed. A study showed that, the wet – land vegetation in lakes could cut 25% ~ 62% of sediment and other pollutants carried by input flow, purifying the water quality. Similarly, the ecological spur – dikes with artificial wet – lands may also affect the land – source polluted flows. The suspended sediment particles, which absorbed pollutants such as heavy metals, would be especially intercepted in the vegetation. Therefore, a laboratory experiment was designed to quantitatively analyze the pollutant interception effect of this kind of ecological spur – dikes. Then the improvement of projects and relative parameters can be further studied.

2 Laboratory hydrodynamic experiment

2.1 Experiment layout

As shown in Fig. 1, the size of experiment flume is: 65 m (length) ×4 m(width) ×0.5 m (height)。Near the both two ends of the flume, there were brick grids for flow stabilizing. A fence – type gate is set in the downstream end to control the water depth. Water flows in the flume can be recycled after passing through the gate. The water from upstream was supplied by the pumps (the largest discharge is 150 L/s).

The experiment section is put in the middle of the flume, 4 spur – dike models with the interval of 2 m were placed here in order to simulate the real spur – dike group. The distance of experiment section from upstream and downstream were both very far (30 m around), so that the influence of entrance and tail gate can be ignored. The dam – field zone between the middle 2 spur – dikes is the wetland simulation zone where polluted water is diverted into. In the downstream of spur – dikes, a measuring bridge for turbidimeters placement is set in order to monitor the variation of sediment concentration in this cross – section. At the same time another bridge for velocity measurement is set to obtain the flow field. In this experiment, the turbidity processes in the downstream were recorded under different discharge, wetland area and input polluted sediment concentration. Therefore the interception effect of ecological spur – dikes can be tested in different conditions.

Electromagnetic flow velocimeters which can detect 2D instantaneous velocity and fluctuation were applied for the measurement of velocity and flow field. Sediment concentration in water was measured by laser turbidimeters with the accuracy of 0. 001 g/L. The wooden solid spur – dike models were 100 cm in length and 6 cm in height, with the side slope of 1: 1. 25 and head slope of 1: 1. 3. According to several trial run we selected fly – ash for the simulation of polluted sediment particles. The vegetation in artificial wetland were simulated by plastic grasses which distributed in the interval of 5 cm around.

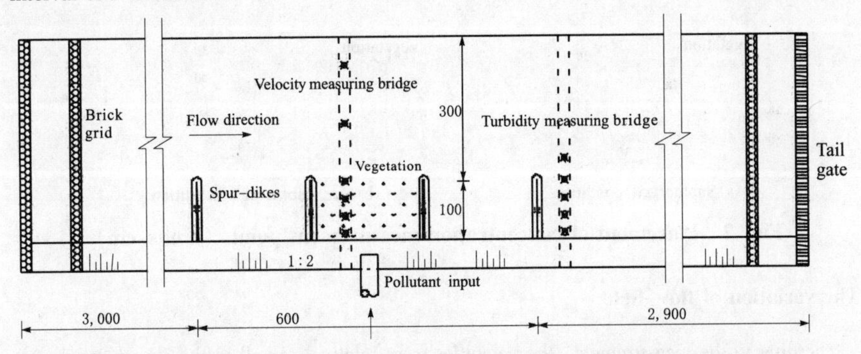

Fig. 1 The layout of laboratory hydrodynamic experiment (Units: cm)

2.2 Experiment conditions and steps

The prototype of this experiment is the spur – dike regulation section between Guiping and Wuzhou in the Pearl River. According to the local hydrodynamic conditions, it was decided to use undistorted model and the horizontal and vertical scales were about $\lambda_L = \lambda_H = 100$, the velocity scale is$\lambda_u = \lambda_H^{1/2} = 10$.

Two typical input discharge values, the regulation stage discharge and flood discharge, were selected. Then the corresponding discharges in model were calculted by scales. Thus in this experiment there are two conditions: non – submerged (regulation stage discharge) and submerged (flood discharge), as shown in Tab. 1.

Tab. 1 Two input flow conditions

Submerged condition	Discharge (L/s)	Depth (cm)
Non - submerged	21.6	6
Submerged	102	10

"Polluted water" of 6 different pollutant concentrations, 1 g/L, 2 g/L, 4 g/L, 6 g/L, 8 g/L and 10 g/L respectively, were poured into the artificial wetland, in order to test the response of vegetation interception. For the convenience of comparison, the total volume of every experiment run was the same 50 L. In the aspect of artificial wetland, 4 schemes of different vegetation proportions were determined. They were named no - vegetation, 1/3 vegetation (from bank to 1/3 spur - dike length), 2/3 vegetation (from bank to 2/3 spur - dike length) and full vegetation (see Fig. 3).

Considering the combination of discharge, input polluted sediment concentration and vegetation proportion, there were 2 × 6 × 4 = 48 experiment runs.

Experiment step: First prepare the vegetation and polluted water of different concentration; Adjust the flow to steady flow of required discharge and water depth by the control of pump and tail gate; Velocity measurement by the 3 - points methods (0.2 h, 0.6 h, 0.8 h), a cross - section for each time, complete all cross - sections by moving the measuring bridge; Discharge the polluted water into the artificial wetland, record the downstream sediment point - concentration (see Fig. 2); Repeat the experiment run after changing working conditions.

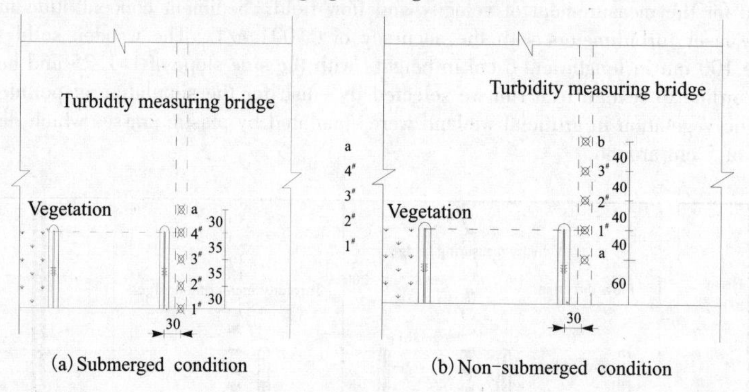

Fig. 2 Placement of concentration measurement points (Units: cm)

3 The variation of flow field

According to the measurement, the velocities were relatively small in the zone between spur - dikes compared with that in the main stream. Especially under the non - submerged condition, the value was only 1/10 ~ 1/5 of the main stream velocity. The further analysis showed that, the main flow had fixed direction. And its velocity was only affected by the input discharge. Then we only need to discuss the variation of flow field in the spur - dike zone:

3.1 Vegetation

The increase of vegetation may decrease the overall velocity or just increase the near - bank back flow. Thus the polluted sediment could be intercepted more easily.

3.2 Submerged condition

The velocity under submerged flow is generally larger than that under non - submerged flow.

As for the flow direction, it was almost a single – direction flow field under submerged flow, except for some local back – flow in the near – bank, near – bed districts. While the flow field under non – submerged flow showed a vortex feature with obvious strong back flow.

3.3 Depth layer

When submerged, the vegetation could only affect the near – bed velocity. But given the non – submerged condition, the influence of vegetation could reach the surface flow field.

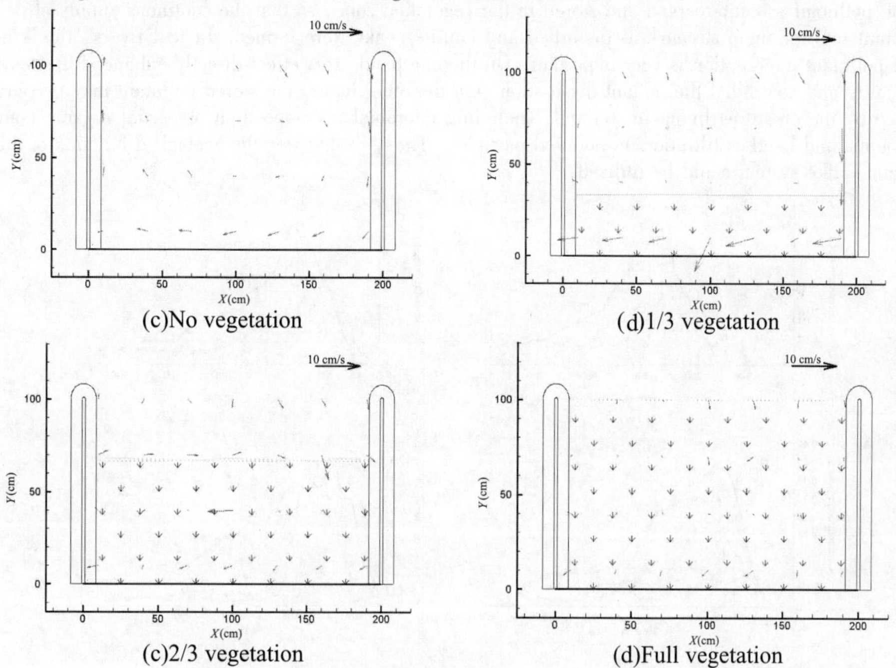

Fig. 3 Flow field between the spur – dikes (Non – submerged, 0.2depth from bed)

4 The variation of downstream concentration

In Fig. 4 and Fig. 5, some downstream concentration process curves were ploted under the non – submerged condition. And Fig. 6, Fig. 7 are the similar curves under the submerged condition. The positions of all measuring points are in Fig. 2. In Fig. 2(b), the value obtained at points a and b were zero in most occasions, indicating that the diffusion of pollutant would not exceed the range of a ~ b when non – submerged ($1^{\#}$ ~ a when submerged). The length of the ranges was around 130 ~ 160 cm.

4.1 Non – submerged condition

(1) The concentration downstream was much smaller than the upstream input concentration due to water dilution. When the input concentration was 1 g/L, the peak value on the downstream cross section was only about 0.1 g/L. And the two values would change proportionally: the corresponding peak value of 10 g/L (input concentration) was a little smaller than 0.9 g/L, almost 10 times of 0.1 g/L. When the vegetation area became larger, the downstream peak value declined to a certain extent. The peak value under full vegetation condition was just 2/3 of the corresponding value under no

vegetation condition. That is to say, the vegetation showed the effect of peak clipping.

(2) According to the shape of concentration process curve, every point – value start to rise nearly at the same time, which is the time when polluted water reach the measuring cross section. Under the no vegetation condition, the concentration process curve was a distinct single – peak type (see Fig. 4). It implied that, without any interception the pollutant entered the main stream continuously until the downstream concentration reaching peak value, then the value start to decline since the remained pollutant became less and less after input stop. However, after adding more vegetation a multi – peaks feature appears in the concentration process curve (see Fig. 5). This means that the pollutant was intercepted and stored in the vegetation zone, so that the continous supply of pollutant into the main stream was disturbed and multi – peaks were formed. In real rivers, this kind of pollutant interception is very important. On the one hand, this effect directly enhanced the water quality and controlled the pollant dissipation. On the other hand, the stored pollutant may be purified by the creatures living in wetland, including microbial decomposition of partial organic compounds and benthos filtration on polluted particles. Thus, in that way the ecological function of this spur – dike system could be utilized.

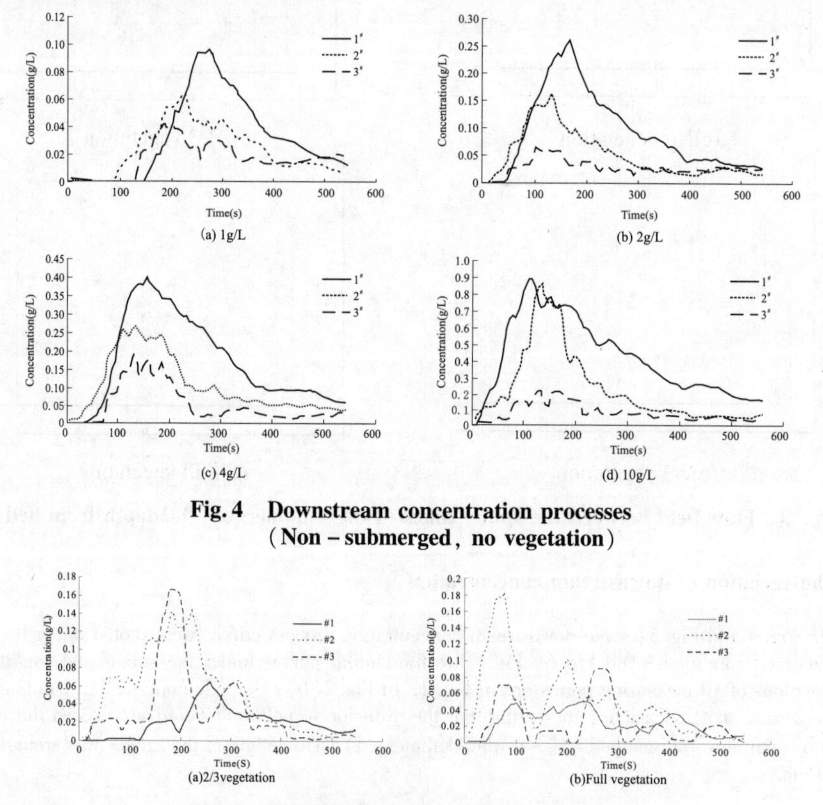

Fig. 4 Downstream concentration processes (Non – submerged, no vegetation)

Fig. 5 Downstream concentration processes (Non – submerged, input concentration 2g/L)

4.2 Submerged condition

(1) The concentration downstream was also much lower than the input ones. And there was a proportional variation feature as well. But the peak clipping effect of vegetation was not that obvious as the non – submerged case.

(2) No matter which vegetation scheme, the concentration of measuring points showed the relationship of $1^{\#} > 2^{\#} > 3^{\#} > 4^{\#}$ (see Fig. 6 and Fig. 7). This result suggested that the polluted current flowed near the bank. Vegetation showed no effect on this phenomenon.

3) As for the curve shape, in all vegeation schemes $1^{\#}$ value was the earliest one to rise compared with other measuring points. That is, a part of the pollutant input from bank was almost instantly carried to the downstream bank by the strong flow. Furthermore, compared with the non – submerged case, the transition from single – peak to multi – peaks along with the vegetation area increase was not that impressive (see Fig. 6 and Fig. 7).

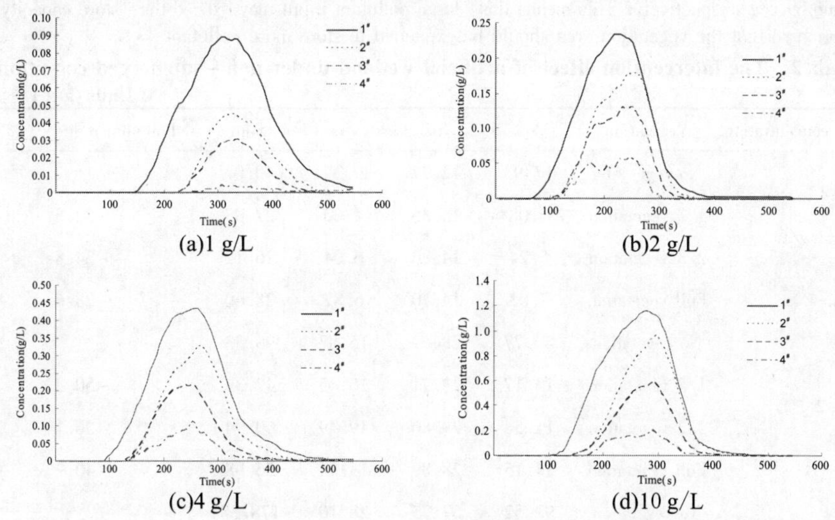

(a)1 g/L (b)2 g/L

(c)4 g/L (d)10 g/L

Fig. 6 Downstream concentration processes (Submerged, no vegetation)

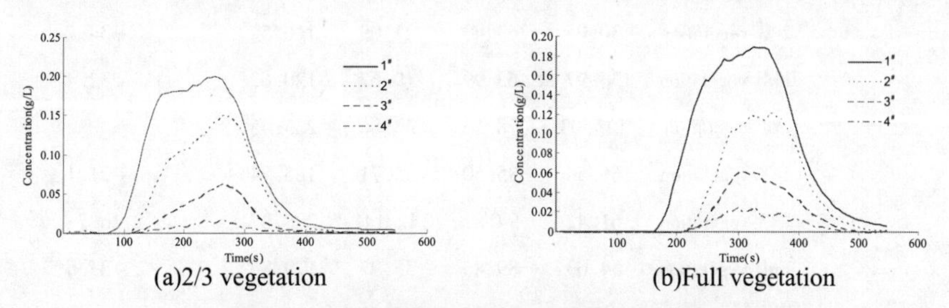

(a)2/3 vegetation (b)Full vegetation

Fig. 7 Downstream concentration processes (Submerged, input concentration 2 g/L)

4.3 Interception calculation

In order to quantify the interception effect of artificial wetland on the polluted sediment discharge, the concentration process curves were integrated by Eq (1) to solve the area S under the curves.

$$S = \int c\mathrm{d}t \approx \sum c\Delta t \tag{1}$$

Where, c is the concentration value on process curves; Δt is the measuring time step of the instrument (6 s here). (It is easy to understand that, given the same discharge from upstream, the sum of pollutant passed a certain measuring point is proportional to the S value at that point. Therefore, the pollutant entering downstream cross section can be estimated indirectly by the variation of S value.

The corresponding S values under non – submerged and submerged conditions were listed in Tab. 2 and Tab. 3 respectively.

The results indicated that, when non – submerged the artificial wetland may impair the S value greatly, reducing 30% or even more. Thus the vegetation can effectively intercept and store the polluted sediment, decreasing the pollutant entering downstream. The interception effect was obvious even with just 1/3 vegetation, but the importance of vegetation area can be revealed along with the increase of input concentration. For example, when inputing 10 g/L polluted water, 1/3 vegetation could only reduce 10.5% of the S value, while the effect of 2/3 vegetation and full vegetation were 21.3% and 39.6% respectively. This means that, large pollutant input may exceed the "store capacity" of vegetation so that the vegetation area should be expanded to store more pollutant.

Tab. 2 The interception effect of artificial wetland under non – submerged condition

(Units: g/(L · s))

Input concentration	Vegetation	$S_{1\#}$	$S_{2\#}$	$S_{3\#}$	Sum	Pollutant reduced(%)
1 g/L	No vegetation	17.93	12.77	9.37	40.07	
	1/3 vegetation	7.08	15.35	4.60	27.03	-32.5
	2/3 vegetation	5.24	14.84	6.04	26.12	-34.8
	Full vegetation	7.03	14.70	6.87	28.60	-28.6
2 g/L	No vegetation	50.77	28.69	15.87	95.33	
	1/3 vegetation	13.17	23.71	10.45	47.33	-50.3
	2/3 vegetation	11.58	29.86	19.49	60.94	-36.1
	Full vegetation	14.15	28.89	13.94	56.98	-40.2
4 g/L	No vegetation	92.52	57.75	28.10	178.38	
	1/3 vegetation	36.58	55.99	27.00	119.58	-33.0
	2/3 vegetation	30.05	63.43	23.09	116.57	-34.7
	Full vegetation	36.97	63.99	19.67	120.62	-32.4
6 g/L	No vegetation	123.03	78.56	33.86	235.45	
	1/3 vegetation	66.54	85.50	33.71	185.74	-21.1
	2/3 vegetation	61.42	94.81	142.14*	298.37	26.7
	Full vegetation	54.03	89.86	35.97	179.86	-23.6
8 g/L	No vegetation	152.81	82.46	86.52	321.79	
	1/3 vegetation	35.39	161.72	59.51	256.62	-20.3
	2/3 vegetation	92.99	90.07	90.63	273.69	-14.9
	Full vegetation	67.34	103.80	42.14	213.29	-33.7
10 g/L	No vegetation	221.13	119.93	42.35	383.41	
	1/3 vegetation	161.52	133.63	47.89	343.05	-10.5
	2/3 vegetation	131.37	112.91	57.56	301.84	-21.3
	Full vegetation	80.04	95.47	56.26	231.77	-39.6

Note: * is Abnormal value.

The effect of artificial wetland on S value was not that significant, reducing less than 20% (see Tab. 3). Considering the vegetation area, the interception effect of 1/3 vegetation and 2/3 vegetation was below 10%, and it can approach 20% only when full vegetation scheme was adopted. This result is relevant to the high original velocity, little influence of vegetation on flow field and sediment tranport above vegetation under submerged condition.

Tab. 3 The interception effect of artificial wetland under submerged condition

(Units: g/(L·s))

Input concentration	Vegetation	$S_{1\#}$	$S_{2\#}$	$S_{3\#}$	$S_{4\#}$	Sum	Pollutant reduced(%)
1 g/L	No vegetation	16.70	6.10	3.77	0.71	27.28	
	1/3 vegetation	14.74	7.32	3.28	1.95	27.29	0.0
	2/3 vegetation	15.81	8.93	3.41	1.74	29.89	9.6
	Full vegetation	14.92	5.45	2.10	1.88	24.35	-10.7
2 g/L	No vegetation	35.24	22.67	16.66	7.19	81.75	
	1/3 vegetation	36.75	20.98	12.41	4.96	75.10	-8.1
	2/3 vegetation	38.53	23.96	8.29	2.76	73.54	-10.0
	Full vegetation	36.25	18.74	8.59	3.49	67.08	-18.0
4 g/L	No vegetation	72.29	43.68	28.11	12.20	156.28	
	1/3 vegetation	76.80	42.29	24.09	5.65	148.82	-4.8
	2/3 vegetation	73.47	47.25	20.90	5.72	147.34	-5.7
	Full vegetation	62.97	38.69	20.42	6.45	128.53	-17.8
6 g/L	No vegetation	107.39	67.74	40.06	10.87	226.06	
	1/3 vegetation	109.48	55.80	36.06	6.55	207.89	-8.0
	2/3 vegetation	109.48	77.29	31.93	8.21	226.91	0.4
	Full vegetation	93.64	50.88	26.31	10.83	181.67	-19.6
8 g/L	No vegetation	147.35	85.42	53.31	20.01	306.09	
	1/3 vegetation	145.33	80.69	47.32	9.39	282.73	-7.6
	2/3 vegetation	150.87	104.13	50.26	12.78	318.05	3.9
	Full vegetation	128.87	76.50	35.96	7.11	248.44	-18.8
10 g/L	No vegetation	180.88	116.80	70.67	23.96	392.32	
	1/3 vegetation	180.28	103.17	61.24	14.25	358.94	-8.5
	2/3 vegetation	182.90	129.38	56.75	13.16	382.19	-2.6
	Full vegetation	177.84	102.66	53.77	15.31	349.58	-10.9

5 Conclusions

In this paper, a laboratory hydrodynamic flume experiment was carried out to investigate the function of artificial wetland in ecological spur-dike system. The results indicated that:

(1) The increase of vegetation may decrease the overall velocity or just increase the near-

bank back flow, especially under non – submerged condition. Thus the polluted sediment could be intercepted more easily. When submerged, the influence of vegetation was restricted only in the near – bed area due to large velocity and depth.

(2) A transition from single – peak to multi – peaks along with the vegetation area increase was shown in the downstream concentration process curves. The peak clipping function of artificial wetland can reduce the downstream concentration, implying that the polluted sediment were intercepted and stored in the ecological spur – dike zone. Furthermore, the water quanlity can be enhanced and the detained pollutant can be biological treated in the real wetland.

(3) The interception effect of artificial wetland on input pollutant was estimated. when non – submerged the reduced proportion could reach 30% or even more. Thus the vegetation can effectively intercept the polluted sediment. This effect was obvious even with just 1/3 vegetation near – bank, but large pollutant input may exceed the "store capacity" of vegetation so that the vegetation area should be expanded to store more pollutant. When submerged, the intercept proportion was relatively small and greatly affected by the vegetation area.

(4) It can be concluded that, the pollutant – interception effect of ecological spur – dikes with artificial wetlands is good. The ecological function is obvious under the normal discharge condition (non – submerged). If the wetland area is restricted by the requirement of flood discharge or navigation, then the near – bank part of wetland can still play the similar function. As for frequently flooded or highly bank – source polluted river sections, the wetland area should be expanded as large as possible. During the floods, some pollutant can directly over flow the spur – dikes. Thus some improverment such as ecological design of the dike body (surface vegetation and so on) can also be considered.

Acknowledgements

This work was financially supported by the National Basic Research 973 Program of China (Grant No. 2012CB417002) and the National Natural Science Foundation of China (Grant No. 51009096).

References

Hohmann J, Konold W. Flussb aum as snahmen an der wutach und ihre bew ertung au s oekologischer sicht [J]. Deutsche Wasserw irtschaft, 1992, 82(9):434 –440.

Lu Yan, Li Shouqian, Lu Yongjun. Ecological effects and stability of an artificial reef spur – dike [C]. Proceedings of the 7th IAHR Symposium on River, Coastal and Estuarine morphodynamics. Beijing, China, 2011: 512 –522.

Wang Peifang, Yang chuanqing, Wang chao. Study on effect of ecological groin on velocity of flow and sediment deposition[C]. Poceedings of the 23rd national Conference on hydrodynamics, Xi'an, 2011:697 –701.

James R T, Martin J, Wool T, et al. A sediment resuspension and water quality model of lake okeechobee[J]. Journal of the American Water Resources Association, 1997, 33: 661 – 680.

Pan G. Adsorption kinetics in natural waters: a generalised ion – exchange model, in adsorption and its application in industry and environmental protection[J]. Studies in Surface Science and Catalysis, 1999, 120: 745 –761.

Study on Key Technology of Ecological Water System Planning of Kaifeng City

Yan Dapeng*, *Jiang Yamin*, *Guo Pengcheng* and *Ma Zhuoluo

Yellow River Engineering Consulting Co. , Ltd. , Zhengzhou, 450003, China

Abstract: The ecological water system of Kaifeng City is an integrated infrastructure of urban water conservancy with integrated functions such as flood control, drainage, water supply, water quality protection, hydrophilic landscape, water ecology and etc. In order to improve the technological level of ecological water system planning, the paper studies on the key technologies in allusion to planning ideas, network layout, water resources allocation mode, water ecology protection, operation protection measures and etc. In the study, the water system planning idea with integrated water, landscape and city is provided by combining with the characteristics of Kaifeng City, a growth type network layout of water system is developed, a multi-level and three-dimensional dynamic equilibrium configuration mode of water resource is formulated, a water ecology protection framework with combined point, line and face is constructed, and a complete winter operation security measure system of artificial wetland, which had successfully applied for a national utility model patent. The achievements of the study provide strong technical support for the ecological water system planning of Kaifeng City, and also provide a useful experience for the efficient, ecological and harmonious formulation of urban water system planning in surrounding similar cities.

Key words: planning idea, network layout, water resources allocation, ecological protection, protection measures

1 Introduction

In the history, Kaifeng was prosperous due to the canal, especially the four rivers (Wuzhang River, Jinshui River, Bian River, Cai River) has created Dongjing city which is a brilliant city in the Northern Song Dynasty. In Qianlong period of Qing Dynasty, Kaifeng is featured with the urban characteristics of "water city, water in the city, the four sides are surrounded by water, and the water is flowing around the city", a famous "North Watertown", and enjoyed the reputation of "a city with full Song Yun and half water". Kaifeng City is an important part of the "integration of Zhengzhou and Kaifeng", "central Henan urban agglomeration" and "Rise of Central China", and undertakes the arduous task of promoting regional economic development. As an important infrastructure of city, the water system of Kaifeng undertakes the tasks of ensuring the safety of flood control and drainage, optimizing the allocation of water resources, building an ecological waterfront space, shaping a pleasant hydrophilic landscape, improving the surrounding people living environment, realizing the recycling of resources, and supporting the urban economic and social sustainable development. However, with the evolution of the natural conditions and the development of economy and society, the potential problems in the development and use of Kaifeng water system is increasingly prominent and has become a bottleneck restricting the urban development of Kaifeng City.

In order to achieve the development goals of "garden city" and "ecological city" with integrated Watertown the Tianlin in Kaifeng City, the planning work for the ecological water system of Kaifeng City is carried out on the basis for the overall urban planning of Kaifeng City. Meanwhile, in order to improve the technological level of ecological water system planning, the paper studies on the key technologies in allusion to planning ideas, network layout, water resources allocation mode, water ecology protection, operation protection measures and etc. , wherein the achievements of the study will be applied to the ecological water system planning of Kaifeng City by combining with the characteristics of the same.

The range for the ecological water system planning of Kaifeng City comprises five districts, i.

e. Drum Tower, Dragon Pavilion, Shunhe, Yuwangtai, and Jinming, and the total area thereof is 546 km^2. In 2020, the planned area of urban construction will be reached to 136 km^2, and the planned population will be 1.46×10^6. The ecological water system of Kaifeng City is an integrated infrastructure of urban water conservancy based on the core of water safety, water system network construction and riparian ecological environment construction, wherein the system integrates flood control, drainage, water supply, water quality protection, hydrophilic landscape, water ecology and other functions to ensure the economic and social sustainable development of the city and effort to achieve the harmony of man and water. According to the general requirements of urban planning and the characteristics for the water system of urban area, the overall concept for the ecological water system planning of Kaifeng City is provided as follows: "safety, health, flowing, clean, water and city blending, and man and water harmony", and the objective thereof is to construct a fused excellent and healthy urban river ecosystem[1] with integrated five aspects, i. e. water security, water environment, water landscape, water culture and water economy. According to the comprehensive consideration for the historical water proportion of Kaifeng City, the local natural environmental conditions, the water resources endowment, the specific requirements of development planning and the like, the appropriate water area of Kaifeng City is about 5% to 10%.

2 Planning idea

In relation to the problem for the protection, inheritance and update of historical and cultural heritage of original water system of ancient city, the paper provides a water system planning idea of integrated water, landscape and city.

Kaifeng, as one of the eight ancient capitals in China, was the ancient capital of seven dynasties, wherein the water system of Song Dynasty is an urban geographical space with unique historical connotations and one of important historic and cultural resources carriers of the city, belonging to resources which are not reproducible and renewable. At the same time, the ancient city where the waster system is located of Song Dynasty belongs to the area with relatively weak urban infrastructure and the long-term problems of water pollution, land occupation, historical infringement and the like, and the water pattern with special historical significance has been damaged in a certain degree, specifically the function of the history water system is gradually shrinking, the landscape is gradually eliminating, and the ecological function is gradually degenerating. The method, taken for protecting or restoring the history water system to continue the history form and pattern while meeting the functions and landscape requirements of modern social development, is one of the difficulties to be faced in the study of water system planning.

In the planning for the water system in Songcheng District of Kaifeng City, the idea of integrated water, landscape and city is provided by basing on the water system pattern of Song Dynasty, combining the water system application and development condition of city evolution, and considering the combination for the historical and cultural environment and natural water landscape environment of ancient city of Song Dynasty, the water system pattern of "city surrounded by water" is planned, and the water system of the ancient city will be injected vitality to restore the historical appearance by combining with ancient city protection, pollution controlling and intercepting, perforating the water system, and enriching the water resources. Therefore, the idea for the protection, continuation and development of historical relics in the planning of ancient historic and cultural city is reflected fully.

3 Planning layout

The study provides a growth type network layout of water system, constructs a communicated and integrated pattern of water system planning, and realizes the harmonious blend of historical culture and modern civilization in the water system planning.

The scope of water system planning in Kaifeng comprises three districts, i. e. new district, Songcheng district, and east district. In the "Overall City Planning of Kaifeng City" (2010 ~ 2020), the three districts are provided with different urban layouts and functional localizations, wherein the Songcheng district represents the historic and cultural relics district of Kaifeng City, and the new district and east district represent the two ends of new economic development and regional communication of Kaifeng City.

At present, the water system of Songcheng is relatively independent, the communication to the water systems of new district and east district is not smooth, the area ratio for the original water of the new district and east district is low, the distribution is uneven, and the requirements of the economic and social development are difficult to be met. The method of how to communicating and blending the new and old, history and modern, heritage and development in the water system planning is another difficulty in the planning.

From the view of historical development, the generation and development of urban water system is like a three with constant growth, i. e. a gradual metabolism process developing from nothing and prosperously developing from small to large. By tracking, the waster system of Kaifeng is grown and developed by moistening and rising with the water of Yellow River. Therefore, the new district is provided with a water system framework of "one lake, two districts, three transverses, and four verticals" and "a narrow strip of water, and fly side by side", the Songcheng district is provided with a water system framework of "city surrounded by water", the east district is provided with a water system framework of "a situation of tripartite confrontation, and seizing the central plain", and the water system of the three districts with the same origins in the cultural connotation are actually and ironically assimilated into the water system planning, and a water system communication pattern with freely storage and discharge, smooth introduction and drainage and frequent water exchange is formed by communicating and sharing the water resources, equally allocating the water, and cohering and integrating the water landscape. Protection means development, the development opportunities are pregnant in the protection, and the protection and development supplement each other to finally realize the coordination and integration of historical heritage and modern civilization development in the water system planning.

4 Allocation of water resources

In order to meet the water quantity and quality requirements of the water system in different times and levels, the paper provides a multi-level and three-dimensional dynamic equilibrium configuration mode of water resource in Kaifeng City, and a water resource strategy of "introducing the water of Yellow River from north side, using rain water, protecting groundwater, and developing reclaimed water" is formulated.

Although Kaifeng is a famous "North Watertown" in the history, the current problem of water shortage therein is also serious with the rapid economic and social development. Due to the uneven allocation of water quantity, the guarantee rate of water supply is low, the water distribution is uneven, the groundwater funnel area of some areas is gradually expanded and etc. Therefore, one of the key factors for the success of the water system planning is the planning and allocation of water resources.

According to the water resource allocation strategy of "introducing the water of Yellow River from north side, using rain water, protecting groundwater, and developing reclaimed water", the surface water, Yellow River water, groundwater, recycled water, rainwater and other water resources are subjected to overall allocation to enrich and broaden the water resources, the allocation and use method of all the water resources of different conditions are designed for meeting the water quantity requirements for the leakage, evaporation, exchange, ecology, landscape and the like of water system in rich, general and dry years as well as preventing the risk of water supply caused by the over-reliance on a single water resource, the utilization ratio and efficiency of water resources are improved by using the water according to the quality, and the dynamical equilibrium for the supplied and required water quantity is finally carried out. By using the flexible and pragmatic strategy above, a multi-level and three-dimensional water allocation mode is constructed, and the water resources allocation for water system in the region with water shortage is provided with a useful reference.

5 Protection of water ecology

The study provides a favorable support for the restoration and construction of water system ecology of Kaifeng by constructing a three-dimensional urban water system ecology protection framework with combined point, line and face.

For a long time, the disorder development activities in the coast of water system of Kaifeng have change the original ecological structure of the rivers and lakes, the water environment has been polluted, the ecological corridor has been destructed and appropriated, and the problems of severe degeneration and even disappearing of aquatic ecosystem, damage of hydrophytic habitat, lack of ecological water requirement, damage of biological habitat, reduction of biodiversity and etc. The restoration and construction of aquatic ecosystem is one of the important problems to be solved in the water system planning, and it directly impacts the health and vitality of the water system. Moreover, a complete three-dimensional urban water system ecology protection framework with combined point, line and face shall be constructed under the premise of the pollution intercepting and controlling.

"Point"—create creature dwellings, stock appropriate aquatic organisms, and create a basic habitat for aquatic organisms;

"Line"—maintain and restore the natural zigzag form of the river during the treatment process, construct a variety of habitats with shallows, deep pools, torrents, waterfalls, Streams, sandbanks, rivers, wetlands and the like by combining with the construction of ecological waterfront shore, and improve the fullness of aquatic ecosystem of the whole planning area by restoring and completing the ecological corridor, connecting and communicating the ecological plaques;

"Face" —establish a water ecological protection zone according to the different water ecological function zones, determine the protection scope, improve recommendations and requirements from the policy, system and management, and establish an ecological accident emergency response mechanism to dispose sudden water environment and water ecological accidents.

6 Operation supporting measures

Artificial wetland is an important engineering measure in the planning of ecological water system, so that the study provides a complete winter operation security measure system of artificial wetland in relation to the problem for the reduction of winter operation effect of artificial wetland.

According to the overall layout of the urban land use, the water system planning of Kaifeng is provided with three artificial wetlands, i. e. Longpan wetland, Wanxihe wetland, and Majiahe wetland to improve the water quality of the rive and improve the function of surrounding environment of the river. As the realization of planned water system function is dependent on the safety control of water environment quality, and the water environment quality is determined according to the exertion for the purification capacity of wetland, the artificial wetland plays a vital role in the overall water system planning of Kaifeng; furthermore, the method of how to ensure the operation effect of artificial wetland in winter will directly determine the success or failure for the overall water system planning of Kaifeng. Thus, the plan firstly provides the complete winter operation security measure system of artificial wetland, and actually applies that in Kaifeng.

The winter operation security measure system of artificial wetland mainly comprises two stages, i. e. design and operation. In the design stage, the pool type, matrix type and wetland plants of the artificial wetland surviving in winter shall be determined according to the local climatic condition, influent water quality, available land area and other influencing factors. Moreover, the normal operation water level, ice water level and operation water level in winter of the artificial wetland shall be determined according to the simulation of the mode while combining with the operation experience of actual project. In the operation stage, the water level shall be adjusted through the water level regulator of artificial wetland according to the actual temperature changing and sowing conditions, and the corresponding winter operation security measures shall be taken in relation to the various conditions presented in the operation process.

7 Planning measures

In the study, Google earth, ArcGIS and other software are applied to the planning of urban water system, and used for enriching and improving the basic information as important auxiliary tools, so that the scientificity and rationality of the planning work are improved, and the planning results are enriched.

Due to the limitation of a variety of objective factors, the problem that the complete and effective topographic map and other information cannot be obtained in time is frequently presented in the early stage of the planning work, which will delay the schedule of the whole project, influence the formulation of the planning proposal and the intuitive understanding and expression of the related elements.

In the plan, Google earth, ArcGIS and other software are used as the important auxiliary tools in the early stage of water system planning, the effective terrain, physiognomy, geographical position and other information are obtained through the public information system, and the problem for the lack of information in the early stage is solved by drawing the topographic map by ourselves. On the basis, we have a preliminary understanding for the current layout status of Kaifeng water system by investigating the site, and primarily analyze the problems for the current overall water system layout, and primarily propose the layout proposal of the water system. Thereafter, the work is gradually detailed with the gradual putting in place of relevant mapping information, wherein the terrain information above is reviewed, the primarily proposed proposal is adjusted and completed, so that the time is saved greatly, the smooth completion of the project is promoted, and the planning work is finally and successfully carried out.

8 Conclusions

The paper explores a method to resolve the major problems of the ecological water system planning in the idea of philosophy, engineering layout, ecological protection, operation security measures and the like by studying the key technologies mentioned above, successfully applies the method to the overall water system planning of Kaifeng City which is a traditional "Northern Watertown", as well as provides a useful reference experience for the efficient, ecological and harmonious development of urban water systems planning in the surrounding cities such as Xi'an, Luoyang, Xuzhou, Fuyang and etc.

References

Wang Peifang, Wang Chao, Hou Jun. Study on Integrated Environment Construction Mode of Urban River Ecosystem [J]. Journal of Hohai University (Natural Science Edition), 2005 (1).

Kang Ying, Chen Zhigang. Simulation Model of Water Resources Allocation in Plain River Network Area [J]. Water Resources Protection, 2007(5).

Peng Xiyuan, Yu Qicheng. Research on Ecological Protection and Restoration Scheme of Wuhan 1 + 8 Urban Agglomeration [J]. Yangtze River, 2010(11).

Liu Zhiguo, Wang Yujian. Analysis of Methods for 3D Models Publishing in Google Earth [J]. IT, 2009(6).

Study on Water Resources Protection and Management of the Yellow River

Li Hao, *Liu Tingting*, *Zhou Yanli* and *Fan Yinqin*

Yellow River Basin Water Resources Protection Bureau, Henan, Zhengzhou, 450004, China

Abstract: The Yellow River resources protection and management is an integrated transaction, and the protection and management requires not only the support of state laws and regulations, but also requires a strong national policy support. Generally, Water Law of the People's Republic of China and other laws had provided water resources management system and mechanism yet. In recent years, the central government clearly defined policies and corresponding measures to implement the most stringent water management system, and put forward a newer and higher requirement. Based on the yellow river basin water resources protection management system and mechanism system, to establish the appropriate institutional mechanism system of the yellow river resources is significant in order to meet the actual needs in the water resources protection and the water quality monitoring.

Key words: water resource, protection, management and system

1 The status of Yellow River Basin water resources protection

The Yellow River is one of the most important rivers in China, and the development is related to national economic and social sustainable development. Within the past thirty years, especially since the beginning of this century, governments at all levels in the Yellow River Basin and Yellow River Conservancy Commission have (YRCC) been strengthening the protection of water resources management in water function zones, management of sewage into the river, drinking water source protection, water quality monitoring, and disposal of major water pollution incidents to make a greater effort to curb the deterioration of the river basin water resources.

At present, China has formulated a series of laws and regulations related to water resources protection, such as Water Law of the People's Republic of China, Law of the People's Republic of China on Prevention and Control of Water Pollution and so on. The basin management as a basic water resource management system has been got much attention from the governments at all levels, but it is still in infancy with the special natural and socio-economic situation in the basin, lots of the policies and managements therefore is insufficient to meet the needs of practical work, and all of these have been seriously restricting the working of the basin water resources protection. At the same time, some rules are difficult to put into operation and practice, which needs to be updated for the Yellow River Basin water resources protection.

2 Problems in yellow river basin water resources protection

Over the years, water resources protection in the yellow river basin is not optimistic even if a lot of manpower and material resources has been invested, and one of the main causes is the management system can not meet the development of community and economic. The management system of the yellow river basin is formed under the planned economy, the initial task is the flood control and development while ignored other aspects which are important from present perspective. The most important reason for the deterioration of water quality in the yellow river basin is a regional effluent out of control. At present, the yellow river water resources protection work lags behind the requirements from the sustainable use of water resources in the national economy development. The technical capacity and means of watershed management still have a long way to go. These problems are mainly manifested in the following areas.

2.1 Limited responsibility and function of the river basin organization

As a basin agencies, YRCC is just an organization under the accredit ion from the Ministry of Water Resources, rather than an level 1 administrative unit, what it can do be only authorized by the Ministry of Water Resources to exercise water administrative power in the basin. In this case, without just authorities and functions, and supporting laws and regulations in the basin, the administrative practices of YRCC are not enforcing enough. Meanwhile, the Water Law of the People's Republic of China /not give the administrative penalties right to management units which have difficult to place unless the great support of subordinate units. Basin within the scope of regional plan should be subject to watershed plan, the law is difficult to ensure that all administrative departments in the real implementation of river basin plan, all that is serious impact on the Yellow River Water Resources Protection's work.

2.2 Settings of sewage inlets on the river

From the beginning of 2006, permitting of effluent to the river has been carried out a in a large scale in the Yellow River Basin had obtained good results. However, because of the seriousness of the basin water pollution and all work at its early stage, there are still some problems. For example, some inlets already set on the river for receiving effluent are not in line with the water function zone management. At the same time, the Yellow River Basin has not yet developed a comprehensive plan for setting the sewage inlets, and doesn't have unified regulation and management for the old pollution sources . In this case, there are no reliable and assuring bases for verifying the effluent permitting and discharge volume when a new project is launched, and this could further exacerbate water pollution of the Yellow River.

2.3 Implementation issues in identifying and supervising source water protection area

At present, the area of the source water protection region is 5,286.9 km^2, and quasi – source water area is 15,225.3 km^2. Among the 176 rivers and lakes in Yellow River Basin, there are 354 water function zones at level 1, of which 160 zones are developed. The developed zones are further divided into 399 regions or in another words, water function zones at level 2, among which 57 regions are for drinkable source water protection. There are 18 water function zones on the Yellow River mainstream, among which 10 are developed that include 14 drinkable source water protection areas. The delineation of those zones has played a very important role for the safety of urban and rural livelihoods, but still there are issues waiting to be solved, for example, the supervision and management system needs to be improved, emergency responding system for security of urban and rural drinking water needs to be updated, compensation is not in place.

2.4 Monitoring system

In the current water regulations, there are two monitoring systems in the water sector which is run by the Ministry of Environmental Protection and the Ministry of Water Resources irrespectively, and the management usually is not cross – department. In this case, the incompatible monitoring systems lead to the different monitoring and evaluation results which is difficult to be shared among the different sectors. The basin agency in this context couldn't get the timely information from the local when water pollution event happened, which is not favorable for the downstream to reducing the losses and avoid the risks. The Yellow River is a sand – laden river with a special hydraulic features which requires a corresponding monitoring method adaptive to it, but related laws and regulations don't authorize the basin agency or YRCC to formulate the special method. This leads to the monitoring data from the different methods can'nt be compared.

3 Discussion on the key management system

In January 12, 2012, the State Council issued the third document, and clearly put forward the guiding ideology and basic principles to implement the most stringent water management system including the main objectives, management measures and safeguards. Therefore, the establishment and implementation of various systems for the Yellow River Water Resources Protection has become the need, and the formulation of these system and regulations could not only promote management, but also be conducive to regulate the Yellow River Basin Management and Development.

3.1 To improve protection and management system for water function area

The water pollution in the Yellow River Basin is very serious, which requires that the Yellow River Basin must quickly establish and implement the management system relating water function. At present, the Yellow River Basin of water function areas has been approved by the State Council, which stresses the improvement of water function area management based on the delineation of water function areas, and strengthening of supervising and managing the basin water function area. The first point is to identify and check water assimilative capacity, to figure out pollution emission control plan including emission volume in different stages while keep the total within the limitation, and to let the related department know by the law. Meanwhile, it must strictly supervise and manage the pollution outfall to the river, and strengthen the water quality monitoring at provincial boundaries and important control cross – section to enforce the monitoring and control of the total amount of sewage into the river, and promptly informed government and relevant departments at all levels of the information. In accordance with the requirements of the "opinions" of the State Council, responsibility system and examination system should be established as soon as possible in managing the water function area.

3.2 The system for capping the pollution emission

In the Yellow River, pollutants discharged into rivers should base on the water function area as a unit to delineate the "red line" of total pollution control volume and the "red line" of water quality control. In this regard, we should actively carry out the supervision and management of the pollution outfall into the river. At the same time, in accordance with the requirements of water quality objectives, the total amount of sewage into the river should be regularly informed, and if necessary, water intake of the relevant areas could be limited or stopped. Also, the sites and area of monitoring both water quality and flow would be gradually expanded, and try to achieve real – time monitor. The objective of environmental planning and the program should be compatible with the total control, and meet the assigned quota. Sometimes, specialized requirements should be developed to make the management through, which include the total amount of monitoring, investigation, measurement and other places at least.

3.3 Strictly implement the system of drinking water source areas protection

Central Document No. 1 in 2011 clearly pointed out to strengthen the protection of water sources, and designate drinking water source protection areas in accordance with the law, strengthen the emergency management of drinking water sources. With the economic and social development of the Yellow River Basin, the people's living standards gradually improved, the urbanization rate is also increased every year, and total water consumption for households in the basin is also increasing year by year. So, to strengthen the protection of drinking water of the Yellow River Basin is more and more important. According to the statistics and projections, in 2010 and 2020, urban domestic water consumption in the Yellow River Basin are 1.94×10^9 m^3 and 26.7 L/(person ·

d) respectively. The average water quota are 106 L and 115 L/ (person · d). Rural household water use of the Yellow River Basin in 2010 and 2020 are 1.37×10^9 m^3 and 1.46×10^9 m^3 respectively, the average water quota, respectively, are 54 L/ (person · d) and 63L /(person · d).

Therefore, in order to guarantee people ' s living water, Yellow River Basin drinking water source protection system must be strictly implemented. According to its economic , social development planning and the actual needs, drink water functions and optimization of the sewage systems should be strengthened. This includes classification management system to further improve the drinking water source protection area, prohibiting in a protection zone expansion (change) to build facilities and projects, to prohibit the protected areas to build new (changed, expanded) construction projects. Also, we should strengthen the protective engineering measures, ecological restoration and protection of engineering measures, sediment and non – point source pollution control measures and quasi – protected areas, pollution control system, improve the management system of the water source, monitoring and control, emergency plans and integrated management to further strengthen the supervision and management of watershed drinking water source protection area.

3.4 Enhance the effective management system of pollution outfalls into the river

In 2006, the Yellow River Conservancy Commission issued the Supervision and Administration Rules of the Implementation of sewage into the river mouth. However, the Water Pollution Control of Yellow River Basin requires not only the joint efforts, but also requires close collaboration of the Ministry of Water Resources and the environmental protection departments and with other departments, as well as strong public support. The "Rules", aiming to help strengthen the supervision and monitoring system of the sewage into the river in both watershed and region, and the management objectives from both macroscopic and microscopic, requires to establish and improve the communication and coordination system among different parts, registration of sewage into the river, the survey data sharing to each other and so on. All these are very important. However, the implementation details is only a normative document, no mandatory binding on environmental protection and other wading departments and local governments, the relevant provisions of its settings in the outfall into the river only as a reference the terms of the guidance related work. Therefore, from the legal system level, the setting system of sewage into the river has been carried out and implemented, but in order to strengthen the effect of the implementation of the system, it is necessary to enhance the effectiveness of the legal position, while improving the content.

4 Conclusions

It must be clearly that the management of yellow river is not the affair of the river basin organization, not just the water administrative department and the affair of the department of environmental protection, it should be a basin – wide duties and obligations of all parties concerned. We must strengthen the social management. Aiming at maintenance of the Yellow River Basin river health, water safety, drinking water safety, environment safety, food security, national security and people ' s lives, watershed organizations, local governments and relevant departments, water users should be mobilized to establish and improve basin water environmental safety, in a timely manner to identification, evaluation, prediction, prevention, control, eliminate and protect the resources and the environment in watersheds to keep lives from the water environment pollution. Meanwhile, the crisis management should be systematic and have the rule of law, policy to follow to ensure the safety of the healthy life of rivers, promote watershed social, economic and environmental comprehensive, coordinated and sustainable development.

References

Li Tingyi, Weng Jianhua . Discussion on the focus of water quality trend analysis and management of the Yellow River[J]. Hydrological,2003(5):16 – 19.

Diao Lifang, Gao Hong, Fan Yinqin. Yellow River Water Resource Protection and Management Disscussion[J]. The Third International Yellow River Forum, 2007.10:119 – 124.

Zhang Shaofeng, Wang Xianfeng. Yellow River water resources protection decision support system [J]. Yellow River, 2005,27(9).